Studies in Systems, Decision and Control

Volume 133

Series editor

Janusz Kacprzyk, Polish Academy of Sciences, Warsaw, Poland
e-mail: kacprzyk@ibspan.waw.pl

The series "Studies in Systems, Decision and Control" (SSDC) covers both new developments and advances, as well as the state of the art, in the various areas of broadly perceived systems, decision making and control- quickly, up to date and with a high quality. The intent is to cover the theory, applications, and perspectives on the state of the art and future developments relevant to systems, decision making, control, complex processes and related areas, as embedded in the fields of engineering, computer science, physics, economics, social and life sciences, as well as the paradigms and methodologies behind them. The series contains monographs, textbooks, lecture notes and edited volumes in systems, decision making and control spanning the areas of Cyber-Physical Systems, Autonomous Systems, Sensor Networks, Control Systems, Energy Systems, Automotive Systems, Biological Systems, Vehicular Networking and Connected Vehicles, Aerospace Systems, Automation, Manufacturing, Smart Grids, Nonlinear Systems, Power Systems, Robotics, Social Systems, Economic Systems and other. Of particular value to both the contributors and the readership are the short publication timeframe and the world-wide distribution and exposure which enable both a wide and rapid dissemination of research output.

More information about this series at http://www.springer.com/series/13304

Viet-Thanh Pham · Sundarapandian Vaidyanathan
Christos Volos · Tomasz Kapitaniak
Editors

Nonlinear Dynamical Systems with Self-Excited and Hidden Attractors

 Springer

Editors
Viet-Thanh Pham
School of Electronics and
 Telecommunications
Hanoi University of Science and Technology
Hanoi
Vietnam

Sundarapandian Vaidyanathan
Research and Development Centre
Vel Tech University
Chennai, Tamil Nadu
India

Christos Volos
Department of Physics
Aristotle University of Thessaloniki
Thessaloniki
Greece

Tomasz Kapitaniak
Division of Dynamics
Faculty of Mechanical Engineering
Lodz University of Technology
Łódź
Poland

ISSN 2198-4182   ISSN 2198-4190   (electronic)
Studies in Systems, Decision and Control
ISBN 978-3-030-10034-6   ISBN 978-3-319-71243-7   (eBook)
https://doi.org/10.1007/978-3-319-71243-7

Printed on acid-free paper

This Springer imprint is published by Springer Nature
The registered company is Springer International Publishing AG
The registered company address is: Gewerbestrasse 11, 6330 Cham, Switzerland

Preface

Recently, there has been an increasing interest in a new classification of nonlinear dynamical systems including two kinds of attractors: self-excited attractor and hidden attractor. Previous research has established that a self-excited attractor has a basin of attraction which is excited from unstable equilibrium point. As a result, classical nonlinear systems such as Lorenz's system, Rössler's system, Chen's system, Lü's system, or Sprott's system are considered as systems with self-excited attractors. Several attempts have been made to study systems with self-excited attractors, which appear in various fields from computer sciences, physics, communications, biology, mechanics, chemistry, to economics and finance. However, there are still different questions which invite more investigation in such systems with self-excited attractors.

In recent years, systems with hidden attractors have received great attention from both a theoretical and a practical viewpoint. There are a number of important differences between self-excited attractors and hidden attractors. Self-excited attractor can be localized straightforwardly by applying a standard computational procedure. By contrast, we have to develop a specific computational procedure to identify a hidden attractor due to the fact that the equilibrium points do not help in their localization. There is evidence that hidden attractors play a crucial role in the fields of oscillators, describing convective fluid motion, model of drilling system, or multilevel DC/DC converter. In addition, hidden attractors are attracting widespread interest because they may lead to unexpected and disastrous responses, for example, in a structure like a bridge or an airplane wing. Therefore, it is useful for engineering students and researchers to know emergent topics of this new classification of attractors. For the past 5 years, although there has been a rapid rise in the discovery of systems with hidden attractors, there is still very little scientific understanding of hidden attractors. For example, to date there has been little discussion on the existence of systems with different families of hidden attractors. Further studies need to be carried out in order to provide insights for hidden attractor.

The aim of this book, *Nonlinear Dynamical Systems with Self-Excited and Hidden Attractors*, is to report the latest advances, developments, research trends, design, and realization as well as practical applications of nonlinear systems with self-excited attractors and hidden attractors. The book consists of 20 contributed chapters of experts who are specialized in these areas. We hope that this book will serve as a reference book about nonlinear systems with self-excited and hidden attractors for researchers and graduate students.

We would like to thank the authors of all chapters submitted to our book. We also wish to thank the reviewers for their contributions in reviewing the chapters. In addition, we would like to express our gratitude to Springer, especially to the book editorial team.

Hanoi, Vietnam   Viet-Thanh Pham
Chennai, India   Sundarapandian Vaidyanathan
Thessaloniki, Greece   Christos Volos
Łódź, Poland   Tomasz Kapitaniak
October 2017

Contents

Part I
Nonlinear Dynamical Systems with Self-Excited Attractors

Bifurcation Analysis and Chaotic Behaviors of Fractional-Order Singular Biological Systems

Komeil Nosrati and Christos Volos

Abstract In this chapter, singular system theory and fractional calculus are utilized to model the biological systems in the real world, some *fractional-order singular (FOS)* biological systems are established, and some qualitative analyses of proposed models are performed. Through the fractional calculus and economic theory, a new and more realistic model of biological systems predator-prey, logistic map and SEIR epidemic system have been extended, and besides some mathematical analysis, the numerical simulations are considered to illustrate the effectiveness of the numerical method to explore the impacts of fractional-order and economic interest on the presented systems in biological contexts. It will be demonstrated that the presence of fractional-order changes the stability of the solutions and enrich the dynamics of system. In addition, singular models exhibit more complicated dynamics rather than standard models, especially the bifurcation phenomena and chaotic behaviors, which can reveal the instability mechanism of systems. Toward this aim, some materials including several definitions and existence theorems of uniqueness of solution, stability conditions and bifurcation phenomena in FOS systems and detailed introductions to fundamental tools for discussing complex dynamical behavior, such as chaotic behavior have been added.

Keywords Fractional-Order singular system · Bifurcation and chaos
Biological systems · Qualitative analysis

K. Nosrati (✉)
Department of Electrical Engineering, Amirkabir University of Technology,
424 Hafez Ave, 15875-4413, Tehran, Iran
e-mail: nosrati_k@aut.ac.ir

C. Volos
Physics Department, Aristotle University of Thessaloniki,
54124 Thessaloniki, Greece
e-mail: chvolos@gmail.com

© Springer International Publishing AG 2018
V.-T. Pham et al. (eds.), *Nonlinear Dynamical Systems with Self-Excited
and Hidden Attractors*, Studies in Systems, Decision and Control 133,
https://doi.org/10.1007/978-3-319-71243-7_1

1 Introduction

Singular systems (differential-algebraic systems, descriptor systems, generalized state space systems, semi-state systems, singular singularly perturbed systems, degenerate systems, constrained systems, etc.), more general kind of equations which have been investigated over the past three decades, are established according to relationships among the variables (Dai 1989). As a valuable tool for system modeling and analysis, singular system theory has been widely utilized in different fields including nonlinear electric and electronic circuits, constrained mechanics, networks and economy (Lewis 1986).

This class of systems, which was introduced first by Luenberger in 1977, can be described as the following form.

$$E(t)\dot{x}(t) = H(x(t), u(t), t),$$
$$y(t) = J(x(t), u(t), t), \tag{1}$$

where H and J are appropriate dimensional vector functions, and the matrix $E(t)$ may be singular.

In 1954, Gordon investigated the economic theory of natural resource utilization in fishing industry and discussed the effects of harvest effort on its ecosystem (Gordon 1954). To study the economic interest of the yield of harvest effort in his theory of a common-property resource, Gordon proposed an algebraic equation to put his idea into practice. Recently, by using this theory of natural resource utilization in industry, the effects of harvest effort on biological systems were studied, and some singular model of these ecosystems were investigated to study the economic interest of the yield of harvest effort. Besides, many qualitative analyses such as stability analysis, presence of bifurcations and chaos and controller design were investigated (Zhang et al. 2010; Chakraborty et al. 2011; Zhang et al. 2012, 2014).

The majority of these works has been carried out in dynamical modeling of biological systems using integer-order differential equations which are valuable in understanding the dynamics behavior. However, the effects of long-range temporal memory and long-range space interactions in these systems are neglected. Due to its ability to provide an exact description of different nonlinear phenomena, inherent relation to various materials and processes with memory and hereditary properties and greater degrees of freedom, fractional-order modeling has recently garnered a lot of attention and gained popularity in the evaluation of dynamical systems (Podlubny 1998; Diethelm 2010; Petras 2011). According to these reasons, fractional-order modeling of many real phenomena such as biological systems has more advantages and consistency rather than classical integer-order mathematical modeling (Rivero et al. 2011).

In this chapter, singular system theory besides fractional calculus is utilized to model the biological systems in the real world which takes the general form

$$E(t)\,D^{\alpha}x(t) = F(t, x(t)), \quad t \geq 0, \tag{2}$$

where $F: \mathbb{R}^n \to \mathbb{R}^n$ is a vector function, $0 < \alpha < 1$, $x(t) \in \mathbb{R}^n$, and $E(t) \in \mathbb{R}^{n \times n}$ is a singular matrix.

Based on this model, some fractional-order singular (FOS) biological systems, such as predator-prey models (Holling-II, Holling-Tanner and food web), logistic map and SEIR epidemic model are established. Then, local stability analysis is performed to investigate the complex dynamical behavior and instability of model systems around the interior equilibrium, which are beneficial to study the coexistence and interaction mechanisms of population in these systems. Furthermore, some qualitative analyses of proposed models such as bifurcation and chaos will be illustrated. These studies can be utilized to design different kinds of controllers with the purpose of stabilizing a model system around the interior equilibrium, to restore the model system to a stable state, which are also theoretical guides to formulate related measures to maintain the sustainable development of population resources in such biological systems.

The remainder of this chapter is organized as follows. Section 2 presents some preliminaries in singular systems theory, and fractional-order integral and derivative definitions will be given. Then, the FOS model will be presented, and some definitions and theorems in solvability and stability conditions will be derived. In Sect. 3, we give some theory for the local bifurcations and chaos of vector fields and maps and extend them to FOS systems. Also, we consider the proposed FOS predator–prey models, logistic map and SEIR epidemic model in Sect. 4, which are followed by some discussions of the local stability, the phenomena of bifurcations and chaos, and numerical simulations to verify the effectiveness of the obtained results. This section will be continued with interpreting of results in biological context, and finally, this chapter ends up by concluding remarks.

2 Preliminary FOS Systems Theory

This section explains the proposed FOS model, beginning with the fractional-order systems and some definitions of fractional integral and derivative operators, and also, stability theorem in Sect. 2.1. Section 2.2 explains a mathematical definition of singular systems and gives their properties, and finally, the FOS model is introduced and established, which is followed by discussion of admissibility and stability conditions.

2.1 Fractional-Order Systems

Fractional calculus as a powerful tool for mathematical modeling has been applied in different fields of sciences such as economics, engineering and biological systems. For instance, it covers the widely known classical fields such as Abel's integral equation and viscoelastic material modeling, and also less reputed fields including feedback amplifiers, description of propagation in plane electromagnetic waves, generalized voltage divider, electro-analytical chemistry, electric conductance of biological systems, neurons modeling, etc. (Podlubny 1998). The increasing number of such applications shows that there is a significant demand for more realistic and adequate mathematical modeling of real phenomena using fractional calculus in which provides one possible approach on this way.

In this section, some basic materials on fractional calculus have been presented, and the *Grunwald-Letnikov (GL)*, *Rienlann-Liouville (RL)* and *Caputo* definitions among many interesting definitions of fractional integral and derivatives will be defined as follows.

Definition 1 (*Podlubny* 1998) Relied on a generalization of classical concept in traditional calculus in which derivatives of integer order can be represented as limits of finite differences, the *GL* fractional derivative operator of order $\alpha \in \mathbb{R}^+$ of a continuous function $f: \mathbb{R}^+ \to \mathbb{R}$ is defined by

$$
{}_a^{GL}D_t^\alpha f(t) = \lim_{h \to 0} \frac{\sum_{r=0}^{\left\lfloor \frac{t-a}{h} \right\rfloor} (-1)^r \binom{\alpha}{r} f(t-rh)}{h^\alpha}, \tag{3}
$$

where ${}_a^{GL}D_t^\alpha$ is the *GL* derivative of fractional order operator, a and t are the lower and upper terminals, respectively.

Definition 2 (*Podlubny* 1998) Based on a generalization of classical concept in integral using Cauchy formula, the *RL* fractional integral operator of order $\alpha \in \mathbb{R}^+$ of a continuous function $f: \mathbb{R}^+ \to \mathbb{R}$ is presented by

$$
{}_a^{RL}I_t^\alpha = \frac{1}{\Gamma(\alpha)} \int_a^t (t-\tau)^{\alpha-1} f(\tau) d\tau, \tag{4}
$$

in which ${}_a^{RL}I_t^\alpha$ is the *RL* integral of fractional-order operator, and $\Gamma(\cdot)$ is the Euler gamma function.

Definition 3 (*Podlubny* 1998) According to the Definition 2, the *RL* fractional derivative operator is expressed by

$$
{}_a^{RL}D_t^\alpha = \frac{d^n}{dt^n} \left\{ {}_a^{RL}I_t^{n-\alpha} f(t) \right\} = \frac{d^n}{dt^n} \left\{ \frac{1}{\Gamma(n-\alpha)} \int_a^t (t-\tau)^{n-\alpha-1} f(\tau) d\tau \right\}, \tag{5}
$$

where $_a^{RL}D_t^\alpha$ is the RL derivative of fractional order operator, and $n = \lceil \alpha \rceil = \min\{x \in Z | x \geq \alpha\}$.

Practical problems require definitions of fractional derivatives allowing the utilization of physically interpretable initial conditions. Unfortunately, the RL approach leads to apply initial conditions which are practically useless, and consequently causes to conflict between the well-established and polished mathematical theory and practical needs. A certain solution to this conflict was Caputo's definition which proposed by Michele Caputo as follows (Caputo 1966):

Definition 4 (*Diethelm* 2010) Based on Definition 2, the Caputo fractional derivative operator is defined by

$$
_a^C D_t^\alpha = {_a^{RL}I_t^{n-\alpha}} \left\{ \frac{d^n}{dt^n} f(t) \right\} = \left\{ \frac{1}{\Gamma(n-\alpha)} \int\limits_a^t (t-\tau)^{n-\alpha-1} \frac{d^n}{dt^n} f(\tau) d\tau \right\}, \qquad (6)
$$

where $_a^C D_t^\alpha$ is the Caputo derivative of fractional order operator and n is defined as same as Definition 3.

All these three approaches provide an interpolation among integer-order derivatives, and their definitions must reach to the same results during steady-state dynamical processes studies.

The initial value problem of a time invariant fractional-order differential equation (FODE) model related to Definition 4 is

$$
\begin{cases} _0^C D_t^\alpha x(t) = F(x(t)), \\ x(t)|_{t=0^+} = X_0, \end{cases} \qquad (7)
$$

in which $x(t) \in \mathbb{R}^n$ and $F: \mathbb{R}^n \to \mathbb{R}^n$. The stability theorem on nonlinear fractional-order system (7) has been introduced below.

Theorem 1 (Petras 2011) *Consider the nonlinear autonomous commensurate fractional-order system (7). The equilibrium points of this system can be calculated by solving the equation $f(x) = 0$. This system is locally asymptotically stable if all eigenvalues λ_i ($i = 1, \cdots, n$) of the Jacobian matrix $J = \partial f / \partial x$ evaluated at the equilibrium points lie in the stable regions of R_s^α (Fig. 1).*

2.2 FOS Systems

Singular systems are general kind of equations, which have been investigated during three past decades and established according to relationships among the variables based on differential or algebraic equations that form the mathematical model of the system. As a valuable tool, the theory of these systems has been widely utilized in different fields including modeling and analysis of nonlinear

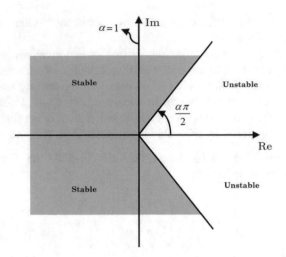

Fig. 1 Stability ($R_s^\alpha: = \{\lambda| \ |\arg(\lambda)| > \alpha\pi/2, \lambda \in \mathbb{C}\}$) and instability ($R_{is}^\alpha: = \{\lambda| \ |\arg(\lambda)| < \alpha\pi/2, \lambda \in \mathbb{C}\}$) regions of a fractional-order system

electric and electronic circuits, constrained mechanics, networks and economy (Podlubny 1998).

Since the 1960s, much research has been extensively focused on analysis of a dynamical system with the state-space variable method as a core feature in modern control theory. The concept of state in a dynamic system refers to a minimum set of variables, and state-space variable method provides us with a completely new method for system analysis and offers us more understanding of systems. Using this method, state-space models of a time-varying nonlinear singular system are obtained as (1). This suitable representation can describe systems that evolve over time, especially; nonlinear singular systems which are the natural outcome of component-based modeling of complex dynamic systems.

If H and J are linear functions of $x(t)$ and $u(t)$, another special form of the system (1) is a time-varying linear singular system as

$$E(t)\dot{x}(t) = A(t)x(t) + B(t)u(t),$$
$$y(t) = C(t)x(t), \tag{8}$$

where $x(t) \in \mathbb{R}^n$, $y(t) \in \mathbb{R}^r$, $u(t) \in \mathbb{R}^m$, and $E(t) \in \mathbb{R}^{n \times n}$, $A(t) \in \mathbb{R}^{n \times n}$, $B(t) \in \mathbb{R}^{n \times m}$, and $C(t) \in \mathbb{R}^{r \times n}$ are time-varying matrices. Here, for singular systems mentioned above, matrix E is considered to be singular, i.e., $rankE = r < n$, otherwise, the system (1) and (8) reduce to a standard (normal) system. In practical system analysis and control system design, many system models may be established in the form of (8), while they could not be described by standard forms (Campbell 1980).

Under the regularity assumption,[1] the state and output responses of singular system is derived, and it has been demonstrated that unlike standard system theory, the singular system (8) has a unique solution only for the consistent initial vector $x(0)$, and for the h times piecewise continuously differentiable input function $u(t)$, where h is the nilpotent index. Also, by using time domain analysis, a fair understanding of the system's structural features and its internal properties such as reachability, controllability, observability, system decomposition, and transfer matrix were obtained (Duan 2010).

Compared with the standard systems, the price paid is that singular systems are more difficult to deal with. However, the advantage they offer over the more often used standard systems is that they are generally easier to formulate and exhibit more complicated dynamics and have been applied widely in different fields of electrical engineering (Ayasun et al. 2004; Marszalek et al. 2005; Yue and Schlueter 2004), aerospace engineering (Masoud et al. 2006), biology (Zhang et al. 2012), chemical processes (Kumar and Daoutidis 1999) and economics (Zhang 1990; Luenberger and Arbel 1997). With the help of singular model for the systems in mentioned fields, complex dynamical behaviors of them, especially the bifurcation phenomena, which can reveal the instability mechanism of systems, have been extensively studied. However, as far as the FOS system theory is concerned, the related research results are few.

Very recently, the study on FOS systems has received much attention due to the fact that the fractional order calculus has contributed great merits, particularly in non-short memory and non-local property of describing physical systems, especially in power systems and biology (Kaczorek and Rogowski 2015, Nosrati and Shafiee 2017). These complicated systems requires considering not only stability, but also regularity and impulse elimination, while the latter do not appear in fractional-order standard ones. Although a number of valuable results and great achievements in the research about FOS systems have been reported in the literature (Yao et al. 2013; N'Doye et al. 2013; Ji and Qiu 2015; Zhang and Chen 2017), there are still many challenging and unsolved problems in the field of stability analysis and controller synthesis.

An initial value problem of a time invariant nonlinear FOS model related to Definition 4 is

$$\begin{cases} E_0^C D_t^\alpha x(t) = F(x(t)), \\ x(t)|_{t=0^+} = X_0, \end{cases} \tag{9}$$

where $x \in \mathbb{R}^n$ and $F: \mathbb{R}^n \to \mathbb{R}^n$, and $E \in \mathbb{R}^{n \times n}$ is a singular matrix ($rankE = r < n$), and $_0^C D_t^\alpha$ denotes the Caputo derivative operator. If F is linear function of $x(t)$, another special form of the system (9) is a time invariant linear FOS system

---

[1] For the singular system (2), if $\det(sE - A) \neq 0$ for some complex number s, then the pair (E, A) is said to be regular (Yang et al. 2012).

$$\begin{cases} E_0^C D_t^\alpha x(t) = Ax(t), \\ x(t)|_{t=0^+} = X_0, \end{cases} \tag{10}$$

where $A = \partial F/\partial x \in \mathbb{R}^{n \times n}$ is the jacobian matrix evaluated at the equilibrium points $(F(x) = 0)$. Parallel to fractional-order standard systems, the concerning basic definitions and relevant facts for the FOS system (10) are given as follows (Yao et al. 2013).

Definition 5 For FOS system (8), the triplet (E, A, α) is called regular if there exists a constant scalar $s_0 \in \mathbb{C}$ such that $|s_0^\alpha E - A| \neq 0$.

Similar to the proof of regularity of integer order singular systems, the triplet (E, A, α) in FOS system (10) is regular if and only if there exist two nonsingular matrices Q and P such that $QEP = diag(I_{n_1}, N)$ and $QAP = diag(A_1, I_{n_2})$ where $n_1 + n_2 = n$, $A_1 \in \mathbb{R}^{n_1 \times n_1}$, $N \in \mathbb{R}^{n_2 \times n_2}$ is nilpotent. Assume the triplet (E, A, α) in FOS system (10) is regular, then, this system can be transformed into

$$\begin{cases} {}_0^C D_t^\alpha x_1(t) = A_1 x_1(t), \\ N_0^C D_t^\alpha x_2(t) = x_2(t), \end{cases} \tag{11}$$

where $\begin{bmatrix} x_1^T(t) & x_2^T(t) \end{bmatrix}^T = P^{-1} x(t)$, $x_1(t) \in \mathbb{R}^{n_1}$ and $x_2(t) \in \mathbb{R}^{n_2}$. The initial state response of FOS system (11) is

$$\begin{bmatrix} x_1(t) \\ x_2(t) \end{bmatrix} = P \begin{bmatrix} E_{\alpha,1}(A_1 t^\alpha) x_1(0) \\ -\sum_{k=1}^{h-1} \delta^{((k-1)\alpha)}(t) N^k x_2(0) \end{bmatrix}, \qquad t \geq 0$$

where $\delta(t)$ is the impulse function, and $E_{\alpha,\beta}(t)$ is the two-parameter Mittag-Leffler function. From the derived response, we know that the triplet (E, A, α) is impulse-free if $N = 0$.

Let $\sigma(E, A, \alpha) = \{\lambda | \lambda \in \mathbb{C}, \lambda \text{ finite}, |\lambda E - A| = 0\}$ denotes the finite pole set for FOS system (10). It can be easily known from Theorem 1 that the system (10) is asymptotically stable, if all the finite dynamic modes lie in the domain R_s^α.

Remark 1 Consider the nonlinear autonomous commensurate FOS system (9). This system is locally asymptotically stable if all eigenvalues λ_i $(i = 1, \cdots, n)$ of the Jacobian matrix $A = \partial F/\partial x$ evaluated at the equilibrium points, satisfy the relation $|\arg(\lambda_i)|_{i=1,\cdots,n} > \alpha\pi/2$. Also, to assess the stability analysis of the system (9), the roots of the equation $|\lambda^\alpha I - A| = 0$ evaluated at the equilibrium points can be checked regarding the imaginary axis.

Definition 6 The generalized eigenvectors v satisfying $Ev = 0$ are defined as:

(1) The infinite eigenvector of order one satisfies $Ev_i^1 = 0$.
(2) The infinite eigenvector of order k satisfies $Ev_i^1 = Av_i^{k-1}$, $k > 1$.

Remark 2 Suppose that $Ev^1 = 0$, then the infinite eigenvalues associated with the generalized principal vectors v^k satisfying $Ev^k = v^{k-1}$ are impulsive modes. The triplet (E, A, α) is impulse-free if and only if there exists no infinite eigenvector of order two.

Definition 7 FOS system (10) is said to be admissible, if the triplet (E, A, α) is regular, impulse-free, and all the finite eigenvalues of triplet (E, A, α) lie in the stable regions of R_s^α.

System (9) is the general or fully-implicit nonlinear time-invariant FOS system. The dynamics of a large class of physical systems, including nonlinear circuits, robotics, and biological system, can be modeled by an important especial case of the system (9), called parameter dependent semi-explicit FOS system of the form

$$\begin{matrix} {}_0^C D_t^\alpha z(t) \end{matrix} = f(z(t), y(t), p) \quad f: \mathbb{R}^{n+m+q} \to \mathbb{R}^n, \tag{12a}$$

$$0 = g(z(t), y(t), p) \quad g: \mathbb{R}^{n+m+q} \to \mathbb{R}^m, \tag{12b}$$

where $z \in Z \subset \mathbb{R}^n$, $y \in Y \subset \mathbb{R}^m$ and $p \in P \subset \mathbb{R}^q$. In the state-space $Z \times Y$, dynamic state variables z and instantaneous state variable y are distinguished. The dynamics of the states z are directly defined by (12a) while the dynamics of the y variable in such that the system satisfies the constraints (12b). The parameters p define a specific system configuration and the operating condition.

As an example, for the predator-prey system, typical dynamic state variables are time dependent values of population densities of the prey and predator, and instantaneous variable is harvest effort performed by a static human population. The parameter space is composed of system parameters such as capture rate, growth rate, carrying capacity, etc., and operating parameters such as net economic profit. The interactions between prey and predator define the f equations and the constraint $g = 0$ is defined by the economic interest equation.

Lemma 1 (Nosrati and Shafiee 2017) *The characteristic polynomial of system (12a, 12b) can be obtained by $|\lambda^\alpha I - J| = 0$, where I is the identity matrix, and*

$$J = \frac{\partial f}{\partial z} + \frac{\partial f}{\partial y} \left(\frac{\partial g}{\partial y}\right)^{-1} \frac{\partial g}{\partial z}. \tag{13}$$

Theorem 2 *The FOS system (12a, 12b) is stable if and only if the fractional degree characteristic polynomial*

$$D(\lambda) = |\lambda^\alpha I - J| = a_0 \lambda^{\alpha_n} + a_1 \lambda^{\alpha_{n-1}} + \ldots + a_n \lambda^{\alpha_0}, \tag{14}$$

with $\alpha_n = \alpha = n\alpha'$, i.e., this polynomial has no zero in the closed right-half of the Riemann complex surface, that is

$$D(\lambda) \neq 0, \text{ for } \operatorname{Re}\lambda \geq 0. \tag{15}$$

It is assumed that the fractional order is commensurate, i.e., $\alpha_i = i\alpha'$, for $i = 0, 1, \ldots, n-1$, and $\alpha \in \mathbb{R}$.

Proof It can be directly derived from Theorem 9.1 in (Kaczorek 2011).

Remark 3 The commensurate degree characteristic polynomial

$$D(\lambda) = a_0 \lambda^{n\alpha'} + a_1 \lambda^{(n-1)\alpha'} + \ldots + a_n \tag{16}$$

is stable if and only if all zeros of this polynomial satisfy the condition (15) or, equivalently, all zeros of the associated natural degree polynomial

$$\tilde{D}(s) = a_0 s^n + a_1 s^{(n-1)} + \ldots + a_1 s + a_n \tag{17}$$

for $s = \lambda^{\alpha'}$, lie in the domain R_s^α.

For the system (12a, 12b), the set of all equilibrium points (*AEP*) and the set of all stable equilibrium points (*SEP*) are defined as

$$AEP = \{(z, y, p) \in Z \times Y \times P; f(z, y, p) = 0, \ g(z, y, p) = 0\}$$

and

$$SEP = \left\{ (z, y, p) \in AEP; \det\left(\frac{\partial g}{\partial y}\right) \neq 0 \text{ and all eigenvalues of matrix } J \text{ lie in } R_s^\alpha \right\},$$

respectively.

Note that the full Jacobian J of the functions f and g in the z and y coordinates is nonsingular for all $(z, y, p) \in SEP$, and therefore, by the implicit function theorem, the equations $f(z, y, p) = 0$ and $g(z, y, p) = 0$ can theoretically be solved uniquely for z and y as functions of the parameter p, locally near any equilibrium point in *SEP*. Hence *SEP* is a p-dimensional submanifold embedded in $AEP \subset Z \times Y \times P$.

Definition 8 Given a stable equilibrium (z_0, y_0) for parameter value p_0, the connected component F of *SEP* which contain (z_0, y_0, p_0) is called the feasibility region of (z_0, y_0, p_0), and its boundary is named feasibility boundary.

This definition provides a convenient mathematical apparatus for analyzing the local stability properties in the nonlinear semi-explicit FOS system (12a, 12b) in light of special nonlinear phenomena that may arise near the equilibrium point. The feasibility boundary for the large system can be solved for the common zeroes (zero sets) of three different sets of functions. These three zero sets are each connected with a special nonlinear property, which are equilibrium points at the singularity, proximity of multiple equilibrium points and birth of limit cycle for the nonlinear system (12a, 12b).

Theorem 3 (Extended Feasibility Boundary Theorem) *For a system defined in (12a, 12b), the feasibility boundary of a feasibility region F consists of three zero sets*

$$\partial F = (\partial F \cap C_{sib}) \cup (\partial F \cap C_{sn}) \cup (\partial F \cap C_H),$$

where

$$C_{sib} = \left\{ (z, y, p) \in AEP;\ \det\left(\frac{\partial g}{\partial y}\right) = 0 \right\},$$

$$C_{sn} = \left\{ (z, y, p) \in AEP;\ \det\left(\frac{\partial g}{\partial y}\right) \neq 0,\ \det(A) = 0 \right\}$$

and

$$C_H = \left\{ (z, y, p) \in AEP;\ \det\left(\frac{\partial g}{\partial y}\right) \neq 0,\ \det(A) \neq 0,\ \det(H_{n-1}(J)) = 0 \right\},$$

where H_{n-1} is the Hurwitz matrix as

$$H_{n-1} = \begin{pmatrix} a_1 & a_3 & a_5 & \cdots & a_{2n-3} \\ a_2 & a_4 & a_6 & \cdots & a_{2n-4} \\ 0 & a_1 & a_3 & \cdots & a_{2n-5} \\ 0 & a_2 & a_4 & \cdots & a_{2n-6} \\ \vdots & \vdots & \vdots & & \vdots \\ 0 & 0 & 0 & \cdots & a_{n-1} \end{pmatrix}$$

corresponding to the coefficient a_i of the following characteristic polynomial

$$D(\lambda) = |\lambda^\alpha I - J| = a_0 \lambda^{\alpha_n} + a_1 \lambda^{\alpha_{n-1}} + \ldots + a_n \lambda^{\alpha_0}.$$

Proof The proof is in the analogous manner with the proof of Theorem 1 in (Venkatasubramanian et al. 1995).

3 Different Bifurcations and Chaos

As we have seen from (12a, 12b), systems of physical interest typically have parameters that appear in the defining systems of equations. As these parameters are varied, changes may occur in the qualitative structure of the solutions for certain parameter values. These changes are called bifurcations and the parameter values are called bifurcation values. The bifurcation theory provides a natural platform for studying the parameter space phenomena by establishing the dynamic mechanisms that effect changes in the system structure upon parameter variations. When the system parameters are varied, the dynamics of system (12a, 12b) changes continuously; however, topologically the structure remains unchanged under small perturbations provided the system is structurally stable. Structurally unstable points then identify the parameter values where the structure of system undergoes changes.

Systems of the form (12a, 12b) typically have singular points where the implicit function theorem for solving the constraint $g = 0$ is not applicable. When the constraint (12b) is absent, it can be shown that the feasibility boundary essentially corresponds to two of three different local bifurcations, namely, the saddle-node bifurcation, transcritical bifurcation, and Hopf bifurcation. For constrained models (12a, 12b), however, the feasibility boundary typically also contains another bifurcation segment named the singularity induced bifurcation (SIB) which occurs when the system equilibrium is at the singularity. When this happens, some of the system eigenvalues may become unbounded. In what follows, we describe four bifurcations of equilibrium points and give some theory for the local bifurcations of vector fields and maps. This section will be ended by some explanations about chaos, other possible types of equilibrium behaviors, which may occur in FOS systems.

3.1 Saddle-Node Bifurcation

The saddle-node bifurcation is well understood mathematically and has been much studied in different type of systems such as power system, biology etc. A saddle-node bifurcation occurs when a system has non-hyperbolic equilibrium with a geometrically simple zero eigenvalue at the bifurcation point and additional transversality conditions are satisfied (Sotomayor 1973). By definition, the points in the set C_{sn} are not singular, i.e., $\det(\partial g/\partial y) \neq 0$. Therefore, by the implicit function theorem, we can reduce the system (12a, 12b) to fractional-order system

$$\,_0^C D_t^\alpha z(t) = f_R(z(t), p) \tag{18}$$

locally near (z_0, y_0, p_0) for a suitable and unique function f_R.

Then points in C_{sn} are saddle-node bifurcations if the following conditions are satisfied:

(a) The matrix $\frac{\partial f_R}{\partial z} = J = \frac{\partial f}{\partial z} + \frac{\partial f}{\partial y} \left(\frac{\partial g}{\partial y}\right)^{-1} \frac{\partial g}{\partial z}$ has a geometrically simple zero eigenvalue with right eigenvector v and left eigenvector w and there is no other eigenvalue on the imaginary axis.

(b) $w^T \left(\frac{\partial f_R}{\partial p}\right) = w^T \left(\frac{\partial f}{\partial p} + \frac{\partial f}{\partial y} \left(\frac{\partial g}{\partial y}\right)^{-1} \frac{\partial g}{\partial p}\right) \neq 0.$

(c) $w^T \left(\frac{\partial^2 f_R}{\partial z^2} (v, v)\right) \neq 0.$

At this type of bifurcations, stable and unstable equilibrium points meet and disappear in the feasibility boundary, resulting in a loss of equilibrium points locally near the bifurcation point on the wrong side of the feasibility boundary. As an example, this can be represented by the differential equation $\dot{x} = p - x^2$ which depends on a single parameter p. The bifurcation diagram for this equation is depicted in Fig. 2a.

Hypotheses (b) and (c) are the transversality conditions that control the non-degeneracy of the behavior with respect to the parameter and the dominant effect of the quadratic nonlinear term. The results obtained from the conditions above are limited in two different ways. On the one hand, it is possible that more quantitative information about the flows near bifurcation can be extracted. The

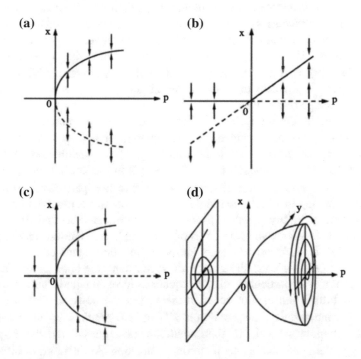

Fig. 2 Different bifurcations; **a** Saddle-node bifurcation **b** Transcritical bifurcation **c** Pitchfork bifurcation (supercritical) bifurcation **d** Hopf bifurcation

second limitation is that there may be global changes in a phase portrait associated with a saddle-node bifurcation.

3.2 Transcritical and Pitchfork Bifurcation

The importance of the saddle-node bifurcation is that all bifurcations of one parameter families at an equilibrium with a zero eigenvalue can be perturbed to saddle-node bifurcations. Thus, one expects that the zero eigenvalue bifurcations encountered in applications will be saddle-nodes. If they are not, then there is probably something special about the formulation of the problem that restricts the context so as to prevent the saddle-node from occurring. The transcritical bifurcation is one example that illustrates how the setting of the problem can rule out the saddle-node bifurcation (Hartman 2002; Kielhoefer 2004).

In classical bifurcation theory, it is often assumed that there is a trivial solution from which bifurcation occurs. Thus, the reduced system (18) is assumed to satisfy $f_R(0, p) = 0$ for all p, so that $z = 0$ is an equilibrium for all parameter values. Since the saddle-node families contain parameter values for which there are no equilibria near the point of bifurcation, this situation is qualitatively different. To formulate the appropriate transversality conditions, we look at the one-parameter families that satisfy the constraint that $f_R(0, p) = 0$ for all p. This prevents hypothesis (b) from being satisfied. If we replace this condition with the requirement that $w^T(\partial^2 f_R / \partial p \partial z)(v) \neq 0$, then the phase portraits of the family near the bifurcation will be topologically equivalent to those of Fig. 2b and we have a transcritical bifurcation or exchange of stability.

As an example, the transcritical bifurcation can be represented by the normal form $\dot{x} = px - x^2$, which depend on a single parameter p (Fig. 2b). This kind of bifurcation can be considered as an unfolding of the saddle-node bifurcation because if we apply the transformation $\zeta = x - p/2$ in the normal form, we obtain $\zeta = p^2/4 - \zeta^2$ is a normal form of saddle-node bifurcation parameterized by p.

A second setting in which the saddle-node does not occur involves systems that have a symmetry. Many physical problems are formulated so that the equations defining the system do have symmetries of some kind. The reduced fractional-order system (18) is symmetric with respect to the symmetry $z \rightarrow -z$ if $f_R(-z, p) = -f_R(z, p)$. Thus, the symmetric vector fields are ones for which f_R is an odd function of z. In particular, all such equations have an equilibrium at zero. The transcritical bifurcation cannot occur in these systems, however, because an odd function f_R cannot satisfy the condition $\partial^2 f_R / \partial z^2 \neq 0$ required by the transcritical bifurcation hypothesis (c). If this condition is replaced by the hypothesis $\partial^3 f_R / \partial z^3 \neq 0$, then one obtains the pitchfork bifurcation. At the point of bifurcation, the stability of the trivial equilibrium changes, and a new pair of equilibrium points appear to one side of the point of bifurcation in parameter space, as in Fig. 2c. (The pitchfork bifurcation can be represented by the normal form $\dot{x} = px - x^3$).

3.3 Hopf Bifurcation

A Hopf bifurcation occurs at points where the system has non-hyperbolic equilibrium connected with a pair of purely imaginary eigenvalues, but no zero eigenvalues, and additional transversality conditions are met (Guckenheimer and Holmes 1983). As an example, the Hopf bifurcation can be represented by the following normal form, which depends on a single parameter p (Fig. 2d).

$$\begin{cases} \dot{x} = -y + x(p - (x^2 + y^2)), \\ \dot{y} = x + y(p - (x^2 + y^2)). \end{cases}$$

One of the basic differences between dynamical behavior of fractional-order systems and integer-order systems is that the limit set of a trajectory of integer-order system as the limit cycle of this system is a solution for this system, but in the fractional-order case, the limit set of a trajectory of fractional-order system can be, not a solution for this system (Tavazoei et al. 2009a, b). In (Tavazoei et al. 2009a, b), the authors claimed there are no periodic orbits in fractional order systems, and in (Tavazoei 2010), the authors gave an example where the solutions of the system are not periodic, but they converge to periodic signals. In (Abdelouahab et al. 2012), the authors were interested about the final state of trajectory, and it has been demonstrated that chaos, as well as the other usual nonlinear dynamic phenomena, occur in this system with mathematical order less than three. The largest Lyapunov exponents and the bifurcation diagrams show the period-doubling bifurcation and the transformation from periodic to chaotic motion through the fractional-order and confirm the justness of the proposed fractional Hopf bifurcation conditions.

Let consider the following the reduced fractional-order commensurate system (18). Suppose that $z \in \mathbb{R}^3$, and e^* is an equilibrium point of this system. In the integer case ($\alpha = 1$), the stability of e^* is related to the sign of $\mathrm{Re}(\lambda_i)$, $i = 1, 2, 3$, where λ_i are the eigenvalues of Jacobian matrix $A = \partial f_R / \partial z|_{e^*}$. The conditions of system (18) with $\alpha = 1$, to undergo a Hopf bifurcation at the equilibrium point e^* when $p = p^*$, are

(a) The Jacobian matrix has two complex-conjugate eigenvalues $\lambda_{1,2}(p) = \theta(p) \pm i\omega(p)$, and one real $\lambda_3(p)$ (this can be expressed by $D(P_{e^*}(p^*)) < 0$, where D is the discriminant of characteristic equation $P(\lambda) = |\lambda I - A|$).
(b) $\theta(p^*) = 0$, and $\lambda_3(p^*) \neq 0$.
(c) $\omega(p^*) \neq 0$.
(d) $d\theta/dp|_{p=p^*} \neq 0$.

But in the fractional case, the stability of e^* is related to the sign of $m_i(\alpha, p) = \alpha \pi / 2 - |\arg(\lambda_i(p))|$, $i = 1, 2, 3$. If $m_i(\alpha, p) < 0$ for all $i = 1, 2, 3$, then e^* is locally asymptotically stable. If there exist i such that $m_i(\alpha, p) > 0$, then e^* is unstable. So, the function $m_i(\alpha, p)$ has a similar effect as the real part of eigenvalue in integer systems, therefore, we extend the Hopf bifurcation conditions to the fractional systems by replacing $\mathrm{Re}(\lambda_i)$ with $m_i(\alpha, p) < 0$ as follows:

(a) $D(P_{e^*}(p^*)) < 0$.
(b) $m_{1,2}(\alpha, p^*) = 0$, and $\lambda_3(p^*) \neq 0$.
(c) $dm/dp|_{p=p^*} \neq 0$.

Hopf bifurcations are especially interesting for the large system because they signal the birth or the annihilation of periodic orbits for the system (12a, 12b) which are otherwise impossible to observe by purely numerical means.

3.4 Singularity Induced Bifurcation (SIB)

SIB is a new type of bifurcation which has been characterized by a singular system and refers to a stability change of the singular system possessing some eigenvalues which diverges to infinity (Venkatasubramanian et al. 1995). The result is impulse phenomenon of the singular system which may cause to the collapse of this system.

An SIB occurs when an equilibrium point e^* crosses the singular surface

$$S := \left\{ (z, y, p) \in \mathbb{R}^{n+m+q}; \ g(z, y, p) = 0, \ \Delta(z, y, p) := \det\left(\frac{\partial g}{\partial y}\right) = 0 \right\},$$

that is a point in the zero set C_{sib}, and certain additional transversality conditions are satisfied at (z, y, p):

(a) $\partial g/\partial y$ has an algebraically simple zero eigenvalue, and

$$trace\left(\frac{\partial f}{\partial y} adj\left(\frac{\partial g}{\partial y}\right) \frac{\partial g}{\partial z}\right)\Bigg|_{e^*} \neq 0.$$

(b) The following two matrices are nonsingular in e^*.

$$\begin{bmatrix} \frac{\partial f}{\partial z} & \frac{\partial f}{\partial y} \\ \frac{\partial g}{\partial z} & \frac{\partial g}{\partial y} \end{bmatrix}, \quad \begin{bmatrix} \frac{\partial f}{\partial z} & \frac{\partial f}{\partial y} & \frac{\partial f}{\partial p} \\ \frac{\partial g}{\partial z} & \frac{\partial g}{\partial y} & \frac{\partial g}{\partial p} \\ \frac{\partial \Delta}{\partial z} & \frac{\partial \Delta}{\partial y} & \frac{\partial \Delta}{\partial p} \end{bmatrix}.$$

Suppose the above conditions are satisfied at $(0, 0, p_0)$, then there exists a smooth curve of equilibrium points in $\mathbb{R}^{n+m+1}$ that passes through this point and is transversal to the singular surface at $(0, 0, p_0)$. When p increases through p_0, one eigenvalue of the system moves from $\mathbb{C}^+$ to $\mathbb{C}^+$ if $B/C > 0$ (respectively, from $\mathbb{C}^+$ to $\mathbb{C}^+$ if $B/C < 0$) along the real axis by diverging through infinity. The other

$(n-1)$ eigenvalues remain bounded and stay away from the origin. The constants B and C can be computed by

$$B = - trace\left(\frac{\partial f}{\partial y} adj\left(\frac{\partial g}{\partial y}\right)\frac{\partial g}{\partial z}\right),$$

and

$$C = \left(\frac{\partial \Delta}{\partial p} - \begin{pmatrix} \frac{\partial \Delta}{\partial z} & \frac{\partial \Delta}{\partial y} \end{pmatrix}\begin{pmatrix} \frac{\partial f}{\partial z} & \frac{\partial f}{\partial y} \\ \frac{\partial g}{\partial z} & \frac{\partial g}{\partial y} \end{pmatrix}^{-1}\begin{pmatrix} \frac{\partial f}{\partial p} \\ \frac{\partial g}{\partial p} \end{pmatrix}\right).$$

3.5 Chaotic Behavior

The asymptotic behavior of an autonomous dynamical system is uniquely specified by their initial conditions. Equilibrium point, limit cycle, torus and chaos are four possible types of equilibrium behaviors. A chaotic system is a deterministic system that exhibits irregular and unpredictable behavior (Giannakopoulos et al. 2002). Chaos occurs in many nonlinear systems, and its main characteristic is that system does not repeat its past behavior. In spite of their irregularity, chaotic dynamical systems follow deterministic equations (Baker and Gollub 1990). The unique characteristic of chaotic systems is dependence on the initial conditions sensitively. Slightly different initial conditions result in very different orbits. There are various methods for detecting chaos such as Poincare maps and Lyapunov exponents.

One-dimensional bifurcation diagrams of Poincare maps present information about the dependence of the dynamics on a certain parameter to gain preliminary insight into the properties of the dynamical system. The analysis reveal the type of attractor to which the dynamics will ultimately settle down after passing the initial transient phase and within which the trajectory will then remain forever. The dynamical behavior on a Poincare surface of section can be described by a discrete map whose phase-space dimension is less than that of the original continuous flow.

Moreover, the Lyapunov exponent is another approach to detect chaos, and it is a measure of the speeds at which initially nearby trajectories of the system diverge. The Lyapunov exponent is related to the predictability of the system, and the largest Lyapunov exponent of a stable system does not exceed zero. However, a chaotic system has at least one positive Lyapunov exponent, and the more positive the largest Lyapunov exponent, the more unpredictable the system is. Consistent with the idea that the chaotic attractor is globally stable, thus the sum of all Lyapunov exponents of a chaotic system will be negative.

4 Bifurcation Analysis and Chaotic Behaviors of FOS Biological Models

As it mentioned in Sect. 1, fractional-order modeling has recently garnered a lot of attention and gained popularity in the evaluation of dynamical systems due to its ability to provide an exact description of different nonlinear phenomena and inherent relation to various materials and processes with memory and hereditary properties. It allows greater degrees of freedom in the model and is closely related to fractals which are abundant in integer-order descriptions of biological systems and describes the whole time domain for a physical process, while the integer-order derivative is related to the local properties of a certain position and indicates a variation or certain attribute at particular time. According to these reasons, fractional-order modeling of many real phenomena especially biological systems has more advantages and consistency rather than classical integer-order mathematical modeling.

In 1954, Gordon investigated the economic theory of natural resource utilization in fishing industry, and discussed the effects of harvest effort on its ecosystem (Gordon 1954). The harvest can be affected by numerous factors such as seasonality, revenue, market demand and harvest cost, and then, it's reasonable to consider the harvest effort as a variable from the real point of view, and consequently harvest function $h(t)$ can be expressed by $h(t) = x(t)y(t)$, where $x(t)$ is the harvest effort performed by a static human population, and $y(t)$ is a harvested specious in a considered ecosystem. Finally, he proposed the following algebraic equation to study the economic interest of the yield of harvest effort in his theory of a common-property resource:

$$ph(t) - cx(t) = m \tag{19}$$

where m represents the net economic profit, $ph(t)$ is total revenue and $cx(t)$ is total cost, where p and c are the price of a unit of the harvested biomass and the cost of a unit of the effort, respectively (Gordon 1954).

In line with this theory, differential-algebraic (singular) integer-order biological systems were proposed, and dynamic behaviors analysis was investigated to design some control strategies (Chakraborty et al. 2011; Zhang et al. 2012). Combining the economic theory of fishery resource with fractional calculus, some FOS biological economic models such as predator-prey models (Holling-II, Holling-Tanner and food web), logistic map and SEIR epidemic model will be introduced as follows, and their qualitative behaviors such as bifurcation and chaos will be illustrated.

4.1 Predator-Prey Models

The last few decades have been active in the development of different kinds of predator–prey model within the traditional territory of population biology.

Most studies of generalists have focused on their functional response, and many authors have explored the dynamics of predator-prey systems based on type-II, Holling-Tanner and Leslie-Grower functional responses. In recent years, there was a growing interest in the research field of the predator-prey with multi-species (especially one predator and two prey) which is called food web systems, and rich dynamical behavior has been found in such a system (Gakkhar and Singh 2007; Gakkhar and Naji 2003). Here, we explain the most popular predator-prey model with Holling type-II functional response, and its FOS model will be investigated in details. We only introduce the FOS models of Holling-Tanner and food web and neglect their detail analysis which can be expressed in an analogous manner.

4.1.1 Model Formulation and Qualitative Analysis

Freedman introduced the most popular predator-prey model with the Holling type-II functional response $\beta x_1(t)x_2(t)/(1+\sigma x_1(t))$, where x_1 and x_2 are the population densities of the prey and predator, respectively (Freedman 1980). β is the feeding rate, and σ is a positive constant that explains the effects of capture rate. The interactions between prey and predator take the form with the following ordinary differential equations:

$$\begin{aligned}
\frac{dx_1(t)}{dt} &= rx_1(t)\left(1 - \frac{x_1(t)}{K}\right) - \frac{\beta x_1(t)x_2(t)}{1+\sigma x_1(t)}, \\
\frac{dx_2(t)}{dt} &= \frac{\beta x_1(t)x_2(t)}{1+\sigma x_1(t)} - ax_2(t),
\end{aligned} \tag{20}$$

where a is a positive real number and the logistic growth $rx_1(t)(1-x_1(t)/K)$ is assumed to be the prey host population with carrying capacity K and a specific growth rate constant r.

Using the fractional calculus and the economic theory, the integer-order standard predator-prey model (20) can be extended based on the algebraic economic interest Eq. (19), and accordingly, the proposed FOS model of the predator-prey system which consists of two fractional-order differential equations and one algebraic equation can take the following form (Nosrati and Shafiee 2017):

$$\begin{cases}
{}_0^C D_t^\alpha x_1(t) = rx_1(t)\left(1 - \frac{x_1(t)}{K}\right) - \frac{\beta x_1(t)x_2(t)}{1+\sigma x_1(t)}, \\
{}_0^C D_t^\alpha x_2(t) = \frac{\beta x_1(t)x_2(t)}{1+\sigma x_1(t)} - ax_2(t) - h(t), \\
0 = x_3(t)(px_2(t) - c) - m, \quad t \geq 0.
\end{cases} \tag{21}$$

The system (21) can also be written as the FOS system (9), where $F: \mathbb{R}^3 \to \mathbb{R}^3$, $x(t) \in \mathbb{R}^3$ and the matrix $E \in \mathbb{R}^{3 \times 3}$ have the following forms:

$$x(t) = \begin{bmatrix} x_1(t) \\ x_2(t) \\ x_3(t) \end{bmatrix}, \quad E = \begin{bmatrix} 1 & 0 & 0 \\ 0 & 1 & 0 \\ 0 & 0 & 0 \end{bmatrix},$$

$$F = \begin{bmatrix} f_1 \\ f_2 \\ f_3 \end{bmatrix} = \begin{bmatrix} rx_1(t)\left(1 - \frac{x_1(t)}{K}\right) - \frac{\beta x_1(t)x_2(t)}{1+\sigma x_1(t)} \\ \frac{\beta x_1(t)x_2(t)}{1+\sigma x_1(t)} - ax_2(t) - x_3(t)x_2(t) \\ x_3(t)(px_2(t) - c) - m \end{bmatrix}.$$

As seen, the system (21) is in form of the semi-explicit FOS system (12a, 12b) in which $z(t) = [x_1 \quad x_2]^T$, $y(t) = x_3(t)$, $f = [f_1 \quad f_2]^T$, $g = f_3$.

Theorem 4 (Nosrati and Shafiee 2017) *The FOS model of predator-prey system (21) is solvable if $x_2 \neq c/p$.*

The main objective is to investigate the local stability of the system (21) based on singular system, bifurcation theories and the effects of economic profit on dynamics of this system in which will be discussed in the region $R_+^3 = \{(x_1, x_2, x_3) | x_i \geq 0, \ i = 1, 2, 3\}$ as an admissible space.

When $m = 0$, there exist following six equilibrium points $X_i^* = \left({}_ix_1^* \quad {}_ix_2^* \quad {}_ix_3^* \right)^T$ ($i = 1, 2, \ldots, 6$) for the system (21):

$$X_1^* = \begin{pmatrix} 0 \\ 0 \\ 0 \end{pmatrix}, X_2^* = \begin{pmatrix} K \\ 0 \\ 0 \end{pmatrix}, X_3^* = \begin{pmatrix} \frac{a}{\beta - a\sigma} \\ \frac{-r(a - k(\beta - a\sigma))}{k(\beta - a\sigma)} \\ 0 \end{pmatrix}, X_4^* = \begin{pmatrix} 0 \\ \frac{c}{p} \\ -a \end{pmatrix}, X_5^* = \begin{pmatrix} {}_5x_1^* \\ \frac{c}{p} \\ {}_5x_3^* \end{pmatrix}, X_6^* = \begin{pmatrix} {}_6x_1^* \\ \frac{c}{p} \\ {}_6x_3^* \end{pmatrix}$$

where ${}_5x_1^*$ and ${}_6x_1^*$ (${}_5x_1^* \leq {}_6x_1^*$) are roots of the equation $pr\sigma x_1^2 + pr(1 - k\sigma)x_1 + K(\beta c - pr) = 0$, and also, ${}_5x_3^* = -a + \beta {}_5x_1^*/(1 + \sigma_5 x_1^*)$ and ${}_6x_3^* = -a + \beta {}_6x_1^* /(1 + \sigma_6 x_1^*)$.

Regarding any positive parameters and admissible space definition, all these points can be admissible except X_4^* which is always negative. To assess the stability analysis of the system (21), using Remark 3 and Lemma 1, the argument eigenvalues of Jacobian matrix J evaluated at the admissible equilibrium points will be checked respect to $a\pi/2$.

Obviously, the equilibrium point X_1^* is saddle node. The eigenvalues of system (21) at equilibrium point X_2^* are $\lambda_1 = -r$ and $\lambda_2 = -a + \beta K/(1 + \sigma K)$. Using Remark 3, λ_1 is always stable, since $|\arg(\lambda_1)| = \pi > a\pi/2$, and the stability of λ_2 changes under parameter variation:

$$|\arg(\lambda_2)| = \begin{cases} 0 < \frac{a\pi}{2} & \text{if } \beta K > a(1 + \sigma K) \quad \Rightarrow \quad \text{unstable} \\ \pi > \frac{a\pi}{2} & \text{if } \beta K < a(1 + \sigma K) \quad \Rightarrow \quad \text{stable} \end{cases}.$$

According to the analysis illustrated above, the stability of equilibrium point X_2^* changes from stable to unstable when β increases through $a(1 + \sigma K)/K$.

Then, β can be regarded as a bifurcation parameter, and the following theorem can be extracted:

Theorem 5 (Nosrati and Shafiee 2017) *The system (21) undergoes transcritical bifurcation at the equilibrium point X_2^* when bifurcation parameter $\mu = \beta$ is increased through $a(1 + \sigma K)/K$.*

Based on the results derived in Subsect. 3.2, it is adequate to check the following statements to prove the theorem:

(1) $\left. \frac{\partial f_{1,2}}{\partial x_{1,2}} \right|_{X_2^*} = \begin{bmatrix} -r & -a \\ 0 & 0 \end{bmatrix}$, then $|\lambda I - \partial f_{1,2}/\partial x_{1,2}|_{X_2^*}$ has a simple zero eigenvalue

with right eigenvector $v = \begin{pmatrix} 1 & -r/a \end{pmatrix}^T$ and left eigenvector $w = \begin{pmatrix} 0 & 1 \end{pmatrix}$.

(2) $w \left(\left. \frac{\partial^2 f_{1,2}}{\partial \mu \partial x_{1,2}} \right|_{X_2^*} \right) v \neq 0.$

(3) $w \left(\left. \frac{\partial^2 f_{1,2}}{\partial^2 x_{1,2}} \right|_{X_2^*} \right) (v, v) \neq 0.$

At the equilibrium point X_3^*, it is easy to check under different parameter values, this equilibrium point can be stable focus or node. The equilibrium points X_5^* and X_6^* are at the singularity, and after that, the matrix J is not well defined because $\partial f_3 / \partial x_3$ is singular. Therefore, the matrix J might have some unbounded eigenvalues, and subsequently, the system (21) may show SIB behavior. Based on following theorem, the system (21) has a SIB at equilibrium points X_5^* and X_6^* when the bifurcation parameter m is zero. If m increases through zero, one eigenvalue of the system (21) evaluated at these equilibrium points will move from an open complex half plane to other open complex half plane along the real axis by diverging into infinity. The other eigenvalue remains bounded and stays away from the origin.

Theorem 6 (Nosrati and Shafiee 2017) *Assume $\partial f_1 / \partial x_1|_{X_5^*, X_6^*} \neq 0$. The FOS model of predator-prey system (21) has an SIB at the equilibrium points X_5^* and X_6^* when the bifurcation parameter m increases through zero. Besides, the stability of the equilibrium points varies from stable to unstable.*

Suppose $\Upsilon = \partial f_3 / \partial x_3 = p x_2(t) - c$. According to the results, we have

(1) $\Upsilon|_{X_5^*, X_6^*}$ has a simple zero eigenvalue, and

$$(2) \quad \begin{bmatrix} \frac{\partial f_{1,2}}{\partial x_{1,2}} & \frac{\partial f_{1,2}}{\partial x_3} \\ \frac{\partial f_3}{\partial x_{1,2}} & \frac{\partial f_3}{\partial x_3} \end{bmatrix}\Bigg|_{X_5^*,X_6^*} = c(r - \frac{2r_{5,6}x_1^*(t)}{K} - \frac{\beta c}{p(1+\sigma_{5,6}x_1^*(t))^2})5,6x_3^*(t)(t) \neq 0$$

and

$$\begin{bmatrix} \frac{\partial f_{1,2}}{\partial x_{1,2}} & \frac{\partial f_{1,2}}{\partial x_3} & \frac{\partial f_{1,2}}{\partial \mu} \\ \frac{\partial f_3}{\partial x_{1,2}} & \frac{\partial f_3}{\partial x_3} & \frac{\partial f_3}{\partial \mu} \\ \frac{\partial Y}{\partial x_{1,2}} & \frac{\partial Y}{\partial x_3} & \frac{\partial Y}{\partial \mu} \end{bmatrix}\Bigg|_{X_5^*,X_6^*} = c(r - \frac{2r_1 x_{5,6}^*(t)}{K} - \frac{\beta c}{p(1+\sigma_1 x_{5,6}^*(t))^2}) \neq 0.$$

Therefore, there exists stability change of the equilibrium points X_5^* and X_6^* when m increases through zero; i.e., one eigenvalue of the system (eigenvalue of Jacobian matrix J evaluated along the equilibrium locus related to X_5^* and X_6^*) moves from one half plane to other half plane. On the other hand, $B = px_2(t)x_3(t)|_{X_5^*,X_6^*}$, and $C = (1/px_3(t))|_{X_5^*,X_6^*}$. Regarding the admissibility space, $B > 0$ and $C > 0$. After that, when μ increases through zero, this eigenvalue of the system moves from left half plane to right half plane along the real axis by diverging into infinity because $B/C > 0$. The other eigenvalue maintains bounded and stays away from the origin in left half plane. Thus, the stability of system (21) changes from stable to unstable at the equilibrium points X_5^* and X_6^* when the economic profit increases through zero. This completes the proof.   □

A complete analysis on this system under positive economic profit can be seen in (Nosrati and Shafiee 2017).

4.1.2 Numerical Simulation

In order to solve (21), the method introduced by Atanackovic and Stankovic can be used. Atanackovic and Stankovic showed that for a function $f(t)$, the Caputo fractional derivative of order α may be expressed as

$$_0^C D_t^\alpha f(t) \simeq \Omega(\alpha, t, M)\dot{f}(t) + \Phi(\alpha, t, M)f(t) + \sum_{n=2}^{M} A(\alpha, t, M)\frac{v_n(f)(t)}{t^{n-1+\alpha}}, \qquad (22)$$

where

$$A(\alpha, t, M) = -\frac{\Gamma(n-1+\alpha)}{\Gamma(2-\alpha)\Gamma(\alpha-1)(n-1)!}, \quad \Omega(\alpha, t, M) = \frac{1}{\Gamma(2-\alpha)t^{\alpha-1}} + \sum_{n=1}^{M}\frac{A(\alpha, t, n)}{nt^{\alpha-1}}, \quad \Phi(\alpha, t,$$

$$M) = \frac{1-\alpha}{t^\alpha \Gamma(2-\alpha)} + \sum_{n=2}^{M}\frac{A(\alpha, t, n)}{t^\alpha} \quad \text{and} \quad v_n(f)(t) = -(n-1)\int_o^t \tau^{n-2}f(\tau)d\tau, \quad n = 2, 3, \ldots$$

(Atanackovic and Stankovic 2004). Thus, the system (21) can be expressed by

$$E'\dot{x}'(t,n) = F'(x'(t,n)), \quad 0 < \alpha < 1, \quad n = 2, 3, \ldots, M, \quad t \geq 0, \quad (23)$$

where $x'(t,n) = [\, x_1(t) \quad w^n(t) \quad x_2(t) \quad u^n(t) \quad x_3(t) \,]^T$, and also, $F' : \mathbb{R}^5 \to \mathbb{R}^5$ and $E' \in \mathbb{R}^{5 \times 5}$ have the following forms:

$$E = \begin{bmatrix} I_4 & 0 \\ 0 & 0 \end{bmatrix}, F' = \begin{bmatrix} f_1' \\ f_2' \\ f_3' \\ f_4' \\ f_5' \end{bmatrix} = \begin{bmatrix} \frac{1}{\Omega(\alpha,t,M)}\left[f_1 - \Phi(\alpha,t,M)x_1(t) - \sum_{n=2}^{M} A(\alpha,t,M)\frac{w_n(t)}{t^{n-1+\alpha}} \right] \\ \quad - (n-1)t^{n-2}x_1(t) \\ \frac{1}{\Omega(\alpha,t,M)}\left[f_2 - \Phi(\alpha,t,M)x_2(t) - \sum_{n=2}^{M} A(\alpha,t,M)\frac{u_n(t)}{t^{n-1+\alpha}} \right] \\ \quad - (n-1)t^{n-2}x_2(t) \\ f_3 \end{bmatrix}$$

Now, numerical solution of the singular ordinary differential system (23) will be considered to derive orbits of the FOS predator-prey system (21) for different set of parameters. For convenience, the simulation will be implemented using the fixed parameter values $r = 0.2$, $K = 5$, $\beta = 0.2$, $\sigma = 0.01$, $a = 0.2$, $p = 1.5$, $c = 1$, $M = 10$ and m will be varied.

Numerical values of prey and predator, and also, phase portrait of the system (21) are presented in Figs. 3 and 4 for the set parameter values and two different values of β. As seen in Fig. 3, the trajectories of the system converge to the equilibrium point X_3^* in steady state, and the equilibrium point X_2^* is unstable because $\beta = 0.2 > 0.042$. In Fig. 4, the system (21) is simulated for $\beta = 0.041 < 0.042$. In this case, the equilibrium point X_3^* is unstable, and the trajectories of the system converge to the equilibrium point X_2^* in steady state which verifies the existence of transcritical bifurcation (Theorem 5). In all numerical runs,

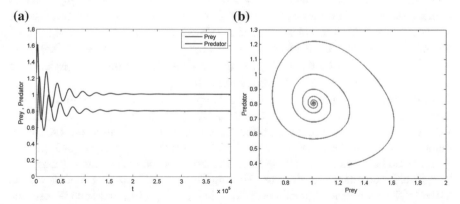

(a)   **(b)**

Fig. 3 **a** Numerical value of $x_1(t)$ and $x_2(t)$ respect to time. **b** Phase portrait of system (21) ($\alpha = 0.8$, $\beta = 0.2$, $x_1(0) = 1.3$, $x_2(0) = 0.4$, $x_3(0) = 0.00025$, $m = 0$)

(a)   (b)

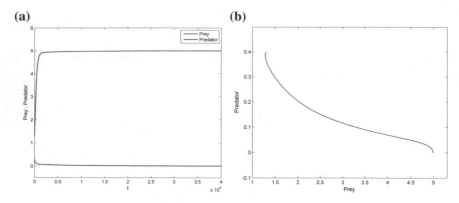

Fig. 4 **a** Numerical value of $x_1(t)$ and $x_2(t)$ respect to time. **b** Phase portrait of system (21) ($\alpha = 0.8$, $\beta = 0.041$, $x_1(0) = 1.3$, $x_2(0) = 0.4$, $x_3(0) = 0.00025$, $m = 0$)

the solution has been approximated using the parameter values given in the captions of the figures.

The admissible equilibrium point X_5^* is at the singularity. When economic profit $m = -0.0001$, the eigenvalues are $\lambda_1 = -193.8$ and $\lambda_2 = -0.067$, and then become $\lambda_1 = 191.4$ and $\lambda_2 = -0.066$ when the parameter value $m = 0.0001$. Obviously, λ_2 remains almost constant and λ_1 moves from the open complex left half plane to the open complex right half plane along the real axis by diverging through infinity. This verifies the Theorem 6 and demonstrates that the system (21) has an SIB at the equilibrium point X_5^* when the bifurcation parameter $m = 0$.

Numerical values of the system (21) are presented in Figs. 8 and 9 for two different economic profit values $m = -0.0001$ and $m = 0.0001$. When $m = -0.0001$, the equilibrium point X_5^* is stable and the trajectories of system (21) converge to X_5^* (Fig. 5). Besides the admissible equilibrium point X_5^*, there is another stable equilibrium point X_6^* (related to $m < 0$) when the initial condition is varied. This equilibrium point is not admissible because the trajectory $x_3(t)$ converges to a negative point in steady state (Fig. 6). Also, when $m = 0.0001$, the stability of the equilibrium point X_5^* changes to unstable, and therefore, trajectories of the system converge to X_6^* (related to $m > 0$) (Fig. 7).

Furthermore, to explain the oscillation damping properties, the phase portrait of system (21) for three different values of α are given in Fig. 8, for two different values of m. The results show that, the fractional derivative damps the oscillation behavior of the model when α decreases, which leads to improve the stability.

It should be noted that using the fractional calculus and the economic theory, the integer-order standard predator-prey Holling-Tanner and web food models can be extended based on the algebraic economic interest Eq. (19), and accordingly, the proposed FOS model of these systems can take the (24) and (25), respectively.

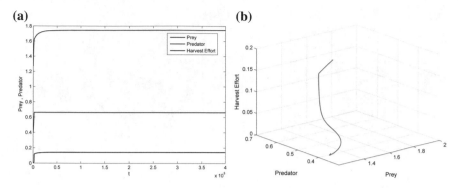

Fig. 5 **a** Numerical value of $x_1(t)$, $x_2(t)$ and $x_3(t)$ respect to time. **b** Phase portrait of system (21) ($\alpha = 0.8$, $\beta = 0.2$, $x_1(0) = 1.3$, $x_2(0) = 0.4$, $x_3(0) = 0.00025$, $m = -0.0001$)

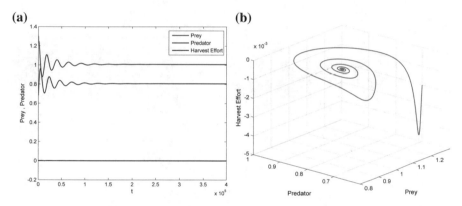

Fig. 6 **a** Numerical value of $x_1(t)$, $x_2(t)$ and $x_3(t)$ respect to time. **b** Phase portrait of system (21) ($\alpha = 0.8$, $\beta = 0.2$, $x_1(0) = 1.3$, $x_2(0) = 0.7$, $x_3(0) = -0.002$, $m = -0.0001$)

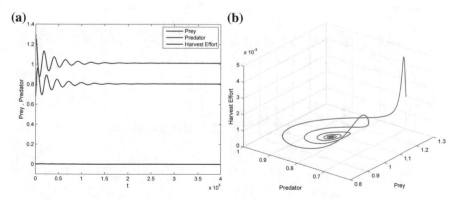

Fig. 7 **a** Numerical value of $x_1(t)$, $x_2(t)$ and $x_3(t)$ respect to time. **b** Phase portrait of system (21) ($\alpha = 0.8$, $\beta = 0.2$, $x_1(0) = 1.3$, $x_2(0) = 0.7$, $x_3(0) = 0.002$, $m = 0.0001$)

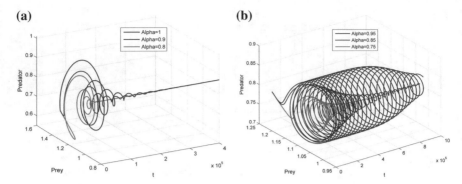

Fig. 8 (Oscillation damping property). Phase portrait of system (21) respect to time **a** $m = 0$ **b** $m = 0.002$ ($\beta = 0.2$, $x_1(0) = 1.3$, $x_2(0) = 0.7$, $x_3(0) = 0.0025$)

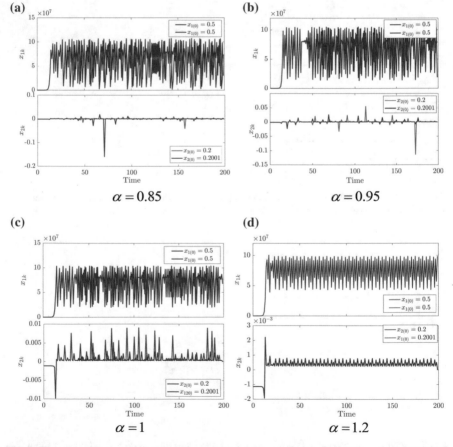

Fig. 9 The x_{1k} and x_{2k} graphs for two nearby initial conditions $(0.5, 0.2)$ and $(0.5, 0.2001)$ ($r = 2.8$)

$$\begin{cases} {}_0^C D_t^\alpha x_1(t) = x_1(t)(1 - x_1(t)) - \dfrac{x_1(t)x_2(t)}{\gamma + x_1(t)} + \lambda(1 - cos\theta t)x_1(t), \\[3mm] {}_0^C D_t^\alpha x_2(t) = x_2(t)(\delta - \beta\dfrac{x_2(t)}{x_1(t)}), \\[3mm] 0 = x_3(t)(px_2(t) - c) - m, \quad t \geq 0. \end{cases} \tag{24}$$

$$\begin{cases} {}_0^C D_t^\alpha x_1(t) = x_1(t)\left(1 - x_1(t) - \dfrac{w_2 x_3(t)}{1 + w_3 x_1(t) + w_4 x_2(t)}\right), \\[3mm] {}_0^C D_t^\alpha x_2(t) = x_2(t)\left((1 - x_2(t))w_5 - \dfrac{w_7 x_3(t)}{1 + w_4 x_2(t) + w_3 x_1(t)}\right), \\[3mm] {}_0^C D_t^\alpha x_3(t) = x_3\left(t\left(\dfrac{w_8 x_1(t) + w_9 x_2(t)}{1 + w_3 x_1(t) + w_4 x_2(t)} - w_{10}\right) - w_{11}x_3(t) - x_3(t)x_4(t), \\[3mm] 0 = x_4(t)x_3(t)(\dfrac{a}{b + x_4(t)x_3(t)} - \dfrac{d}{x_3(t)}) - m, \quad t \geq 0. \end{cases} \tag{25}$$

where the parameters interpretation is mentioned in (Zhang et al. 2012). The systems (24) and (25) can also be written as the FOS system (9), and therefore, their analysis can be studied in a same analogous as the system (21) which are omitted in this study.

4.2 Logistic Map

The logistic model is widely used to investigate the growth law of various biological ecosystems such as some kind of single-cell, marine population, and birds and insects populations on continent (Clark 1990). Although many discussions have been applied on the behavior of integer-order standard logistic map (Alligood et al. 1997), fewer efforts have been contributed to the behaviors of the fractional-order and singular cases. For the famous logistic map

$$x_{k+1} = r x_k \left(1 - \frac{x_k}{K}\right) \tag{26}$$

popularized by May in (1976), the system exhibits chaotic behaviors for most values of the growth coefficient r. For the system (26), $x(k) > 0$ represents population density, $r > 0$ represents the intrinsic growth rate, and $K > 0$ represents the environment capacity.

In (Zhang et al. 2012), a discrete singular logistic system was proposed, and its dynamics were discussed. It was demonstrated that the model system bifurcates into periodical orbits and finally admits chaotic behavior under parameter variations. Also, in some literatures, it has been demonstrated that there is a discrete fractional logistic map which has a generalized chaos behavior (Munkhammar 2013; Wu and

Baleanu 2014; Guckenheimer and Holmes 1983). These studies introduced a fractional discrete logistic map using the fractional-order difference in different senses. Compared with the one of the integer order, the fractional model has a discrete memory and a fractional difference order. When the difference order changes in the numerical results, new chaotic behaviors of the logistic map are observed. It has been demonstrated that the chaotic zones not only depends on the coefficients r but the difference order. Although the chaos theory for discrete maps is well understood, how it is related to fractional calculus phenomena is perhaps less clarified and need a further investigation.

In this subsection, a discrete fractional-order singular logistic system is proposed, and the dynamics of the model system, especially chaotic behavior, are discussed.

4.2.1 Model Formulation

The growth law of various biological species is usually described by the classic logistic model (26). Compared with the continuous model, the dynamics of the discrete logistic model with one dimension are abundant. There are two fixed points for this system: $x_1^* = 0$ and $x_2^* = K$. Using these two real equilibrium points and the eigenvalues of the corresponding Jacobian matrix, the behavior of the system can be evaluated, and its rich dynamics can be derived when the parameter r changes.

According to the Gordon theory and fractional calculus, the following discrete FOS system is proposed to investigate the dynamics of the logistic system and the economic interest of the harvest effort on its population:

$$E_0^{GL} \Delta_{k+1}^\alpha x_{k+1} = F(x_k) \tag{27}$$

where $_0^{GL}\Delta_k^\alpha$ denotes the GL difference operator, and $F: \mathbb{R}^2 \to \mathbb{R}^2$, $x_k \in \mathbb{R}^2$ and the matrix $E \in \mathbb{R}^{2 \times 2}$ have the following forms:

$$x_k = \begin{bmatrix} x_{1k} \\ x_{2k} \end{bmatrix}, \quad E = \begin{bmatrix} 1 & 0 \\ 0 & 0 \end{bmatrix}, \quad F = \begin{bmatrix} f_1 \\ f_2 \end{bmatrix} = \begin{bmatrix} rx_{1k}\left(1 - \frac{x_{1k}}{K}\right) - x_{1k}x_{2k} \\ x_{2k}(px_{1k} - c) - m \end{bmatrix}.$$

As seen, the system (27) is in the discrete form of the semi-explicit FOS system (12a, 12b) in which $z_k = [x_{1k} \quad x_{2k}]^T$, $y_k = x_{2k}$, $f = f_1$, $g = f_2$.

The fractional order GL difference is given by

$$_0^{GL}\Delta_k^\alpha x_k = \frac{1}{h^\alpha} \sum_{j=0}^{k} (-1)^j \binom{\alpha}{j} x_{k-j} \tag{28}$$

where $\alpha = diag\{\alpha_1, \ldots, \alpha_n\} \in \mathbb{R}^n$ is the real orders of the fractional difference, h is the sampling interval, k is the number of samples for which the derivative is calculated, and the coefficient

$$\binom{\alpha}{j} = diag\left(\binom{\alpha_1}{j}, \cdots, \binom{\alpha_n}{j}\right)$$

is the extended form of integer-valued binomial coefficient developed by the gamma function idea, with

$$\binom{\alpha_i}{j} = \begin{cases} 1 & \text{if } j=0 \\ \frac{\Gamma(\alpha_i+1)}{\Gamma(\alpha_i+1-j).\Gamma(j+1)} & \text{if } j>0 \end{cases},$$

for $i=1, \ldots, n$. According to this definition, discrete equivalent of the fractional-order derivative and integration can be obtained when α is positive and negative, respectively.

For the FOS logistic system (27), x_{1k}, r and K share the same biological interpretations as in (26), and x_{2k} represents the harvest effort on population, p is the unit price of the harvested population, and c is the united cost of the harvest effort.

4.2.2 Numerical Simulation

From (28), we can obtain the following equivalent difference equation form of the FOS logistic model (27):

$$x_{1(k+1)} = h^\alpha\left(rx_{1k}\left(1-\frac{x_{1k}}{K}\right) - x_{1k}x_{2k} - \frac{1}{h^\alpha}\sum_{j=1}^{k+1}(-1)^j\binom{\alpha}{j}x_{k+1-j}\right)$$

$$0 = x_{2k}(px_{1k}-c)-m \tag{29}$$

Compared with the map of the integer order (25), the fractionalized one (27) has a discrete kernel function. As seen from (29), the state x_{1k} depends on the past information $x_{1(k-1)}, x_{1(k-2)}, \ldots, x_{1(0)}$. As a result, the memory effects of the discrete maps mean that their present state of evolution depends on all past states.

Assume $K = 8.05 \times 10^7$, $p = 5 \times 10^{-3}$, $c = 8.75 \times 10^4$, $m = 100$ and $h = 1$. We can derive the numerical solutions x_k using the Matlab. In what follows, Figs. 9 and 10 show the numerical solutions for different r and α, for two slightly different initial conditions, one at $(0.5, 0.2)$ and the other at $(0.5, 0.2001)$. These graphs are nearly identical for a certain time period, but then they differ considerably. No matter how close two solutions start, they always move apart in this manner when they are close to the attractor. This is sensitive dependence on initial conditions, one of the main features of a chaotic system. In Fig. 9, for a fixed parameter $r = 2.8$, when the order α decreases, period doubling event occurs, and finally system undergoes to a chaotic behavior. Especially, Fig. 9c verifies the results obtained by the analysis of the integer-order singular logistic map introduced by Zhang et al. (2012) in which the proposed system showed the chaotic behavior when $r = 2.8$. Also, in Fig. 10,

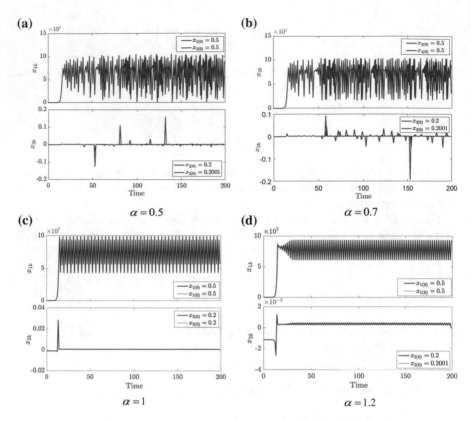

Fig. 10 The x_{1k} and x_{2k} graphs for two nearby initial conditions $(0.5, 0.2)$ and $(0.5, 0.2001)$ $(r = 2.5)$

this simulation is repeated for a fixed parameter $r = 2.5$, and again we can see that when the order α decreases, period doubling event occurs, and again system undergoes to a chaotic behavior but this happens for a smaller order α. Unlike Fig. 9c, as it can be seen from Fig. 10c, there is no chaotic behavior when $r = 2.5$ and $\alpha = 1$, which verifies the previous results obtained in literatures.

To discuss more precisely under parameters variations, the bifurcation diagrams of Poincare for model system (27) against variation of parameters α and r are depicted in Figs. 11 and 12, respectively. As seen, the bifurcation diagram against variation parameter α moves from right to left of plane when the parameter value r decreases (Fig. 11). It means that chaotic behavior happens for a smaller value of α when the parameter r decreases. This can be seen from Fig. 13a in which the Lyaponuv exponent against variation parameter α gets positive for a smaller parameter value r. Also, the bifurcation diagram against variation parameter r moves from left to right of plane when the parameter value α increases (Fig. 12). It means that chaotic behavior happens for a bigger value of r when the parameter α

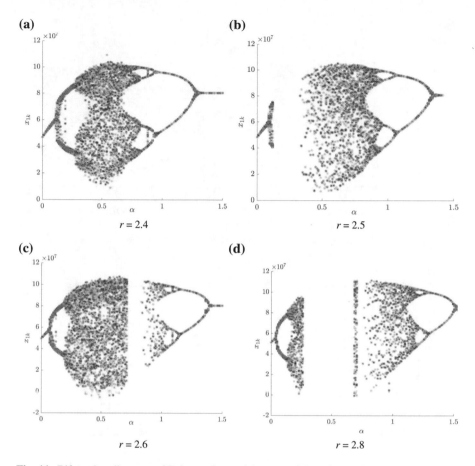

Fig. 11 Bifurcation diagrams of Poincare for model system (27) against variation of parameter α

increases. This can be seen from Fig. 13b in which the Lyaponuv exponent against variation parameter r gets positive for a smaller parameter value α.

There is a strange evolution in the bifurcation diagrams of the system (27) when the order α decreases (Fig. 11). This system exhibits stable equilibrium point behavior at first and undergoes to period doubling route to chaos and eventually enters to a chaotic space. When we continue the simulation while decreasing the order α, the system (27) behaves in a reverse treatment and undergoes to inverse period doubling and is finished by stable equilibrium point. Figure 14 shows the x_{1k} and x_{2k} graphs for two nearby initial conditions and two parameter order values $\alpha = 0.05$ and $\alpha = 1.4$, and also, $r = 2.5$. It shows that system is stable out of a certain band of the parameter values α, and when the order α increases (from left) and decreases (from right), from both sides the system undergoes to period doubling route to chaos in a reverse treatment.

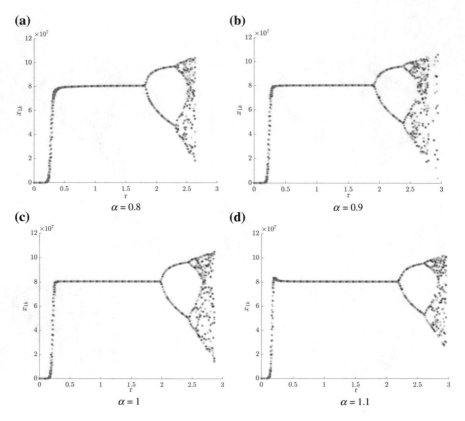

Fig. 12 Bifurcation diagrams of Poincare for model system (27) against variation of parameter r

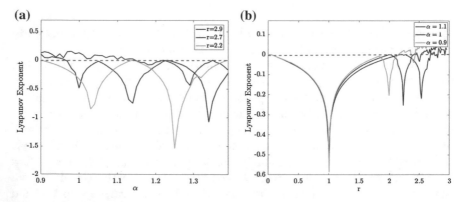

Fig. 13 Lyapunov Exponent diagram of the system (27) against variations of parameters **a** α and **b** r

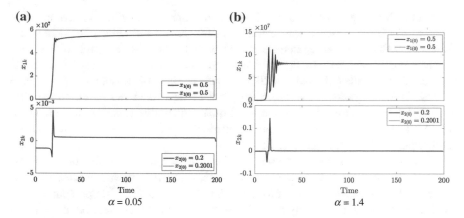

Fig. 14 The x_{1k} and x_{2k} graphs for two nearby initial conditions ($r = 2.5$)

4.3   SEIR Epidemic System

Modeling of the population dynamics of infectious diseases has been playing an important role in better understanding epidemiological patterns, and many epidemic models have been proposed and analyzed in recent years to control of disease for a long time (Kermack and McKendrick 1927; Kot 2001; Li et al. 2001; May and Oster 1976). The primary models, which customarily called an SIR (susceptible-infectious-recovered) or SIRS (susceptible-infectious-recovered-susceptible) system, assumes that the disease incubation can be negligible that, once infected, each susceptible individual (in class S) becomes infectious instantaneously (in class I) and later recovers (in class R) with a permanent or temporary acquired immunity (Glendinning and Perry 1997; Greenhalgh et al. 2004).

To study the role of incubation in disease transmission, the systems that are more general than SIR or SIRS types need to be studied. Thus, the resulting models are of SEIR (susceptible-exposed-infectious-recovered) or SEIRS (susceptible-exposed-infectious-recovered-susceptible) types, respectively, depending on whether the acquired immunity is permanent or not, and many analysis such as the stability, bifurcation and chaos behavior of these epidemic systems have been studied (Kuznetsov and Piccardi 1994; Sun et al. 2007; Xu et al. 2005).

Although many epidemic systems were described by differential and algebraic equations (Zhang et al. 2007, Zhang and Zhang 2007), they were studied by reducing the dimension of epidemic models to differential systems, accordingly, the dynamical behaviors of the whole system were not better described. Via analysis of the whole system by singular model, one can find more complex dynamical behaviors if the SEIR epidemic system. In (Zhang et al. 2014), integer-order singular SEIR epidemic system with seasonal forcing in transmission rate was discussed, and hyper-chaotic behavior of this system and its control with the aim of elimination of the disease was illustrated.

Although a large amount of work has been done in modeling the dynamics of epidemiological diseases, it was restricted to integer-order differential equations and few works discussed an epidemics model with fractional-order case. It has been demonstrated the great properties of fractional calculus are very useful to model epidemics problems (Liu and Lu, 2014; Goufo et al. 2014; Rostamy and Mottaghi 2016). In recent years, it has turned out that SEIR system can be described very successfully by the model using fractional-order differential equations in which help us to reduce the errors arising from the neglected parameters in modeling (Ozalp and Demirci 2011; Area et al. 2015).

However, no literature discusses fractional-order singular SEIR epidemic system. To the best of our knowledge, chaotic behavior first appears in these systems based on this subsection. In what follows, the FOS model of SEIR epidemic system will be introduced, and the dynamical behaviors of the model will be analyzed.

4.3.1   Model Formulation

The Fractional-order singular SEIR epidemic model with nonlinear transmission rate is introduced as follows. At time t, the population of size $N(t)$ is divided into four subpopulation containing susceptible $S(t)$, exposed $E(t)$, infectious $I(t)$, and recovers $R(t)$. It is assumed that death and birth occur with the same constant rate, i.e. the population size is constant. Thus, the host total population is $N(t) = S(t) + I(t) + R(t) + E(t)$ at any time t. In addition, it is assumed that immunity is permanent and recovered individuals do not revert to the susceptible class, and also, all newborns are susceptible and there is a uniform birth rate.

The following fractional-order singular SEIR system is derived based on the basic assumptions:

$$
\begin{cases}
{}_0^C D_t^\alpha x_1(t) = bx_5(t) - dx_1(t) - \beta \dfrac{x_1(t)x_3(t)}{x_5(t)} \\[2mm]
{}_0^C D_t^\alpha x_2(t) = \beta \dfrac{x_1(t)x_3(t)}{x_5(t)} - (\zeta + d)x_2(t) \\[2mm]
{}_0^C D_t^\alpha x_3(t) = \zeta x_2(t) - (\gamma + d)x_3(t) \\[2mm]
{}_0^C D_t^\alpha x_4(t) = \gamma x_3(t) - dx_4(t) \\[2mm]
\quad\quad 0 = x_1(t) + x_2(t) + x_3(t) + x_4(t) - x_5(t)
\end{cases}
\tag{30}
$$

where $x_1(t)$, $x_2(t)$, $x_3(t)$, $x_4(t)$ and $x_5(t)$ are the population $S(t)$, $E(t)$, $I(t)$, $R(t)$ and $N(t)$, respectively. Also, the parameter $b > 0$ is the rate for natural birth and $d > 0$ is the rate for natural death. The parameter $\zeta > 0$ is the rate at which the exposed individuals become infectious, and $\gamma > 0$ is the rate of recovery. The force of infection is $\beta x_3(t)/x_4(t)$, where $\beta > 0$ is effective per capita contact rate of infectious individuals and the incidence rate is $\beta x_1(t)x_3(t)/x_4(t)$.

The FOS system (30) can describe the whole behavior of certain epidemic spreads in a certain area. The first to fourth fractional-order differential equations of this system describe whole dynamical behaviors of every dynamic element and the last algebraic equation describes restriction of every dynamic element of system.

The transmission rate with seasonal forcing can be considered as $\beta = \beta_0(1 + \beta_1 \cos(2\pi t))$, where β_0 is the base transmission rate, and $0 \le \beta_1 \le 1$ measures the degree of seasonality. By utilizing some transformation as

$$x_1' = \frac{x_1}{x_5}, \ x_2' = \frac{x_2}{x_5}, \ x_3' = \frac{x_3}{x_5}, \ x_4' = \frac{x_4}{x_5},$$

the system (30) can be attacked by studying the following subsystem:

$$\begin{cases} {}_0^C D_t^\alpha x_1'(t) = b - dx_1'(t) - \beta x_1'(t)x_3'(t) \\ {}_0^C D_t^\alpha x_2'(t) = \beta x_1'(t)x_3'(t) - (\zeta + d)x_2'(t) \\ {}_0^C D_t^\alpha x_3'(t) = \zeta x_2'(t) - (\gamma + d)x_3'(t) \\ \quad 0 = x_1'(t) + x_2'(t) + x_3'(t) + x_4'(t) - 1 \end{cases} \tag{31}$$

The variable x_4' is described by the fractional-order differential equation $\gamma x_3'(t) - dx_4'(t)$ as well as algebraic equation $x_4'(t) = 1 - x_1'(t) - x_2'(t) - x_3'(t)$, and there is no the variable x_4' in the first to third equations of system (30). That is why the forth equation is removed.

The system (31) can also be written as the FOS system (9), where $F: \mathbb{R}^4 \to \mathbb{R}^4$, $x(t) \in \mathbb{R}^4$ and the matrix $E \in \mathbb{R}^{4 \times 4}$ have the following forms:

$$x(t) = \begin{bmatrix} x_1(t) \\ x_2(t) \\ x_3(t) \\ x_4(t) \end{bmatrix}, \quad E = \begin{bmatrix} I_3 & 0 \\ 0 & 0 \end{bmatrix}, \quad F = \begin{bmatrix} f_1 \\ f_2 \\ f_3 \\ f_4 \end{bmatrix} = \begin{bmatrix} b - dx_1'(t) - \beta x_1'(t)x_3'(t) \\ \beta x_1'(t)x_3'(t) - (\zeta + d)x_2'(t) \\ \zeta x_2'(t) - (\gamma + d)x_3'(t) \\ x_1'(t) + x_2'(t) + x_3'(t) + x_4'(t) - 1 \end{bmatrix}.$$

As seen, the system (30) is in form of the semi-explicit FOS system (12a, 12b) in which $z(t) = [x_1 \ \ x_2 \ \ x_3]^T$, $y(t) = x_4(t)$, $f = [f_1 \ \ f_2 \ \ f_3]^T$ and $g = f_4$.

4.3.2 Numerical Simulation

In this subsection, we consider two case of varying parameters β_1 and α, and discuss the behaviors of the system (31) under these variations. In order to solve the proposed system, the method introduced by Atanackovic and Stankovic can be used similar to Sect. 4.1.2. The numerical results show that there is chaotic dynamical behavior for the FOS SEIR system (31) with $\beta_1 = 0.28$ when the order α is equal to one. Also, for the case of varying parameter α, the dynamical behaviors of system

(31) are analyzed by simulation results, and it will be showed that the chaotic behavior occurs under different parameter value β_1.

Let β_1 be a varying parameter of (31), and the remaining parameters are as follows: $b = d = 0.02$, $\xi = 35.84$, $\gamma = 100$, and $\beta_0 = 1800$, respectively (Olsen and Schaffer 1990). Figure 15 shows the $x_1(t)$ and $x_2(t)$ coordinates of two solutions that start out nearby, one at $(0.016, 0.006, 0.012, 0.02)$ and the other at $(0.016001, 0.006, 0.012, 0.02)$. From Fig. 15a, when $\alpha = 1$ and $\beta_1 = 0.28$, these graphs are nearly identical for a certain time period, but then they differ considerably. No matter how close two solutions start, they always move apart in this manner when they are close to the attractor. This is sensitive dependence on initial conditions, one of the main features of a chaotic system. Also, when the order α

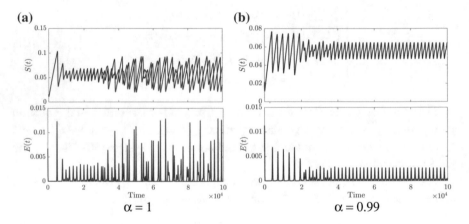

Fig. 15 The $x_1(t)$ and $x_2(t)$ graphs for two nearby initial conditions and $\beta_1 = 0.28$

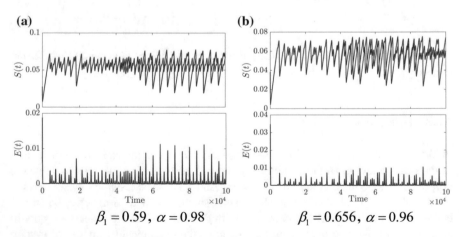

Fig. 16 The $x_1(t)$ and $x_2(t)$ graphs for two nearby initial conditions and different parameter values β_1 and α

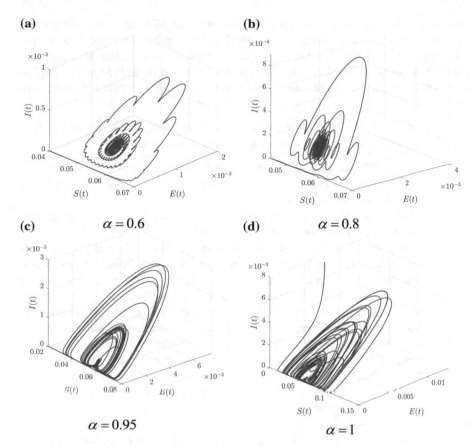

Fig. 17 FOS SEIR attractors in for $\beta_1 = 0.28$ and different parameter values of α

decreases insignificantly, the behavior of system changes and gets periodic (Fig. 15b). Furthermore, it can be demonstrated from simulation results that the system (31) exhibit chaotic behavior at $\alpha < 1$ when the parameter β_1 increases (Fig. 16). Figure 17 depicts the phase portrait for model system (31) in the case of $\beta_1 = 0.28$. As seen, the system (31) undergoes to period doubling when α increases and gets chaotic at $\alpha = 1$. The mathematical analysis of the system (31) will be illustrated in further investigations.

5 Conclusions and Discussions

In this chapter, some fractional-order singular (FOS) biological systems were established to investigate the impacts of economic profit and fractional derivative on the dynamic behaviors of these ecosystems. Our study extended previous models of biological systems predator-prey, logistic map and SEIR epidemic system, and proposed new and more realistic biological systems using fractional calculus and singular theory. Besides some mathematical analysis, the numerical simulations were considered to illustrate the effectiveness of the numerical method to explore the following impacts of fractional-order and singular modeling on the presented systems:

1. *The effect of fractional derivative:* It has been demonstrated that using fractional derivative can have following influences on the proposed models:

   - It reduces the errors arising from the neglected parameters in modeling of the memory-based biological systems which leads to derive the exact dynamical behavior of species interactions.
   - It acts as a time lag in ordinary differential model and causes to notably increase in the complexity of the observed behavior.
   - It takes less time for predator and prey and infectious diseases population to be settled as the fractional order decreases. Also, it will take the maximum time for the standard motion, i.e., $\alpha = 1$. In logistic map, we encountered with a strange behavior. When the order α decreases, this system exhibits stable equilibrium point behavior at first and undergoes to period doubling route to chaos and eventually enters to a chaotic space. Continuing this process, the system behaves in a reverse treatment and undergoes to inverse period doubling and is finished by stable equilibrium point.
   - The combination of fractional derivative and economic profit in singular form may change the stability of the system and cause the population and capture capability to be more sustainable.
   - The fractional derivative in the presented models damps the oscillation behavior and improves the stability of the solutions. In addition, the fractional order can impress the switching time from stability to instability. In recent case, the persistence and sustainable development of the ecosystem can be attained.

2. *The effect of singular modeling:* It is found that singular models exhibit more complicated dynamics rather than standard models, especially the bifurcation phenomena and chaotic behaviors, which can reveal the instability mechanism of systems. The most derived features are as follows:

   - Through the theoretical analysis and numerical simulation in predator-prey model, it has been demonstrated that there is a phenomenon of singularity induced bifurcation due to variation of economic interest of harvesting. This brings impulse phenomenon and causes a rapid growth of the species population. If this phenomenon prolongs a period of time, the species population will be out of the carrying capacity of the environment, and the collapse of the ecosystem may be happened.
   - It has been shown the predator-prey model exhibits another bifurcation phenomenon called transcritical bifurcation which varies the stability of the system and leads to extinct the predator population.

- Using singular modeling, high dimension chaotic attractor was occurred in SEIR epidemic models. Biologic signification of these types of attractor in epidemic models is that the epidemic disease will break out suddenly and spread gradually in a region at the period of the high incidence of the epidemic disease, and accordingly, many people in the region would be infected by disease. Also, singular modeling of logistic map can affect its behavior and enrich its dynamical properties.

All results show that extinction, speciation and stability of the biological ecosystems can be affected by fractional derivative and economic interest in singular form, and with considering the constraints imposed on the ecosystem, persistence and sustainable development of the ecosystem can be attained.

The future directions of research include:

- *Mathematical analysis*: During FOS modeling of logistic map and SEIR epidemic system, these models exhibited equilibrium point, period doubling and chaotic behaviors. These results were derived from simulation point of view, and more detailed analysis and synthesis of these new extended models need further investigations.
- *Application to medicine and engineering:* Biologic signification of chaotic behavior in proposed ecosystems is that the extinction, speciation and stability of the biological ecosystems can be affected and the species population will break out suddenly. Many species would be infected by disease, for example, and some of them would even lose their lives. Nevertheless, there exists uncertain prediction for the low period of the incidence of these events. Therefore, it is important to control chaos of the biological models, which need further investigations. Further, characteristics of deterministic chaos are greatly affecting basic concepts of engineering such as prediction, control, computation, information, and optimization. New introduced chaotic models can be applied in this field of science.
- *Extending to other systems:* In this chapter, we proposed only three FOS models of biological systems and investigated their qualitative behaviors. This new era of modeling can be extended to other biological systems and even other systems from different fields of science such as power system, robotic, economics and so on.

References

Abdelouahab MS, Hamri NE, Wang J (2012) Hopf bifurcation and chaos in fractional-order modified hybrid optical system. Nonlin Dyn 69(1):275–284

Alligood K, Sauer T, Yorke J (1997) An introduction to dynamical systems. Springer, New York

Area I, Batarfi H, Losada J, Nieto JJ et al (2015) On a fractional order Ebola epidemic model. Adv Differ Equ 278(1):1–12

Atanackovic TM, Stankovic B (2004) An expansion formula for fractional derivatives and its application. Fract Calculus Appl Anal 7(3):365–378

Ayasun S, Nwankpa CO, Kwatny HG (2004) Computation of singular and singularity induced bifurcation points of differential-algebraic power system model. IEEE Trans Cir Syst I 51(8): 1525–1537

Baker GL, Gollub JP (1990) Chaotic dynamics; an introduction. Cambridge University Press, Cambridge

Campbell SL (1980) Singular systems of differential equations. Priman, London

Caputo M (1966) Linear models of dissipation whose Q is almost frequency independent. Ann Geophys 19(4):383–393

Chakraborty K, Das S, Kar TK (2011) Optimal control of effort of a stage structured prey–predator fishery model with harvesting. Nonlin Anal Real World Appl 12(6):3452–3467

Clark CW (1990) Mathematical bioeconomics: the optimal management of renewable resource. Wiley, New York

Dai L (1989) Singular control system. Springer, New York

Diethelm K (2010) The analysis of fractional differential equations. Springer, Berlin

Doungmo Goufo EF, Maritz R, Munganga J (2014) Some properties of the Kermack-McKendrick epidemic model with fractional derivative and nonlinear incidence. Adv Differ Equ 278(1):1–9

Duan GR (2010) Analysis and design of descriptor linear systems. Springer, New York

Freedman HI (1980) Deterministic mathematical models in population ecology. Marcel Dekker, New York

Gakkhar S, Naji RK (2003) Existence of chaos in two-prey, one-predator system. Chaos Soli Frac 17(4):639–649

Gakkhar S, Singh B (2007) The dynamics of a food web consisting of two preys and a harvesting predator. Chaos Soli Frac 34(4):1346–1356

Giannakopoulos K, Deliyannis T, Hadjidemetriou J (2002) Means for detecting chaos and hyperchaos in nonlinear electronic circuits. In: 14th international conference on digital signal processing, Santorini, Greece, 1–3 July 2002

Glendinning P, Perry LP (1997) Melnikov analysis of chaos in a simple epidemiological model. J Math Biol 35(3):359–373

Gordon H (1954) The economic theory of a common property resource: the fishery. J Polit Econ 62(2):124–142

Greenhalgh D, Khan QJA, Lewis FI (2004) Hopf bifurcation in two SIRS density dependent epidemic models. Math Comp Model 39(11):1261–1283

Guckenheimer J, Holmes P (1983) Nonlinear oscillations, dynamical systems, and bifurcations of vector fields. Springer, New York

Hartman P (2002) Ordinary differential equations. Cambridge University Press, Cambridge

Kaczorek T (2011) Selected problems of fractional systems theory. Springer, London

Kaczorek T, Rogowski K (2015) Fractional linear systems and electrical circuits. Springer, Bialystok

Kermack WO, McKendrick AG (1927) A contribution to the mathematical theory of epidemics. In: Proceedings of the royal society, London

Kielhoefer H (2004) Bifurcation theory: an introduction with applications to PDEs. Springer, New York

Kot M (2001) Elements of mathematical biology. Cambridge University Press, Cambridge

Kumar A, Daoutidis P (1999) Control of nonlinear differential-algebraic equation systems with applications to chemical process. CRC Press, London

Kuznetsov YA, Piccardi C (1994) Bifurcation analysis of periodic SEIR and SIR epidemic models. Math Bio 32(2):109–121

Lewis FL (1986) A survey of linear singular systems. Circuits Syst Signal Proc 5(1):3–36

Li XZ, Gupur G, Zhu GT (2001) Threshold and stability results for an age-structured SEIR epidemic model. Comp Math Appl 42(6):883–907

Liu Z, Lu P (2014) Stability analysis for HIV infection of CD4 + T-cells by a fractional differential time-delay model with cure rate. Adv Differ Equ 1:1–20

Luenberger DG (1977) Dynamic Equations in Descriptor Form. IEEE Trans Automat Control 22(3):312–321

Luenberger DG, Arbel A (1997) Singular dynamic Leontief systems. Econometrica 45:991–995

Marszalek W, Trzaska ZW (2005) Singularity-induced bifurcations in electrical power system. IEEE Trans Pow Syst 20(1):302–310

Masoud M, Masoud S, Caro L et al (2006) Introducing a new learning method for fuzzy descriptor systems with the aid of spectral analysis to forecast solar activity. J Atmo Sol-Terr Phy 68(18):2061–2074

May RM (1976) Simple mathematical models with very complicated dynamics. Nature 261 (5560):459–467

May RM, Oster GF (1976) Bifurcation and dynamic complexity in simple ecological models. Amer Nat 110(974):573–599

Munkhammar J (2013) Chaos in a fractional order logistic map. Fract Calc Appl Anal 16(3): 511–519

N'Doye I, Darouach M, Zasadzinski M et al (2013) Robust stabilization of uncertain descriptor fractional-order systems. Automatica 49(6):1907–1913

Nosrati K, Shafiee M (2017) Dynamic analysis of fractional-order singular Holling type-II predator–prey system. Appl Math Comput 313:159–179

Olsen LF, Schaffer WM (1990) Chaos versus periodicity: alternative hypotheses for childhood epidemics. Science 249:499–504

Ozalp N, Demirci E (2011) A fractional order SEIR model with vertical transmission. Math Comput Model 54(1):1–6

Petras I (2011) Fractional-order nonlinear systems: Modeling, analysis and simulation. Springer, New York

Podlubny I (1998) Fractional differential equations: An introduction to fractional derivatives, fractional differential equations, to methods of their solution and some of their applications. Academic Press, California

Rivero M, Trujillo JJ, Vazquez L et al (2011) Fractional dynamics of population. Appl Math Comput 218(3):1089–1095

Rostamy D, Mottaghi E (2016) Stability analysis of a fractional-order epidemics model with multiple equilibriums. Adv Differ Equ. https://doi.org/10.1186/s13662-016-0905-4

Sotomayor J (1973) Generic bifurcations of dynamical systems. Dynamical Systems. Academic Press, New York

Sun CJ, Lin YP, Tang SP (2007) Global stability for a special SEIR epidemic model with nonlinear incidence rates. Chaos Soli Frac 33(1):290–297

Tavazoei MS, Haeri M, Attari M, Bolouki S et al (2009a) More details on analysis of fractional-order Van der Pol oscillator. J Vib Control 15(6):803–819

Tavazoei MS, Haeri M, Attari M (2009b) A proof for non existence of periodic solutions in time invariant fractional order systems. Automatica 45(8):1886–1890

Tavazoei MS (2010) A note on fractional-order derivatives of periodic functions. Automatica 46(5):945–948

Venkatasubramanian V, Schaettler H, Zaborszky J (1995) Local bifurcations and feasibility regions in differential-algebraic systems. IEEE Trans Auto Contr 40(12):1992–2013

Wu GC, Baleanu D (2014) Discrete fractional logistic map and its chaos. Nonlin Dyn 75(1): 283–287

Xu WB, Liu HL, Yu JY et al (2005) Stability results for an age-structured SEIR epidemic model. J Sys Sci Inf 3(3):635–642

Yao YU, Zhuang JIAO, Chang-Yin SUN (2013) Sufficient and necessary condition of admissibility for fractional-order singular system. Acta Autom Sin 39(12):2160–2164

Yang C, Zhang Q, Zhou L (2012) Stability analysis and design for nonlinear singular systems. Springer, Berlin

Yude, J, Qiu J (2015) Stabilization of fractional-order singular uncertain systems. ISA Trans 56:53-64

Yue M, Schlueter R (2004) Bifurcation subsystem and its application in power system analysis. IEEE Trans Pow Syst 19(4):1885–1893

Zhang JS (1990) Singular system economy control theory. Tsinghua Press, Beijing

Zhang Y, Zhang QL, Zhao LC et al (2007) Tracking control of chaos in singular biological economy systems. J Nor Uni 28(2):157–164

Zhang Y, Zhang QL (2007) Chaotic control based on descriptor bioeconomic systems. Cont Dec 22(4):445–452

Zhang G, Zhu L, Chen B (2010) Hopf bifurcation and stability for a differential-algebraic biological economic system. Appl Math Comput 217(1):330–338

Zhang Q, Liu C, Zhang X (2012) Complexity, analysis and control of singular biological systems. Springer, London

Zhang Y, Zhang Q, Yan XG (2014) Complex dynamics in a singular Leslie-Gower predator–prey bioeconomic model with time delay and stochastic fluctuations. Phys A 404:180–191

Zhang X, Chen Y (2017) Admissibility and robust stabilization of continuous linear singular fractional order systems with the fractional order α: the $0 < \alpha < 1$ case. ISA Trans. https://doi.org/10.1016/j.isatra.2017.03.008

Chaos and Bifurcation in Controllable Jerk-Based Self-Excited Attractors

Wafaa S. Sayed, Ahmed G. Radwan and Hossam A. H. Fahmy

Abstract In the recent decades, utilization of chaotic systems has flourished in various engineering applications. Hence, there is an increasing demand on generalized, modified and novel chaotic systems. This chapter combines the general equation of jerk-based chaotic systems with simple scaled discrete chaotic maps. Two continuous chaotic systems based on jerk-equation and discrete maps with scaling parameters are presented. The first system employs the scaled tent map, while the other employs the scaled logistic map. The effects of different parameters on the type of the response of each system are investigated through numerical simulations of time series, phase portraits, bifurcations and Maximum Lyapunov Exponent (MLE) values against all system parameters. Numerical simulations show interesting behaviors and dependencies among these parameters. Analogy between the effects of the scaling parameters is presented for simple one-dimensional discrete chaotic systems and the continuous jerk based chaotic systems with more complicated dynamics. The impacts of these scaling parameters appear on the effective ranges of other main system parameters and the ranges of the obtained solution. The dependence of equilibrium points on the sign of one of the scaling parameters results in coexisting attractors according to the signs of the parameter and the initial point. In addition, switching can be used to generate double-scroll attractors. Moreover, bifurcation and chaos are studied for fractional-order of the derivative.

W. S. Sayed (✉) · A. G. Radwan
Faculty of Engineering, Engineering Mathematics and Physics Department,
Cairo University, Giza 12613, Egypt
e-mail: wafaa.s.sayed@eng.cu.edu.eg

A. G. Radwan
Nanoelectronics Integrated Systems Center, Nile University, Cairo 12588, Egypt
e-mail: agradwan@ieee.org

H. A. H. Fahmy
Faculty of Engineering, Electronics and Communications Engineering Department,
Cairo University, Giza 12613, Egypt
e-mail: hfahmy@alumni.stanford.edu

© Springer International Publishing AG 2018
V.-T. Pham et al. (eds.), *Nonlinear Dynamical Systems with Self-Excited and Hidden Attractors*, Studies in Systems, Decision and Control 133,
https://doi.org/10.1007/978-3-319-71243-7_2

45

Keywords Coexisting attractors · Discrete maps · Double-scroll attractors
Encryption applications · Fractional calculus · Maximum Lyapunov exponent
Scaling parameters

1 Introduction

Chaos theory studies the capability of generating aperiodic sequences, which are
unpredictable on the long term, from deterministic relations. Chaotic generators are
characterized by their sensitive dependence on initial conditions. Hence, they are
widely utilized in many applications, which belong to various fields such as: biology, chemistry, physics (Moaddy et al. 2012; Strogatz 2014), circuit theory (Radwan
2012, 2013b; Radwan et al. 2008a, b), control and synchronization (Henein et al.
2016; Radwan et al. 2013, 2014a, 2017; Sayed et al. 2017b, 2016a), communication and cryptography (Abd-El-Hafiz et al. 2016, 2015, 2014; Abdelhaleem et al.
2014; Barakat et al. 2013, 2011; Gan et al. 2016; Hua et al. 2017; Kocarev and Lian
2011; Li et al. 2016; Lin et al. 2016; Radwan and Abd-El-Hafiz 2013, 2014; Radwan
et al. 2014b, 2015b; Sayed et al. 2017a, 2015a, b; Radwan et al. 2007b; Wang et al.
2016; Zidan et al. 2011).

The two main categories of chaotic systems are discrete-time maps and continuous differential equations. Continuous chaotic systems based on differential equations overpass discrete chaotic systems based on difference equations or iterative
maps because the former are characterized by more complicated dynamics. Various implementations of discrete and continuous chaotic systems on electronic platforms have been presented (Radwan et al. 2004, 2007a; Radwan 2013a; Radwan
et al. 2003; Sayed et al. 2017d; Zidan et al. 2012). The recent decades witness an
increasing demand on generalized, modified and novel chaotic systems to satisfy
the requirements of modeling and random number generation. Researches ranged
between continuous and discrete chaotic systems, but they rarely combined ideas
from both and investigated the results.

Fractional calculus has also flourished in the last few decades and found its way to
real world applications in various fields including electromagnetics (Shamim et al.
2011), bioengineering (Magin 2006), chaotic systems (Petras 2011; Radwan et al.
2011a), image encryption (Ismail et al. 2015; Radwan et al. 2012, 2015a), circuits,
modeling and control (AbdelAty et al. 2017; Fouda and Radwan 2015; Fouda et al.
2016; Psychalinos et al. 2016; Radwan and Fouda 2013; Radwan et al. 2016, 2011b;
Semary et al. 2016; Soltan et al. 2012, 2015, 2017). In addition, advances in numerical methods for solving fractional-order systems and their electronic implementations have been reported (Caponetto 2010; Gorenflo and Mainardi 1997; Semary
et al. 2017; Tolba et al. 2017). Fractional calculus is more suitable for modeling
the continuous non-standard behaviors of nature due to the flexibility offered by
the extra degrees of freedom and including memory effects. Recently, most of the
chaotic dynamical systems based on integer-order calculus have been extended into

the fractional-order domain (AboBakr et al. 2017) to fit the experimental data much precisely than the their integer-order counterparts.

Leading researches investigated the possibility of introducing novel continuous chaotic systems, which are algebraically simple and, hence, suitable for hardware realization (Sprott 1994, 2000a, 2007). The well-established examples of chaotic flows occur in nonlinear systems with self-excited attractors, which have one or more saddle points (Alligood et al. 1996). In a self-excited attractor, the trajectories starting at some of the initial values converge to a saddle equilibrium point while the others diverge from it. If most of the initial values diverge from the saddle point, then it is called unstable and this set of initial values is called the unstable manifold of the saddle. Consequently, the computational procedure for locating strange attractors is carried out through choosing an initial value on the unstable manifold in the vicinity of the saddle point (Jafari et al. 2015).

Simple chaotic systems based on the jerk equation gained the interest of the scientific community. The jerk-equation (Sprott 2000a) is given by:

$$\dddot{x} = -\dot{x} - r\ddot{x} + f(x), \tag{1}$$

where $f(x)$ is a nonlinear function. The assignments

$$\dot{x} = y, \quad \dot{y} = z, \quad \dot{z} = -y - rz + f(x), \tag{2}$$

transform the third order differential equation (1) into an autonomous system of three first order differential equations with a single nonlinear term. The choice of the type of the nonlinearity $f(x)$ yields new jerk-based systems. Several researches discussed jerk-based chaotic systems including analysis, simulations, implementation and applications (Elwakil et al. 2000; Mansingka et al. 2013; Sayed et al. 2017c; Sprott 1997, 1994, 2000b, 2011; Vaidyanathan 2015; Vaidyanathan et al. 2015b, 2014, 2015c).

Recently, generalized tent and logistic maps with signed control parameter and added scaling parameters have been analyzed in (Radwan and Abd-El-Hafiz 2013; Sayed et al. 2015a, b) studying their effects on the chaotic properties. The advantages of these generalizations include the added control capabilities on the ranges of parameters and outputs and the extra degrees of freedom increasing the keyspace in encryption applications. While the generalized tent map represents a piece-wise nonlinearity, the generalized logistic map represents a quadratic nonlinearity.

This chapter aims at utilizing both the generalized scaled discrete-time tent and logistic maps in chaotic systems based on differential equations. Two continuous chaotic systems with self-excited attractors are presented. The two systems are based on the jerk-equation and discrete maps with scaling parameters in the form of piece-wise nonlinearity and quadratic nonlinearity. The effects of different parameters on the type of the response of each system are studied. Time series, phase portraits, bifurcation diagrams and Maximum Lyapunov Exponent (MLE) are investigated against all system parameters. It is shown that the role of each parameter is related to its role in the corresponding case of discrete maps.

Section 2 reviews the related previous works on jerk-based chaotic systems and the two nonlinearities utilized in the two proposed systems. Section 3 presents the two systems, their equations, attractor diagrams and Lyapunov exponents for specific parameter values. Section 4 discusses the type of system response against the two main system parameters and the scaling parameters through bifurcation diagrams and Maximum Lyapunov Exponent (MLE) plots. In addition, it studies the effects of the scaling parameters on the ranges of the main system parameters and output responses. Moreover, it illustrates the possibility of generating coexisting and double-scroll attractors from the proposed systems. Section 5 studies the behavior of the systems in the factional-order domain. Finally, Sect. 6 summarizes the main contributions of the chapter and suggests possible approaches for future work.

2 Survey of the Related and Utilized Continuous and Discrete Chaotic Systems

2.1 Jerk-Based Chaotic Attractors

Chaotic time series starting at close initial conditions begin to diverge, as time passes, leading to completely different long term behaviors as mentioned in Sect. 1. This divergence property is quantified using Lyapunov exponents, which characterize the divergence and convergence properties of an attractor. The number of Lyapunov exponents equals to the number of orthogonal directions of divergence or convergence in the phase space (Strogatz 2014). To have a dissipative dynamical system, the values of all Lyapunov exponents should sum to a negative number. For this system to be chaotic, the maximum Lyapunov exponent should be finite positive. Continuous flows expressed in terms of ordinary differential equations can have numerous types of post transient solution(s). An attractor or phase portrait is defined as the set of points approached by the orbit as the number of iterations increase to infinity representing its long term behavior.

Many researches presented the analysis of new jerk-based chaotic systems, their implementations and applications. Sprott (1994) presented general three dimensional autonomous ordinary differential equations with quadratic nonlinearities, which are composed of either five terms and two nonlinearities or six terms and one nonlinearity. Sprott (1997) presented systems with cubic nonlinearities, which were recently employed in Vaidyanathan et al. (2015a). Elwakil et al. (2000) presented a very simple jerk-based system with piecewise nonlinearity generated by a signum function. Sprott (2000a) discussed fourteen systems of the general form $\dddot{x} + A\ddot{x} + \dot{x} = f(x)$ where $f(x)$ is a nonlinearity in the form of a piece-wise linear, quadratic, cubic, sinusoidal, or hyperbolic tangent non-linear function. The systems posses an unstable equilibrium point at the zero of $f(x)$. Two of the systems with piece-wise and quadratic nonlinearities, respectively, are given in Table 1 because they have some similarities in common with the systems which will be proposed in Sect. 3.

Table 1 Sprott's chaotic systems based on the jerk-equation (1) at $r = 0.6$ and their properties, where C is a scaling parameter

$f(x)$	$-B|x| + C$	$-B(\frac{x^2}{C} - C)$
B	1	0.58
MLE	0.036	0.078
Phase portrait		

Vaidyanathan et al. (2014) presented a six-term three dimensional novel jerk chaotic system with two hyperbolic sinusoidal nonlinearities. Vaidyanathan (2015) presented a seven-term three dimensional novel jerk chaotic system with two quadratic nonlinearities. Vaidyanathan et al. (2015c) presented a six-term three dimensional jerk chaotic system with two exponential nonlinearities. Vaidyanathan et al. (2015b) presented a four dimensional novel hyperchaotic hyperjerk system. Synchronization applications and electronic circuit realization were also presented in Vaidyanathan (2015); Vaidyanathan et al. (2015b, 2014). Mansingka et al. (2013) presented fully digital implementations of four different systems in the third order jerk-equation based chaotic family using Euler approximation.

2.2 Two Modified Non-linearities

This subsection reviews generalized forms of two well-known discrete-time chaotic maps, which will be utilized as the nonlinear function of the jerk-equation in Sect. 3. The two generalizations are the scaled tent map with piece-wise nonlinearity and the scaled logistic map with quadratic nonlinearity. The complete bifurcation diagram using negatively valued parameters in tent and logistic maps has been recently analyzed in Sayed et al. (2015b, 2016b). The new parameter range provides a controlling capability resulting in a wider output range.

2.2.1 Piece-Wise Nonlinearity: Scaled Tent Map

Scaled tent map (Sayed et al. 2015a) with piece-wise nonlinearity is given by:

$$f(x) = \begin{cases} \mu \, sgn(b)x, \; x \le \frac{a}{b+sgn(b)} \\ \mu(a - bx), \; x > \frac{a}{b+sgn(b)} \end{cases}, \tag{3}$$

where μ, a and b are parameters, $a \in R^+$, $b \in R - \{0\}$ and $sgn(b)$ is the sign function or signum function which is an odd mathematical function that extracts the sign of b as follows

$$sgn(b) = \begin{cases} -1, \; b < 0 \\ 1, \;\; b > 0 \end{cases} \tag{4}$$

The forms of the scaled tent map can be classified into positive, mostly positive, negative, and mostly negative maps named after the sign of the obtained output range. Figure 1 shows the graphs of the map equation for the first two forms, in which $b > 0$, expressing the output ranges in terms of the map parameters.

(a)   **(b)**

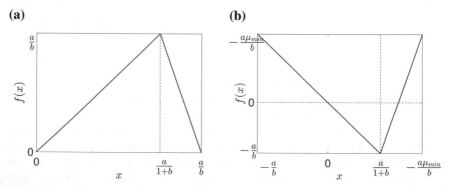

Fig. 1 **a** Scaled positive tent map and **b** Scaled mostly positive tent map, where $\mu_{min} = -\left(1 + \frac{1}{b}\right)$

For a discrete-time map represented as a recurrence relation, the bifurcation diagram is a plot of its steady state solution versus the control parameter(s) of the map. Plotting bifurcation diagrams is one of the approaches towards identifying the effective range of parameters through which the system exhibits bounded responses. In addition, it is used to classify the corresponding qualitative type of the post-transient solution into stable, periodic or chaotic. Figure 2 shows the general schematic of the bidirectional bifurcation diagram of the scaled tent map, which changes its shape as the parameter b exceeds 1. The figure shows the main bifurcation points and the ranges of the parameter μ and the output x. The effective range of the parameter μ, in which the output is bounded, depends on the scaling parameter b in an inverse proportionality relation. The output range depends on both scaling parameters, where it widens as the value of the parameter a increases and/or the value of the parameter b decreases. These effects can be further inferred from the three-dimensional snapshots of bifurcation diagrams against the main system parameter μ for different values of the scaling parameters a and b, which are shown in Fig. 3a, b respectively.

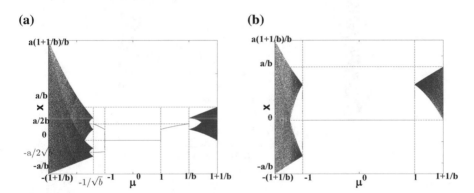

(a) **(b)**

Fig. 2 General schematic of the bidirectional bifurcation diagram of the scaled tent map in both sides of μ **a** $b < 1$, **b** $b > 1$

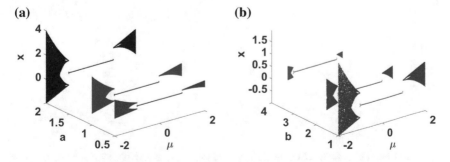

(a) **(b)**

Fig. 3 3D snapshots of the bifurcation diagrams of the scaled tent map **a** at $b = 1$ and $a = \{0.5, 1, 2\}$ and **b** at $a = 1$ and $b = \{1, 2, 4\}$

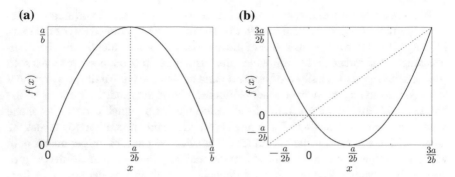

Fig. 4 **a** Scaled positive logistic map and **b** Scaled mostly positive logistic map

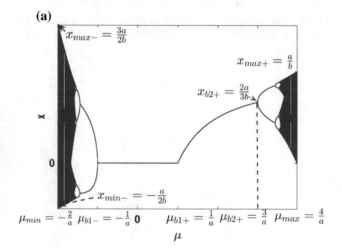

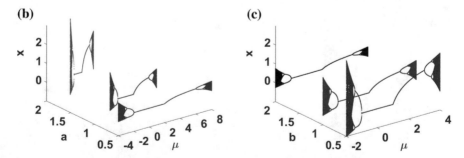

Fig. 5 **a** General bifurcation diagram of the scaled logistic map and 3D snapshots **b** at $b = 1$ and $a = \{0.5, 1, 2\}$ and **c** at $a = 1$ and $b = \{0.5, 1, 2\}$

2.2.2 Quadratic Nonlinearity: Scaled Logistic Map

Similarly, scaled logistic map (Sayed et al. 2015b) with quadratic nonlinearity is given by:

$$f(x) = \mu \, sgn(b)x(a - bx), \tag{5}$$

resulting in four forms similar to the scaled tent map. Figures 4 and 5 show the graphs of two map versions and their bifurcation diagrams. The dependence of the range of the output x on the scaling parameters is similar to the scaled tent map. However, the effective range of the parameter μ depends on the scaling parameter a in an inverse proportionality relation. Bifurcation diagrams against the scaling parameters and more detailed analyses of the different aspects of the scaled tent and logistic maps can be found in Sayed et al. (2015a, b).

3 Proposed Systems and Their Properties

Substituting either the scaled tent map (3) or the scaled logistic map (5) in the jerk-system (2) as $f(x)$ yields the piece-wise nonlinearity system and the quadratic non-linearity system, respectively. For both systems, the equations, attractor diagrams in the three-dimensional space and different projections, and lyapunov exponents at the specified parameter values are shown in Table 2. The attractor diagrams and the projections of the two systems resemble those of the two systems with similar nonlinearities which were introduced in Sprott (2000a) and reviewed in Sect. 2.1. However, they do not exhibit the same ranges of the three state space variables x, y and z. The obtained values for Lyapunov exponents for the two systems are in the same range obtained for the similar systems (Sprott 2000a). Both systems belong to the dissipative systems category because the sum of the three Lyapunov exponents for each system is negative (Strogatz 2014). Moreover, they exhibit chaotic strange attractors since the maximum Lyapunov exponent is finite positive.

For a continuous system of differential equations, the equilibrium points are defined to be those points at which all time derivatives equal zero. The equilibrium points are $(x^*, 0, 0)$ where $x^* = \{x | f(x) = 0\}$. For both forms of $f(x)$, $x^* = 0$, a/b. Hence, there are two equilibrium points $(0, 0, 0)$ and $(a/b, 0, 0)$. Hence, the sign of the x-coordinate of the nontrivial equilibrium point, x^*, depends on the sign of the parameter b and some consequences of this property will be discussed in Sect. 4.3.

The linear stability of each of the obtained points can be determined by calculating the eigen values of the linearized Jacobian matrix (Sprott 1994) (Routh-Hurwitz criterion). Specifically, if all the eigenvalues have real parts that are negative, then the system is stable near the equilibrium point. If any eigenvalue has a positive real part, then the point is unstable. If the matrix has at least one eigenvalue with positive real part, at least one with negative real part, and no eigenvalues with zero real part, then the point is called a saddle (Alligood et al. 1996). The Jacobian matrix of the proposed jerk-based systems is given by:

Table 2 Proposed systems and their properties

	The piece-wise nonlinearity system	The quadratic nonlinearity system
Nonlinearity	$f(x) = \begin{cases} \mu \, sgn(b)x, & x \le \frac{a}{b+sgn(b)} \\ \mu(a - bx), & x > \frac{a}{b+sgn(b)} \end{cases}$	$f(x) = \mu \, sgn(b)x(a - bx)$
Parameter Values	$\mu = 1$ $a = 1$ $b = 1$ $r = 0.6$	$\mu = 1$ $a = 1$ $b = 1$ $r = 0.5$
Attractor Diagram		
Lyapunov Exponents	$(0.038, 0, -0.64)$	$(0.092, 0, -0.59)$

$$\begin{pmatrix} 0 & 1 & 0 \\ 0 & 0 & 1 \\ f'(x) & -1 & -r \end{pmatrix} \tag{6}$$

4 Simulation Results in Integer-Order Domain

For the two systems presented in Sect. 3, the type of response obtained at the different values of the four parameters and the sensitivity to parameter variation need to be studied. This study can be carried out in a discrete manner, where the phase portrait and the time series are plotted at chosen values of each parameter fixing the other parameters. Continuous bifurcation diagrams provide a better representation of the systems behavior, which is also more consistent with the continuous description where parameters vary in narrow steps. For continuous chaotic systems, the bifurcation diagram versus a chosen parameter is generated through plotting the value of x every time it reaches a local maximum, where the time series is sampled as shown in Fig. 6, revealing whether the time series is stable, periodic or chaotic.

As previously detailed, Lyapunov Exponents (LE) measure the sensitivity to initial conditions through the exponential divergence of nearby trajectories. The Maximum Lyapunov Exponent (MLE) exhibits finite positive values for parameter ranges which correspond to chaotic behavior. To further indicate which parameter ranges exhibit chaotic behavior, MLE values are plotted against each studied parameter.

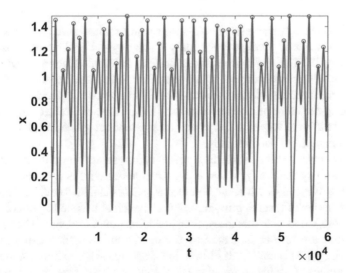

Fig. 6 Time series sampling to decide the type of system response

Table 3 Responses versus the parameter r at $a = b = \mu = 1$

	$r = 0.55$	$r = 0.57$	$r = 0.7$	$r = 1.1$
Attractor diagram				Single point
Time series				
Response type	Divergent	Chaotic	Periodic	Stable

4.1 Sensitivity to Main System Parameters

To study the effect of parameters r and μ, the scaling parameters a and b are kept constant $a = b = 1$ corresponding to the unity scaling case. For the piece-wise nonlinearity system, Tables 3 and 4 show its responses at different values of r and μ, respectively. The post-transient attractor diagrams, time series, and the obtained response type at different values of the parameter r within a chosen interval are plotted in Table 3 fixing the other parameter values to 1. The value of r is fixed at 0.6 to study responses at different values of the parameter μ, which are given in Table 4. Negative values of μ can be studied similarly.

Figure 7 shows the bifurcation diagrams of both systems versus the system parameter r. For both systems, at $\mu = a = b = 1$, chaotic behavior is reported starting at a critical value of r, below which no bounded responses can be found and the solution diverges. A series of reverse bifurcations from the chaotic state to periodic orbits is noticed as the value of r increases, then stable responses prevail. The results discussed earlier are further indicated by the maximum Lyapunov exponent (MLE) plots, which appear below each bifurcation diagram. MLE exhibits finite positive values for ranges of r which correspond to chaotic behavior, whereas it is negative in the regions of stable solution. It roughly equals zero for ranges of r which correspond to periodic responses.

Table 4 Responses versus the parameter μ at $a = b = 1$ and $r = 0.6$

	$\mu = 0.9$	$\mu = 0.95$	$\mu = 1$	$\mu = 1.1$
Attractor diagram				
Time series				
Response type	Periodic	Chaotic	Chaotic	Divergent

Fixing r at 0.6 for the piece-wise nonlinearity system and 0.5 for the quadratic nonlinearity system and studying the effect of μ yields the diagrams shown in Table 5. Bounded responses are reported when the value of the parameter μ belongs to a given interval, where around the middle of the interval, stable responses are obtained. Then, the response type changes gradually to periodic in a series of period doubling bifurcations as $|\mu|$ increases. Afterwards, the response becomes chaotic as μ approaches the lower and upper bounds. The possibility of bounded responses and the generation of chaotic sequences at both positive and negative values of μ in a double sided bifurcation are analogous to the behavior in the discrete domain (Sayed et al. 2015a, b).

4.2 Sensitivity to Scaling Parameters

This section studies the effects of scaling parameters a and b on the system responses. For the piece-wise nonlinearity system, Table 6 shows the responses at different values of the parameter b, which was noticed to be related to the variation of the parameter μ. In addition, Table 7 shows the continuous bifurcation diagrams and MLE values against both scaling parameters a and b. The response type does not change as the value of the parameter a increases. The response is chaotic for almost all values

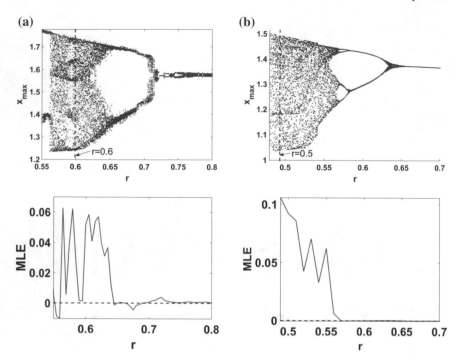

Fig. 7 Bifurcation diagram and MLE against the parameter r for **a** the piece-wise nonlinearity system at $\mu = a = b = 1$ and **b** the quadratic non-linearity system at $\mu = a = b = 1$

of a, where the range of the obtained solution gets wider as the value of a increases. MLE value is almost kept constant when varying the value of the parameter a. The parameter a acts only as a scaling parameter that widens the range of the solution, which can be further inferred from Fig. 8, where increasing the value of a increases the size of the attractor diagram. Table 7 shows that b is a signed parameter and that the system response exhibits double sided period doubling bifurcations when varying the value of b. In addition, the bifurcation diagram is limited by a value b_{max} controlled by the value of μ analogous to discrete scaled tent map case (Sayed et al. 2015a). The corresponding MLE plot exhibits values that match the response types shown in the bifurcation diagram.

For the quadratic nonlinearity system, bifurcation diagrams and MLE versus the scaling parameters are shown in Table 8, which can be described similar to the piece-wise nonlinearity system. The effects of the scaling parameters a and b on the output range remain the same, where the range of the system output increases as a increases. In addition, the bifurcation diagram is limited by a value a_{max} controlled by the value of μ. The parameter b acts only as a scaling parameter, where as $|b|$ increases the output ranges and the attractor size decrease as shown in Fig. 9.

Figure 10a shows the dependence of the effective range of the parameter μ on the value of b. For $b > 0$, the range of μ that yields bounded responses decreases

Table 5 Summary of the sensitivity to the system parameter μ and the similarities with the discrete scaled tent and logistic maps

Bifurcation and MLE	Properties

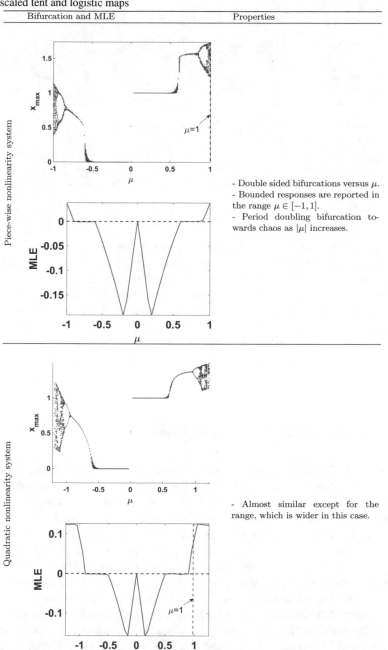

- Double sided bifurcations versus μ.
- Bounded responses are reported in the range $\mu \in [-1, 1]$.
- Period doubling bifurcation towards chaos as $|\mu|$ increases.

- Almost similar except for the range, which is wider in this case.

Table 6 Piece-wise nonlinearity system responses for different combinations of the parameters b and μ at $a = 1$ and $r = 0.6$

	$b = 2, \mu = 0.6$	$b = -2, \mu = 0.8$	$b = 2, \mu = -0.8$	$b = 0.5, \mu = -1.3$
Attractor diagram				
Time series				

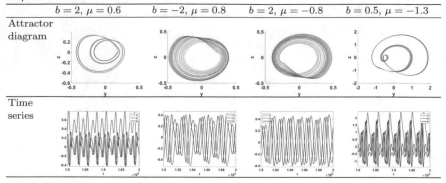

as b increases and sometimes no chaotic behavior can be reported. The dependence between μ and b resemble their dependence for discrete scaled tent map (Sayed et al. 2015a, c). Figure 10a also shows that the range of the solution shrinks as the absolute value of b increases. On the other hand, the parameter a does not affect the range of μ. The effects of a and b on the quadratic nonlinearity system differ from their effects on the piece-wise nonlinearity system from the viewpoint of the effective range of μ, where the roles are exchanged. The effective range of μ is affected by the value of a analogous to the discrete scaled logistic map (Sayed et al. 2015b, c), where it decreases as a increases as shown in Fig. 10b.

4.3 Co-existing and Multi-scroll Attractors

Each of the two studied systems can exhibit co-existing attractors for different signs of the parameter b, which controls the sign of the x-coordinate of the equilibrium point as mentioned before as shown in Fig. 11. Two different attractor diagrams can be obtained at distinct values and/or signs of b. In addition, if the parameter b varies dynamically with time and switches its value and sign as time advances, then multi-scroll attractors can be generated similar to the procedure given in Elwakil et al. (2002). Figure 12 shows various examples in which parameter switching is used to generate multi-scroll chaotic attractors from the two systems. In this case, the attractor diagrams are joined and undergo switching from side to side throughout the simulation time. Figure 12a generates a four-scroll attractor from the piece-wise nonlinearity system through switching the parameter values from $a = b = 1$, to $a = 2$ and $b = -0.7$, followed by $a = 3$ and $b = 1$, and finally $a = 4$ and $b = -0.7$, each case for quarter the simulation time, respectively, where $r = 0.6$ and $\mu = 1$. Figure 12b generates a double-scroll attractor from the quadratic nonlinearity system through

Table 7 Summary of the sensitivity of the piece-wise nonlinearity system to the scaling parameters a and b and the similarities with the discrete scaled tent map

Bifurcation and MLE	Properties

- The response is chaotic for almost all values of a, where the range of the obtained solution gets wider as the value of a increases. MLE values are positive and slightly vary versus a.

- The range of μ decreases as b increases. The bifurcation diagram is limited by a value b_{max} controlled by the value of μ.

- a acts only as a scaling parameter, where as a increases the output ranges and the size of the attractor diagram increases.

- Double sided period doubling bifurcations towards chaos exist as $|b|$ increases.

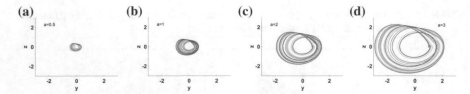

Fig. 8 Scaled chaotic responses of the piece-wise nonlinearity system for different values of the parameter a at $b = \mu = 1, r = 0.6$

switching the value of the parameter b from 1 to -0.6 after half of the simulation time passes, where $r = 0.5$ and $a = \mu = 1$.

5   Sensitivity to Fractional-Order Parameters

Consider the fractional-order differential equation

$$D^\alpha x(t) = f(t, x). \tag{7}$$

Grünwald-Letnikov method of approximation (Hussian et al. 2008) is defined as follows:

$$D^\alpha x(t) = \lim_{h \to 0} h^{-\alpha} \sum_{j=0}^{t/h} (-1)^j \begin{pmatrix} \alpha \\ j \end{pmatrix} x(t - jh), \tag{8}$$

where h is the step size. This equation can be discretized as follows:

$$\sum_{j=0}^{n+1} c_j^\alpha x(t - jh) = f(t_n, x(t_n)), \qquad j = 1, 2, 3, \ldots \tag{9}$$

where $t_n = nh$ and c_j^α are the Grünwald-Letnikov coefficients defined as:

$$c_j^\alpha = \left(1 - \frac{1+\alpha}{j}\right) c_{j-1}^\alpha, \quad j = 1, 2, 3, \ldots, \quad c_0^\alpha = h^{-\alpha}. \tag{10}$$

The NSFD discretization technique is based on replacing the step size h by a function $\phi(h)$ (Hussian et al. 2008; Moaddy et al. 2012) and applying it with (9) to solve (7). Same algebraic manipulation can be applied to a system of three fractional-order differential equations. Tables 9 and 10 show the time series of the three phase space dimensions x, y and z as well as the post-transient attractor diagram illustrating the obtained type of solution for different values of the fractional-order.

Table 8 Summary of the sensitivity of the quadratic nonlinearity system to the scaling parameters a and b and the similarities with the discrete scaled logistic map

Bifurcation and MLE	Properties

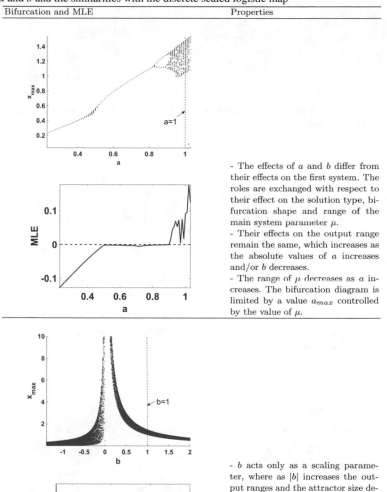

- The effects of a and b differ from their effects on the first system. The roles are exchanged with respect to their effect on the solution type, bifurcation shape and range of the main system parameter μ.
- Their effects on the output range remain the same, which increases as the absolute values of a increases and/or b decreases.
- The range of μ decreases as a increases. The bifurcation diagram is limited by a value a_{max} controlled by the value of μ.

- b acts only as a scaling parameter, where as $|b|$ increases the output ranges and the attractor size decreases.

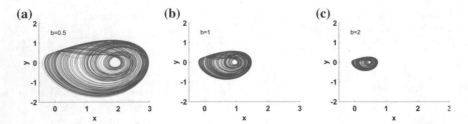

Fig. 9 Scaled chaotic responses of the quadratic nonlinearity system for different values of the parameter b at $a = \mu = 1$ and $r = 0.5$

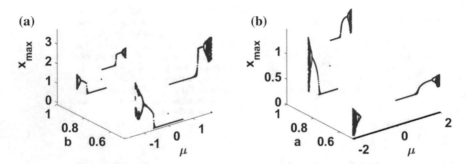

Fig. 10 Bifurcation diagrams versus μ of **a** the piece-wise nonlinearity system at $b = \{0.5, 1\}$ and **b** the quadratic nonlinearity system at $a = \{0.5, 1\}$

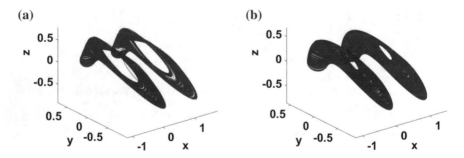

Fig. 11 Coexisting attractor diagrams at $b > 0$ (blue) and $b < 0$ (red) for **a** the piece-wise nonlinearity system and **b** the quadratic nonlinearity system

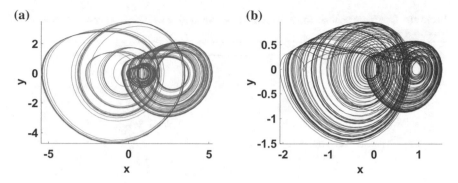

Fig. 12 **a** Four-scroll attractor using the piece-wise nonlinearity system and **b** Double-scroll attractor using the quadratic nonlinearity system

Table 9 Piece-wise nonlinearity system responses versus the fractional-order α at parameter values $a = b = \mu = 1$ and $r = 0.6$

$\alpha = 0.9$	$\alpha = 0.95$	$\alpha = 0.99$
Stable	Periodic	Periodic

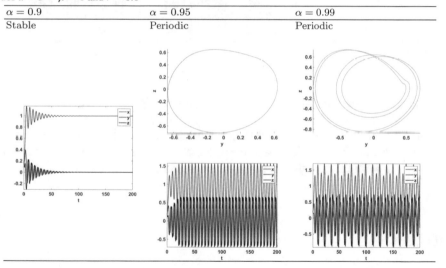

6 Conclusions

In this chapter, controllable jerk-based chaotic systems with extra parameters were presented. The systems utilized generalized forms of well-known discrete time chaotic maps as the nonlinear function of the jerk equation. While the piece-wise nonlinearity system employs the scaled tent map, the quadratic nonlinearity system employs the scaled logistic map. An analogy exists between the effects of the scaling parameters a and b in simple one-dimensional discrete chaotic maps and their effects in continuous jerk-based chaotic systems with more complicated dynamics.

Table 10 Quadratic nonlinearity system responses versus the fractional-order α at parameter values $a = b = \mu = 1$ and $r = 0.5$

$\alpha = 0.9$	$\alpha = 0.98$	$\alpha = 0.99$
Stable	Periodic	Chaotic

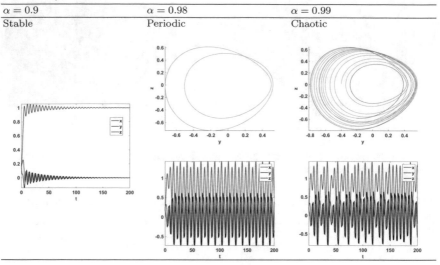

The impacts of these scaling parameters appear on the effective ranges of the parameter μ and the ranges of the obtained solution. Similar research ideas can be extended from the discrete domain to the continuous domain and combined to produce new systems. Increased nonlinearity and extra degrees of freedom can be added to the systems through using more complicated maps (e.g., Radwan 2013a; Sayed et al. 2017a).

References

Abd-El-Hafiz SK, AbdElHaleem SH, Radwan AG (2016) Novel permutation measures for image encryption algorithms. Opt Lasers Eng 85:72–83

Abd-El-Hafiz SK, Radwan AG, AbdEl-Haleem SH (2015) Encryption applications of a generalized chaotic map. Appl Math Inform Sci 9(6):3215

Abd-El-Hafiz SK, Radwan AG, AbdelHaleem SH, Barakat ML (2014) A fractal-based image encryption system. IET Image Process 8(12):742–752

AbdelAty AM, Soltan A, Ahmed WA, Radwan AG (2017) On the analysis and design of fractional-order Chebyshev complex filter. Circ Syst Signal Process pp 1–24

Abdelhaleem SH, Radwan AG, Abd-El-Hafiz SK (2014) A chess-based chaotic block cipher. In: IEEE 12th international new circuits and systems conference (NEWCAS), 2014. IEEE, pp 405–408

AboBakr A, Said LA, Madian AH, Elwakil AS, Radwan AG (2017) Experimental comparison of integer/fractional-order electrical models of plant. AEU-Int J Electron Commun

Alligood KT, Sauer TD, Yorke JA (1996) Chaos: an introduction to dynamical systems. Springer

Barakat ML, Mansingka AS, Radwan AG, Salama KN (2013) Generalized hardware post-processing technique for chaos-based pseudorandom number generators. ETRI J 35(3):448–458

Barakat ML, Radwan AG, Salama KN (2011) Hardware realization of chaos based block cipher for image encryption. In: IEEE international conference on microelectronics (ICM), pp 1–5

Caponetto R (2010) Fractional order systems: modeling and control applications, vol 72. World Scientific

Elwakil A, Salama K, Kennedy M (2000) A system for chaos generation and its implementation in monolithic form. In: IEEE international symposium on circuits and systems (ISCAS), vol 5. IEEE, pp 217–220

Elwakil AS, Ozoguz S, Kennedy MP (2002) Creation of a complex butterfly attractor using a novel Lorenz-type system. IEEE Trans Circ Syst I: Fundam Theory Appl 49(4):527–530

Fouda ME, Radwan AG (2015) Fractional-order memristor response under DC and periodic signals. Circ Syst Signal Process 34(3):961–970

Fouda ME, Soltan A, Radwan AG, Soliman AM (2016) Fractional-order multi-phase oscillators design and analysis suitable for higher-order PSK applications. Analog Integr Circ Sig Process 87(2):301–312

Gan Q, Yu S, Li C, Lü J, Lin Z, Chen P (2016) Design and arm-embedded implementation of a chaotic map-based multicast scheme for multiuser speech wireless communication. Int J Circ Theory Appl

Gorenflo R, Mainardi F (1997) Fractional calculus. Springer

Henein MMR, Sayed WS, Radwan AG, Abd-El-Hafiez SK (2016) Switched active control synchronization of three fractional order chaotic systems. In: 13th international conference on electrical engineering/electronics, computer, telecommunications and information technology

Hua Z, Yi S, Zhou Y, Li C, Wu Y (2017) Designing hyperchaotic cat maps with any desired number of positive Lyapunov exponents. IEEE Trans Cybern

Hussian G, Alnaser M, Momani S (2008) Non-standard discretization of fractional differential equations. In: Proceeding of 8th seminar of differential equations and dynamical systems in Isfahan, Iran

Ismail SM, Said LA, Radwan AG, Madian AH, Abu-ElYazeed MF, Soliman AM (2015) Generalized fractional logistic map suitable for data encryption. In: 2015 International conference on science and technology (TICST). IEEE, pp 336–341

Jafari S, Sprott JC, Nazarimehr F (2015) Recent new examples of hidden attractors. Eur Phys J

Kocarev L, Lian S (2011) Chaos-based cryptography: theory, algorithms and applications, vol 354. Springer

Li X, Li C, Lee I-K (2016) Chaotic image encryption using pseudo-random masks and pixel mapping. Signal Process 125:48–63

Lin Z, Yu S, Li C, Lü J, Wang Q (2016) Design and smartphone-based implementation of a chaotic video communication scheme via WAN remote transmission. Int J Bifurcat Chaos 26(09):1650158

Magin RL (2006) Fractional calculus in bioengineering. Begell House Redding

Mansingka AS, Zidan MA, Barakat ML, Radwan AG, Salama KN (2013) Fully digital jerk-based chaotic oscillators for high throughput pseudo-random number generators up to 8.77 Gbits/s. Microelectron J 44(9):744–752

Moaddy K, Radwan AG, Salama KN, Momani S, Hashim I (2012) The fractional-order modeling and synchronization of electrically coupled neuron systems. Comput Math Appl 64(10):3329–3339

Petras I (2011) Fractional-order nonlinear systems: modeling, analysis and simulation. Springer Science & Business Media

Psychalinos C, Elwakil AS, Radwan AG, Biswas K (2016) Guest editorial: fractional-order circuits and systems: theory, design, and applications. Circ Syst Signal Process 35:1807–1813

Radwan A (2012) Stability analysis of the fractional-order $RL_\beta C_\alpha$ circuit. J Fract Calc Appl 3(1): 1–15

Radwan A, Moaddy K, Hashim I (2013) Amplitude modulation and synchronization of fractional-order memristor-based Chua's circuit. In: Abstract and applied analysis, vol 2013. Hindawi Publishing Corporation

Radwan A, Moaddy K, Salama KN, Momani S, Hashim I (2014a) Control and switching synchronization of fractional order chaotic systems using active control technique. J Adv Res 5(1):125–132

Radwan A, Soliman A, El-Sedeek A (2004) MOS realization of the modified Lorenz chaotic system. Chaos, Solitons Fractals 21(3):553–561

Radwan A, Soliman AM, Elwakil AS (2007a) 1-D digitally-controlled multiscroll chaos generator. Int J Bifurcat Chaos 17(01):227–242

Radwan AG (2013a) On some generalized discrete logistic maps. J Adv Res 4(2):163–171

Radwan AG (2013b) Resonance and quality factor of the fractional circuit. IEEE J Emerg Sel Top Circ Syst 3(3):377–385

Radwan AG, Abd-El-Hafiz SK (2013) Image encryption using generalized tent map. In: IEEE 20th international conference on electronics, circuits, and systems (ICECS). IEEE, pp 653–656

Radwan AG, Abd-El-Hafiz SK (2014) The effect of multi-scrolls distribution on image encryption. In: 2014 21st IEEE international conference on electronics, circuits and systems (ICECS). IEEE, pp 435–438

Radwan AG, Abd-El-Hafiz SK, AbdElHaleem SH (2012) Image encryption in the fractional-order domain. In: 2012 International conference on engineering and technology (ICET). IEEE, pp 1–6

Radwan AG, Abd-El-Hafiz SK, AbdElHaleem SH (2014b) An image encryption system based on generalized discrete maps. In: IEEE 21st international conference on electronics, circuits and systems (ICECS). IEEE, pp 283–286

Radwan AG, Abd-El-Hafiz SK, AbdElHaleem SH (2015a) Image encryption based on fractional-order chaotic generators. In: 2015 international symposium on nonlinear theory and its applications NOLTA'2015, Kowloon, Hong Kong, China, 1–4 Dec 2015. IEEE, pp 688–691

Radwan AG, AbdElHaleem SH, Abd-El-Hafiz SK (2015b) Symmetric encryption algorithms using chaotic and non-chaotic generators: a review. J Adv Res

Radwan AG, Elwakil AS, Soliman AM (2008a) Fractional-order sinusoidal oscillators: design procedure and practical examples. IEEE Trans Circ Syst I: Regul Pap 55(7):2051–2063

Radwan AG, Fouda ME (2013) Optimization of fractional-order RLC filters. Circ Syst Signal Process 32(5):2097–2118

Radwan AG, Maundy BJ, Elwakil AS (2016) Fractional-order oscillators. Oscillator Circ Front Des Anal Appl 32:25

Radwan AG, Moaddy K, Momani S (2011a) Stability and non-standard finite difference method of the generalized Chuas circuit. Comput Math Appl 62(3):961–970

Radwan AG, Sayed WS, Abd-El-Hafiz SK (2017) Control and synchronization of fractional-order chaotic systems. In: Fractional order control and synchronization of chaotic systems. Springer, pp 325–355

Radwan AG, Shamim A, Salama KN (2011b) Theory of fractional order elements based impedance matching networks. IEEE Microwave Wirel Compon Lett 21(3):120–122

Radwan AG, Soliman AM, El-Sedeek A-L (2003) An inductorless CMOS realization of Chuas circuit. Chaos, Solitons Fractals 18(1):149–158

Radwan AG, Soliman AM, Elwakil AS (2007b) 1-D digitally-controlled multiscroll chaos generator. Int J Bifurcat Chaos 17(01):227–242

Radwan AG, Soliman AM, Elwakil AS (2008b) First-order filters generalized to the fractional domain. J Circ Syst Comput 17(01):55–66

Sayed WS, Fahmy HA, Rezk AA, Radwan AG (2017a) Generalized smooth transition map between tent and logistic maps. Int J Bifurcat Chaos 27(01):1730004

Sayed WS, Henein MM, Abd-El-Hafiz SK, Radwan AG (2017b) Generalized dynamic switched synchronization between combinations of fractional-order chaotic systems. Complexity

Sayed WS, Radwan AG, Abd-El-Hafiez SK (2016a) Generalized synchronization involving a linear combination of fractional-order chaotic systems. In: 13th international conference on electrical engineering/electronics, computer, telecommunications and information technology

Sayed WS, Radwan AG, Fahmy HA (2015a) Design of a generalized bidirectional tent map suitable for encryption applications. In: 11th international computer engineering conference (ICENCO). IEEE, pp 207–211

Sayed WS, Radwan AG, Fahmy HA (2015b) Design of positive, negative, and alternating sign generalized logistic maps. Discrete Dyn Nat Soc

Sayed WS, Radwan AG, Fahmy HA (2016b) Double-sided bifurcations in tent maps: analysis and applications. In: 3rd international conference on advances in computational tools for engineering applications (ACTEA). IEEE, pp 207–210

Sayed WS, Radwan AG, Fahmy HA (2017c) Chaotic systems based on jerk equation and discrete maps with scaling parameters. In: 6th international conference on modern circuits and systems technologies (MOCAST). IEEE, pp 1–4

Sayed WS, Radwan AG, Fahmy HAH, Hussien AE (2015c) Scaling parameters and chaos in generalized 1D discrete time maps. In: international symposium on nonlinear theory and its applications (NOLTA). pp 688–691

Sayed WS, Radwan AG, Rezk AA, Fahmy HA (2017d) Finite precision logistic map between computational efficiency and accuracy with encryption applications. Complexity

Semary MS, Hassan HN, Radwan AG (2017) Controlled picard method for solving nonlinear fractional reaction-diffusion models in porous catalysts. Chem Eng Commun 204(6):635–647

Semary MS, Radwan AG, Hassan HN (2016) Fundamentals of fractional-order LTI circuits and systems: number of poles, stability, time and frequency responses. Int J Circ Theory Appl 44(12):2114–2133

Shamim A, Radwan AG, Salama KN (2011) Fractional Smith chart theory. IEEE Microwave Wirel Compon Lett 21(3):117–119

Soltan A, Radwan AG, Soliman AM (2012) Fractional order filter with two fractional elements of dependant orders. Microelectron J 43(11):818–827

Soltan A, Radwan AG, Soliman AM (2015) Fractional order Sallen-Key and KHN filters: stability and poles allocation. Circ Syst Signal Process 34(5):1461–1480

Soltan A, Soliman AM, Radwan AG (2017) Fractional-order impedance transformation based on three port Mutators. AEU-Int J Electron Commun

Sprott J (1997) Some simple chaotic jerk functions. Am J Phys 65(6):537–543

Sprott JC (1994) Some simple chaotic flows. Phys Rev E 50(2):R647

Sprott JC (2000a) A new class of chaotic circuit. Phys Lett A 266(1):19–23

Sprott JC (2000b) Simple chaotic systems and circuits. Am J Phys 68(8):758–763

Sprott JC (2007) A simple chaotic delay differential equation. Phys Lett A 366(4):397–402

Sprott JC (2011) A new chaotic jerk circuit. IEEE Trans Circ Syst II: Express Briefs 58(4):240–243

Strogatz SH (2014) Nonlinear dynamics and chaos: with applications to physics, biology, chemistry, and engineering. Westview press

Tolba MF, AbdelAty AM, Soliman NS, Said LA, Madian AH, Azar AT, Radwan AG (2017) FPGA implementation of two fractional order chaotic systems. AEU-Int J Electron Commun 78:162–172

Vaidyanathan S (2015) Analysis, control and synchronization of a 3-D novel jerk chaotic system with two quadratic nonlinearities. Kyungpook Math J 55:563–586

Vaidyanathan S, Idowu BA, Azar AT (2015a) Backstepping controller design for the global chaos synchronization of Sprott's jerk systems. In: Chaos modeling and control systems design. Springer, pp 39–58

Vaidyanathan S, Volos C, Pham V-T, Madhavan K (2015b) Analysis, adaptive control and synchronization of a novel 4-D hyperchaotic hyperjerk system and its spice implementation. Arch Control Sci 25(1):135–158

Vaidyanathan S, Volos C, Pham V-T, Madhavan K, Idowu BA (2014) Adaptive backstepping control, synchronization and circuit simulation of a 3-D novel jerk chaotic system with two hyperbolic sinusoidal nonlinearities. Arch Control Sci 24(3):375–403

Vaidyanathan S, Volos CK, Kyprianidis I, Stouboulos I, Pham V (2015c) Analysis, adaptive control and anti-synchronization of a six-term novel jerk chaotic system with two exponential nonlinearities and its circuit simulation. J Eng Sci Technol Rev 8(2):24–36

Wang Q, Yu S, Li C, Lü J, Fang X, Guyeux C, Bahi JM (2016) Theoretical design and FPGA-based implementation of higher-dimensional digital chaotic systems. IEEE Trans Circuits Syst I Regul Pap

Zidan, MA, Radwan, AG, Salama, KN (2011) Random number generation based on digital differential chaos. In: IEEE 54th international midwest symposium on circuits and systems (MWSCAS), 2011. pp 1–4

Zidan MA, Radwan AG, Salama KN (2012) Controllable V-shape multiscroll butterfly attractor: system and circuit implementation. Int J Bifurcat Chaos 22(06):1250143

Self-Excited Attractors in Jerk Systems: Overview and Numerical Investigation of Chaos Production

Wafaa S. Sayed, Ahmed G. Radwan and Salwa K. Abd-El-Hafiz

Abstract Chaos theory has attracted the interest of the scientific community because of its broad range of applications, such as in secure communications, cryptography or modeling multi-disciplinary phenomena. Continuous flows, which are expressed in terms of ordinary differential equations, can have numerous types of post transient solutions. Reporting when these systems of differential equations exhibit chaos represents a rich research field. A self-excited chaotic attractor can be detected through a numerical method in which a trajectory starting from a point on the unstable manifold in the neighborhood of an unstable equilibrium reaches an attractor and identifies it. Several simple systems based on jerk-equations and different types of nonlinearities were proposed in the literature. Mathematical analyses of equilibrium points and their stability were provided, as well as electrical circuit implementations of the proposed systems. The purpose of this chapter is double-fold. First, a survey of several self-excited dissipative chaotic attractors based on jerk equations is provided. The main categories of the included systems are explained from the viewpoint of nonlinearity type and their properties are summarized. Second, maximum Lyapunov exponent values are explored versus the different parameters to identify the presence of chaos in some ranges of the parameters.

Keywords Jerk equation · Maximum Lyapunov exponent · Phase portraits
Time series

W. S. Sayed (✉) · A. G. Radwan · S. K. Abd-El-Hafiz
Faculty of Engineering, Engineering Mathematics and Physics Department,
Cairo University, Giza 12613, Egypt
e-mail: wafaa.s.sayed@eng.cu.edu.eg

A. G. Radwan
e-mail: agradwan@ieee.org

S. K. Abd-El-Hafiz
e-mail: salwa@computer.org

A. G. Radwan
Nanoelectronics Integrated Systems Center, Nile University, Cairo 12588, Egypt

© Springer International Publishing AG 2018   71
V.-T. Pham et al. (eds.), *Nonlinear Dynamical Systems with Self-Excited
and Hidden Attractors*, Studies in Systems, Decision and Control 133,
https://doi.org/10.1007/978-3-319-71243-7_3

1   Introduction

Nonlinear dynamical systems with chaotic or strange attractors are characterized by the sensitivity to initial conditions, which is a required property for many applications (Layek 2015; Schöll 2001; Strogatz 2014). Chaos theory, dating back to Lorenz (1963), has attracted the interest of the scientific community and took part in many engineering applications such as dynamical modeling, pseudo-random number generation for secure communication and cryptography applications (Abd-El-Hafiz et al. 2014, 2015, 2016; Abdelhaleem et al. 2014; Barakat et al. 2013; Chien and Liao 2005; Frey 1993; Kocarev and Lian 2011; Lau and Tse 2003; Radwan and Abd-El-Hafiz 2013, 2014; Radwan et al. 2012, 2015a, b; Radwan AG et al. 2014; Sayed et al. 2015a, b, 2017a; Tolba et al. 2017) and control and synchronization (Azar and Vaidyanathan 2015, 2016; Azar et al. 2017; Henein et al. 2016; Martínez-Guerra et al. 2015; Radwan et al. 2013, 2017; Radwan A et al. 2014; Sayed et al. 2016, 2017b). Consequently, chaotic systems have been implemented in several numerical and electronic forms (Petras 2011; Radwan et al. 2003, 2004, 2007; Radwan 2013; Sayed et al. 2017d; Zidan et al. 2012).

Continuous flows expressed in terms of ordinary differential equations can have numerous types of post-transient solutions. An attractor is defined as the set of points approached by the orbit as the number of iterations increases to infinity representing its long term behavior. For a continuous system of differential equations, the equilibrium points are defined to be those points at which all time derivatives equal zero. The linear stability of each of the obtained points can be determined by Routh-Hurwitz stability criterion (Sprott 1994). The eigenvalues of the linearized Jacobian matrix are calculated. If all eigenvalues have negative real part, then the system is stable near the equilibrium point. If any eigenvalue has a real part that is positive, then the point is unstable. If the matrix has at least one eigenvalue with positive real part, at least one with negative real part, and no eigenvalues with zero real part, then the point is called a saddle (Alligood et al. 1996).

Furthermore, nearby trajectories diverge on strange attractors, giving rise to the butterfly effect in chaotic dynamical systems. This divergence is exponential and may be quantified using characteristic exponents known as Lyapunov exponents (Addison 1997). The number of Lyapunov exponents is equal to the number of phase space dimensions, or the order of the system of differential equations. They are arranged in a descending order and if the maximum Lyapunov exponent is positive, then the system is chaotic. The sum of Lyapunov exponents represents the average contraction rate of volumes in phase space. The sum is less than zero in dissipative dynamical systems, as the post-transient solutions lie on attractors with zero phase volume. Dissipative systems exhibit chaos for most initial conditions in a specified range of parameters. On the other hand, a conservative system exhibits periodic and quasi-periodic solutions for most values of parameters and initial conditions, and can exhibit chaos for special values only. Consequently, dissipative systems usually appear in most applications of chaos theory such as chaos-based communication, physical and financial modeling. Conservative systems have another different set of applications that study the development of chaos in some kinds of systems. They are useful in describing certain dynamical systems where there is no dissipation, or it is so slight that it can be ignored, e.g., models of the solar system. Another important classification of chaotic attractors is either self-excited or hidden. A self-excited attractor has a basin of attraction that is associated with or excited from unstable equilibria. On the other hand, a hidden attractor has a basin of attraction that does not intersect with small neighborhoods of any equilibrium points. From a computational point of view, a self-excited chaotic attractor can be detected through a numerical method in which a trajectory started from a point on the unstable manifold in the neighborhood of an unstable equilibrium reaches an attractor and identifies it. Hidden attractors cannot be found by this method (Leonov and Kuznetsov 2013).

Introducing novel chaotic systems requires a system of, at least, three differential equations involving, at least, one nonlinear term. A system of three or more first order ordinary differential equations that contain one or more nonlinear term(s) is constructed with the tendency to be as simple as possible. Some systematic numerical search methods have been developed for detecting the presence of chaotic solutions for new systems that contain multiple parameters. These parameters mainly appear as the coefficients of each term in the system of differential equations. Those numerical methods aim at setting many coefficients to zero with the others set to ± 1 if possible or otherwise to a small integer or decimal fraction with the fewest possible digits (Sprott 1994). These systems, with the least number of existing coefficients and nonlinear terms, should exhibit chaotic properties of aperiodic bounded long-time evolution and sensitive dependence on initial conditions for some ranges of parameters. Reporting the parameter ranges for which systems of differential equations exhibit chaos or a strange attractor represents a rich research field.

Many researches focused on coming up with novel chaotic systems that, in the simplest form, involve a differential equation of at least third order $\dddot{x} = G(\ddot{x}, \dot{x}, x)$ and a nonlinearity. Differential equations of this form are called jerk equations because they involve third derivatives. The word "jerk" refers to the rate of change of acceleration, i.e., the derivative of acceleration with respect to time, the second derivative of velocity, and the third derivative of position. The mathematically simple jerk equation, which is equivalent to a system of three first-order ordinary non-linear differential equations, was shown to have solutions that exhibit chaotic behavior (Gottlieb 1996). Moreover, the simple circuit implementation of jerk-based systems suggests their utilization in secure communications and broadband signal generation. Systems involving a fourth or higher derivative are accordingly called hyperjerk systems.

Several simple systems based on the jerk-equation and different types of nonlinearities were proposed in the literature (Elwakil et al. 2000; Sayed et al. 2017c; Sprott 1994, 1997, 2000a, b, 2011; Vaidyanathan 2015; Vaidyanathan et al. 2014, 2015b, c). Mathematical analysis of equilibrium points and their stability were provided, as well as electrical circuit implementation of the proposed systems. Jerk-based chaotic systems express a third order ordinary differential equation as a system of three simultaneous first-order ordinary differential equations. Hence, they are considered as one of the simplest types of continuous chaotic systems. Consequently, they have been utilized in many applications including control and synchronization.

This chapter focuses on dissipative chaotic systems with self-excited attractors because they are not easily driven away from chaotic behavior when correctly adjusting the parameters and varying the initial conditions. All the reviewed systems are based on jerk equations and were shown to be chaotic for specific values of the parameters in the original papers which introduced them. The main properties of the selected systems, which have different types of nonlinear terms, are reviewed. The associated phase portraits and Maximum Lyapunov Exponent (MLE) values are tabulated in Sect. 2. Section 3 explores the responses in wider ranges of parameters to investigate the possibility of chaos production using MLE as a chaotic measure. Section 4 summarizes the contributions of the chapter.

2 Review of Some Self-Excited Jerk-Based Attractors

Early researches presented different variations on chaotic systems such that their equations look simpler or more "elegant" (Sprott 1994). Sprott (2000a) discussed several systems of the general form $\dddot{x} + A\ddot{x} + \dot{x} = f(x)$, where $f(x)$ is a nonlinear function satisfying some conditions that guarantee boundedness. The equation is redefined as $\dot{x} = y$, $\dot{y} = z$ and $\dot{z} = -Az - y + f(x)$. An electrical circuit implementation has been suggested for cases in which $f(x)$ is a piecewise linear function. Several cases in which $f(x)$ is a piecewise linear, quadratic, cubic, sinusoidal or hyperbolic tangent nonlinear function are illustrated in Fig. 1.

The systems presented in (Sprott 2000a) are listed as fourteen systems in Tables 1, 2, 3 and 4. Systems (1) to (14) are self-excited attractors that posses an unstable equilibrium point at the zero of $f(x)$. These systems are elementary, both in the sense of having the algebraically simplest autonomous Ordinary Differential Equation (ODE) and in the form of the nonlinearity. The first five systems, which are discussed in Table 1, represent the simplest cases with piece-wise nonlinearity. Their governing equations are the easiest to implement on electronic platforms. They represent a class of chaotic electrical circuit that is simple to construct, analyze, and scale over a wide

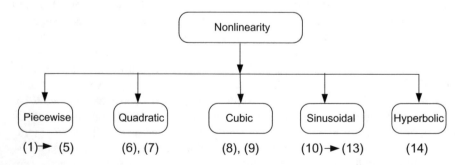

Fig. 1 Categorization of the reviewed dissipative chaotic systems with self-excited attractors from the viewpoint of the type of nonlinearity

Table 1 Systems with piece-wise linear function

Equations	Attractor
(1) $f(x) = B|x| - C$ $B = 1$ MLE = 0.036	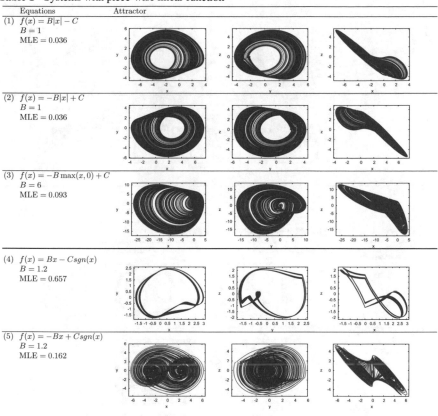
(2) $f(x) = -B|x| + C$ $B = 1$ MLE = 0.036	
(3) $f(x) = -B\max(x,0) + C$ $B = 6$ MLE = 0.093	
(4) $f(x) = Bx - C sgn(x)$ $B = 1.2$ MLE = 0.657	
(5) $f(x) = -Bx + C sgn(x)$ $B = 1.2$ MLE = 0.162	

Table 2 Systems with quadratic nonlinearity

Equations	Attractor
(6) $f(x) = B(\frac{x^2}{C} - C)$ $B = 0.58$ MLE = 0.078	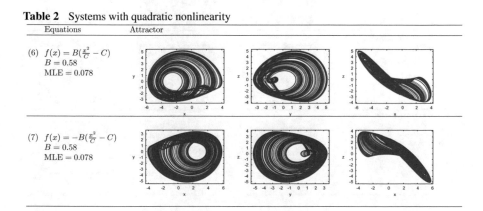
(7) $f(x) = -B(\frac{x^2}{C} - C)$ $B = 0.58$ MLE = 0.078	

Table 3 Systems with cubic nonlinearity

Equations	Attractor
(8) $f(x) = Bx(\frac{x^2}{C} - 1)$ $B = 1.6$ MLE $= 0.103$	
(9) $f(x) = -Bx(\frac{x^2}{C} - 1)$ $B = 0.9$ MLE $= 0.126$	

Table 4 Systems with sinusoidal or hyperbolic nonlinearity

Equations	Attractor
(10) $f(x) = B\sin(Cx)/C$ $B = 2.7$ MLE $= 0.069$	
(11) $f(x) = -B\sin(Cx)/C$ $B = 2.7$ MLE $= 0.069$	
(12) $f(x) = B\cos(Cx)/C$ $B = 2.7$ MLE $= 0.069$	
(13) $f(x) = -B\cos(Cx)/C$ $B = 2.7$ MLE $= 0.069$	
(14) $f(x) = -B[x - 2\tanh(Cx)/C]$ $B = 2.2$ MLE $= 0.221$ Hyperbolic tangent	

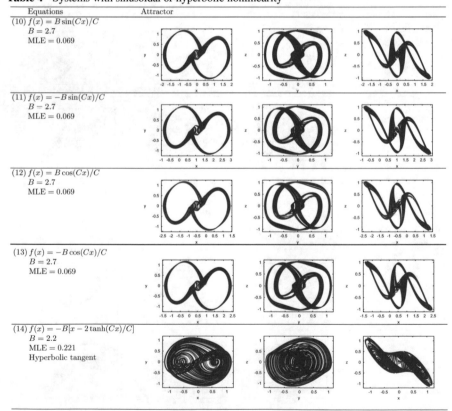

range of frequencies. In addition, it does not involve analog multiplication and uses only resistors, capacitors, diodes and operational amplifiers. The rest of the systems range between quadratic, cubic, sinusoidal and hyperbolic nonlinearities. Although they are more complicated, they are still good candidates for detailed quantitative analysis of bifurcation theory and other chaotic properties to be compared with simulation or implementation results.

For all the systems, the parameter $A = 0.6$, while C can be arbitrarily chosen as it acts as a scaling factor for the size of the attractor diagram. Each table provides the nonlinear function $f(x)$ and the specific value of the parameter B that produces chaos. In addition, the attractor diagrams or phase portraits are shown with the corresponding positive value of MLE, base-e, both indicating chaotic behavior. The diagrams have been plotted using Economics and Finance (E&F) chaos software (Diks et al. 2008) at the specified parameter values and the MLE values were given in (Sprott 2000a). Systems with similar equations exhibit similar attractor diagrams, e.g., systems (1) and (2) and systems (10) to (13).

Several other papers presented jerk-based chaotic attractors (Elwakil et al. 2000; Sprott 1994, 1997, 2000b, 2011). General three dimensional autonomous ordinary differential equations with quadratic nonlinearities were examined in (Sprott 1994). The resulting simple chaotic systems are composed of either five terms and two nonlinearities or six terms and one nonlinearity. Systems with cubic nonlinearities were presented in (Sprott 1997) and employed in (Vaidyanathan et al. 2015a). A very simple jerk-based system with piecewise nonlinearity generated by a signum function was presented in (Elwakil et al. 2000).

Several recent researches presented new jerk based systems as part of their work (Sayed et al. 2017c; Vaidyanathan 2015; Vaidyanathan et al. 2014, 2015b, c). A six-term three dimensional novel jerk chaotic system with two hyperbolic sinusoidal nonlinearities was presented in (Vaidyanathan et al. 2014). An adaptive backstepping controller was designed to stabilize the system with two unknown parameters. In addition, synchronization of two systems with two unknown parameters was achieved. Moreover, an electronic circuit realization of the novel jerk chaotic system using Spice was presented. A four-dimensional novel hyperchaotic hyperjerk system was proposed in (Vaidyanathan et al. 2015b) associated with control, synchronization and electronic circuit realization. A six-term three-dimensional jerk chaotic system with two exponential nonlinearities was presented in (Vaidyanathan et al. 2015c). A seven-term three-dimensional novel jerk chaotic system with two quadratic nonlinearities was presented in (Vaidyanathan 2015). Adaptive backstepping control of the proposed system and synchronization of two identical entities with unknown parameters were also proposed. Generalized forms of two well-known discrete-time chaotic maps were utilized as the nonlinear function of the jerk-equation in (Sayed et al. 2017c). The two maps are the scaled tent map with piece-wise nonlinearity and the scaled logistic map with quadratic nonlinearity.

Fully digital implementations of four different systems in the third order jerk-equation based chaotic family using Euler approximation were presented in (Mansingka et al. 2013). The systems ranged between absolute value, signum, quadratic and cubic nonlinearities. The high performance metrics of the digitally implemented systems as pseudo-random number generators were verified and shown to be suitable for communication systems and hardware encryption applications.

3 Sensitivity to Parameter Variations

This section provides some extra results and simulations for a selected set of the systems summarized in the previous section. The results mainly focus on plotting the phase portraits or strange attractors at values of parameters around those specified in the original paper (Sprott 2000a). A simulation-based procedure for specifying parameter ranges of chaos production around the specified values is discussed through plotting MLE versus the different parameters.

In general, the basin of attraction is the set of initial conditions which leads to a particular post-transient solution. Parameter values can control whether chaotic behavior is exhibited or not. While the parameter values that drive the system into chaos are called parameter basin of attraction of the chaotic attractor, the initial points that converge to a chaotic orbit are called its basin of attraction. There are two reasons for the importance of parameter basin of attraction. First, to test the robustness of the solution or its sensitivity to small parameter changes. Second, to have an estimation of the allowed parameter space and which values produce chaos.

Plots of phase portraits and MLE versus parameters have been generated by the aid of E&F software (Diks et al. 2008). In addition, some of the calculations of Lyapunov exponents were carried out by Lyapunov Exponent Toolbox (LET) (Siu 1998) and a MATLAB-based program for dynamical system investigation (MATDS) (Govorukhin 2004). Calculations of Lyapunov exponents have been carried out for 10, 000 iterations up to accuracy of four decimal places. These choices are made to ensure reaching a post-transient value of Lyapunov exponents. Lyapunov exponent calculations at the specified parameters and initial conditions should satisfy the two conditions for chaos production. First, their summation should be less than zero since they are dissipative dynamical systems. Second, the MLE should be positive which accounts for chaotic behavior.

Several numerical simulations are performed to get the ranges of parameters, rather than specific values only, that produce chaos. The procedure makes use of the calculated Lyapunov exponents in determining approximate ranges of parameters that produce chaos. The systems are shown to satisfy the condition of dissipative systems by using the Lyapunov exponent calculation function of MATDS. MLE is plotted versus different system parameters using E&F chaos software. For the parameter values specified in (Sprott 2000a), the neighborhood of each parameter value is explored while fixing the other parameters. The approximate ranges of parameters that exhibit positive values of MLE are recorded. Visualizing phase portraits is also used as a check of chaos production.

The parameter values that correspond to maximum chaos (largest MLE) are specified. In addition, other values of parameters are shown to drive the response out of chaos and generate other types of solutions. The flows of most dynamical systems with respect to parameter variation exhibit the following pattern of different types of solutions: stable or fixed followed by periodic then quasi-periodic and afterwards chaotic and finally unstable or divergent, whether in the direction of increasing the parameter or vice versa. Lyapunov exponents can be used to determine the type of the attractor as follows, where three dimensional phase space is assumed for simplicity. Lyapunov exponents with signs $(+, 0, -)$ correspond to chaos or strange attractor, $(-, -, -)$ to fixed point, $(0, -, -)$ to limit cycle and $(0, 0, -)$ to quasi-periodic torus (Addison 1997).

3.1 A Dissipative Self-Excited Attractor with Quadratic Nonlinearity: System (6)

System (6) in Table 2 represents a sample for quadratic nonlinearity. Figure 2 shows the ranges of parameters A, B and C that can produce chaos in system (6). Wider ranges were investigated using E&F chaos software, but the chaotic range is focused as shown in the figure. Figure 2a shows that for approximately $0.6 \leq A \leq 0.675$, the value of MLE varies but remains positive through almost the whole interval. This indicates that the system exhibits chaotic behavior in this range of the parameter A.

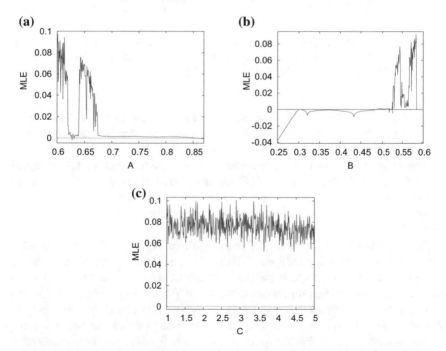

Fig. 2 Ranges of parameters that produce chaos for system (6) **a** MLE versus A, **b** MLE versus B, and **c** MLE versus C

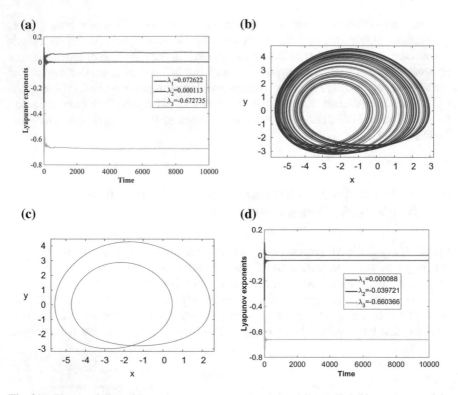

Fig. 3 **a** Time evolution of Lyapunov exponents at $A = 0.6$ and $B = 0.58$, **b** Phase portrait of the chaotic attractor at $A = 0.62$ and $B = 0.55$, **c** Non-chaotic phase portrait, and **d** Time evolution of Lyapunov exponents at $A = 0.7$ and $B = 0.58$ for system (6)

For A slightly less than 0.6, MLE diverges corresponding to unstable system, while for A slightly greater than 0.675, MLE is around zero or negative corresponding to periodic responses.

Regarding the effect of the parameter B, Fig. 2b shows that for approximately $0.525 \leq B \leq 0.585$, the value of MLE varies but remains positive. For B slightly greater than 0.585, MLE diverges corresponding to unstable system, while for B slightly less than 0.525, MLE is around zero or negative corresponding to periodic responses.

Figure 2c shows that the value of C does not affect the type of the behavior, which conforms to its description in (Sprott 2000a) as a scaling factor for the attractor size.

The values chosen in (Sprott 2000a) to produce chaos are $A = 0.6$ and $B = 0.58$ corresponding to the attractor diagrams shown in Table 2. Figure 3a shows the time evolution of Lyapunov exponents at these values of parameters, which satisfy the conditions for chaotic behavior. The MLE value approaches the value given in Table 2 as time advances. Furthermore, it is shown in Fig. 3b that other combinations of A and B, which belong to the intervals specified in this section, can yield a chaotic attractor too. Values of parameters outside the specified ranges drive the system out of chaos and can yield periodic responses as shown in Fig. 3c. Such responses exhibit $(0, -, -)$ values for the three Lyapunov exponents as shown in Fig. 3d indicating a limit cycle.

3.2 A Dissipative Self-Excited Attractor with Cubic Nonlinearity: System (8)

The range of the parameter A for system (8), with cubic nonlinearity, to exhibit chaotic behavior is limited to roughly about $0.6 \leq A < 0.65$ as shown in Fig. 4a with the largest MLE occurring at $A = 0.6$. For A slightly less than 0.6, the response diverges, while for A slightly greater than 0.65, periodic responses start to appear.

For the parameter B, chaos is produced in the approximate interval $1.5 < B < 1.65$, preceded by periodic responses and followed by divergent ones as shown in Fig. 4b. The conditions on the values of Lyapunov exponents at the parmeter values given in Table 3 can be illustrated similar to the previous case. Furthermore, Fig. 4c shows the phase portrait of a chaotic attractor at other parameter values that belong to the intervals defined in this section.

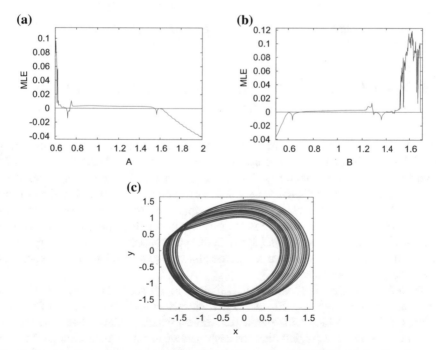

Fig. 4 Ranges of parameters that produce chaos for system (8) **a** MLE versus A, **b** MLE versus B, and **c** Phase portrait at $A = 0.62$ and $B = 1.64$

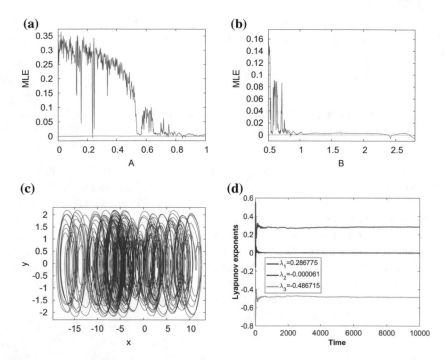

Fig. 5 Ranges of parameters that produce chaos for system (10) **a** MLE versus A, **b** MLE versus B, **c** Phase portrait, and **d** Time evolution of Lyapunov exponents at $A = 0.2$ and $B = 2.7$

3.3 A Dissipative Self-Excited Attractor with Sinusoidal Nonlinearity: System (10)

System (10) is studied as a sample of systems with sinusoidal nonlinearity. Various values that belong to the approximate intervals $0 < A < 0.7$ and $0.5 < B < 2.75$ correspond to chaotic behavior as shown in Fig. 5a and b. The intervals are not continuous, i.e., some exceptional values that correspond to non-chaotic behavior are found in between. The ranges of parameters which correspond to chaotic responses for the systems with sinusoidal nonlinearity are wider than the previous systems. Values other than those stated in (Sprott 2000a) can produce chaos with larger values of MLE. This is illustrated through the phase portrait and Lyapunov exponents in Fig. 5c and d, respectively.

Table 5 summarizes the main results obtained in this section for three continuous dissipative chaotic systems with self-excited attractors and various types of nonlinearities. A combination of the parameter values, which produce chaos, was given in (Sprott 2000a) as a single value for each parameter rather than a range. Table 5 shows the attractor digram of each system at the specified parameter values and wider ranges of parameters that produce chaos, which were not mentioned in the original

Table 5 Summary of the results obtained for the selected systems

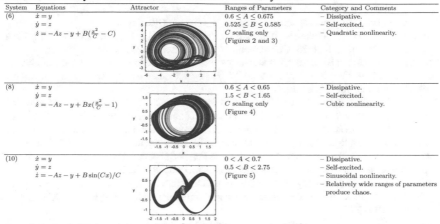

System	Equations	Attractor	Ranges of Parameters	Category and Comments
(6)	$\dot{x} = y$ $\dot{y} = z$ $\dot{z} = -Az - y + B(\frac{x^2}{C} - C)$		$0.6 \leq A \leq 0.675$ $0.525 \leq B \leq 0.585$ C scaling only (Figures 2 and 3)	– Dissipative. – Self-excited. – Quadratic nonlinearity.
(8)	$\dot{x} = y$ $\dot{y} = z$ $\dot{z} = -Az - y + Bx(\frac{x^2}{C} - 1)$		$0.6 \leq A < 0.65$ $1.5 < B < 1.65$ C scaling only (Figure 4)	– Dissipative. – Self-excited. – Cubic nonlinearity.
(10)	$\dot{x} = y$ $\dot{y} = z$ $\dot{z} = -Az - y + B\sin(Cx)/C$		$0 < A < 0.7$ $0.5 < B < 2.75$ (Figure 5)	– Dissipative. – Self-excited. – Sinusoidal nonlinearity. – Relatively wide ranges of parameters produce chaos.

paper. Moreover, the main category to which each system belongs and comments on its behavior are included.

4 Conclusions

A review of dissipative jerk-based continuous chaotic systems with self-excited attractors has been presented. The systems posses various types of nonlinearities: piecewise, quadratic, cubic, sinusoidal and hyperbolic. The parameter values and chaotic properties of each system have been validated through phase portraits and MLE values. Using numerical simulations, wider ranges of parameters that correspond to chaotic behavior have been defined and shown to exhibit positive value of MLE. In addition, either periodic or divergent responses corresponding to values of parameters outside these ranges have been included.

References

Abd-El-Hafiz SK, AbdElHaleem SH, Radwan AG (2016) Novel permutation measures for image encryption algorithms. Opt Lasers Eng 85:72–83

Abd-El-Hafiz SK, Radwan AG, AbdEl-Haleem SH (2015) Encryption applications of a generalized chaotic map. Appl Math Inf Sci 9(6):3215

Abd-El-Hafiz SK, Radwan AG, AbdelHaleem SH, Barakat ML (2014) A fractal-based image encryption system. IET Image Process 8(12):742–752

Abdelhaleem, SH, Radwan AG, Abd-El-Hafiz SK (2014) A chess-based chaotic block cipher. In: IEEE 12th international new circuits and systems conference (NEWCAS). IEEE, pp 405–408

Addison PS (1997) Fractals and chaos: an illustrated course. CRC Press

Alligood KT, Sauer TD, Yorke JA (1996) Chaos: an introduction to dynamical systems. Springer

Azar AT, Vaidyanathan S (2015) Chaos modeling and control systems design. Springer

Azar AT, Vaidyanathan S (2016) Advances in chaos theory and intelligent control, vol 337. Springer

Azar AT, Vaidyanathan S, Ouannas A (2017) Fractional order control and synchronization of chaotic systems, vol 688. Springer

Barakat ML, Mansingka AS, Radwan AG, Salama KN (2013) Generalized hardware post-processing technique for chaos-based pseudorandom number generators. ETRI J 35(3):448–458

Chien T-I, Liao T-L (2005) Design of secure digital communication systems using chaotic modulation, cryptography and chaotic synchronization. Chaos Solitons Fractals 24(1):241–255

Diks C, Hommes C, Panchenko V, Van Der Weide R (2008) E&F chaos: a user friendly software package for nonlinear economic dynamics. Comput Econ 32(1–2):221–244

Elwakil A, Salama K, Kennedy M (2000) A system for chaos generation and its implementation in monolithic form. In: IEEE international symposium on circuits and systems (ISCAS), vol 5. IEEE, pp 217–220

Frey DR (1993) Chaotic digital encoding: an approach to secure communication. IEEE Trans Circuits Syst II Analog Digital Signal Process 40(10):660–666

Gottlieb H (1996) Question 38. what is the simplest jerk function that gives chaos? Am J Phys 64(5):525–525

Govorukhin VN (2004) MATDS: MATLAB based program for dynamical systems investigation

Henein MMR, Sayed WS, Radwan AG, Abd-El-Hafiez SK (2016) Switched active control synchronization of three fractional order chaotic systems. In: 13th international conference on electrical engineering/electronics, computer, telecommunications and information technology

Kocarev L, Lian S (2011) Chaos-based cryptography: theory, algorithms and applications, vol 354. Springer

Lau F Tse CK (2003) Chaos-based digital communication systems. Springer

Layek G (2015) An introduction to dynamical systems and chaos. Springer

Leonov GA, Kuznetsov NV (2013) Hidden attractors in dynamical systems. from hidden oscillations in Hilbert-Kolmogorov, Aizerman, and Kalman problems to hidden chaotic attractor in Chua circuits. Int J Bifurcat Chaos 23(01):1330002

Lorenz EN (1963) Deterministic nonperiodic flow. J Atmos Sci 20(2):130–141

Mansingka AS, Zidan MA, Barakat ML, Radwan AG, Salama KN (2013) Fully digital jerk-based chaotic oscillators for high throughput pseudo-random number generators up to 8.77 Gbits/s. Microelectron J 44(9):744–752

Martínez-Guerra R, Pérez-Pinacho CA, Gómez-Cortés GC (2015) Synchronization of integral and fractional order chaotic systems: a differential algebraic and differential geometric approach with selected applications in real-time. Springer

Petras I (2011) Fractional-order nonlinear systems: modeling, analysis and simulation. Springer Science & Business Media

Radwan A, Moaddy K, Hashim I (2013) Amplitude modulation and synchronization of fractional-order memristor-based Chua's circuit. In: abstract and applied analysis, vol 2013. Hindawi Publishing Corporation

Radwan A, Moaddy K, Salama KN, Momani S, Hashim I (2014) Control and switching synchronization of fractional order chaotic systems using active control technique. J Adv Res 5(1):125–132

Radwan A, Soliman A, El-Sedeek A (2004) MOS realization of the modified Lorenz chaotic system. Chaos Solitons Fractals 21(3):553–561

Radwan A, Soliman AM, Elwakil AS (2007) 1-D digitally-controlled multiscroll chaos generator. Int J Bifurcat Chaos 17(01):227–242

Radwan AG (2013) On some generalized discrete logistic maps. J Adv Res 4(2):163–171

Radwan AG, Abd-El-Hafiz SK (2013) Image encryption using generalized tent map. In: IEEE 20th international conference on electronics, circuits, and systems (ICECS). IEEE, pp 653–656

Radwan AG, Abd-El-Hafiz SK (2014) The effect of multi-scrolls distribution on image encryption. In: 21st IEEE international conference on electronics, circuits and systems (ICECS). IEEE, pp 435–438

Radwan AG, Abd-El-Hafiz SK, AbdElHaleem SH (2012) Image encryption in the fractional-order domain. In: International conference on engineering and technology (ICET). IEEE, pp 1–6

Radwan AG, Abd-El-Hafiz SK, AbdElHaleem SH (2014) An image encryption system based on generalized discrete maps. In: IEEE 21st international conference on electronics, circuits and systems (ICECS). IEEE, pp 283–286

Radwan AG, Abd-El-Hafiz SK, AbdElHalee SH (2015a) Image encryption based on fractional-order chaotic generators. In: 2015 international symposium on nonlinear theory and its applications NOLTA2015, Kowloon, Hong Kong, China, December 1-4, 2015. IEEE, pp 688–691

Radwan AG, AbdElHaleem SH, Abd-El-Hafiz SK (2015b) Symmetric encryption algorithms using chaotic and non-chaotic generators: a review. J Adv Res

Radwan AG, Sayed WS, Abd-El-Hafiz SK (2017) Control and synchronization of fractional-order chaotic systems. Fractional order control and synchronization of chaotic systems. Springer, pp 325–355

Radwan AG, Soliman AM, El-Sedeek A-L (2003) An inductorless CMOS realization of Chuas circuit. Chaos Solitons Fractals 18(1):149–158

Sayed WS, Fahmy HA, Rezk AA, Radwan AG (2017a) Generalized smooth transition map between tent and logistic maps. Int J Bifurcat Chaos 27(01):1730004

Sayed WS, Henein MM, Abd-El-Hafiz SK, Radwan AG (2017b) Generalized dynamic switched synchronization between combinations of fractional-order chaotic systems. Complexity

Sayed WS, Radwan AG, Abd-El-Hafiez SK (2016) Generalized synchronization involving a linear combination of fractional-order chaotic systems. In: 13th international conference on electrical engineering/electronics, computer, telecommunications and information technology

Sayed WS, Radwan AG, Fahmy HA (2015a) Design of a generalized bidirectional tent map suitable for encryption applications. In: 11th international computer engineering conference (ICENCO). IEEE, pp 207–211

Sayed WS, Radwan AG, Fahmy HA (2015b) Design of positive, negative, and alternating sign generalized logistic maps. Discrete Dyn Nat Soc

Sayed WS, Radwan AG, Fahmy HA (2017c) Chaotic systems based on jerk equation and discrete maps with scaling parameters. In: 6th international conference on modern circuits and systems technologies (MOCAST). IEEE, pp 1–4

Sayed WS, Radwan AG, Rezk AA, Fahmy HA (2017d) Finite precision logistic map between computational efficiency and accuracy with encryption applications. Complexity

Schöll E (2001) Nonlinear spatio-temporal dynamics and chaos in semiconductors, vol 10. Cambridge University Press

Siu S (1998) Lyapunov exponent toolbox. MATLAB central file exchange, file ID 233

Sprott J (1997) Some simple chaotic jerk functions. Am J Phys 65(6):537–543

Sprott JC (1994) Some simple chaotic flows. Phys Rev E 50(2):R647

Sprott JC (2000a) A new class of chaotic circuit. Phys Lett A 266(1):19–23

Sprott JC (2000b) Simple chaotic systems and circuits. Am J Phys 68(8):758–763

Sprott JC (2011) A new chaotic jerk circuit. IEEE Trans Circuits Syst II Express Briefs 58(4):240–243

Strogatz SH (2014) Nonlinear dynamics and chaos: with applications to physics, biology, chemistry, and engineering. Westview press

Tolba MF, AbdelAty AM, Soliman NS, Said LA, Madian AH, Azar AT, Radwan AG (2017) FPGA implementation of two fractional order chaotic systems. AEU Int J Electron Commun 78:162–172

Vaidyanathan S (2015) Analysis, control and synchronization of a 3-D novel jerk chaotic system with two quadratic nonlinearities. Kyungpook Math J 55:563–586

Vaidyanathan S, Idowu BA, Azar AT (2015a) Backstepping controller design for the global chaos synchronization of Sprott's jerk systems. Chaos modeling and control systems design. Springer, pp 39–58

Vaidyanathan S, Volos C, Pham V-T, Madhavan K (2015b) Analysis, adaptive control and synchronization of a novel 4-D hyperchaotic hyperjerk system and its spice implementation. Arch Control Sci 25(1):135–158

Vaidyanathan S, Volos C, Pham V-T, Madhavan K, Idowu BA (2014) Adaptive backstepping control, synchronization and circuit simulation of a 3-D novel jerk chaotic system with two hyperbolic sinusoidal nonlinearities. Arch Control Sci 24(3):375–403

Vaidyanathan S, Volos CK, Kyprianidis I, Stouboulos I, Pham V (2015c) Analysis, adaptive control and anti-synchronization of a six-term novel jerk chaotic system with two exponential nonlinearities and its circuit simulation. J Eng Sci Technol Rev 8(2):24–36

Zidan MA, Radwan AG, Salama KN (2012) Controllable V-shape multiscroll butterfly attractor: system and circuit implementation. Int J Bifurcat Chaos 22(06):1250143

Synchronization Properties in Coupled Dry Friction Oscillators

Michał Marszal and Andrzej Stefański

Abstract Self-excited vibrations in friction oscillators are known as stick-slip phenomenon. The non-linearity in the friction force characteristics introduces insta-bility to the steady frictional sliding. The self-excited friction oscillator consists of the mass pushed horizontally on the surface, elastic element (spring) and a drive (convey or belt). Described system serves as a classic toy model for representation of stick-slip motion. Synchronization is an interdisciplinary phenomenon and can be defined as correlation in time of at least two different processes. This chapter focuses on synchronization thresholds in networks of oscillators with dry friction oscillators coupled by linear springs. Oscillators are connected in the nearest neighbour fash-ion into topologies of open and closed ring. In course of the numerical modelling we are interested in identification of complete and cluster synchronization regions. The thresholds for complete synchronization are determined numerically using brute force numerical integration and by means of the master stability function (MSF). Estimation of the MSF is conducted using approach called two-oscillator probe. Moreover, we perform a parameter study in two-dimensional space, where differ-ent cluster synchronization configurations are explored. The results indicate that the MSF can be applied to non-smooth system such as stick-slip oscillator. Synchro-nization thresholds determined using MSF occur to be in line with the one obtained numerically.

1 Introduction

Synchronization phenomenon draws attention of scientists in different disciplines of science, e.g. biology, social science, engineering, physics. Word "synchronization"

M. Marszal (✉) · A. Stefański
Division of Dynamics, Lodz University of Technology,
ul. Stefanowskiego 1/15, 90-924 Lodz, Poland
e-mail: michal.marszal@p.lodz.pl

A. Stefański
e-mail: andrzej.stefanski@p.lodz.pl

© Springer International Publishing AG 2018
V.-T. Pham et al. (eds.), *Nonlinear Dynamical Systems with Self-Excited and Hidden Attractors*, Studies in Systems, Decision and Control 133,
https://doi.org/10.1007/978-3-319-71243-7_4

has Greek origins and is combined of two parts: *syn*—common and *chronos*—time, which together mean happening at the same time. Synchronization may be defined as adjustments of rhythms of oscillating objects due to their weak interaction (Pikovsky et al. 2003).

Dutch scientist Christiaan Huygens was a pioneer of research in the field of synchronization, when back in the 17th century he observed synchronization of two pendula hanging on a common support (Huygens 1673). In the 19th century Sir John William Strutt (Lord Rayleigh) described synchronization in organ pipes (Rayleigh 1896). Pipes with the same pitch, placed side by side cause the sound to quench. Beginning of the 20th century brought observation of synchronization in electric engineering, when Eccles and Vincent (1920) discovered synchronization property of triode generator. The experiment they proposed proved that coupling generator forces common frequency of system vibration (current frequency of single generator depends on electric properties circuit elements). This idea was later developed by Appleton (1922), van der Pol (1927).

In the second half of the 20th century the synchronization phenomenon was reported in biological systems (Mirollo and Strogatz 1990; Winfree 1967). John and Elisabeth Buck investigated the synchronization phenomenon among fireflies is south-east Asia (Buck and Buck 1968), where males emit synchronous light flashes to attract female during the mating season. Phenomenon of swarm behaviour in groups of animals (e.g. fish school, flock of birds) is addressed in Heppner and Grenander (1990), Reynolds (1987). Existence of synchronization is found in pacemakers cells (Jalife 1984; Michaels et al. 1987), adjustments of menstrual cycle among women (Graham and McGrew 1980), rhythmic applause in concert halls (Néda et al. 2000). An example of synchronization in civil engineering is the case of the Millennium Bridge in London. The just opened footbridge started to vibrate unexpectedly after reaching a threshold number of pedestrian. The lateral forces exerted by pedestrians induced the bridge vibrations, which forced the walkers to move in synchronized step, which additionally amplify the lateral oscillations of the bridge (Dallard et al. 2001a, b; Eckhardt et al. 2007; Lenci and Marcheggiani 2012; Strogatz et al. 2005).

Friction is an ubiquitous force in mechanics, responsible for the resistance of contacting surfaces to relative motion. One can distinguish two types of friction: dry friction—when two solid surfaces are in contact and viscous friction—when the contact occurs through a layer of fluid (e.g. lubricant). Friction dissipates the energy of contacting interfaces into heat and can be the source of self-excited vibrations. These can be heard as squeal sound in various devices (e.g. breaks, machining tools, chalk on blackboard, string and bow in violin) (Ghazaly et al. 2013; Patitsas 2010; Warmiński et al. 2003). Proper understanding of friction phenomenon is crucial in control engineering (Gogoussis and Donath 1987; Saha et al. 2010).

The word "friction" is of Latin origin—*fricare*. One of the first scholars studying the properties of friction was da Vinci (1518), who formulated two theories. He reported that friction is directly proportional to the normal load applied on the friction interface. Additionally he stated that friction is independent from the apparent contact area (Dowson 1979; Hutchings 2016; Wojewoda 2008). Works of da

Vinci were unpublished until they had been rediscovered by Amontons (1669) and today are now known as Amontons' laws of friction. Euler (1750, 1761) distinguished static and kinetic friction. He also found the relation between inclination angle of inclined plane and friction coefficient $\mu = \tan \alpha$ (Meyer et al. 1998). French physicist Charles Coulomb, further developed Amontons' ideas (Coulomb 1821). He concluded that kinetic friction is independent of the relative velocity between contacting surface, which is known as Coulomb's law of friction. A basic friction model is named after him (Coulomb friction model). However, the Coulomb model despite robustness in simple case fails in more complex applications. In the beginning of the 20th century German engineer Stribeck (Stribeck 1902) investigated the non-linearity between the friction force and relative velocity, which is known as Stribeck effect. The change between static and kinetic friction is not gradually, but follows non-linear dependency called Stribeck curve. Should the relative velocity between surface of contact be small, the friction force smoothly decrease from the static friction level, converging at kinetic friction level. The difference between static and kinetic friction in systems with energy source leads to self-excited vibration of the investigated mass.

Nowadays a variety of friction models has been proposed, which can be divided into two groups: static models (Armstrong-Helouvry 1991; Bo and Pavelescu 1982; Hess and Soom 1990; Popp and Stelter 1990) and dynamical (Al-Bender et al. 2004; de Wit et al. 1995; Dahl 1968; Stefański et al. 2003; Wojewoda et al. 2008), where friction depends on many variables. The dynamical models have even internal states described by ordinary differential equations.

In this chapter we deal with the synchronization properties and synchronization properties of coupled dry friction oscillators. The research presented in this chapter is a continuation of author's previous research in Marszal (2017), Marszal et al. (2016), Marszal and Stefański (2017). The chapter is organized as follows. Section 2 presents theoretical background in the field of synchronization. Section 3 introduces the mathematical model and the concept of self-excited vibrations in single friction oscillator. In Sect. 4 friction oscillators are coupled forming oscillator networks. Section 5 discusses the result of numerical simulations. Finally conclusions and possible future development are shown in Sect. 6.

2 Synchronization

Let us consider a dynamical system, consisting of N oscillators, which can be described using following matrix-form equation (Stefański 2009), where $\mathbf{x} = (x_1, \ldots, x_N) \in \mathfrak{R}^N$ is a state vector and $\mathbf{F}(\mathbf{x}) = (f_1(x_1), \ldots, f_N(x_N))$ is a function describing the local dynamics of the system, which is independent of the coupling.

$$\dot{\mathbf{x}} = \mathbf{F}(\mathbf{x}) + \sigma(\mathbf{G} \otimes \mathbf{H})\mathbf{x}, \tag{1}$$

The second term in (1) describes the coupling properties. $\mathbf{G}$ is a connectivity matrix, $\mathbf{H} : \mathfrak{R}^N \to \mathfrak{R}^N$ linking functions, $\mathbf{H}$ a linking matrix, σ coupling coefficient, $\otimes$ denotes the Kronecker product of two matrices. For the general case the properties of $\mathbf{G}$ and $\mathbf{H}$ matrices can be arbitrary.

2.1 Types of Synchronization

Synchronization is a complex phenomenon. For the case of this chapter let us limit our consideration to few types of synchronization, namely, complete synchronization, imperfect complete synchronization and cluster synchronization.

Let us restrict the area of considerations to a network of identical oscillators $\mathbf{F}(\mathbf{x})$ and linking functions $\mathbf{H}$. In such a system it is possible to obtain complete synchronization (CS), called also full synchronization. According to Pecora and Carroll (1990) complete synchronization can be observed when two trajectories converge to the same value and later hold that conditions. Stefański (2009) proposes following definition of complete synchronization.

Definition 1 The complete synchronization of two dynamical systems represented with their phase plane trajectories $\mathbf{x}(t)$ and $\mathbf{y}(t)$, respectively, takes place when for all $t > 0$, the following relation is fulfilled:

$$\lim_{t \to \infty} \|\mathbf{x}(t) - \mathbf{y}(t)\| = 0. \tag{2}$$

In practical applications it may be difficult to have identical oscillator nodes in network. Should there be a mismatch between oscillators or coupling properties, differences between their respective trajectories converge to zero with some small tolerance ε. Such a situation is called imperfect complete synchronization (ICS). Stefański (2009) defines ICS as follows.

Definition 2 The imperfect complete synchronization of two dynamical systems represented with their phase plane trajectories $\mathbf{x}(t)$ and $\mathbf{y}(t)$, respectively, occurs when for all $t > 0$, the following inequality is fulfilled:

$$\lim_{t \to \infty} \|\mathbf{x}(t) - \mathbf{y}(t)\| < \varepsilon, \tag{3}$$

where ε is a small parameter.

Supposing the system consists of $N > 2$ identical oscillators one may distinguish two or more subsets for which the particular oscillators are in sync with each other and out of sync with the members of the other subset. Subsets of synchronized oscillators are called clusters. It is important to mention that we can talk about cluster synchronization when whole system is not in complete synchronization. The motion of different cluster may be uncorrelated or one can observe a shift phase between them. Existence of clusters is connected with the existence and stability of synchronization manifold (Perlikowski 2007). The topic of clusters can be found in literature in Belykh et al. (2000, 2001), Dahms et al. (2012), Kaneko (1990), Wu et al. (2009), Yanchuk et al. (2001).

2.2 Synchronous State Stability

A power mathematical tool used in assessing the stability of the synchronous state is the concept of master stability function (MSF) introduced by Pecora and Carroll (1998). Master stability function enables to divide the problem of the synchronous state stability into two parts: (i) the topological part, where we need to calculate the eigenvalues of the connectivity matrix, and (ii) local dynamic part, where one need to calculate Lyapunov exponents of variational equation. The classic approach to estimate the MSF is to calculate transversal Lyapunov exponents (TLE) of the Eq. (7) derived below. Let us begin with obtaining variational equation of Eq. (1):

$$\dot{\xi} = \left[\mathbf{1_N} \otimes D\mathbf{F} + \sigma\mathbf{G} \otimes D\mathbf{H}\right] \xi, \tag{4}$$

where ξ_i is the variation of the ith node, $\xi = (\xi_1, \xi_2, ...\xi_N)$ is variation vector, $D\mathbf{F}$ is the Jacobian of any node, $D\mathbf{H}$ is the Jacobian of the linking function. Diagonalization of Eq. (4) yields to uncoupling the variational Eq. (4) into N block having a form of:

$$\dot{\xi}_k = \left[D\mathbf{F} + \sigma\gamma_k D\mathbf{H}\right] \xi_k, \tag{5}$$

where γ_k is the kth eigenvalue of the $\mathbf{G}$, $i = 0, 1, 2, ..., N - 1$, ξ_k is transverse mode of perturbation from the synchronous state. In case of $k = 0$ eigenvalue is $\gamma_0 = 0$, and consequently Eq. (5) is reduced to

$$\dot{\xi}_0 = D\mathbf{f}\xi_0, \tag{6}$$

which is associated with the longitudinal direction located within the synchronization manifold. The other kth eigenvalues correspond to transverse eigenvectors (Pecora and Carroll 1998). In MSF concept the tendency to synchronization is a function of eigenvalues γ_k. Let us substitute $\sigma\gamma = \alpha + i\beta$ in Eq. (5), where α and β are respective real and imaginary part of eigenvalues.

$$\dot{\xi} = [D\mathbf{F} + (\alpha + i\beta) D\mathbf{H}] \xi, \tag{7}$$

where ξ is an arbitrary transverse mode.

Condition for the existence of invariant synchronization manifold is the zero row sum connectivity matrix $\mathbf{G}$ (Pecora and Carroll 1998). All the real parts of eigenvalues, which correspond to transversal modes, are negative ($\text{Re}(\gamma_{k\neq0}) < 0$). The spectrum of eigenvalues has the descending form, i.e., $\gamma_0 \geq \gamma_1 \geq ... \geq \gamma_{N-1}$. In general case, (Pecora and Carroll 1998) defines MSF as the largest transversal Lyapunov exponent λ_T surface, computed basing on Eq. (7), on a complex numbers plane (α, β). Should the interaction between the nodes be mutual (e.g. mechanical systems), then the eigenvalues have only real part and then MSF is represented only by a curve describing the largest TLE as a function of real number α, defined as

$$\alpha = \sigma\gamma. \tag{8}$$

The synchronous state of dynamical system is stable when all eigenmodes of the discrete eigenvalue spectrum $\sigma\gamma_k$ lay in ranges of the largest negative TLE (see Fig. 1a). Supposing even only one eigenvalue is in the range, where $\lambda_T > 0$ (see Fig. 1b), the global synchronization is unstable, however, cluster synchronization is still possible.

The method mentioned in previous section is robust for time continuous systems, given by smooth equations, where the computation of TLE is relatively easy. However, when dealing with non-smooth dynamical systems, such as dry friction oscillators, the computation of TLE requires special care and algorithms. In such a case, techniques called *three-oscillator universal probe* (Fink et al. 2000) and *two-oscillator probe* (Wu 2001) come to our rescue. Oscillator probe is based on estimating the MSF on the complex plane by direct detection of the complete synchronization in numerical calculations or in the experiment. The methodology is simple, but yet efficient. When calculating MSF in three-oscillator probe for system containing N oscillators, one initially investigates the reference probe of three oscillators. The area on the complex plane (α, β) where the complete synchronization or imperfect complete synchronization occurs is the equivalent of the area of negative transversal Lyapunov exponents. The two-oscillator probe can be applied for mechanical systems, where due to mutual interaction between the nodes, the eigenvalues of the

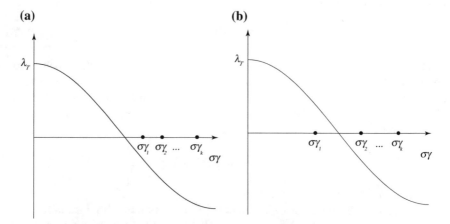

Fig. 1 Examples of the MSF with spectrum of eigenvalue of connectivity matrix for cases of potential stable (**a**) and unstable (**b**) complete synchronization state of all network oscillators

connectivity matrix **G** are real and the MSF is reduced to a curve, which is also the case for real only eigenvalues when using TLE as MSF.

Let us consider system of $N = 2$ coupled oscillators, given by Eq. (1), with connectivity matrix **G**:

$$\mathbf{G} = \begin{pmatrix} -c & c \\ c & -c \end{pmatrix}, \tag{9}$$

where c is the real coupling factor. One can formulate following variational equation of the considered system.

$$\begin{pmatrix} \dot{\xi}_1 \\ \dot{\xi}_2 \end{pmatrix} = \left[\mathbf{I_2} \otimes D\mathbf{f} + \sigma \begin{pmatrix} -c & c \\ c & -c \end{pmatrix} \otimes D\mathbf{H} \right] \begin{pmatrix} \xi_1 \\ \xi_2 \end{pmatrix}, \tag{10}$$

with real eigenvalues $\gamma_0 = 0$, $\gamma_1 = -2c$. This yields to generic variational equation for MSF determination for the two-oscillator probe system

$$\dot{\xi} = (D\mathbf{f} - 2\sigma c D\mathbf{H})\,\xi. \tag{11}$$

If the all non-zero eigenvalues of the connectivity matrix of the system of N oscillators are in the surface or range of complete (imperfect complete) synchronization for the reference probe, then the complete synchronization (imperfect complete synchronization) is possible for the the system in question. When comparing Eq. (11) with Eq. (7) one can notice that $\alpha = 2\sigma c$. Hence, the multiplier 2 has to be taken under consideration when the MSF (e.g. from Fig. 1a) is replaced by two oscillators probe as shown in Fig. 2.

In order to estimate the MSF using two-oscillator probe, it is necessary to couple two oscillators with real coupling and perform numerical or experimental determination of synchronous ranges. The synchronous ranges can be indicated by average synchronization error for two-oscillator probe $\langle e_{II} \rangle = 0$ and are equivalents of stable synchronous region of MSF using the largest TLE. We can project this representative two-oscillator probe for any number of coupled oscillators with arbitrary structure of connection between them via eigenvalues of the connectivity matrix **G** (Fink et al. 2000; Marszal and Stefański 2017; Pecora and Carroll 1998; Stefański 2009; Wu 2001).

One can distinguish three different regimes of synchronous intervals in the MSF $\langle e_{II} \rangle (\alpha)$ analysis: (i) bottom-limited (α_1, ∞), (ii) upper-limited $(\alpha_1 = 0, \alpha_2 > 0$ but of finite value) and (iii) double-limited $(\alpha_1, \alpha_2 > 0$, but of finite value). Values of α_1 and α_2 denote upper and lower ends (Fig. 3) of the synchronous range, respectively (Stefański 2009; Stefański et al. 2007). In this chapter let us focus on the second and third cases. In the double-limited case, two transverse eigenmodes have influence on the synchronization thresholds, i.e., the longest spatial-frequency mode, which corresponds to the largest eigenvalue γ_1, and the shortest spatial frequency, which corresponds to the smallest eigenvalue γ_{N-1}. They determine the size of the synchronous state interval. The loss of stability can be caused by two desynchronization bifurcations. Decrease of σ triggers a long-wavelength bifurcation, as the longest

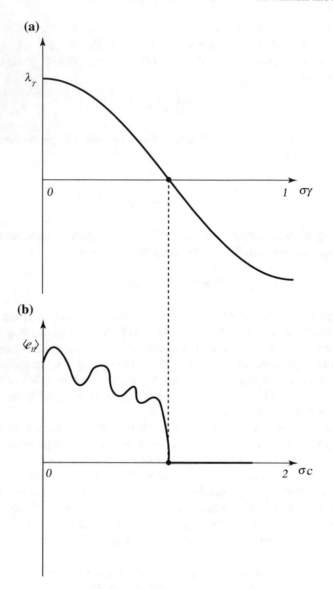

Fig. 2 Equivalence of the MSF (**a**) and two-oscillator probe (**b**)

wavelength mode ξ_1 becomes unstable. Contrary, the increase of σ may lead to short-wavelength bifurcation, because the shortest wavelength mode ξ_{N-1} becomes unstable (Marszal and Stefański 2017; Stefański et al. 2007). One can formulate condition for the existence of the synchronous interval as

$$\frac{\gamma_{N-1}}{\gamma_1} < \frac{\alpha_2}{\alpha_1}, \tag{12}$$

which implies the existence of the maximum number of oscillators, for which the investigated system can be in CS. The increase of N follows the increase of γ_{N-1}/γ_1 ratio. Hence, the inequality in (12) cannot be fulfilled (Barahona and Pecora 2002; Nishikawa et al. 2003; Pecora et al. 2000; Pecora 1998). The discussed case of double limited synchronous interval is depicted in the Fig. 3. For the Fig. 3a the condition in (12) is fulfilled and synchronous intervals overlap with cross-hatched area marking the synchronous range (σ_1, σ_2). Contrary, in the Fig. 3b the synchronous intervals do not overlap and there is no synchronous range for network of N oscillators (Marszal and Stefański 2017; Stefański 2009). In case of upper-limited synchronous interval the synchronization regions depend on the smallest eigenvalue γ_{N-1}. The increase of N forces to narrow the synchronization interval towards the origin of the coordinates system.

As an additional effect of double-limited synchronous interval, the phenomenon of the so called ragged synchronizability can be observed, i.e., alternately occurring synchronous and desynchronous windows (Stefański et al. 2007) (e.g. Fig. 3a).

3 Single Self-excited Friction Oscillator

Consider a single, classic, dry friction, self-excited oscillator, depicted in Fig. 4. The system consists of a drive—conveyor belt, elastic element—spring and moving oscillating mass with a frictional interface. The equation of motion of the system can be formulated as follows:

$$m\frac{d^2x}{dt^2} = F_N f\left(v_r\right) - kx, \tag{13}$$

where: m—mass of the oscillator x—displacement of the oscillator, k—stiffness constant, v_b—velocity of the belt, v_r—relative velocity between the contacting surfaces $\left(v_r = v_b - \frac{dx}{dt}\right)$, F_N—normal load (the weight of oscillator is included in F_N), $f(v_r)$—function describing friction characteristics.

Let us non-dimensionalise the Eq. (13) by applying $\omega_0 = \sqrt{\frac{k}{m}}$, non-dimensional time $\tau = \omega_0 t$ and characteristic constant $x_0 = \frac{g}{\omega_0^2}$, which results in

$$\ddot{\chi} = \epsilon f(\vartheta_r) - \chi. \tag{14}$$

The non-dimensional variables are formulated as follows: $\epsilon = \frac{F_N}{mg}$—load coefficient, $\vartheta_b = \frac{v_b}{x_0 \omega_0}$—non-dimensional velocity of the belt, $\vartheta_r = \vartheta_b - \dot{\chi}$—non-dimensional relative velocity between contacting surfaces, $\chi = \frac{x}{x_0}$—non-dimensional displacement. The overdots stand for the respective derivatives with

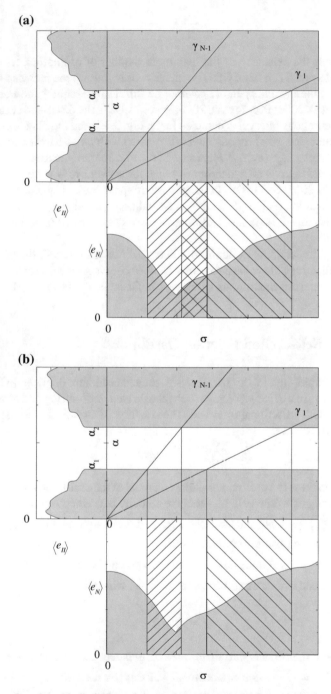

Fig. 3 Examples of double-limited synchronous intervals: **a** synchronous ranges of shortest and longest frequency modes are partly overlapping, yielding to stable synchronous interval (σ_1, σ_2); **b** synchronous ranges of both modes are disconnected—CS is not possible. Gray corresponds to desynchronous regions

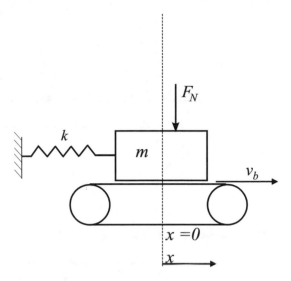

Fig. 4 Single stick-slip dry friction oscillator

respect to τ. The relationships between dimensional and non-dimensional displacement and its derivatives is given by: $\frac{d^2x}{dt^2} = x_0\omega_0^2\ddot{\chi}$, $\frac{dx}{dt} = x_0\omega_0\dot{\chi}$.

Stribeck friction model with the exponential non-linearity is applied as the basis for friction modelling in this work. The Stribeck friction model with the exponential non-linearity is given by formula.

$$f(\vartheta_r) = \left(\mu_k + \left(\mu_s - \mu_k \right) e^{-a|\vartheta_r|} \right) \operatorname{sgn}\vartheta_r, \tag{15}$$

where μ_s—static friction coefficient, μ_k—kinetic friction coefficient, a is constant defining the shape of the friction—relative velocity curve. The friction force f as a function of relative velocity ϑ_r based on the aforementioned model is depicted in the Fig. 5. Note the negative slope of friction force—relative velocity curve, which is essential for the occurrence of self-excited vibrations (Ding 2010).

Figure 6 illustrates a limit-cycle to which all +. The segment of trajectory with horizontal line corresponds to the sticking phase when $\vartheta_r = 0$. The mass moves along with the belt and accumulates the potential energy. The value of friction force adjust itself to maintain the equilibrium with the spring force. When the maximum value of friction force is reached, the friction force cannot balance the spring and the mass begins to slide. Friction is then responsible for the dissipation of energy into heat. Eventually, the velocity of the mass decreases to the level of the velocity of the belt and the mass sticks with it. Such kind of motion form a sawtooth wave (see Fig. 8 with grey regions corresponding to stick phase). Hence, the sound of objects subjected to stick-slip phenomenon is not pleasant to our ears. In Fig. 7 a phase diagram with stick-slip limit cycle is depicted.

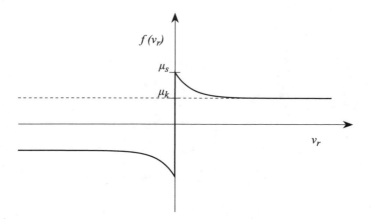

Fig. 5 Friction force f as a function of relative velocity ϑ_r

Fig. 6 Phase portrait of single friction oscillator with self-excited stick-slip vibrations. Multiple trajectories approach the stick-slip limit cycle. System parameters: $\vartheta_b = 0.1$, $\mu_s = 0.3$, $\mu_k = 0.15$, $a = 2.5$, $\epsilon = 2$. (Marszal 2017)

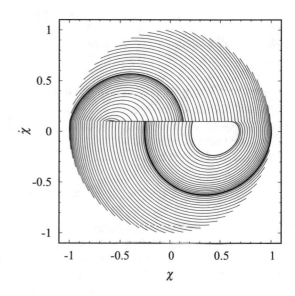

The presented, 1-DOF system, can be treated as a toy model for more complicated systems used in engineering application, e.g. disc brake (Wei et al. 2016). Here mass corresponds to brake pad, belt—to disc of the brake, while spring—to the stiffness of the system, which is subjected to external excitation (Popp et al. 1995).

4 Oscillators Network

Let us now consider an array of N identical oscillators described above, which are coupled using linear springs of stiffness k_C, as shown in the Fig. 9. Additional

Fig. 7 Phase diagram of single self-excited friction oscillator. System parameters: $\vartheta_b = 0.1$, $\mu_s = 0.3$, $\mu_k = 0.15$, $a = 2.5$, $\epsilon = 2$

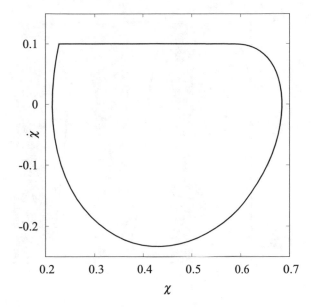

excitation $u \cos \omega \tau$ is applied to each oscillator. The equation of motion for the coupled system can be written in matrix form, where the $\sigma = k_C/k$ stands for the coupling coefficient and determines the strength of the coupling:

$$\begin{Bmatrix} \ddot{\chi}_1 \\ \vdots \\ \ddot{\chi}_N \end{Bmatrix} = - \begin{Bmatrix} \chi_1 \\ \vdots \\ \chi_N \end{Bmatrix} + \sigma \mathbf{G}_N \begin{Bmatrix} \chi_1 \\ \vdots \\ \chi_N \end{Bmatrix} + \begin{Bmatrix} \epsilon f\left(\vartheta_{r_1}\right) + u \cos \omega \tau \\ \vdots \\ \epsilon f\left(\vartheta_{r_N}\right) + u \cos \omega \tau \end{Bmatrix}. \tag{16}$$

Matrix $\mathbf{G_N}$ is connectivity matrix and represents the connection topology of the network. In this work, two distinct topologies are considered, namely open and close ring of oscillators connected in nearest neighbour fashion. The scheme of the topologies is depicted in the Fig. 10. The connectivity matrices for both topologies are presented by Eq. (17) ($\mathbf{G_{N_O}}$—open ring) and Eq. (18) ($\mathbf{G_{N_C}}$—closed ring) respectively.

$$\mathbf{G}_{N_o} = \begin{pmatrix} -1 & 1 & 0 & \cdots & 0 & 0 \\ 1 & -2 & 1 & \ddots & \cdots & 0 \\ 0 & 1 & \ddots & \ddots & \ddots & \vdots \\ \vdots & \ddots & \ddots & \ddots & 1 & 0 \\ 0 & \cdots & \ddots & 1 & -2 & 1 \\ 0 & 0 & \cdots & 0 & 1 & -1 \end{pmatrix}, \tag{17}$$

(a)

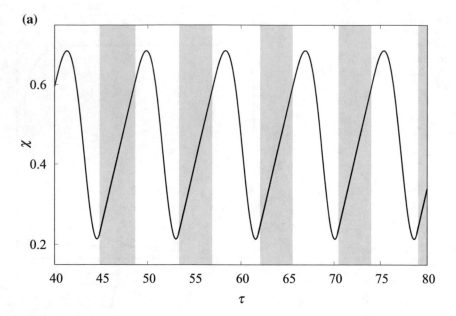

(b)

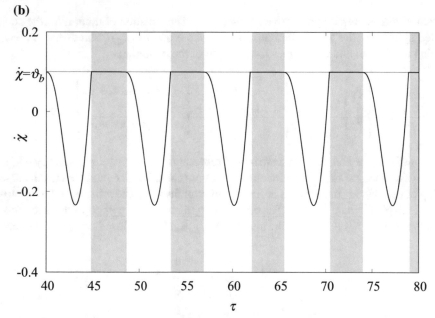

Fig. 8 Time diagram of single self-excited friction oscillator: **a** position of the oscillator, **b** velocity of the oscillator. Gray regions correspond to stick phase. System parameters: $\vartheta_b = 0.1$, $\mu_s = 0.3$, $\mu_k = 0.15$, $a = 2.5$, $\epsilon = 2$

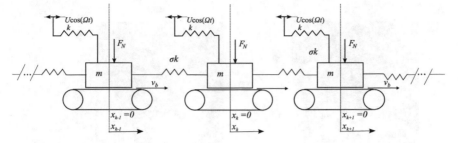

Fig. 9 Array of coupled N stick-slip oscillators

(a)   **(b)**

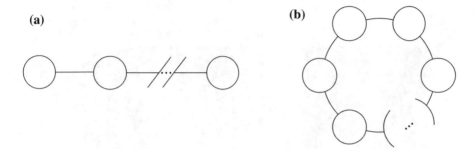

Fig. 10 Different connection topologies for network of N oscillators: **a** open ring **b** closed ring

$$G_{N_C} = \begin{pmatrix} -2 & 1 & 0 & \cdots & 0 & 1 \\ 1 & -2 & 1 & \ddots & \cdots & 0 \\ 0 & 1 & \ddots & \ddots & \ddots & \vdots \\ \vdots & \ddots & \ddots & \ddots & 1 & 0 \\ 0 & \cdots & \ddots & 1 & -2 & 1 \\ 1 & 0 & \cdots & 0 & 1 & -2 \end{pmatrix}. \tag{18}$$

Note that for both matrices there are zero sum rows, which is caused by mutual interaction in mechanical systems. The eigenvalues of the connectivity matrices are used later in the master stability function to determine the synchronization thresholds.

5 Results

In this Section we present the results of numerical studies, based on the numerical model described in Sects. 3 and 4. Additionally, we present the usage of master stability function and two-oscillator probe for determining the synchronization thresholds.

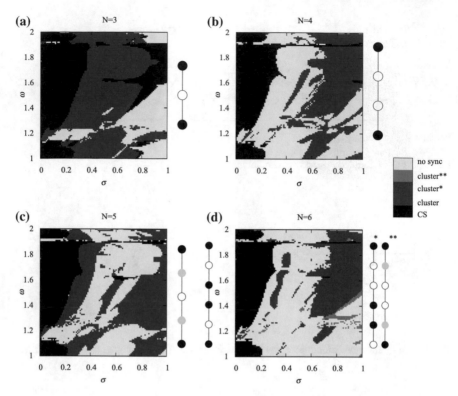

Fig. 11 Synchronization regions for open ring of N oscillators in two-parameter space (σ versus ω), $\epsilon = 1$, on sides of scheme of observed cluster layouts: **a** $N = 3$, **b** $N = 4$, **c** $N = 5$, **d** $N = 6$

The numerical simulations are based on author's own program written in C++ using on Boost Odeint library (Ahnert and Mulansky 2011) as numerical engine. Following non-dimensional parameters are used in simulations: $\vartheta = 0.1$, $\mu_s = 0.35$, $\mu_k = 0.2$, $a = 2.5$, $u = 0.1$. If different values are used, information is placed in figure caption or legend respectively. A transient time equal to 1000 excitation periods ($\Delta\tau_t = 1000 \cdot 2\pi/\omega$) is applied, which is followed by measurement of average synchronization error for time interval corresponding to 200 excitation periods. The investigated systems are started from the initial conditions, when the complete synchronization is slightly perturbed. Should the synchronous state be stable, the trajectories return to synchronous state after time $\Delta\tau_t$.

We perform a study of parameters in (σ, ω) two-dimensional parameter space with goal to detect complete and cluster synchronization regions in open and closed ring connection topology, for different values of the network size N. Based on previous studies (Marszal et al. 2016; Marszal and Stefański 2017), we choose the following ranges of parameters: coupling coefficient $\sigma \in [0, 1]$ and angular frequency of excitation $\omega \in [1, 2]$. The parameter space is discretised into grid with grid element size $\Delta\sigma = \Delta\omega = 0.01$, giving 10 201 elements in total. For each element of the grid aver-

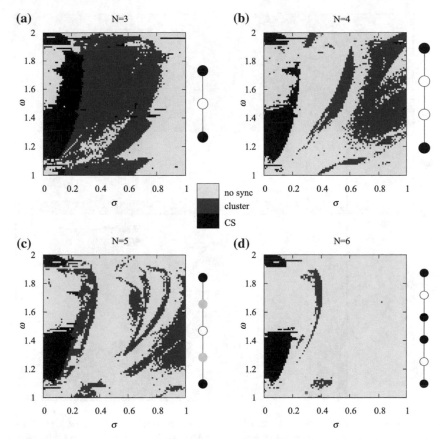

Fig. 12 Synchronization regions for open ring of N oscillators in two-parameter space (σ versus ω), $\epsilon = 1$, on sides of scheme of observed cluster layouts: **a** $N = 3$, **b** $N = 4$, **c** $N = 5$, **d** $N = 6$

age global e (19) and cluster $e_{i,j}$ (20) synchronization errors are computed. Finally a type of synchronization is classified according to definitions in Sect. 2.1. If the respective synchronization error is bellow 10^{-3}, element of the grid is classified as synchronized.

$$e = \sum_{i=2}^{N} \sqrt{\left(\chi_1 - \chi_i\right)^2 + \left(\dot{\chi}_1 - \dot{\chi}_i\right)^2} \tag{19}$$

$$e_{i,j} = \sqrt{\left(\chi_i - \chi_j\right)^2 + \left(\dot{\chi}_i - \dot{\chi}_j\right)^2}. \tag{20}$$

In Figs. 11 and 12 results of the parameter study for the open ring topology are presented. The systems in question are checked for two different values of normal load coefficient: $\epsilon = 1$ (Fig. 15), $\epsilon = 1.5$ (Fig. 12). Black colour depicts the com-

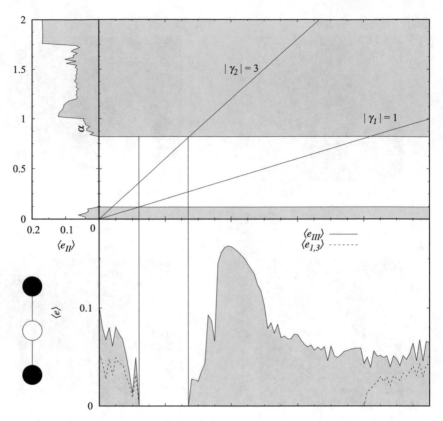

Fig. 13 MSF $\langle e_{II} \rangle (\alpha)$ projected onto respective average synchronization errors $\langle e_3 \rangle$, $\langle e_{1,3} \rangle$ for N coupled oscillators via eigenvalues of connectivity matrix $\mathbf{G}_3$. $N = 3$, $\omega = 1.6$, $\epsilon = 1.5$ initial conditions $\chi = [0.2913, 0, 0.2945, 0, 0.2922, 0]^T$

plete synchronization region, while the yellow lack of synchronization. The other colours correspond to different cluster synchronization layouts. The complete synchronization region is larger for lower values of the normal load coefficient, which yields to lower friction force between the contacting interfaces. Scheme of cluster layouts are placed on side of respective diagrams. Note that for the case of ($\epsilon = 1$, $N = 6$) three cluster layouts are observed (Fig. 11d). For other systems (Fig. 11a–c) only one cluster layout is observed. In all cases the complete synchronization occurs rather for weak coupling. The increase of coupling strength destroys the complete synchronization, however cluster synchronization regions emerges.

Analysis of the eigenvectors of respective eigenvalues enables us to explain the shapes of the clusters (Perlikowski et al. 2010; Yanchuk et al. 2001). Consider the case of three oscillators depicted in one parameter space in the Fig. 13. Here we have following values of eigenvalues: $\gamma_{0(3)} = 0$, $\gamma_{1(3)} = -1$, $\gamma_{2(3)} = -3$; together with corresponding eigenvectors: $\mathbf{v}_{0(3)} = [1, 1, 1]^T$, $\mathbf{v}_{1(3)} = [-1, 0, 1]^T$, $\mathbf{v}_{2(3)} = [1, -2, 1]^T$. The desynchronization process from the complete synchronization state is governed

Table 1 Non-zero eigenvalues and corresponding eigenvectors for the connectivity matrices $\mathbf{G}_N$ for open ring networks.

N=3		N=4		
$\gamma_{1(3)} = -1\ \mathbf{v}_{1(3)}$	$\gamma_{2(3)} = -3\ \mathbf{v}_{2(3)}$	$\gamma_{1(4)} = \sqrt{2}-2$ $\mathbf{v}_{1(4)}$	$\gamma_{2(4)} = -2\ \mathbf{v}_{2(4)}$	$\gamma_{3(4)} = -\sqrt{2}-2$ $\mathbf{v}_{3(4)}$
-1	1	-1	1	-1
0	-2	$1-\sqrt{2}$	-1	$1+\sqrt{2}$
1	1	$-1+\sqrt{2}$	-1	$-1-\sqrt{2}$
		1	1	1

N=5			
$\gamma_{1(5)} =$ $\left(\sqrt{5}-3\right)/2$ $\mathbf{v}_{1(5)}$	$\gamma_{2(5)} =$ $\left(\sqrt{5}-5\right)/2$ $\mathbf{v}_{2(5)}$	$\gamma_{3(5)} =$ $\left(-\sqrt{5}-3\right)/2$ $\mathbf{v}_{3(5)}$	$\gamma_{4(5)} = \left(-\sqrt{5}-5\right)/2\ \mathbf{v}_{4(5)}$
-1	1	-1	1
$\left(1-\sqrt{5}\right)/2$	$\left(-3+\sqrt{5}\right)/2$	$\left(1+\sqrt{5}\right)/2$	$-\left(3+\sqrt{5}\right)/2$
0	$1-\sqrt{5}$	0	$1+\sqrt{5}$
$-\left(1-\sqrt{5}\right)/2$	$\left(-3+\sqrt{5}\right)/2$	$-\left(1+\sqrt{5}\right)/2$	$-\left(3+\sqrt{5}\right)/2$
1	1	1	1

N=6				
$\gamma_{1(6)} = \sqrt{3}-2$ $\mathbf{v}_{1(6)}$	$\gamma_{2(6)} = -1\ \mathbf{v}_{2(6)}$	$\gamma_{3(6)} = -2\ \mathbf{v}_{3(6)}$	$\gamma_{4(6)} = -3\ \mathbf{v}_{4(6)}$	$\gamma_{5(6)} \approx -2-\sqrt{3}$ $\mathbf{v}_{5(6)}$
-1	1	-1	1	-1
$1+\sqrt{3}$	0	1	-2	$1+\sqrt{3}$
$-2+\sqrt{3}$	-1	-1	1	$-2-\sqrt{3}$
$2-\sqrt{3}$	-1	1	1	$2+\sqrt{3}$
$-1+\sqrt{3}$	0	-1	-2	$-1-\sqrt{3}$
1	1	1	1	1

by the $\gamma_{2(3)}$. For this particular eigenvalues the first and the third element of the eigenvector $\mathbf{v}_2$ are equal, leading to the existence of cluster consisting of the first and the third oscillator. The other eigenvector—$\mathbf{v}_{1(3)}$ does not have at least two equal elements, hence it cannot be responsible for the formation of cluster. The eigenvector $\mathbf{v}_{0(3)}$ based on $\gamma_{0(3)}$ corresponds to the direction along synchronization manifold (the global CS state), as all its elements are equal. Table 1 lists non-zero eigenvalues and their eigenvectors for all investigated networks with open ring topology. Note that $\gamma_{0(N)} = 0$ and $\mathbf{v}_{0(N)} = [1, ..., 1]^T$. In the cases of $N = 3, 4, 5$ only one eigenvector pattern can be responsible for the creation of clusters. Thus only single cluster configuration can be observed.

More detailed analysis of global (CS) and cluster synchronization thresholds is performed for certain selection of ω, including verification of the obtained results by means of MSF (see Figs. 13, 14). The MSF is defined as average synchronization

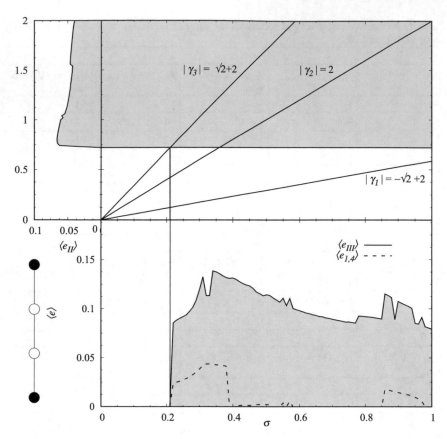

Fig. 14 MSF $\langle e_{II}\rangle(\alpha)$ projected onto average synchronization error between 1st and 4th oscillator $\langle e_{1,4}\rangle$ (dashed line) and average global synchronization error $\langle e_{IV}\rangle$ (solid line) for three coupled oscillators via eigenvalues of connectivity matrix $\mathbf{G_4}$ for excitation angular frequency $\omega = 1.4$. Initial conditions for each value of σ: $\chi_0 = [0.2913, 0, 0.2945, 0, 0.2922, 0]^T$

error for two-oscillator probe $\langle e_{II}\rangle(\alpha)$ as a function of real number α (see Eq. (8)). Next, MSF $\langle e_{II}\rangle(\alpha)$ is projected via eigenvalues of connectivity matrix $\mathbf{G_N}$ onto bifurcation diagrams of average synchronization error for networks consisting of N oscillators with σ as a bifurcation parameter. The complete synchronization for network of N oscillators occurs, provided all eigenvalues spectrum of connectivity matrix $\mathbf{G_N}$ lies within zero $\langle e_{II}(\alpha)\rangle$ function. The areas of zero MSF within the eigenvalues spectrum, as well as complete synchronization regions in networks of N oscillators are marked with grey colour respectively. The method described above is robust for predicting the global CS thresholds and along with the analysis of eigenvectors can be used to explain the cluster synchronizability. However, due to additional coupling factors (i.e., excitation, friction or coexistence of the system attractors) the MSF method indicates only the tendencies of the oscillators to synchronize and might be not always verified in given network configuration.

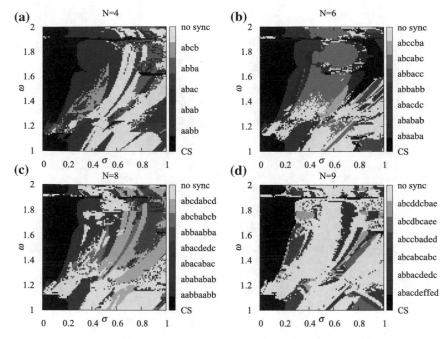

Fig. 15 Synchronization regions for open ring of N oscillators in two-parameter space (σ versus ω), on sides of scheme of observed cluster layouts: **a** $N = 4$, **b** $N = 6$, **c** $N = 8$, **d** $N = 9$

For the case of three oscillators $\epsilon = 1.5, N = 3, \omega = 1.6$ (Fig. 13) the CS region is in range $\sigma \in [0.12, 0.27]$. It is worth mentioning that the first and last oscillator in the network are in cluster synchronization for almost all investigated range of σ. Similar behaviour can be seen in Fig. 12a, marked as red region, wherein cluster synchronization occupies large area of the investigated parameter space. For the case of four oscillators (see Fig. 14) with $\omega = 1.4$, $\epsilon = 1$ CS region is located for $\sigma \in [0, 0.21]$.

In the presented systems it is possible to observe the so called ragged synchronizability phenomenon (Stefański et al. 2007). In Fig. 13 one can observe global ragged synchronizability, where all oscillators in the network are in synchronous or desynchronous state. One can also find for that case, cluster ragged synchronizability, i.e., synchronous and desynchronous regions in clusters.

Similar study as for open ring networks is performed to closed ring network topology in Figs. 15 ($\epsilon = 1$) and 16 ($\epsilon = 1.5$). Table 2 lists non-zero eigenvalues and their eigenvectors for all investigated networks with closed ring topology. Again the parameter space is checked for global and cluster synchronization regions for different length of identical oscillators. The systems are modelled according to Eq. (16) with $\mathbf{G}_{N_C}$ from Eq. (18). The obtained results for different network sizes ($N = 4, 6, 8, 9$) are depicted in the Figs. 15 and 16. The areas of complete synchronization occurs also for low values of coupling. However, in the case of closed ring

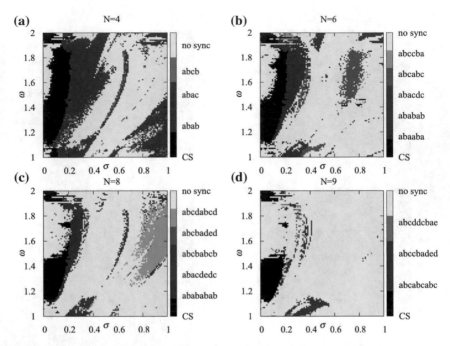

Fig. 16 Synchronization regions for closed ring of N oscillators in two-parameter space (σ versus ω), on sides of scheme of observed cluster layouts: **a** $N = 4$, **b** $N = 6$, **c** $N = 8$, **d** $N = 9$

topology the variety of cluster configuration is richer. This can be explained by the symmetry of the system, which aids the cluster formation.

6 Conclusion

Let us summarize this chapter, which is devoted to the analysis of synchronization properties in dynamical systems with dry friction.

We have performed a parameter study of complete and cluster synchronization properties in two-parameter space (coupling coefficient versus angular frequency of excitation). Numerical investigations involve two different network topologies, i.e., open ring and closed ring. Oscillators are connected in nearest neighbour fashion. The goal was to find synchronization thresholds in various networks of oscillators. The used methodology is based on master stability function. However, MSF is not estimated in a traditional way by TLE but by means of more direct approach, namely, two-oscillator probe.

One needs to bear in mind that MSF describes tendencies of the system to synchronize. It cannot be treated as the final condition for the synchronization. The oscillators in question are coupled also by common excitation and the friction force itself, which may contribute also to the synchronizability of the system. Velocity of the

Table 2 Non-zero eigenvalues and corresponding eigenvectors for the connectivity matrices $\mathbf{G}_N$ for closed ring networks.

N=4			N=6	
i	$\gamma_{i(4)}$	$\mathbf{v}_{i(4)}$	$\gamma_{i(6)}$	$\mathbf{v}_{i(6)}$
1	-2	$[-1,0,1,0]^T$	-1	$[-1,-1,0,1,1,0]^T$
2	-2	$[0,-1,0,1]^T$	-1	$[1,0,-1,-1,0,1]^T$
3	-4	$[-1,1,-1,1]^T$	-3	$[-1,1,0,-1,1,0]^T$
4			-3	$[-1,0,1,-1,0,1]^T$
5			-4	$[-1,1,-1,1,-1,1]^T$

N=8			N=9	
i	$\gamma_{i(8)}$	$\mathbf{v}_{i(8)}$	$\gamma_{i(9)}$	$\mathbf{v}_{i(9)}$
1	$-2+\sqrt{2}$	$[-1,-\sqrt{2},-1,0,1,\sqrt{2},1,0]^T$	-0.468	$[-1,-1.532,-1.347,-0.532,0.532,1.347,1.532,1,0]^T$
2	$-2+\sqrt{2}$	$[\sqrt{2},1,0,-1,-\sqrt{2},-1,0,1]^T$	-0.468	$[1.532,1.347,0.532,-0.532,-1.347,-1.532,-1,0,1]^T$
3	-2	$[-1,0,1,0,-1,0,1,0]^T$	-1.653	$[-1,-0.347,0.879,0.652,-0.652,-0.879,0.347,1,0]^T$
4	-2	$[0,-1,0,1,0,-1,0,1]^T$	-1.653	$[0.347,-0.879,-0.652,0.652,0.879,-0.347,-1,0,1]^T$
5	$-2-\sqrt{2}$	$[-1,\sqrt{2},-1,0,1,-\sqrt{2},1,0]^T$	-3	$[-1,1,0,-1,1,0,-1,1,0]^T$
6	$-2-\sqrt{2}$	$[-\sqrt{2},1,0,-1,\sqrt{2},-1,0,1]^T$	-3	$[-1,0,1,-1,0,1,-1,0,1]^T$
7	-4	$[-1,1,-1,1,-1,1,-1,1]^T$	-3.880	$[-1,1.880,-2.532,2.880,-2.880,2.532,-1.880,1,0]^T$
8			-3.880	$[-1.880,2.532,-2.880,2.880,-2.532,1.880,-1,0,1]^T$

conveyor belt, which is equal for all oscillators as well as the same friction model, provide identity of the parameters, which is necessary condition for the occurrence of CS. Moreover, the common harmonic excitation correlates in time with the driving components of all oscillators and as a consequence facilitates the synchronization.

On the contrary, important factor leading to the desynchronization or appearance of cluster is the coexistence of attractors, which is characteristic and often encountered for the systems with friction and impact oscillators. Coexistence of attractors is a property of non-linear systems, which can occur also in smooth, time-continuous dynamical systems. Therefore, the MSF concept and eigenvectors analysis can be treated only as a tool for estimating the overall, global predisposition of the system of coupled oscillators to synchronize or to cluster. This may explain fact, that for some configuration in closed ring topology, some of the cluster layouts cannot be explained by the eigenvectors or eigenvalues interpretation. The results presented in this chapter also show ragged synchronization phenomenon (i.e., complete synchronization windows).

In general, the synchronization stability criterion given by the MSF does not provide for proper detection of global network synchronization state even in the case of smooth systems described by continuous differential equations. The more this problem occurs in non-smooth systems where the structure of attractors coexistence and their basins of attraction is usually more complex than in smooth systems. Hence, on the basis of our research, we can conclude that for non-smooth dynamical systems the MSF estimated with two oscillators probe can be even more effective than one calculated with use of the TLE, because then we can be sure that the synchronous region was really detected and it is not only a projection of an interval of the negative TLE. Additionally, the numerical results show that the phenomenon of ragged synchronizability concerns also the cluster synchronization case.

Based on the results of the chapter, further research can be conducted in following directions. The first research proposal is to consider the presented model in the framework of earthquake modelling as a version of Burridge-Knopoff model (Burridge and Knopoff 1967). This requires changing the system parameters, as in earthquake modelling transition from static to dynamic behaviour is a crucial property. It is also possible to analyse the presented model in the framework of the so called snaking phenomenon (Papangelo et al. 2017). Another direction of research may concern experimental investigation of the proposed model. This would involve designing and assembling experimental stand with all necessary measurement equipment, which would enable to verify experimentally the synchronizability of the system in question.

Acknowledgements This work has been supported by National Science Center—Poland (NCN) in frame of Project no. DEC-2012/06/A/ST8/00356. This research was supported in part by PL-Grid Infrastructure, in particular by computing resources of ACC Cyfronet AGH. The calculations mentioned in this paper are performed in part using the PLATON project's infrastructure at the Lodz University of Technology Computer Centre.

References

Ahnert K, Mulansky M (2011) Odeint-solving ordinary differential equations in C++. AIP Conf Proc 1389(1):1586–1589

Al-Bender F, Lampaert V, Swevers J (2004) A novel generic model at asperity level for dry friction force dynamics. Tribol Lett 16(1–2):81–93

Amontons G (1669) De la resistance causée dans les machines. In: Histoire de l'Académie royale des sciences avec les mémoires de mathématique et de physique, Imprimerie Royale, Paris, pp 206–227

Appleto EV (1922) Automatic synchronization of triode oscillators. Proc Cambridge Phil Soc 21:231

Armstrong-Helouvry B (1991) Control of machines with friction. Springer, New York

Barahona M, Pecora LM (2002) Synchronization in small-world systems. Phys Rev Lett 89(5):054101

Belykh VN, Belykh IV, Hasler M (2000) Hierarchy and stability of partially synchronous oscillations of diffusively coupled dynamical systems. Phys Rev E 62(5):6332

Belykh VN, Belykh IV, Mosekilde E (2001) Cluster synchronization modes in an ensemble of coupled chaotic oscillators. Phys Rev E 63(3):036216

Bo LC, Pavelescu D (1982) The friction-speed relation and its influence on the critical velocity of stick-slip motion. Wear 82(3):277–289

Buck J, Buck E (1968) Mechanism of rhythmic synchronous flashing of fireflies. Science 159(3821):1319–1327

Burridge R, Knopoff L (1967) Model and theoretical seismicity. Bull Seismol Soc Am 57(3):341–371

Canudas de Wit C, Olsson H, Astrom KJ, Lischinsky P (1995) A new model for control of systems with friction. IEEE Trans Autom Control 40(3):419–425

Coulomb CA (1821) Theorie des machines simples: en ayant egard au frottement de leurs parties et a la roideur des cordages. Bachelier, Paris

da Vinci L (1478-1518) Codex atlanticus. Biblioteca Ambrosiana, Milan

Dahl PR (1968) A solid friction model. Technical report, DTIC Document

Dahms T, Lehnert J, Schöll E (2012) Cluster and group synchronization in delay-coupled networks. Phys Rev E 86(1):016202

Dallard P, Fitzpatrick A, Flint A, Le Bourva S, Low A, Ridsdill Smith R, Willford M (2001a) The London millennium footbridge. Struct Eng 79(22):17–21

Dallard P, Fitzpatrick T, Flint A, Low A, Smith RR, Willford M, Roche M (2001b) London millennium bridge: pedestrian-induced lateral vibration. J Bridge Eng 6(6):412–417

Ding, W. (2010) Stick-slip vibration. In: Self-Excited vibration, Springer, pp 140–166

Dowson D (1979) History of Tribology. Longman, London

Eccles W, Vincent J (1920) On the variations of wave-length of the oscillations generated by three-electrode thermionic tubes due to changes in filament current, plate voltage, grid voltage, or coupling. Proc R Soc Lond A 96(680):455–465

Eckhardt B, Ott E, Strogatz SH, Abrams DM, McRobie A (2007) Modeling walker synchronization on the Millennium Bridge. Phys Rev E 75(2):021110

Euler, L (1750) Sur le frottement des corps solides. Mem Acad Sci Berlin 4:122–132

Euler, L. (1761). De frictione corporum rotantium. Novi Comment Acad Sci Imp Petropol 6:223–270

Fink KS, Johnson G, Carroll T, Mar D, Pecora L (2000) Three coupled oscillators as a universal probe of synchronization stability in coupled oscillator arrays. Phys Rev E 61(5):5080

Ghazaly NM, El-Sharkawy M, Ahmed I (2013) A review of automotive brake squeal mechanisms. J Mech Des Vibr 1(1):5–9

Gogoussis A, Donath M (1987) Coulomb friction joint and drive effects in robot mechanisms. In: 1987 Proceedings of IEEE international conference on robotics and automation, vol 4, pp 828–836

Graham C, McGrew W (1980) Menstrual synchrony in female undergraduates living on a coeducational campus. Psychoneuroendocrino 5(3):245–252

Heppner F, Grenander U (1990) A stochastic nonlinear model for coordinated bird flocks. In: Krasner S (ed) The ubiquity of chaos. AAAS, Washington, pp 233–238

Hess D, Soom A (1990) Friction at a lubricated line contact operating at oscillating sliding velocities. J Tribol 112(1):147–152

Hutchings IM (2016) Leonardo da vinci? s studies of friction. Wear 360:51–66

Huygens C (1673) Horologium oscillatorium. Muget, Paris

Jalife J (1984) Mutual entrainment and electrical coupling as mechanisms for synchronous firing of rabbit sino-atrial pace-maker cells. J Physiol 356(1):221–243

Kaneko K (1990) Clustering, coding, switching, hierarchical ordering, and control in a network of chaotic elements. Physica D 41(2):137–172

Lenci S, Marcheggiani L (2012) On the dynamics of pedestrians-induced lateral vibrations of footbridges. Nonlinear dynamic phenomena in mechanics. Springer, Berlin, pp 63–114

Marszal M (2017) Synchronization effects in systems with dry friction. PhD thesis, Lodz University of Technology

Marszal M, Saha A, Jankowski K, Stefański A (2016) Synchronization in arrays of coupled self-induced friction oscillators. Eur Phys J Spec Top 225(13–14):2669–2678

Marszal M, Stefański A (2017) Parameter study of global and cluster synchronization in array of dry friction oscillators. Phys Lett A 381(15):1286–1301

Meyer E, Gyalog T, Overney RM, Dransfeld K (1998) Nanoscience: friction and rheology on the nanometer scale. World Scientific, Singapore

Michaels DC, Matyas EP, Jalife J (1987) Mechanisms of sinoatrial pacemaker synchronization: A new hypothesis. Circ Res 61(5):704–714

Mirollo RE, Strogatz SH (1990) Synchronization of pulse-coupled biological oscillators. SIAM J Appl Math 50(6):1645–1662

Néda Z, Ravasz E, Vicsek T, Brechet Y, Barabási A-L (2000) Physics of the rhythmic applause. Phys Rev E 61(6):6987

Nishikawa T, Motter AE, Lai Y-C, Hoppensteadt FC (2003) Heterogeneity in oscillator networks: Are smaller worlds easier to synchronize? Phys Rev Lett 91(1):014101

Papangelo A, Grolet A, Salles L, Hoffmann N, Ciavarella M (2017) Snaking bifurcations in a self-excited oscillator chain with cyclic symmetry. Commun Nonlinear Sci Numer Simul 44:108–119

Patitsas A (2010) Squeal vibrations, glass sounds, and the stick-slip effect. Can J Phys 88(11):863–876

Pecora L, Carroll T, Johnson G, Mar D, Fink KS (2000) Synchronization stability in coupled oscillator arrays: solution for arbitrary configurations. Int J Bifurcat Chaos 10(02):273–290

Pecora LM (1998) Synchronization conditions and desynchronizing patterns in coupled limit-cycle and chaotic systems. Phys Rev E 58(1):347

Pecora LM, Carroll TL (1990) Synchronization in chaotic systems. Phys Rev Lett 64(8):821

Pecora LM, Carroll TL (1998) Master stability functions for synchronized coupled systems. Phys Rev Lett 80(10):2109

Perlikowski P (2007) Synchronizacja kompletna sieci nielinoowych układów dynamicznych. PhD thesis, Lodz University of Technology

Perlikowski P, Stefanski A, Kapitaniak T (2010) Discontinuous synchrony in an array of van der pol oscillators. Int J Non Linear Mech 45(9):895–901

Pikovsky A, Rosenblum M, Kurths J (2003) Synchronization: a universal concept in nonlinear sciences, vol 12. Cambridge University Press, Cambridge

Popp K, Hinrichs N, Oestreich M (1995) Dynamical behaviour of a friction oscillator with simultaneous self and external excitation. Sadhana 20(2–4):627–654

Popp K Stelter P (1990) Nonlinear oscillations of structures induced by dry friction. In: Nonlinear dynamics in engineering systems, Springer, pp 233–240

Rayleigh, J. W. S. B. (1896) The theory of sound, vol. 2. Macmillan

Reynolds CW (1987) Flocks, herds and schools: a distributed behavioral model. In: ACM SIG-GRAPH computer graphics, vol. 21, pp 25–34 ACM

Saha A, Bhattacharya B, Wahi P (2010) A comparative study on the control of friction-driven oscillations by time-delayed feedback. Nonlinear Dyn 60(1–2):15–37

Stefański A (2009) Determining thresholds of complete sychronization and application, vol 67. World Scientific

Stefański A, Perlikowski P, Kapitaniak T (2007) Ragged synchronizability of coupled oscillators. Phys Rev E 75(1):016210

Stefański A, Wojewoda J, Wiercigroch M, Kapitaniak T (2003) Chaos caused by non-reversible dry friction. Chaos Soliton Fract 16(5):661–664

Stribeck R (1902) The key qualities of sliding and roller bearings. Z Ver Dtsch Ing 46(38):39

Strogatz SH, Abrams DM, McRobie A, Eckhardt B, Ott E (2005) Theoretical mechanics: crowd synchrony on the millennium bridge. Nature 438(7064):43–44

van der Pol B (1927) Forced oscillations in a circuit with non-linear resistance. (reception with reactive triode). Lond Edinb Dubl Phil Mag 3(13):65–80

Warmiński J, Litak G, Cartmell M, Khanin R, Wiercigroch M (2003) Approximate analytical solutions for primary chatter in the non-linear metal cutting model. J Sound Vib 259(4):917–933

Wei D, Ruan J, Zhu W, Kang Z (2016) Properties of stability, bifurcation, and chaos of the tangential motion disk brake. J Sound Vib 375:353–365

Winfree AT (1967) Biological rhythms and the behavior of populations of coupled oscillators. J Theor Biol 16(1):15–42

Wojewoda J (2008) Efekty histerezowe w tarciu suchym. Politechnika Łódzka

Wojewoda J, Stefański A, Wiercigroch M, Kapitaniak T (2008) Hysteretic effects of dry friction: modelling and experimental studies. Phil Trans R Soc A 366(1866):747–765

Wu CW (2001) Simple three oscillator universal probes for determining synchronization stability in coupled arrays of oscillators. In: 2001 The IEEE international symposium on circuits and systems (ISCAS), vol. 3, IEEE, pp 261–264

Wu W, Zhou W, Chen T (2009) Cluster synchronization of linearly coupled complex networks under pinning control. IEEE Trans Circuits Syst I Regul Pap 56(4):829–839

Yanchuk S, Maistrenko Y, Mosekilde E (2001) Partial synchronization and clustering in a system of diffusively coupled chaotic oscillators. Math Comput Simul 54(6):491–508

Backstepping Control for Combined Function Projective Synchronization Among Fractional Order Chaotic Systems with Uncertainties and External Disturbances

Vijay K. Yadav, Mayank Srivastava and Subir Das

Abstract In the present chapter the combined function projective synchronization among fractional order chaotic systems in the presence of uncertain parameters and external disturbances using backstepping control method is investigated. The chaotic attractors of the systems are found for fractional-order time derivative, which is described in Caputo sense. A new lemma of Caputo derivatives is used to design the controller based on Lyapunov stability theory. During the combined function projective synchronization among the non-identical fractional order systems, the Lorenz, Rossler and Chen systems are taken to illustrate the effectiveness of the considered method. Numerical simulation and graphical results for different particular cases clearly exhibit that the method with this new procedure is easy to implement and reliable for synchronization of non-identical fractional order chaotic systems.

Keywords Fractional order chaotic system · Backstepping method
Lyapunov stability theory · Combined function projective synchronization

V. K. Yadav · S. Das (✉)
Department of Mathematical Sciences, Indian Institute of Technology (BHU), Varanasi
221005, India
e-mail: sdas.apm@iitbhu.ac.in

V. K. Yadav
e-mail: vijayky999@gmail.com

M. Srivastava
Department of Mathematics, Institute of Science, Banaras Hindu University, Varanasi
221005, India
e-mail: mayanksrivastava1983@gmail.com

© Springer International Publishing AG 2018   115
V.-T. Pham et al. (eds.), *Nonlinear Dynamical Systems with Self-Excited
and Hidden Attractors*, Studies in Systems, Decision and Control 133,
https://doi.org/10.1007/978-3-319-71243-7_5

1   Introduction

In last few decades, much attention has been devoted to the study of the fractional calculus and their numerous applications in the area of mathematics, physics and engineering. Fractional differential equations which are generalizations of classical differential equations describe the memory effect, which is the major advantage over integer-order derivatives. It has been extensively applied for modelling of many real problems such in viscoelasticity (Koeller 1984) dielectric polarization (Sun et al. 1984), electromagnetic waves (Heaviside 1971), quantitative finance (Laskin 2000), quantum evolution of complex system (Kunsezov et al. 1999), chaos control of dynamical systems (Chen and Yu 2003; Azar and Vaidyanathan 2015) and the control of fractional order dynamic systems (Hartley and Lorenzo 2002) etc.

It is evident from literature survey that during last few decades the nonlinear phenomena occurring in various areas of scientific fields have gained immense popularity amongst the scientists and engineers who have delivered tireless efforts towards the development of the models using non-linear differential equations. Introduction of fractional calculus in nonlinear models had given a new dimension to the existing problems. The interesting phenomena of nonlinear dynamics are the possibility of chaos. Most of the nonlinear systems reveal chaotic behaviour which is deterministic and has a periodic long-term behaviour, and also exhibit sensitive dependence on initial conditions. A periodic long-term behaviour means that there are trajectories which do not settle down to fixed points, periodic orbits, or quasi-periodic orbits as time approaches to infinity. Deterministic means that the system has no random or noisy inputs. This irregular behaviour arises from the system's nonlinearity, rather than from noisy driving forces. Sensitivity means that a small change in the initial state will lead to progressively larger changes in later system. Hence, an arbitrarily small perturbation of the current trajectory may lead to different future behaviour. The concept of chaos has been used to explain how systems subject to known laws of physics may be predictable in the short term but are apparently random on a longer time scale.

The nonlinear chaotic dynamic system of fractional order has taken care by mathematical and physical communities in the last few years. The chaotic dynamics of fractional order systems are important topics, which are mainly devoted to the chaos synchronization problem in nonlinear dynamical systems. Synchronization of chaos refers to a process wherein two or more identical or non-identical chaotic systems have a common behaviour due to a coupling, which appears to be structurally stable. In other words, synchronization, an important achievement in the research of chaos, means that the trajectories of two systems will converge and they will remain in step with each other. Pecora and Carroll (1990), first introduced a method about synchronization between the drive (master) and response (slave) systems of two identical or non identical systems with different initial conditions, which has important applications in ecological system, physical system, chemical system, modelling brain activity, system identification, pattern recognition phenomena and secure communications etc. Different types of synchronization schemes had already been handled by various researchers for the synchronization of chaotic systems, such as complete synchronization, anti-synchronization lag synchronization, hybrid synchronization, projective synchronization, and function projective synchronization (Agrawal et al. 2012; Yu and Liu 2003; Zhang and Sun 2004; Rosenblum et al. 1997; Srivastava et al. 2013a; Si et al. 2012; Zhou and Zhu 2011) etc. using different types of control scheme such as linear and non linear feedback synchronization, adaptive control, active control, sliding mode control etc.

In the present chapter a new way for combined function projective synchronization among fractional order chaotic systems in the presence of parametric uncertainties and external disturbances is described using backstepping control method. Function projective synchronization (FPS), the generalization of projective synchronization (PS), is one of the synchronization methods where two identical (or different) chaotic systems can synchronize up to a scaling function matrix with different initial values. From literature survey, it is seen that many researchers and scientists have worked on function projective synchronization of fractional order chaotic systems (Yu and Li 2010; Chen and Li 2007; Yadav et al. 2017a). In combination synchronization (Runzi et al. 2011; Yadav et al. 2017b), two or more master systems and one slave system are synchronized. This synchronization scheme has advantages over the usual drive response synchronization, such as being able to provide greater security in secure communication. The influences of the uncertainties during synchronization have been considered late. In the real world applications, such as in secure communication (Vaidyanathan and Volos 2016), the receiver plants will definitely suffer from the various uncertainties including parameter perturbation or external disturbance, which will no doubt influence the accuracy of the communication. Therefore, the synchronization between fractional order chaotic systems with uncertainties and disturbances are tough jobs for researchers. There are possibilities of destroying synchronization with the effects of those parameters (Srivastava et al. 2013b). The synchronization between chaotic systems with uncertainties and disturbances are not easy jobs for researchers since there are always possibilities of destroying synchronization under the effects of those parameters especially for fractional order systems. There are few results about the chaotic systems with uncertainties (Jawaadaa et al. 2012; Chen et al. 2012). Recently, Park (2006), Wu et al. (2009) have shown that the back stepping method is very simple, reliable and powerful for controlling the chaotic behavior and synchronization of chaotic systems. Wang and Ge (2001) proposed the adaptive synchronization of uncertain chaotic systems via backstepping design. In the same year, Lu and Zhang (2001) controlled the Chen's chaotic attractors using backstepping design based on parameters identification. Tan et al. (2003) synchronize the chaotic systems using backstepping design and again in the same year Yu and Zhang (2003) controlled the uncertain behavior of chaotic systems using backstepping design. These have motivated the authors to study on the combined function projective synchronization of fractional order chaotic systems with the presence of parametric uncertainties and external disturbances using backstepping control method. To the best of authors' knowledge the combined function projective synchronization among fractional order chaotic systems in the presence of parametric uncertainties and external disturbances using backstepping control method are few in numbers. Numerical simulation results are displayed graphically which clearly exhibit that the backstepping design control method is effective, easy to implement and reliable for combined function projective synchronizations of two nonlinear fractional order uncertain chaotic systems.

This chapter has been organized as follows. In Sect. 2, problem formulation of the combined function projective synchronization scheme of two different chaotic master systems, and one chaotic response system are presented. Section 3 contains some preliminaries, definition and lemma. In Sect. 4, the system descriptions of Lorenz, Rossler and Chen systems are given. Combined function projective synchronization among fractional order chaotic systems with uncertainties and external disturbances using backstepping control method are discussed in Sect. 5. In Sect. 6, the conclusion of the research work is presented.

2   Problem Formulation

Consider two uncertain fractional order chaotic systems as the master system as

$$D_t^q x = (A_1 + \Delta A_1)x + f_1(x) + d_1, \tag{1}$$

$$D_t^q y = (A_2 + \Delta A_2)y + f_2(y) + d_2, \quad 0 < q < 1 \tag{2}$$

and another uncertain fractional order chaotic system as the slave system as

$$D_t^q z = (A_3 + \Delta A_3)z + f_3(z) + d_3 + u(t), \tag{3}$$

where $x = [x_1, x_2, \ldots x_n]^T \in R^n$, $y = [y_1, y_2, \ldots y_n]^T \in R^n$ and $z = [z_1, z_2, \ldots z_n]^T \in R^n$ are the state vectors, A_1, A_2, $A_3 \in R^{n \times n}$ are constant matrices with proper dimensions, $f_1, f_2, f_3 : R^n \to R^n$ are the nonlinear functions of the systems, $\Delta A_1, \Delta A_2, \Delta A_3 \in R^{n \times n}$ are parametric uncertainties of chaotic systems with $|\Delta A_1| \le \delta_1$, $|\Delta A_2| \le \delta_2$, $|\Delta A_3| \le \delta_3$, where δ_1, δ_2, δ_3 are positive constants and d_1, d_2, d_3 are the external disturbances of uncertain chaotic systems with $|d_1| \le \rho_1$, $|d_2| \le \rho_2$, $|d_3| \le \rho_3$, where ρ_1, ρ_2, $\rho_3 > 0$ and $u(t) \in R^n$ is the control input vector of the uncertain chaotic system (3). Now controller $u(t)$ is to be designed in such a way that the master and slave systems are synchronized through the proper definitions of errors.

If the synchronization error is defined by $e = z - k(y + x)$, where k is the scaling function, then the corresponding error dynamics can be obtained as

$$D_t^q e = (A_3 + \Delta A_3)e - k[(A_1 + \Delta A_1 - A_3 - \Delta A_3)x + (A_2 + \Delta A_2 - A_3 - \Delta A_3)y \\ + f_1(x) + f_2(y) + d_1 + d_2] + f_3(z) + d_3 + u(t)$$ (4)

Therefore, for combined function projective synchronization we use backstepping control method to design the control functions in such a way that the origin becomes asymptotically stable equilibrium point of the error dynamics i.e., $\lim\limits_{t \to \infty} \|z - k(y + x)\| = 0$. The demonstration of backstepping control method is given in Sect. 5.

3   Some Preliminaries, Definition and Lemma

3.1   Fractional Calculus

Fractional calculus is a generalization of integration and differentiation of integer order operator to a non-integer integro-differential operator denoted by $_aD_t^q$ and defined by

$$_aD_t^q = \begin{cases} \frac{d^q}{dt^q}, & R(q) > 0 \\ 1, & R(q) = 0 \\ \int_a^t (d\tau)^{-q}, & R(q) < 0, \end{cases}$$

where q is the fractional order which may be a complex number and $R(q)$ denotes the real part of q and a is the fixed lower terminal and t is the moving upper terminal.

Definition 1 (Kilbas et al. 2006) The Caputo derivative for fractional order q is defined as

$$_a^cD_t^q \phi(t) = \frac{1}{\Gamma(n - q)} \int_a^t \frac{\phi^{(n)}(\tau)}{(t - \tau)^{q + 1 - n}} d\tau, \, t > a,$$

where $q \in R^+$ on the half axis R^+ and $n = \min\{k \in N / k > q\}, q > 0$.

Lemma 1 (Aguila-Camacho et al. 2014) *Let* $x(t) \in R$ *be a continuous and derivable function. Then for any time instant* $t \geq t_0$,

$$\frac{1}{2} {}_{t_0}^cD_t^q x^2(t) \leq x(t) {}_{t_0}^cD_t^q x(t), \forall q \in (0, 1].$$

4 Systems' Description

4.1 Fractional Order Lorenz System

The Lorenz attractor is an example of a non linear dynamical system corresponding to the long term behaviour of the Lorenz oscillation. The Lorenz oscillator is a three dimensional dynamical system that exhibits lemniscates type shaped chaotic flow which shows how the state of dynamical system evolves over time in a complex and non-repeating pattern. The Lorenz equations deal with the stability of fluid flows in the atmosphere. In addition to its interest in the field of non linear mathematics, the Lorenz model has important implications for climate and weather predictions. The case is also applicable for simplified models for lasers (Lorenz 1963) and dynamos (Knobloch 1981).

The fractional order Lorenz system (Wu and Shen 2009; Grigorenko and Grigorenko 2003) is given by

$$
\begin{aligned}
\frac{d^q x_1}{dt^q} &= a_1(y_1 - x_1), \\
\frac{d^q y_1}{dt^q} &= x_1(c_1 - z_1) - y_1, \\
\frac{d^q z_1}{dt^q} &= x_1 y_1 - b_1 z_1,
\end{aligned}
\tag{5}
$$

where a_1 is the Prandtl number, c_1 is the Rayleigh number and b_1 is the size of the region approximated by the system. The phase portraits of Lorenz system is shown through Fig. 1 for the parameters' values $a_1 = 10$, $b_1 = 8/3$, $c_1 = 28$ and initial condition $(0.2, 0, 2)$. The lowest value of fractional order q for which the system remains chaotic is 0.99 (Wu and Shen 2009). The chaotic attractors in the $x_1 - y_1 - z_1$ space, $x_1 - y_1$, $x_1 - z_1$, $y_1 - z_1$ planes are shown in Fig. 1 for order of derivative $q = 0.993$.

The fractional order Lorenz system with uncertain parameters and external disturbances is defined as

$$
\begin{aligned}
\frac{d^q x_1}{dt^q} &= a_1(y_1 - x_1) + 0.11 z_1 - \cos(10d), \\
\frac{d^q y_1}{dt^q} &= x_1(c_1 - z_1) - y_1 - 0.14 x_1 - 2\cos(10d), \\
\frac{d^q z_1}{dt^q} &= x_1 y_1 - b_1 z_1 + 0.23 y_1 - 3\sin(10d),
\end{aligned}
\tag{6}
$$

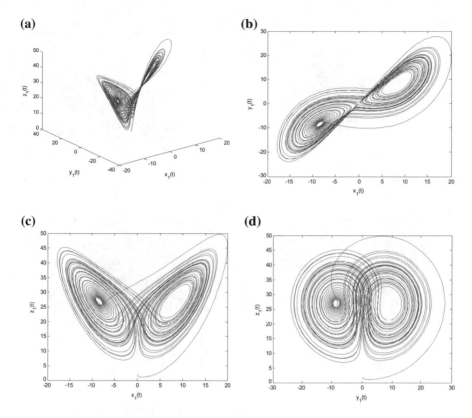

Fig. 1 Phase portraits of the Lorenz system at $q = 0.993$: **a** $x_1 - y_1 - z_1$ space, **b** $x_1 - y_1$ plane, **c** $x_1 - z_1$ plane, **d** $y_1 - z_1$ plane

where uncertain parameter $\Delta A_1 = \begin{bmatrix} 0 & 0 & 0.11 \\ -0.14 & 0 & 0 \\ 0 & 0.23 & 0 \end{bmatrix}$ and disturbance term

$d_1 = \begin{bmatrix} -\cos(10d) \\ -2\cos(10d) \\ -3\sin(10d) \end{bmatrix}$. Figure 2 shows the phase portraits of the fractional order

Lorenz system with uncertainties and disturbances in $x_1 - y_1 - z_1$ space, $x_1 - y_1$, $x_1 - z_1$, $y_1 - z_1$ planes for the order of the derivative $q = 0.993$.

4.2 Fractional Order Rossler Systems

The fractional order Rossler system (Yan and Li 2007; Zhou and Cheng 2008) is given by

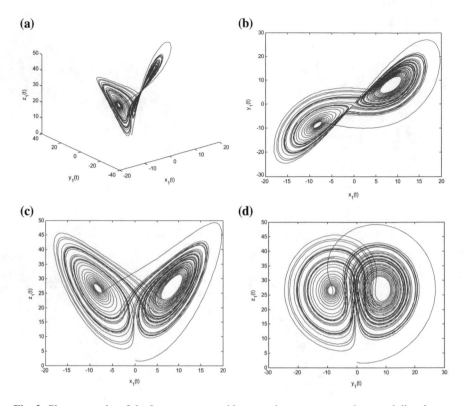

Fig. 2 Phase portraits of the Lorenz system with uncertain parameters and external disturbances at $q = 0.993$: **a** $x_1 - y_1 - z_1$ space, **b** $x_1 - y_1$ plane, **c** $x_1 - z_1$ plane, **d** $y_1 - z_1$ plane

$$\frac{d^q x_2}{dt^q} = -y_2 - z_2,$$
$$\frac{d^q y_2}{dt^q} = x_2 + a_2 y_2,$$
$$\frac{d^q z_2}{dt^q} = b_2 + x_2 z_2 - c_2 z_2,$$
$$(7)$$

For the parameters' values $a_2 = 0.2$, $b_2 = 0.2$, $c_2 = 5.7$ and $q = 0.96$, the system (7) is chaotic. The phase portraits of Rossler system for order of derivative $q = 0.98$ are shown through Fig. 3.

The fractional order Rossler system with uncertain parameters and external disturbances is defined as

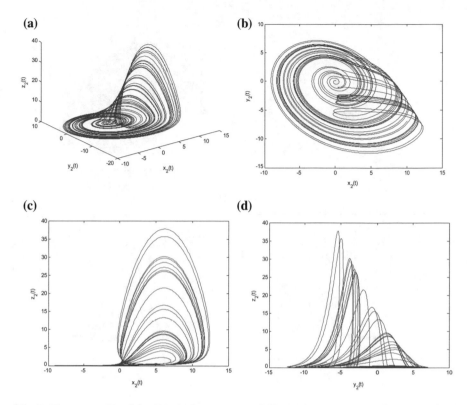

Fig. 3 Phase portraits of the Rossler system at $q = 0.98$: **a** $x_2 - y_2 - z_2$ space, **b** $x_2 - y_2$ plane, **c** $x_2 - z_2$ plane, **d** $y_2 - z_2$ plane

$$\frac{d^q x_2}{dt^q} = -y_2 - z_2 - 0.01x_2 - 0.1\sin(20d),$$

$$\frac{d^q y_2}{dt^q} = x_2 + a_2 y_2 - 0.02z_2 - 0.3\cos(20d), \tag{8}$$

$$\frac{d^q z_2}{dt^q} = b_2 + x_2 z_2 - c_2 z_2 - 0.15y_2 - 0.04\sin(20d),$$

where uncertain parameter $\Delta A_2 = \begin{bmatrix} -0.01 & 0 & 0 \\ 0 & 0 & -0.02 \\ 0 & -0.15 & 0 \end{bmatrix}$ and disturbance

term $d_2 = \begin{bmatrix} -0.1\sin(20d) \\ -0.3\cos(20d) \\ -0.04\sin(20d) \end{bmatrix}$. The phase portraits of fractional order Rossler

system with uncertainties and disturbances in $x_2 - y_2 - z_2$ space, $x_2 - y_2$, $x_2 - z_2$, $y_2 - z_2$ planes are depicted through Fig. 4 at $q = 0.98$.

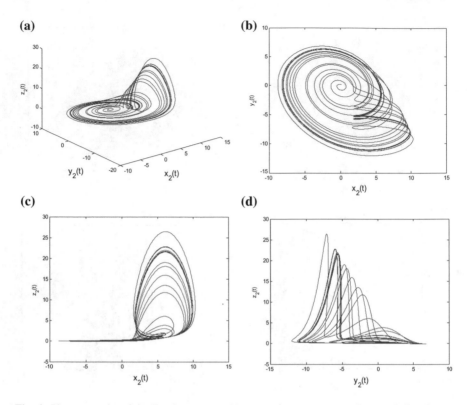

Fig. 4 Phase portraits of the Rossler system with uncertain parameters and external disturbances at $q = 0.98$: **a** $x_2 - y_2 - z_2$ space, **b** $x_2 - y_2$ plane, **c** $x_2 - z_2$ plane, **d** $y_2 - z_2$ plane

4.3   Fractional Order Chen System

The fractional order Chen system (Lu and Chen 2006) is defined as

$$\frac{d^q x_3}{dt^q} = a_3(y_3 - x_3),$$

$$\frac{d^q y_3}{dt^q} = (c_3 - a_3)x_3 - x_3 z_3 + c_3 y_3, \tag{9}$$

$$\frac{d^q z_3}{dt^q} = x_3 y_3 - b_3 z_3,$$

For the parameters' values $a_3 = 35$, $b_3 = 3$, $c_3 = 28$, $q = 0.7$ and initial condition $(3, 4, 6)$, the system (9) shows the chaotic behaviour. The phase portraits of Chen system at $q = 0.90$ are depicted through Fig. 5.

Fractional order Chen system with uncertain parameters and external disturbances is defined as

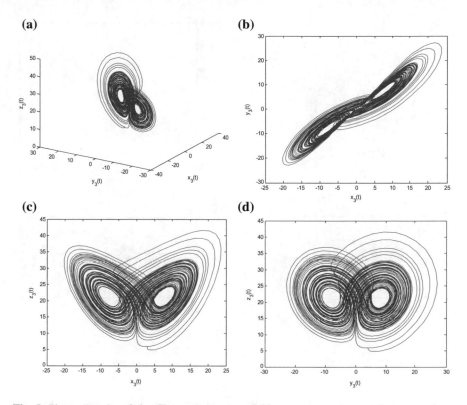

Fig. 5 Phase portraits of the Chen system at $q = 0.90$: **a** $x_3 - y_3 - z_3$ space, **b** $x_3 - y_3$ plane, **c** $x_3 - z_3$ plane, **d** $y_3 - z_3$ plane

$$\frac{d^q x_3}{dt^q} = a_3(y_3 - x_3) - 0.2z_3 + 0.1\sin(100d),$$

$$\frac{d^q y_3}{dt^q} = (c_3 - a_3)x_3 - x_3 z_3 + c_3 y_3 - 0.4z_3 - 0.2\cos(100d), \qquad (10)$$

$$\frac{d^q z_3}{dt^q} = x_3 y_3 - b_3 z_3 + 0.1x_3 - \sin(100d),$$

where uncertain parameter $\Delta A_3 = \begin{bmatrix} 0 & 0 & -0.2 \\ 0 & 0 & -0.4 \\ 0.1 & 0 & 0 \end{bmatrix}$ and disturbance term

$d_3 = \begin{bmatrix} 0.1\sin(100d) \\ -0.2\cos(100d) \\ -\sin(100d) \end{bmatrix}$. The phase portraits of fractional order Chen system with

uncertainties and disturbances in $x_3 - y_3 - z_3$ space, $x_3 - y_3$, $x_3 - z_3$, $y_3 - z_3$ planes are depicted through Fig. 6 at $q = 0.90$.

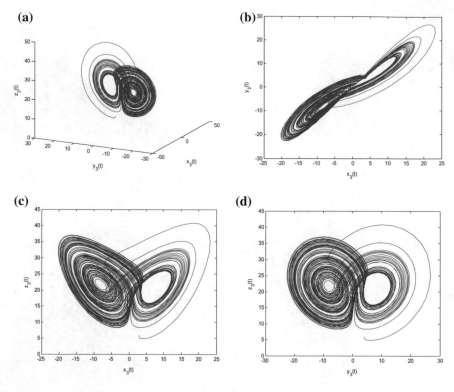

Fig. 6 Phase portraits of the Chen system with uncertain parameters and external disturbances at $q = 0.90$: **a** $x_3 - y_3 - z_3$ space, **b** $x_3 - y_3$ plane, **c** $x_3 - z_3$ plane, **d** $y_3 - z_3$ plane

5   Combined Function Projective Synchronization Among Fractional Order Chaotic Systems with Uncertainties and External Disturbances Using Backstepping Control Method

For the study of combined function projective synchronization among fractional order chaotic systems with uncertain parameters and external disturbances, two systems viz., Lorenz system (6) and Rossler system (8) are considered as drive system-I and drive system-II and Chen system (10) is considered as response system. The response system with the control functions is defined as

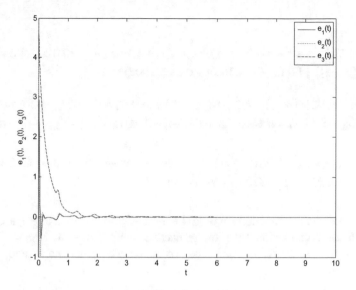

Fig. 7 Evolution of the error functions $e_1(t)$, $e_2(t)$ and $e_3(t)$ for fractional order $q = 0.96$

$$\frac{d^q x_3}{dt^q} = a_3(y_3 - x_3) - 0.2z_3 + 0.1\sin(100d) + u_1(t),$$

$$\frac{d^q y_3}{dt^q} = (c_3 - a_3)x_3 - x_3z_3 + c_3y_3 - 0.4z_3 - 0.2\cos(100d) + u_2(t), \qquad (11)$$

$$\frac{d^q z_3}{dt^q} = x_3y_3 - b_3z_3 + 0.1x_3 - \sin(100d) + u_3(t),$$

where $u(t) = [u_1(t),\ u_2(t),\ u_3(t)]^T$ is the control functions to be deigned later.
Defining the error functions as

$$e_1 = x_3 - k_1(x_2 + x_1)$$

$$e_3 = y_3 - k_2(y_2 + y_1)$$

$$e_3 = z_3 - k_3(z_2 + z_1),$$

we obtain the error system as

$$\frac{d^q e_1}{dt^q} = a_3(e_2 - e_1) - 0.2e_3 + \phi_1 + u_1(t),$$

$$\frac{d^q e_2}{dt^q} = (c_3 - a_3)e_1 + c_3e_2 - 0.4e_3 + \phi_2 + u_2(t), \qquad (12)$$

$$\frac{d^q e_3}{dt^q} = -b_3e_3 + 0.1e_1 + \phi_3 + u_3(t),$$

where

$$\phi_1 = -0.2k_3(z_2 + z_1) + a_3k_2(y_2 + y_1) - k_1[a_3(x_2 + x_1) - y_2 - z_2 - 0.01x_2 - 0.1\sin(20d)$$
$$+ a_1(y_1 - x_1) + 0.11z_1 - \cos(10d)] + 0.1\sin(100d)$$

$$\phi_2 = (c_3 - a_3)k_1(x_2 + x_1) - k_2[-c_3(y_2 + y_1) + x_2 + a_2y_2 - 0.02z_2 - 0.3\cos(20d)$$
$$+ x_1(c_1 - z_1) - y_1 - 0.14x_1 - 2\cos(10d)] - 0.4k_3(z_2 + z_1) - x_3z_3 - 0.2\cos(100d)$$

$$\phi_3 = 0.1k_1(x_2 + x_1) - k_3[b_3(z_2 + z_1) + b_2 + x_2z_2 - c_2z_2 - 0.15y_2 - 0.04\sin(20d) + x_1y_1$$
$$- b_1z_1 + 0.23y_1 - 3\sin(10d)] + x_3y_3 - \sin(100d).$$

Now the control functions would be properly designed using backstepping approach for combination function projective synchronization among fractional order chaotic systems in presence of uncertain parameters and external disturbances.

Theorem 1 *If the control functions are chosen as*

$$u_1(t) = 0.2e_3 - \phi_1,$$

$$u_2(t) = -c_3w_1 - w_2 - c_3w_2 - \phi_2,$$

$$u_3(t) = -0.1w_1 + 0.4w_2 - \phi_3,$$

where $w_1 = e_1$, $w_2 = e_2$, $w_3 = e_3$, *the systems (6) and (8) will be synchronized with the system (10).*

Proof To achieve control functions, we use active backstepping procedure through following three steps.

Step I: Considering $w_1 = e_1$, the fractional derivative of w_1 is

$$\frac{d^q w_1}{dt^q} = \frac{d^q e_1}{dt^q} = a_3(e_2 - w_1) - 0.2e_3 + \phi_1 + u_1(t), \tag{13}$$

where $e_2 = \alpha_1(w_1)$ is regarded as an virtual controller. To stabilize w_1-subsystem, we define the Lyapunov function V_1 as

$$V_1 = \frac{1}{2}w_1^2.$$

Fractional derivative of V_1 is

$\frac{d^q V_1}{dt^q} = \frac{1}{2}\frac{d^q w_1^2}{dt^q} \leq w_1 \frac{d^q w_1}{dt^q}$ (Using Lemma 1)

$$\text{i.e.,} \qquad \leq w_1[a_3(\alpha_1 - w_1) - 0.2e_3 + \phi_1 + u_1(t)].$$

Taking $\alpha_1(w_1) = 0$ and $u_1(t) = 0.2e_3 - \phi_1$, we get $\frac{d^q V_1}{dt^q} \leq -a_3 w_1^2 < 0$, negative definite, which implies that w_1-subsystem (13) is asymptotically stable. For the virtual control function $\alpha_1(w_1)$, we define a variable w_2 between e_2 and $\alpha_1(w_1)$ as

$$w_2 = e_2 - \alpha_1(w_1).$$

Then, (w_1, w_2) subsystem is obtained as

$$\frac{d^q w_1}{dt^q} = a_3(w_2 - w_1), \tag{14}$$
$$\frac{d^q w_2}{dt^q} = (c_3 - a_3)w_1 + c_3 w_2 - 0.4e_3 + \phi_2 + u_2(t).$$

Let $e_3 = \alpha_2(w_1, w_2)$ is an virtual controller.

Step II: In this step to stabilize (w_1, w_2)-subsystem (14), define the Lyapunov function V_2 as

$$V_2 = V_1 + \frac{1}{2}w_2^2 = \frac{1}{2}w_1^2 + \frac{1}{2}w_2^2.$$

Now

$$\frac{d^q V_3}{dt^q} = \frac{1}{2}\frac{d^q w_1^2}{dt^q} + \frac{1}{2}\frac{d^q w_2^2}{dt^q}$$
$$\leq w_1 \frac{d^q w_1}{dt^q} + w_2 \frac{d^q w_2}{dt^q}$$

$$\text{i.e.,} \quad \leq -a_3 w_1^2 + w_2[c_3 w_1 + c_3 w_2 - 0.4\alpha_2(w_1, w_2) + \phi_2 + u_2(t)],$$

If $\alpha_2(w_1, w_2) = 0$ and $u_2(t) = -c_3 w_1 - w_2 - c_3 w_2 - \phi_2$, then $\frac{d^q V_2}{dt^q} \leq -a_3 w_1^2 - w_2^2$ < 0 makes the subsystem (14) asymptotically stable.

Considering $w_3 = e_3 - \alpha_2(w_1, w_2)$, we get the following (w_1, w_2, w_3)-subsystem as

$$\frac{d^q w_1}{dt^q} = a_3(w_2 - w_1),$$
$$\frac{d^q w_2}{dt^q} = -a_3 w_1 - w_2 - 0.4w_3 \tag{15}$$
$$\frac{d^q w_3}{dt^q} = -b_3 w_3 + 0.1w_1 + \phi_3 + u_3(t),$$

Step III: In order to stabilize (w_1, w_2, w_3)-subsystem (15), choosing the Lyapunov function as

$$V_3 = V_2 + \frac{1}{2}w_3^2 = \frac{1}{2}w_1^2 + \frac{1}{2}w_2^2 + \frac{1}{2}w_3^2,$$

we get

$$\frac{d^q V_3}{dt^q} = \frac{1}{2}\frac{d^q w_1^2}{dt^q} + \frac{1}{2}\frac{d^q w_2^2}{dt^q} + \frac{1}{2}\frac{d^q w_3^2}{dt^q}$$

$$\leq w_1 \frac{d^q w_1}{dt^q} + w_2 \frac{d^q w_2}{dt^q} + w_3 \frac{d^q w_3}{dt^q},$$

i.e., $$\leq -a_3 w_1^2 - w_2^2 - 0.4 w_2 w_3 - b_3 w_3^2 + w_3 [0.1 w_1 + \phi_3 + u_3(t)].$$

If $u_3(t) = -0.1 w_1 + 0.4 w_2 - \phi_3$, then $\frac{d^q V_3}{dt^q} \leq -a_3 w_1^2 - w_2^2 - b_3 w_3^2 < 0$ negative definite. In view of $w_1 = e_1$, $w_2 = e_2 - \alpha_1(w_1) = e_2$, $w_3 = e_3 - \alpha_2(w_1, w_2) = e_3$, the error states will converge to zero after a finite period of time, and thus the combined function projective synchronization among Lorenz, Rossler and Chen systems in the presence of uncertain parameters and external disturbances will be achieved.

5.1   Numerical Simulation and Results

In numerical simulation, the parameters of Lorenz system, Rossler system and Chen system are taken as $a_1 = 10$, $b_1 = 8/3$, $c_1 = 28$; $a_2 = 0.2$, $b_2 = 0.2$, $c_2 = 5.7$ and $a_3 = 35$, $b_3 = 3$, $c_3 = 28$ respectively. Time step size is taken as 0.005. The initial condition of two master systems and one slave system are taken as (0.1, 0.1, 0.1), (0.2, 0, 2) and (3, 4, 6) respectively. Thus according to definition of error functions, the initial errors are (2.85, 3.96, 4.95).

During the combined function projective synchronization the scaling functions are taken as periodic function as

$$k_1 = a_{11} \cos(a_{12} x_1) + a_{13}$$

$$k_2 = a_{21} \cos(a_{22} y_1) + a_{23}$$

$$k_3 = a_{31} \cos(a_{32} z_1) + a_{33}.$$

For the values of parameters $a_{11} = 0.4, a_{12} = 0.1, a_{13} = 0.1, a_{21} = 0.1, a_{22} = 0.2, a_{23} = 0.3, a_{31} = 0.3, a_{32} = 0.3, a_{33} = 0.2$ it is seen from Fig. 7 that the error functions asymptotically converge to zero as time becomes large for the order of the derivatives $q = 0.96$, which shows that the master systems (6) and (8) are synchronized with the slave system (10).

6 Conclusion

The contribution of the present chapter is the investigation of the combined function projective synchronization among different fractional order chaotic systems with uncertainties and external disturbances using backstepping method. Based on Lyapunov stability theory, the synchronization with function scaling factor of chaotic systems through the proper design of control functions is achieved. The components of error state tend to zero as time becomes large help to get the time requires for combined synchronization among the systems. Numerical simulation results demonstrate that the method is reliable, convenient and effective for the combined function projective synchronization even for fractional order chaotic systems.

References

Agrawal SK, Srivastava M, Das S (2012) Synchronization of fractional order chaotic systems using active control method. Chaos, Solitons Fractals 45:737–752

Aguila-Camacho N, Duarte-Mermoud MA, Gallegos JA (2014) Lyapunov functions for fractional order systems. Commun Nonlinear Sci Numer Simul 19:2951–2957

Azar AT, Vaidyanathan S (2015) Chaos modeling and control systems design. In: Studies in computational intelligence, vol 581. Springer, Germany

Chen GR, Yu XH (2003) Chaos control: theory and applications. Springer, Berlin

Chen M, Wu Q, Jiang C (2012) Disturbance observer based robust synchronization control of uncertain chaotic systems. Nonlinear Dyn 70:2421–2432

Chen Y, Li X (2007) Function projective synchronization between two identical chaotic systems. Int J Mod Phys C 18:883–888

Grigorenko I, Grigorenko E (2003) Chaotic dynamics of the fractional order Lorenz system. Phys Rev Lett 91:4–034101

Hartley TT, Lorenzo CF (2002) Dynamics and control of initialized fractional-order systems. Nonlin. Dyn. 29:201–233

Heaviside O (1971) Electromagnetic theory. Chelsea, New York

Jawaadaa W, Noorani MSM, Al-sawalha MM (2012) Robust active sliding mode anti-synchronization of hyperchaotic systems with uncertainties and external disturbances. Nonlinear Anal Real World Appl 13:2403–2413

Kilbas AA, Srivastava HM, Trujillo JJ (2006) Theory and applications of fractional differential equations. Elsevier

Knobloch E (1981) Chaos in the segmented disc dynamo. Phys Lett A 82:40–439

Koeller RC (1984) Application of fractional calculus to the theory of viscoelasticity. J Appl Mech 51:299–307

Kunsezov D, Bulagc A, Dang GD (1999) Quantum levy processes and fractional kinetics. Phys Rev Lett 82:1136–1139

Laskin N (2000) Fractional market dynamics. Phys A 287:482–492

Lorenz EN (1963) Deterministic nonperiodic flow. J Atmos Sci 20:130–141

Lu JG, Chen G (2006) A note on the fractional-order Chen system. Chaos, Solitons Fractals 27:685–688

Lu JH, Zhang SC (2001) Controlling Chen's chaotic attractor using backstepping design based on parameters identification. Phys Lett A 286:52–148

Park JH (2006) Synchronization of Genesio chaotic system via backstepping approach. Chaos, Solitons Fractals 27:1369–1375

Pecora LM, Carroll TL (1990) Synchronization in chaotic systems. Phys Rev Lett 64:24–821

Rosenblum MG, Pikovsky AS, Kurths J (1997) From phase to lag synchronization in coupled chaotic oscillators. Phys Rev Lett 78:4193–4196

Runzi L, Yinglan W, Shucheng D (2011) Combination synchronization of three classic chaotic systems using active backstepping design. Chaos 21:043114

Si G, Sun Z, Zhang Y, Chen W (2012) Projective synchronization of different fractional-order chaotic systems with non-identical orders. Nonlinear Anal Real World Appl 13:1761–1771

Srivastava M, Agrawal SK, Das S (2013a) Adaptive projective synchronization between different chaotic systems with parametric uncertainties and external disturbances. Pramana J Phy 81:417–437

Srivastava M, Das S, Leung AYT (2013b) Hybrid phase synchronization between identical and non-identical three dimensional chaotic systems using active control method. Nonlinear Dyn 73:2261–2272

Sun HH, Abdelwahed AA, Onaral B (1984) Linear approximation for transfer function with a pole of fractional order. IEEE Trans Autom Control 29:441–444

Tan XH, Zhang JY, Yang YR (2003) Synchronization chaotic systems using backstepping design. Chaos, Solitons Fractals 16:37–45

Vaidyanathan S, Volos C (2016) Advances and applications in chaotic systems. Springer, Berlin

Wang C, Ge SS (2001) Adaptive synchronization of uncertain chaotic systems via backstepping design. Chaos, Solitons Fractals 12:206–1199

Wu XJ, Shen SL (2009) Chaos in the fractional-order Lorenz system. Int J Comp Math 86:82–1274

Wu Y, Zhou X, Chen J, Hui B (2009) Chaos synchronization of a new 3D chaotic system. Chaos, Solitons Fractals 42:1812–1819

Yadav VK, Bhadauria BS, Das S, Singh AK, Srivastava M (2017a) Stability analysis, chaos control of fractional order chaotic chemical reactor system and its function projective synchronization with parametric uncertainties. Chin J Phy 55:594–605

Yadav VK, Prasad G, Som T, Das S (2017b) Combined synchronization of time-delayed chaotic systems with uncertain parameters. Chin J Phy 55:457–466

Yan JP, Li CP (2007) On chaos synchronization of fractional differential equations. Chaos, Solitons Fractals 32:725–735

Yu IIJ, Liu YZ (2003) Chaotic synchronization based on stability criterion of linear systems. Phys Lett A 314:292–298

Yu Y, Li HX (2010) Adaptive generalized function projective synchronization of uncertain chaotic systems. Nonlinear Anal 11:2456–2464

Yu YG, Zhang SC (2003) Controlling uncertain system using backstepping design. Chaos, Solitons Fractals 15:897–902

Zhang Y, Sun J (2004) Chaotic synchronization and anti-synchronization based on suitable separation. Phys Lett A 330:442–447

Zhou P, Cheng X (2008) Synchronization between different fractional order chaotic systems. In: Proceeding of the 7th world congress on intelligent control and automation, pp 25–27

Zhou P, Zhu W (2011) Function projective synchronization for fractional-order chaotic systems. Nonlinear Anal. Real World Appl. 12:811–816

Chaotic Business Cycles within a Kaldor-Kalecki Framework

Giuseppe Orlando

Abstract This chapter, after providing some background on business cycles, Kaldor's original model and related literature, presents an original specification Orlando (Math Comput Simul 125:83–98, 2016) which adds to the cyclical behaviour some peculiar characteristics such as an asymmetric investment and consumption function, lagged investments and integration of economic shocks. A further section proves the chaotic behaviour of the model and adds some insights derived from recurrence quantification analysis. The final part draws some concluding remarks and makes some suggestions for future research. This work investigates chaotic behaviours within a Kaldor-Kalecki framework. This can be achieved by an original specification of the functions describing the investments and consumption as variants of the hyperbolic tangent function rather than the usual arctangent. Therefore fluctuations of economic systems (i.e. business cycles) can be explained by the shape of the investment and saving functions which, in turn, are determined by the behaviour of economic agents. In addition it is explained how the model can accommodate those cumulative effects mentioned by Kaldor which may have the effect of translating the saving and investment functions. This causes the so-called shocks which may be disruptive to the economy or that may have the effect of helping the system to recover from a crisis.

Keywords Numerical chaos · Applications to economics · Economic dynamics

G. Orlando (✉)
Department of Economics and Mathematical Methods,
Università degli Studi di Bari Aldo Moro, Via C. Rosalba 53, 70124 Bari, Italy
e-mail: giuseppe.orlando@uniba.it; giuseppe.orlando@unicam.it

G. Orlando
School of Science and Technologies, Università degli Studi di Camerino,
Via M. delle Carceri 9, 62032 Camerino, Italy

© Springer International Publishing AG 2018   133
V.-T. Pham et al. (eds.), *Nonlinear Dynamical Systems with Self-Excited
and Hidden Attractors*, Studies in Systems, Decision and Control 133,
https://doi.org/10.1007/978-3-319-71243-7_6

1 Introduction

The seminal work of Kaldor (1940) on the business cycle is one of the most fruitful for researching on non-linear dynamics in economics. In that paper the author's intention, contrary to the traditional Keynesian multiplier-accelerator concept, was to explain from a macroeconomic viewpoint the fundamental reasons for cyclical phenomena. After several years the Kaldor model re-gained the attention of some scholars interested in non-linear phenomena in economics such as Morishima (1959), Yasui (1953) and Ichimura (1955a, b) that were the first in investigating the existence, stability and uniqueness of limit cycles in a nonlinear trade cycle model. Among other notable contributors were Hicks (1950), Goodwin (1951) and particularly Kalecki (1966). The latter divided the investment process into three steps where the first is the decision, the second the time needed for the production and the last is the delivery of the capital good. In such a way the dynamic of capital stock in the economy is described by a non-linear difference-differential equation which exhibits a complex behaviour (including chaos) and, as a result, oscillations of capital induce fluctuations of other economic variables. Chiarella (1990) showed how the model could adjust to adaptive expectation of inflation. Krawiec and Szydlowski (1999, 2001) analyzing the Kaldor-Kalecki model of business cycle found a Hopf bifurcation leading to a limit cycle. Last but not least Pham et al. (2017) observed that the presence of time delay, such as those hypothesized by Kalecki, could induce unexpected oscillations, therefore time-delay systems may be suitable to introducing chaotic dynamics.

Further we applied recurrence plots (RPs) and their quantitative description provided by recurrence quantification analysis (RQA) to detect relevant changes in the dynamic regime of business time series. RQA aims at a direct and quantitative appreciation of the amount of deterministic structure of time series and has been proven to be an efficient and relatively simple tool in non-linear analysis of a wide class of signals. The technique allows for the identification of sudden phase-changes possibly pointing to underlying phenomena. Therefore RQA may be suitable for studying business cycles as well as identifying possible signals of changes in the economy.

The remainder of the chapter is organized as follows: Sect. 2 contains the definition of business cycle and summarizes the literature on recurrence quantification analysis and its applications to economics and finance. Section 3 presents some results on the applicability of RQA to economic time series. Section 4 illustrates the model along with its parameters and peculiarities. Section 5 shows the numerical analysis performed and the results obtained. Finally Sect. 6 consolidates the ideas and presents concluding remarks.

2 Literature Review

2.1 On the Business Cycles

A definition of business cycles can be found in Burns and Mitchell (1946)

> Business cycles are a type of fluctuation found in the aggregate economic activity of nations that organize their work mainly in business enterprises: a cycle consists of expansions occurring at about the same time in many economic activities, followed by similarly general recessions, contractions, and revivals which merge into the expansion phase of the next cycle.

and it has been used for:

1. Creation of composite leading, coincident, and lagging indices based on the consistent pattern of comovement among various variables over the business cycle (e.g. Shishkin 1961).
2. The identification within the business cycles of separate phases or regimes.

The National Bureau of Economic Research (NBER) defines a recession as "a significant decline in economic activity spread across the economy, lasting more than a few months, normally visible in real GDP, real income, employment, industrial production, and wholesale-retail sales. A recession begins just after the economy reaches a peak of activity and ends as the economy reaches its trough. Between trough and peak, the economy is in an expansion. Expansion is the normal state of the economy; most recessions are brief and they have been rare in recent decades." Figure 1 illustrates the business cycle where recession (trough) follows expansion (peak) and Fig. 2 shows the financial and business cycle in the United States.

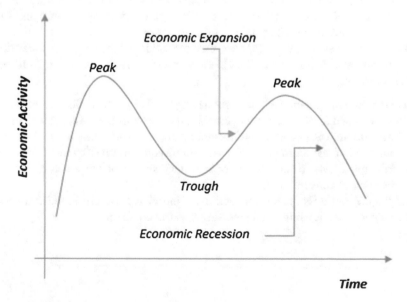

Fig. 1 The business cycle

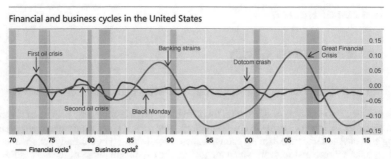

Financial and business cycles in the United States

¹ The financial cycle as measured by frequency-based (bandpass) filters capturing medium-term cycles in real credit, the credit-to-GDP ratio and real house prices; Q1 1970 = 0.   ² The business cycle as measured by a frequency-based (bandpass) filter capturing fluctuations in real GDP over a period from one to eight years; Q1 1970 = 0.

Sources: M Drehmann, C Borio and K Tsatsaronis, "Characterising the financial cycle: don't lose sight of the medium term!", *BIS Working Papers*, no 380, June 2012; BIS calculations.

Fig. 2   BIS 85th annual report 2015

Schumpeter (1954) mentioned four stages linking together production, stock exchange, confidence, demand, interest rates and prices:

1. Expansion (increase in production and prices and low interest rates)
2. Crisis (stock exchanges crash and multiple bankruptcies of firms occur)
3. Recession (drops in prices and in production and high interest rates)
4. Recovery (stocks recover because of the fall in prices and incomes)

In addition he suggested that each business cycles has its own typology according to the periodicity so a number of cycles were named after their discoverers: see Korotayev and Sergey 2010: Kitchin (1923) or inventory cycle (3–5 years long), Juglar (1862) cycle (7–11 years long), Kuznets (1930) cycle (15–25 years long) and Kondratiev (1935) technological cycle (45–60 years long).

Theories on business cycles expound on the volatility of economies and may differ (for a review see Hillinger and Sebold-Bender 1992; Zarnowitz 1992; Mullineux 1984) depending on:

1. Their ability to explain cycle without having to rely on outside forces/shocks and they are called respectively endogenous and exogenous business cycle theories.
2. The assumption of a general equilibrium framework (neoclassical theories) or the assumption of market imperfections and/or disequilibrium (Keynesian theories).
3. Attributing cycles to real shocks or monetary shocks or too much or too little investment or consumption.
4. Explaining business cycles from actions of individuals (micro-founded theories) economic units or using aggregate variables (macro theories).

2.2 On the Recurrence Quantification Analysis

Recurrence Quantification Analysis (RQA) in economics started relatively recently see Zbilut (2005), Crowley and Schultz (2010), Karagianni and Kyrtsou (2011), Chen (2011), Moloney and Raghavendra (2012). Fabretti and Ausloos (2005) found cases where RQA could detect a warning before a crash. In accordance to that Addo et al. (2013) assert "the usefulness of recurrence plots in identifying, dating and explaining financial bubbles and crisis". Strozzi et al. (2007) claim that determinism and laminarity change "more clearly than standard deviation and then they provide an alternative measure of volatility". Finally, to quote Piskun and Piskun (2011), laminarity (LAM) "is the most suitable measure, sensitive to critical events on markets".

The ability of RQA to predict catastrophic changes stems from the fact that RQA is based upon the change in correlation structure of the observed phenomenon that is known to precede the actual event in many different systems, ranging from physiology Zimatore et al. (2011) and geophysics Zimatore et al. (2017) to economy Crowley (2008). Gorban et al. (2010) found out that even before crisis correlation increases as well as variance (and volatility) increases too. In particular their dataset composed of thirty largest companies from British stock market within the period 2006–2008 supports the hypothesis of increasing correlations during a crisis and, therefore, that correlation (or equivalently determinism) increases when the market goes down (respectively decreases when it recovers).

3 Recurrence Plot

As mentioned by Eroglu et al. (2014) "recurrence plots (RPs) have been shown to be a powerful technique to uncover statistically many characteristic properties" of

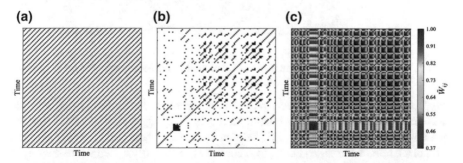

Fig. 3 (Color online) Recurrence plot of logistic map for **a** r = 3.5 (periodic regime) and **b** r = 4.0 (chaotic regime) and **c** weighted recurrence plot of logistic map for r = 4.0 (chaotic regime). *Source* Eroglu et al. (2014)

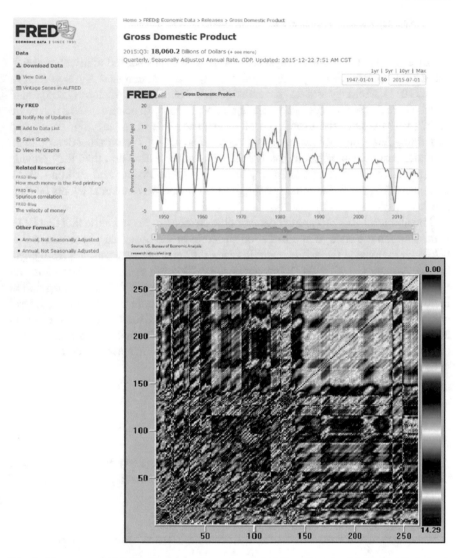

Fig. 4 Changes in US GDP (above) and its unthreshold RP or Distance Matrix (DM) (below). Period: 01-01-1947–01-07-2015. Compare the RP of the logistic Fig. 3c with the RP of business cycles. *Source* St. Louis Fed, FRED database

complex dynamical systems. "In a given m-dimensional phase space, two points are considered to be recurrent if their state vectors lie in a neighbourhood characterized by a threshold ε". Therefore in a RP, "elements $R_{i,j} \equiv 1$ (recurrence) are usually said to be black dots, whereas $R_{i,j} \equiv 0$ (no recurrence) are usually called white dots" (see Fig. 3a and b).

In order to have some indications whether business cycles are chaotic and to study recessions from the point of view of a phase transition of non-linear phenomena we applied RQA on time series extracted from Federal Reserve Economic Data

(FRED)—St. Louis Fed. In Fig. 4 the recurrence plot of USA GDP% variation is shown right below the FRED graph. Greyed areas correspond to periods of economic recessions as reckoned by FRED. For the unthreshold RP or Distance Matrix (DM) it is possible to observe the anticipating transitions to turbulent phases. The noteworthy results consist in a correspondence between vertical lines in DM (i.e. chaos to chaos transitions) and grey lines (recession periods) as well as in the apparent resemblance between the RP of the chaotic logistic Fig. 3c with the RP of business cycles Fig. 4.

4 The Model

The discretized Kaldor model is

$$
\begin{aligned}
Y_{t+1} - Y_t &= \alpha(I_t - S_t) = \alpha[I_t - (Y_t - C_t)] \\
K_{t+1} - K_t &= I_t - \delta K_t
\end{aligned}
\tag{1}
$$

where Y, I, S, K define respectively income, investment, saving and capital, α is the "speed" by which the output responds to excess investment and δ represents the depreciation rate of capital. It is worth mentioning that a key feature for Kaldor is that $I = I(Y, K)$ and $S = S(Y, K)$ are non-linear functions of income and capital.

The author's original variant is (for a detailed description see Orlando 2016)

$$
\begin{aligned}
Y_{t+1} - Y_t &= \alpha \left[f_1 \left(g \left(\tfrac{Y_{t-1}-Y_{t-2}}{Y_{t-2}} - \tfrac{K_{t-1}-K_{t-2}}{K_{t-2}} \right) \right) + f_2 \left(g \left(\tfrac{Y_t-Y_{t-1}}{Y_{t-1}} - \tfrac{C_t-C_{t-1}}{C_{t-1}} \right) \right) - Y_t \right] \\
K_{t+1} - K_t &= f_1 \left(g \left(\tfrac{Y_{t-1}-Y_{t-2}}{Y_{t-2}} - \tfrac{K_{t-1}-K_{t-2}}{K_{t-2}} \right) \right) - \delta K_t
\end{aligned}
\tag{2}
$$

Where the parameters $\alpha, \delta, \tau_1, \tau_2, \rho, \hat{c}, k$ have the following meaning:

1. α is the savings adjustment speed with regard to investment. Its reciprocal in physics is called delay and measures the time necessary for the adjustment.
2. δ is a percentage that determines the fixed capital which is lost during the productive process (due to obsolescence or actual consumption).
3. τ determines the f function knee; it is, therefore, a measure of the reactivity of the function to the variation in its argument.
4. ρ measures the maximum possible level (in capital terms) of investment. This value changes according to the economic system (pre-industrial, industrial, post-industrial) and the type of investment (i.e. high or low capital intensity).
5. $\hat{c} = 1 - c$ multiplied by actual income, determines the minimum level of consumption, therefore it is also called the base level. Its complement to 1, is c and represents the average level of consumption.
6. The k parameter changes according to the economic development.

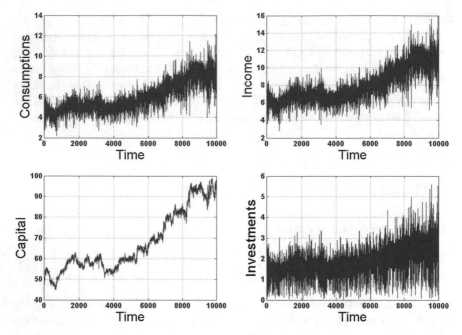

Fig. 5 A simulation displaying a steady growth

In Figs. 5 and 6 are two examples of the system's dynamic which differ only because a different initial starting point.

4.1 Shocks in the Economy

Kaldor explained that there are some factors which affect consumption and saving and which have the effect of shifting the functions in one direction or the other. In the model this translation can be easily achieved and operates when the capital or the income changes negatively (with the ultimate effect to help the system recover from a crisis).

4.2 Consumption, Saving and Economic Recessions

The idea that an increase in the disposable income (which is the basis of fiscal stimulus such as tax rebates that are supposed to encourage consumption, and hence aggregate demand), automatically translates into an increase in the aggregate demand, can be fallacious as it neglects to consider the state of health of the economy and

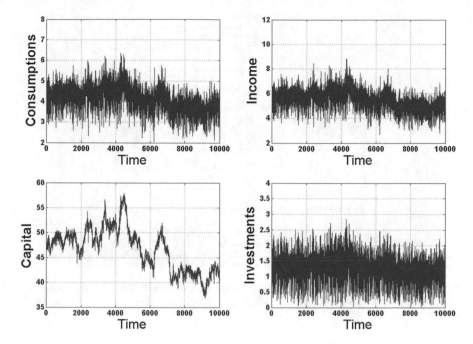

Fig. 6   A simulation displaying a steady fall

therefore the confidence in it. In fact, if confidence in the economy is low, it could be that people may reduce their consumption during the recession years: consumers will continue the process of deleveraging (they use the money to pay off debt and save more) because of uncertainty in the future.

For the above mentioned reasons, in the model, the change in consumption is linked to the change of income as follows

$$w = \frac{\Delta Y}{Y} - \frac{\Delta C}{C} \tag{3}$$

which we believe describes correctly the behaviour of consumption (Figs. 7 and 8).

4.3   Modelling Investment and Consumption

In the usual set-up, modified versions of Kaldor's approach adopt the trigonometric investment function arctg (see Mircea et al. 1963; Kaddar and Alaoui 2009; Agliari et al. 2007; Januario et al. 2005, 2009; Bischi et al. 2001 etc.). We have decided, instead, to use a different functional form: the hyperbolic tangent. The reasons why this specific functional form has been chosen are that there is no particular

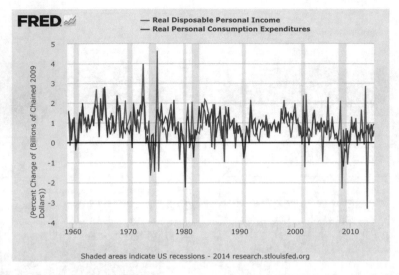

Fig. 7 Changes in US real disposable personal income (blue—DSPIC96) and real personal consumption expenditures (red—PCECC96) 1959 (Q1)—2014 (Q2). *Source* St. Louis Fed, FRED database

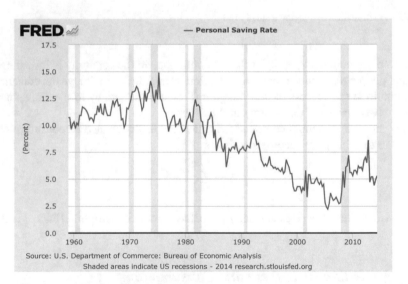

Fig. 8 US personal savings rate (PSAVERT) 1959 (Q1)—2014 (Q2). *Source* St. Louis Fed. FRED Database

justification to prefer the arctg,[1] whilst there are several to prefer the tanh. A good reason, for example, is that across multiple sciences growth and decay are better modelled by such exponential-like functions as tanh instead of trigonometric function such as arctg. For some applications to physics (such as radioactive decay, capacitor discharge, damped oscillations, etc.), to chemistry and biology (such as first order reaction rates and population and cancer growth), to actuarial science and finance (e.g. Gompertz-Makeham law of mortality, compound interest, etc.) Stewart (2010), Benson (2008), Reger et al. (2010), Purves et al. (2003), Wheldon (1988), etc. Moreover the arctg tends to its asymptotes quite slowly compared to the hyperbolic tangent whilst we wanted to design a framework in which economic agents can adjust quickly (how quickly depends on some parameters as τ) to changes. Given this framework the suggested model could link up, for example, to the classic Solow-Swan growth model in which labour and knowledge are represented by exponential functions.

Finally it should be noted that while consumption and investment are similarly function of a difference (i.e., respectively, between the growth rates of income and capital and the growth rates of income and consumption) their timing differs. In fact, *à la* Kalecki, we suggest that the investment process has different timing than does consumption hence the difference in the considered time lags (see Eq. 2).

5 Numerical Analysis

Up to now the sensitive dependence on initial conditions and the irregular trend of variables over time has only been shown graphically. Naturally the experimental evidence is not sufficient to prove the chaoticity of a system. Therefore we must use some numeric instruments in order to have a better insight into the nature of the system. Specifically, we will report the results obtained by the spectral analysis as well as the calculation of the correlation integral, the Lyapunov exponents, the Kolmogorov entropy and the embedding dimension.

5.1 Spectral Analysis

Spectral analysis has the aim of determining the spectral content of a time series by decomposing a given time series into different harmonic series with different frequencies and, by doing that, identifies the contribution of each series to the overall signal Stoica and Moses (2005).

Spectral analysis may aid in identifying chaos for a given time series as well as helping in discovering hidden periodicities in data. In the following three pictures of

---

[1]On some issues related to the arctg see for example Bradford and Davenport (2002), Collicott (2012), Walter (2010), Gonnet and Scholl (2009).

logistic map are presented in order to illustrate how the power spectra change with the parameter μ. The first box (top left) shows the cobweb diagram, the second (bottom left) is for the orbits, the third (top right) depicts the power spectrum obtained with a rectangular (sometimes called boxcar or Dirichlet) window and the last box shows the power spectrum with a Hamming window. Figures 9, 10 and 11 illustrate, respectively, very regular orbits with a single frequency peak, regular orbits with two frequency peaks and irregular orbits (chaos) with several frequency peaks.

Even though the technique is not conclusive if the system has "many hidden degrees of freedom of which the observer is unaware" (Moon 1987 p. 45) "chaotic time series are known to have aperiodic cycles of many lengths, so it seems reasonable to assume that if they are present in the candidate time series they should have been observed" (McBurnett 1996 p. 50). In any case, as the proposed model is by construction deterministic, the spectral analysis can definitively help in understanding whether the system shows chaotic dynamics.

Following this reasoning we have run a simulation in order to show that for the generated time series there is no peak that clearly dominates all other peaks (power spectrum for C, K, I, Y with rectangular and Hamming windows are shown in Figs. 12 and 13, respectively).

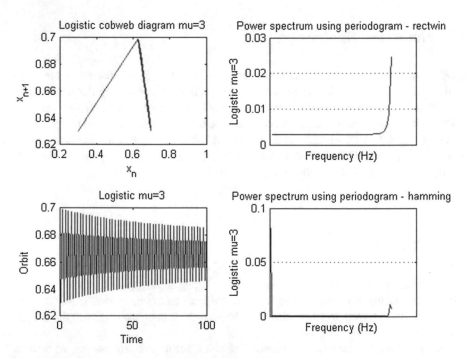

Fig. 9 Logistic map, $\mu = 3$ cobweb diagram and periodogram

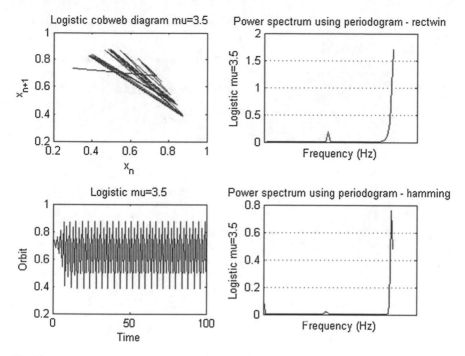

Fig. 10 Logistic map, $\mu = 3.5$ cobweb diagram and periodogram

5.2 Embedding Dimension

"The embedding dimension is the smallest dimension required to embed an object (a chaotic attractor for instance). In other words, this is the minimum dimension of the space in which you reconstruct a phase portrait starting from your measurements and in which the trajectory does not cross itself, that is, in which the determinism is verified. Of course, this is a statistical measure, meaning that you may have some "rare" self-crossings. When a global model is attempted, this is the minimum dimension your model must have" Letellier (2013).

Cao (1997) has suggested an algorithm based on the work of Kennel et al. (1992) for estimating the embedding dimension (see Takens 1981; Adachi 1993; Whitney 1992) through E1(d) and E2(d) functions, where d denotes the dimension.[2] The function E1(d) stops changing when d is greater than or equal to the embedding dimension staying close to 1. The function E2(d), instead, is used to distinguish deterministic from stochastic signals. If the signal is deterministic, there exist some d such

---

[2]Which has the following advantages: (a) does not require any subjective parameters except for the time-delay for the embedding; (b) does not strongly depend on the number of data points; (c) is able to distinguish between deterministic and stochastic signals; (d) it is computationally efficient and works well in presence of high-dimensional attractors.

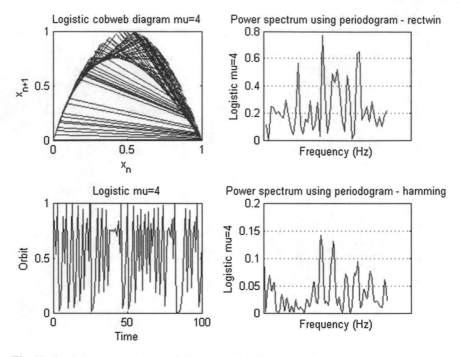

Fig. 11 Logistic map, $\mu = 4$ cobweb diagram and periodogram

that E2(d)! = 1 whilst if the signal is stochastic E2(d) is approximately 1 for all the values (see also Arya 1993; Arya et al. 1998).

For example, in Figs. 14 and 15, analogously to Cao (2002) we report the embedding dimension for the FX British Pound/US Dollar showing the values of E1 and E2 for a time series of 1,008 data points. In "looking at the results of the quantity E2 whose values are very close to 1 with some oscillations when the dimensions are small, this implies that the time series is likely a random time series, comparing with the case of random colored noise shown in Cao (1997). Given the oscillation behaviour away from 1 when the embedding dimensions are small, the time series should contain some determinism although the determinism may be weak".[3]

By applying the same analysis to our model, it is possible to observe a similar behaviour in the following figures where the E1 and E2 are calculated on consumption (Fig. 16), income (Fig. 17), capital (Fig. 18) and investment (Fig. 19). In fact we can observe that E2, on the four macroeconomic variables that by construction are not deterministic, is not 1 for all values but approaches it for $d \geq 10$.

---

[3]By contrast in the proposed model determinism is ensured by construction.

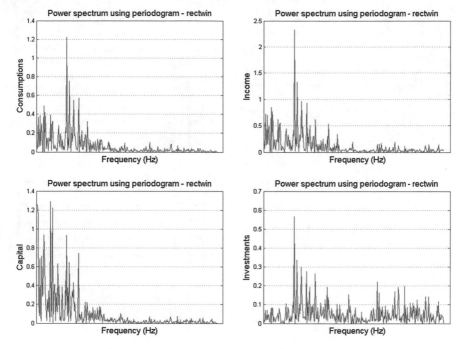

Fig. 12 Power spectrum with rectangular window

5.3 Correlation Integral

This kind of correlation is called spatial correlation and it aims to measure the degree of "relationship" between the different points on the strange attractor. The function which performs this task is called *correlation integral* $C(r, m)$ (see Lorenz 1993) which represents a direct arithmetic average of the pointwise mass function Theiler (1990) and it is defined as

$$C(r, m) = \lim_{m \to \infty} \lim_{r \to 0} \frac{1}{\vartheta} ln \frac{C(r, m)}{C(r, m + 1)} \tag{4}$$

with m embedding dimension and r radius (i.e. the space contraction or stretching).

In the following Figs. 20 and 21 we can observe that its trend (or its logarithm) is a function of r the radius and when it depicts a regular growth it confirms that the system is deterministic.

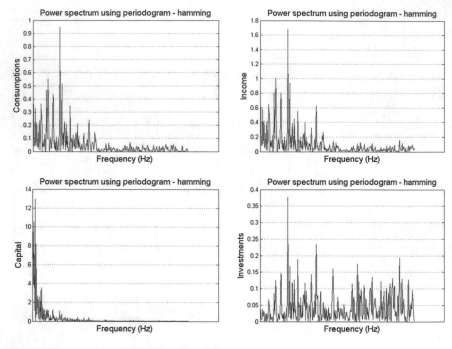

Fig. 13 Power spectrum with Hamming window

Fig. 14 Embedding
dimension for USD/EUR FX
rates (tao = 1, data points =
260)

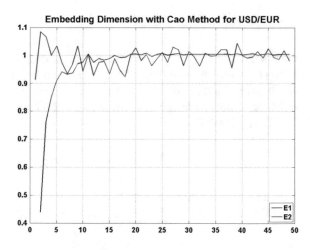

Fig. 15 Embedding dimension for USD/GBP FX rates (tao = 1, data points = 260)

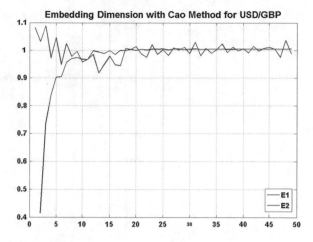

Fig. 16 Embedding dimension for consumption (tao = 1, data points = 10,000)

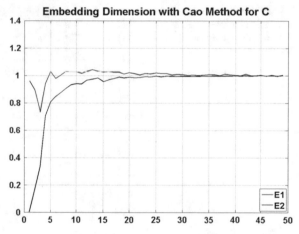

Fig. 17 Embedding dimension for income (tao = 1, data points = 10,000)

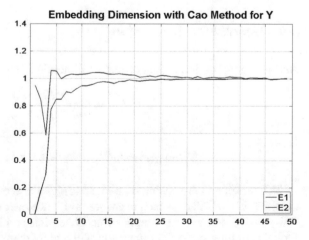

Fig. 18 Embedding
dimension for capital
(tao = 1, data points
= 10,000)

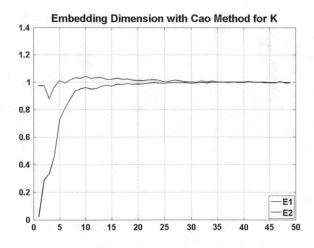

Fig. 19 Embedding
dimension for investments
(tao = 1, data points =
10,000)

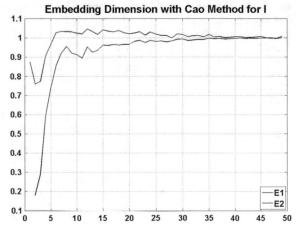

5.4 Correlation Dimension

Another useful notion is the correlation dimension defined as

$$D^C(m) = \lim_{r \to 0} \frac{ln(C(r,m))}{ln(r)} \tag{5}$$

hence

$$D^C(m)ln(r) \approx ln(C(r,m)) \ i.e. \ r^{D^C(m)} \approx C(r,m). \tag{6}$$

The correlation dimension is intended to measure the information content "where the limit of small size is taken to ensure invariance over smooth coordinate changes. This small-size limit also implies that dimension is a local quantity and that any global definition of fractal dimension will require some kind of averaging" Theiler

Fig. 20 Correlation integral
trend versus r

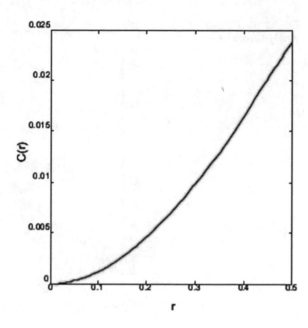

Fig. 21 Log-log plot

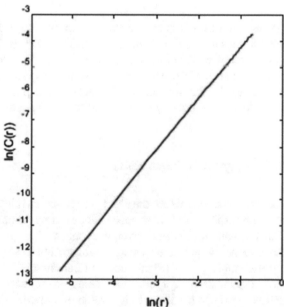

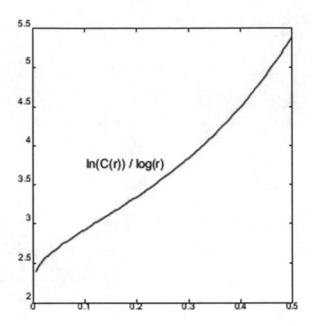

Fig. 22 Correlation dimension when $r \to 0$

(1990). Moreover it is shown that the correlation dimension D^C is an approxima-tion of Hausdorff's dimension D^H with $D^C \le D^H$. In Fig. 22 it can be seen that the dimension of correlation is noninteger. As D^C is a "more relevant measure of the attractor than D^H because it is sensitive to the *dynamical* process of the coverage of the attractor" (Grassberger and Procaccia 1983c p. 348), we can say that the system is fractal (see Grassberger and Procaccia 1983a; Grassberger 1986).

5.5 *Lyapunov Exponents*

"Lyapunov exponents are the average exponential rates of divergence or convergence of nearby orbits in the phase space. Since nearby orbits correspond to nearly identical states, exponential orbital divergence means that systems whose initial differences that may not be possible to resolve will soon behave quite differently, i.e. predictive ability is rapidly lost" (Sivakumar and Berndtsson 2010 p. 424). Dynamical systems have a spectrum of Lyapunov exponents, one for each dimension of the phase space and, similarly to the largest eigenvalue of a matrix, the largest Lyapunov exponent determines the dominant behaviour of a system. The sensible dependence on initial conditions, hence, can be restated as follows:

$$\|\delta x(t)\| \approx e^{\lambda t}\|\delta x_0\| \tag{7}$$

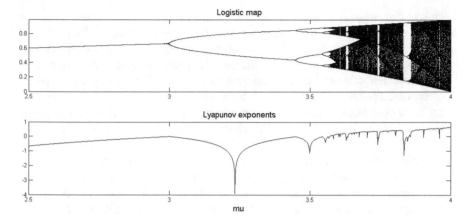

Fig. 23 Logistic map with Lyapunov exponents

Table 1 Lyapunov exponents

	Min	Max	Mean
Consumption	6.22	11.399	10.885
Income	12.8338	19.6440	13.3534
Capital	7.3165	14.594	12.999
Investment	5.511	11.969	11.049

Calculated Lyapunov's exponents for 10,000 points time series of C, Y, K and I

"where λ, the mean rate of separation of trajectories of the system, is called the leading Lyapunov exponent. In the limit of infinite time the Lyapunov exponent is a global measure of the rate at which nearby trajectories diverge, averaged over the strange attractor" Cvitanovic et al. (2012).

Figure 23 shows the Lyapunov exponents of the Logistic Map for comparison (see Schuster 1988).

Lyapunov exponents are used to measure the rate at which nearby trajectories of a dynamical system diverge (see for example Sivakumar and Berndtsson 2010; Cvitanovic et al. 2012). A dynamic dissipative system is chaotic if its biggest Lyapunov exponent is a positive number (see Lorenz 1993). In our simulations, by adopting the Wolf algorithm (see Wolf et al. 1985; Wolf 1986), we have found out that the biggest Lyapunov exponent has always been positive (see Table 1).

5.6 Entropy

To supplement the above mentioned analysis it could be useful to refer to the entropy K of Kolmogorov-Sinai (see Farmer 1982; Schuster 1988). We know it must converge to a positive finite value in order for the time series to be defined as chaotic because "Kolmogorov entropy is the mean rate of information created by the system. It is important in characterizing the average predictability of a system of which it represents the sum of the positive Lyapunov exponents. The Kolmogorov entropy quantifies the average amount of new information on the system dynamics brought by the measurement of a new value of the time series. In this sense, it measures the rate of information produced by the system, being zero for periodic or quasiperiodic (i.e. completely predictable) time series, and infinite for white noise (i.e. unpredictable by definition), and between the two for chaotic system" (Sivakumar and Berndtsson 2010 p. 424). In fact, according to the Pesin's theorem, Pesin (1977), the sum of all the positive Lyapunov exponents gives an estimate of the Kolmogorov-Sinai entropy. So, if $K > 0$, then the biggest Lyapunov exponent is bigger than zero and the system is chaotic. In order to measure K we used the approximation K_2 as defined by Grassberger and Procaccia (1983b) and we found that it was positive (e.g. 21.34561).

5.7 Symplectic Geometry Method

In addition to the embedding dimension, the symplectic geometry method (see Lei et al. 2002; Xie et al. 2005; Lei and Meng 2011) is used as a consistency check to verify the appropriate embedding dimension from a scalar time series. Symplectic similarity transformation is nonlinear and has measure-preserving properties i.e. time series remain unchanged when performing symplectic similarity transformation. For this reason symplectic geometry spectra (SGS) is preferred to singular value decomposition (SVD) (which is by nature a linear method that can bring distorted and misleading results Palus and Dvorak 1992).

Figures 24 and 25 show respectively the embedding dimension of the Logistic (which is deterministic) and the Gauss white noise and they can be be used as reference for comparing the embedding dimension obtained for consumption (Fig. 26), income (Fig. 27), capital (Fig. 28) and investment (Fig. 29) depicted.

5.8 Correlation Integral and Embedding Dimension

As we have repeatedly shown the system behaves stochastically but by construction is deterministic. In Table 2 we report the correlation integral versus the embedding dimension for our variables. As it can be observed the correlation integral does not increase with the embedding dimension confirming the validity of this analysis and that the system is deterministic (see Lorenz 1993 p. 213).

Fig. 24 Embedding dimension symplectic geometry method for logistic, $\mu = 3.9$ (data points = 1,000, abscissa is d, ordinate is $log(\frac{\sigma_i}{tr(\sigma_i)})$)

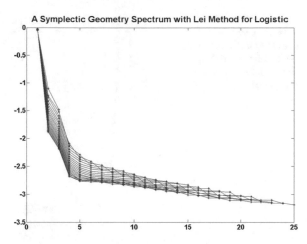

Fig. 25 Embedding dimension symplectic geometry method for consumption (data points = 10,000, abscissa is d, ordinate is $log(\frac{\sigma_i}{tr(\sigma_i)})$)

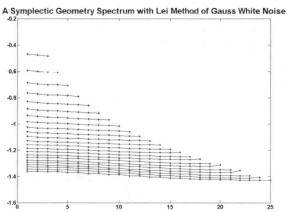

Fig. 26 Embedding dimension symplectic geometry method for capital (data points = 10,000, abscissa is d, ordinate is $log(\frac{\sigma_i}{tr(\sigma_i)})$)

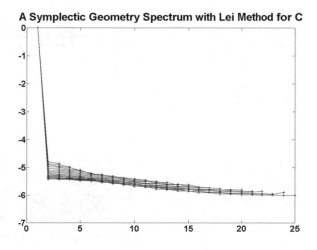

Fig. 27 Embedding
dimension symplectic
geometry method for income
(data points = 10,000,
abscissa is d, ordinate is
$log(\frac{\sigma_i}{tr(\sigma_i)})$)

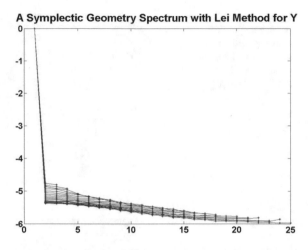

Fig. 28 Embedding
dimension symplectic
geometry method for capital
(data points = 10,000,
abscissa is d, ordinate is
$log(\frac{\sigma_i}{tr(\sigma_i)})$)

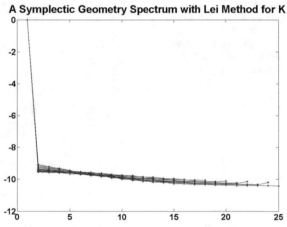

Fig. 29 Embedding
dimension symplectic
geometry method for
investments (data points =
10,000, abscissa is d,
ordinate is $log(\frac{\sigma_i}{tr(\sigma_i)})$)

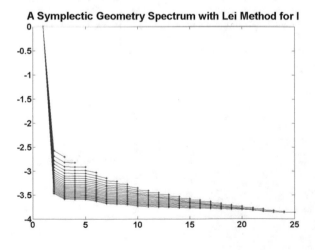

Table 2 Correlation integral versus embedding dimension

Correlation integral for		Embedding dimension						
		2	3	4	5	6	7	8
Consumptions	Min	0.053543	0.021604	0.0092081	0.0043155	0.0022121	0.0012546	0.00079854
	Max	0.71333	0.70895	0.70449	0.69998	0.69658	0.69372	0.69127
	Mean	0.30254	0.2599	0.23142	0.21153	0.19686	0.1859	0.17797
Income	Min	0.053543	0.021604	0.0092081	0.0043155	0.0022121	0.0012546	0.00079854
	Max	0.71333	0.70895	0.70449	0.69998	0.69658	0.69372	0.69127
	Mean	0.30254	0.2599	0.23142	0.21153	0.19686	0.1859	0.17797
Capital	Min	0.053543	0.021604	0.0092081	0.0043155	0.0022121	0.0012546	0.00079854
	Max	0.71333	0.70895	0.70449	0.69998	0.69658	0.69372	0.69127
	Mean	0.30254	0.2599	0.23142	0.21153	0.19686	0.1859	0.17797
Investments	Min	0.053543	0.021604	0.0092081	0.0043155	0.0022121	0.0012546	0.00079854
	Max	0.71333	0.70895	0.70449	0.69998	0.69658	0.69372	0.69127
	Mean	0.30254	0.2599	0.23142	0.21153	0.19686	0.1859	0.17797

Correlation integral versus embedding dimension for 10,000 points time series of C, Y, K and I. As shown the correlation integral is quite stable for $m = 2, \ldots, 8$.

6 Conclusions

We share the view that "economists will be led, as natural scientists have been led, to seek in nonlinearities an explanation of the maintenance of oscillation" Goodwin (1951). In this paper, with the help of RP, we have first shown that business cycles may be chaotic in nature and then we have proposed a non-linear model has a chaotic behaviour. But while the latter result has been achieved by many, we suggested a different functional form (i.e. the hyperbolic tangent) instead of the usual arctangent. Moreover the originality of this work lies in the specification of consumption and investment as a function of the difference, respectively, between the growth rates of income and capital and the growth rates of income and consumption. This has been obtained by considering, à la Kalecki, that the investment process has different timing than does consumption, hence the difference in the considered time lags. Last but not least the model can accommodate such external perturbations as shocks by a translation of the argument of the function f.

In future we will research the calibration of the model to real economy as well as study further its features with the help of RQA (which, possibly, may result to identifying some leading indicators of economic crashes). Moreover here we stress that such non-linear behaviour has some implications which could be potentially considered as being of some interest, e.g. what should be the rules set by a regulator (such as the central bank) within a chaotic framework?

For example if the system is not predictable, not reachable and then not observable but nevertheless controllable (see Romieras et al. 1992; Grebogi and Laib 1997; Calvo and Cartwright 1998; Pettini 2005), can one set up a system of controls that is able to drive the economy?

Acknowledgements The author is grateful to the editors, the referees and his colleagues. Special thanks go to Giovanna Zimatore (CNR-IDASC, Institute of Acoustics and Sensors O. M. Corbino, Rome) for the support given on the RQA analysis, to Edward Bace (Middlesex University London), Carlo Lucheroni (School of Science and Technologies, University of Camerino) and to Michele Mininni (Department of Economics and Mathematical Methods, University of Bari) for his guidance, patience and invaluable time.

References

Adachi M (1993) Embeddings and immersions. American Mathematical Society

Addo PM, Billio M, Guegan D (2013) Nonlinear dynamics and recurrence plots for detecting financial crisis. N Am J Econ Finan 26:416–435

Agliari A, Dieci R, Gardini L (2007) Homoclinic tangles in a Kaldor-like business cycle model. J Econ Behav Organ 62:324–347

Arya S, Mount DM (1993) Approximate nearest neighbor searching. In: Proceedings of 4th annual ACM-SIAM symposium on discrete algorithms (SODA'93), pp 271–280

Arya S, Mount DM, Netanyahu NS, Silverman R, Wu AY (1998) An optimal algorithm for approximate nearest neighbor searching. J ACM 45(6):891–923

Benson H (2008) University physics. Wiley

Bischi GI, Dieci R, Rodano G, Saltari E (2001) Multiple attractors and global bifurcations in a kaldor-type business cycle model. J Evolut Econ 11:527–554

Bradford R, Davenport JH (2002) Towards better simplification of elementary functions. In: ISSAC '02 Proceedings of the 2002 international symposium on symbolic and algebraic computation, New York. ACM, pp 16–22

Burns AF, Mitchell WC (1946) Measuring business cycles. In: National bureau of economic research

Calvo O, Cartwright JHE (1998) Fuzzy control of chaos. Int J Bifurc Chaos 8:1743–1747

Cao L (1997) Practical method for determining the minimum embedding dimension of a scalar time series. Phys D 110(1–2):43–50

Cao L (2002) Determining minimum embedding dimension from scalar time series. In: Soofi A, Cao L (eds) Modelling and forecasting financial data, vol 2. Studies in computational finance. US, Springer, pp 43–60

Chen W-S (2011) Use of recurrence plot and recurrence quantification analysis in taiwan unemployment rate time series. Phys A Stat Mech Appl 390(7):1332–1342

Chiarella C (1990) The elements of a nonlinear theory of economic dynamic. Springer, Berlin-Heidelberg-New York

Collicott SH (2012) Never trust an arctangent. https://engineering.purdue.edu/~collicot/NTAA_files/Chapter1.pdf

Crowley PM (2008) Analyzing convergence and synchronicity of business and growth cycles in the euro area using cross recurrence plots. Eur Phys J Spec Topics 164(1):67–84

Crowley PM, Schultz A (2010) A new approach to analyzing convergence and synchronicity in growth and business cycles: cross recurrence plots and quantification analysis. Bank of Finland research discussion paper (16)

Cvitanović P, Artuso R, Mainieri R, Tanner G, Vattay G (2012) Lyapunov exponents. In: Chaos: classical and quantum, chapter 6. Niels Bohr Institute, Copenhagen. http://ChaosBook.org/version14ChaosBook.org/version14

Eroglu D, Peron TKD, Marwan N, Rodrigues FA, Costa LdF, Sebek M, Kiss, IZ, Kurths J (2014) Entropy of weighted recurrence plots. Phys Rev E 90(4):042919

Fabretti A, Ausloos M (2005) Recurrence plot and recurrence quantification analysis techniques for detecting a critical regime. Examples from financial market indices. Int J Mod Phys C 16(05):671–706

Farmer D (1982) Chaotic attractors of an infinite-dimensional dynamical system. Phys D 4:366–93

Gonnet GH, Scholl R (2009) Scientific computation. Cambridge University Press, Cambridge

Goodwin RM (1951) The nonlinear accelerator and the persistence of business cycle. Econometrica 19(1)

Gorban AN, Smirnova EV, Tyukina TA (2010) Correlations, risk and crisis: from physiology to finance. Phys A Stat Mech Appl 389(16):3193–3217

Grassberger P (1986) Estimating the fractal dimension and entropies of strange attractors. In: Holden AV (ed) Chaos. Manchester University Press, Manchester, pp 291–311

Grassberger P, Procaccia I (1983a) Characterization of strange attractors. Phys Rev Lett 50:346

Grassberger P, Procaccia I (1983b) Estimation of the Kolmogorov entropy from a chaotic signal. Phys Rev A 28:2591–2593

Grassberger P, Procaccia I (1983c). Measuring the strangeness of strange attractors. Physica D 9:189–208

Grebogi C, Laib YC (1997) Controlling chaotic dynamical systems. Syst Control Lett 31(5):307–312

Hicks JR (1950) A contribution to the theory of the trade cycle. Clarendon Press, Oxford

Hillinger C, Sebold-Bender M (1992) Cyclical growth in market and planned economies. Oxford University Press, Oxford

Ichimura S (1955a) Notes on non-linear business cycle theories. Osaka economic papers

Ichimura S (1955b) Toward a general nonlinear macrodynamic theory of economic fluctuations. In: Kurihara KK (ed) Post-Keynesian economics, chapter 8. George Allen & Unwin Ltd., London, pp 192–226

Januário C, Grácio C, Duartea J (2009) Measuring complexity in a business cycle model of the Kaldor type. Chaos, Solitons Fractals 42(5):2890–2903

Januário C, Grácio C, Ramos JS (2005) Chaotic behaviour in a two-dimensional business cycle model. In: Elaydi S, Cushing J, Lasser R, Ruffing A, Papageorgiou V, Assche WV (eds) Proceedings of the international conference, difference equations, special functions and orthogonal polynomials, pp 294–304, Munich

Kaddar A, Alaoui HT (2009) Global existence of periodic solutions in a delayed Kaldor-Kalecki model. Nonlinear Anal Model Control 14(4):463–472

Kaldor N (1940) A model of trade cycle. Econ J 50(197):78–92

Kalecki M (1966) Studies in the theory of business cycles, 1933–1939. New York, A.M, Kelley

Karagianni S, Kyrtsou C (2011) Analysing the dynamics between US inflation and Dow Jones index using non-linear methods. Stud Nonlinear Dyn Econom 15(2)

Kennel MB, Brown R, Abarbanel HDI (1992) Determining embedding dimension for phase-space reconstruction using a geometrical construction. Phys Rev A 45(6):3403–3411

Korotayev AV, Sergey TV (2010) A spectral analysis of world gdp dynamics: Kondratieff waves, Kuznets swings, Juglar and Kitchin cycles in global economic development, and the 2008–2009 economic crisis. Struct Dyn 4(1)

Krawiec A, Szydlowski M (1999). The Kaldor-Kalecki business cycle model. Ann Oper Res, 89–100

Krawiec A, Szydlowski M (2001) On nonlinear mechanics of business cycle model. Regul Chaotic Dyn 6(1):101–118

Lei M, Meng G (2011) Symplectic principal component analysis: a new method for time series analysis. Math Probl Eng 2011. Article ID 793429, 14 p

Lei M, Wang Z, Feng Z (2002) A method of embedding dimension estimation based on symplectic geometry. Phys Lett A 303(2–3):179–189

Letellier C (2013) Estimating the minimum embedding dimension

Lorenz HW (1993) Nonlinear dynamical economics and chaotic motion, 2nd edn. Springer, Berlin-Heidelberg-New York

McBurnett M (1996) Probing the underlying structure in dynamical systems: an introduction to spectral analysis, chapter 2. The University of Michigan Press, pp 31–51

Mircea G, Neamt M, Opris D (1963) The Kaldor and Kalecki stochastic model of business cycle, nonlinear analysis: modelling and control. J Atmos Sci 16(2):191–205

Moloney K, Raghavendra S (2012) A linear and nonlinear review of the arbitrage-free parity theory for the cds and bond markets. In: Topics in numerical methods for finance. Springer, pp 177–200

Moon FC (1987) Chaotic vibrations: an introduction for applied scientists and engineers. Wiley, New York

Morishima M (1959) A contribution to the nonlinear theory of the trade cycle. Zeitschrift für Nationalökonomie 18(4):166–170

Mullineux AW (1984) The business cycle after Keynes. Wheatsheaf Books Ltd, Brighton, Sussex

Orlando G (2016) A discrete mathematical model for chaotic dynamics in economics: Kaldor's model on business cycle. Math Comput Simul 125:83–98

Palus M, Dvorak I (1992) Singular-value decomposition in attractor reconstruction: pitfalls and precautions. Phys D Nonlinear Phenom 55(1–2):221–234

Pesin YB (1977) Characteristic Lyapunov exponents and smooth ergodic theory. Rus Math Surv 32:55–114

Pettini M (2005) Controlling chaos through parametric excitations. In: Dynamics and stochastic processes theory and applications. Lecture notes in physics, vol 355. Springer, Berlin-Heidelberg-New York, pp 242–250

Pham V-T, Volos C, Vaidyanathan S (2017) A chaotic time-delay system with saturation nonlinearity. Int J Syst Dyn Appl (IJSDA) 6(3):111–129

Piskun O, Piskun S (2011) Recurrence quantification analysis of financial market crashes and crises. arXiv:1107.5420

Purves WK, Orians GH, Sadava D, Heller HC (2003) Life: the science of biology, vol 3. Macmillan

Reger D, Goode S, Ball D (2010) Chemistry: principles and practice. Brooks/Cole, 3rd edn

Romieras FJ, Ott E, Grebogi C, Daiawansa WP (1992) Controlling chaotic dynamical systems. Physica D, 58:165–192

Schumpeter JA (1954) History of economic analysis. George Allen & Unwin, London

Schuster H (1988) Deterministic chaos—an introduction. VcH Verlagsgesellschaft mbH

Shishkin J (1961) Signals of recession and recovery. NBER Occasional Paper n 77

Sivakumar B, Berndtsson R (2010) Advances in data-based approaches for hydrologic modeling and forecasting, chapter 9. World Scientific, pp 411–461

Stewart J (2010) Single variable calculus, 4th edn. Brooks/Cole Publishing Company

Stoica P, Moses R (2005) Spectral analysis of signals. Prentice Hall

Strozzi F, Gutierrez E, Noè C, Rossi T, Serati M, Zaldivar J (2007) Application of non-linear time series analysis techniques to the nordic spot electricity market data. Libero istituto universitario Carlo Cattaneo

Takens F (1981) Dynamical systems and turbulence. Lecture notes in mathematics, chapter Detecting strange attractors in turbulence, vol 898. Springer, Berlin-Heidelberg-New York, pp 366–381

Theiler J (1990) Estimating fractal dimension. J Opt Soc Am A 7:1055–1073

Walter FS (2010) Waves and oscillations: a prelude to quantum mechanics. Oxford University Press, Oxford

Wheldon TE (1988) Mathematical models in cancer research. Taylor & Francis

Whitney H (1992) Hassler Whitney collected papers. In: Eells J, Toledo D (eds) Hassler Whitney collected papers, volume I II of contemporary mathematicians. Birkhäuser Verlag, Basel-Boston-Stuttgart

Wolf A (1986) Quantifying chaos with Lyapunov exponents. In: Holden AV (ed) Chaos. Manchester University Press, Manchester, pp 273–290

Wolf A, Swift JB, Swinney HL, Vastano JA (1985) Determining Lyapunov exponents from a time series. Physica D 16:285–317

Xie H, Wang Z, Huang H (2005) Identification determinism in time series based on symplectic geometry spectra. Phys Lett A 342(1–2):156–161

Yasui E (1953) Non-linear self-excited oscillations and business cycles. Cowles Comm Discuss Paper 2063:1–20

Zarnowitz V (1992) *Business cycles: theory, history, indicators, and forecasting*. National bureau of economic research studies in business cycles, vol 27. The University of Chicago Press, Chicago and London

Zbilut JP (2005) Use of recurrence quantification analysis in economic time series. In: Economics: complex windows. Springer, pp 91–104

Zimatore G, Fetoni AR, Paludetti G, Cavagnaro M, Podda MV, Troiani D (2011) Post-processing analysis of transient-evoked otoacoustic emissions to detect 4 khz-notch hearing impairment—a pilot study. Med Sci Monit Int Med J Experimental Clin Res, 17(6):MT41

Zimatore G, Garilli G, Poscolieri M, Rafanelli C, Terenzio Gizzi F, Lazzari M (2017) The remarkable coherence between two Italian far away recording stations points to a role of acoustic emissions from crustal rocks for earthquake analysis. Chaos: An Interdisciplinary. J Nonlinear Sci 27(4):043101

Analysis of Three-Dimensional Autonomous Van der Pol–Duffing Type Oscillator and Its Synchronization in Bistable Regime

Gaetan Fautso Kuiate, Victor Kamdoum Tamba
and Sifeu Takougang Kingni

Abstract This chapter proposes a three-dimensional autonomous Van der Pol-Duffing (VdPD) type oscillator which is designed from a nonautonomous VdPD two-dimensional chaotic oscillator driven by an external periodic source through replacing the external periodic drive source with a direct positive feedback loop. The dynamical behavior of the proposed autonomous VdPD type oscillator is investigated in terms of equilibria and stability, bifurcation diagrams, Lyapunov exponent plots, phase portraits and basin of attraction plots. Some interesting phenomena are found including for instance, period-doubling bifurcation, symmetry recovering and breaking bifurcation, double scroll chaos, bistable one scroll chaos and coexisting attractors. Basin of attraction of coexisting attractors is computed showing that is associated with an unstable equilibrium. So the proposed autonomous VdPD type oscillator belongs to chaotic systems with self-excited attractors. A suitable electronic circuit of the proposed autonomous VdPD type oscillator is designed and its investigations are performed using ORCAD-PSpice software. Orcard-PSpice results show a good agreement with the numerical simulations. Finally, synchronization of identical coupled proposed autonomous VdPD type oscillators in bistable regime is studied using the unidirectional linear feedback methods. It is found from the numerical simulations that the quality of synchronization depends on the coupling coefficient as well as the selection of coupling variables.

G. F. Kuiate
Department of Physics, Higher Teacher Training College, University of Bamenda,
PO Box 39, Bamenda, Cameroon
e-mail: fautsokuiate@yahoo.co.uk

V. K. Tamba
Department of Telecommunication and Network Engineering, IUT-Fotso
Victor of Bandjoun, University of Dschang, P. O. Box: 134, Bandjoun, Cameroon
e-mail: victorkamdoum@yahoo.fr

S. T. Kingni (✉)
Department of Mechanical and Electrical Engineering, Institute of Mines
and Petroleum Industries, University of Maroua, P.O. Box 46, Maroua, Cameroon
e-mail: stkingni@gmail.com

© Springer International Publishing AG 2018

V.-T. Pham et al. (eds.), *Nonlinear Dynamical Systems with Self-Excited
and Hidden Attractors*, Studies in Systems, Decision and Control 133,
https://doi.org/10.1007/978-3-319-71243-7_7

Keywords Van der Pol-Duffing oscillator · Chaos · Coexisting attractors
Bistable regime · Electronic circuit implementation · Synchronization

1 Introduction

In recent years, the implementation and study of chaotic systems have grown up in many fields and attracts many scientists because chaos was found useful with great potential in many fields, including liquid mixing with low power consumption, human brain and heartbeat regulation, and secure communications (Otto 1993; Hilborn 2006; Xiaofan and Guanrong 2000; Liu et al. 2011). Chaotic oscillations can be generated in the third-order or higher-order autonomous nonlinear differential equations. In the case of non-autonomous differential equations, i.e. nonlinear damped systems driven by external periodic signal, the minimal order of the differential equations can be reduced to two. Among the periodically forced autonomous (or self-excited) oscillators, one of the most extensively studied examples is the VdPD oscillator. The nonautonomous VdPD oscillator can be used as a basic model for describing periodically self-excited oscillators in physics, engineering, electronics, biology, neurology and many other disciplines (Kapitaniak 1998). In the literature, nonautonomous VdPD oscillator has been widely investigated because it is a classical example of a nonlinear dynamical system exhibiting complex behaviors such as chaos (Ueda and Akamatsu 1981; Rudowski and Szemplinska-Stupnicka 1997; Kozlov et al. 1999). Maccari (2008) investigated the dynamics and the vibration amplitude of a VdPD oscillator under time-delayed position and velocity feedbacks. Asymptotic perturbation method is used to derive two slow-flow equations for the amplitude and phase of the fundamental resonance response. The author of reference (Maccari 2008) proved that the introduction of the control term guarantees the stability of the periodic solutions. In reference (Cui et al. 2016), the stability analysis of periodic solutions of the forced VdPD oscillator is reported. Vincent and coworkers (Vincent et al. 2011) investigated the dynamics of VdPD circuit driven by an external periodical signal. They showed that driven VdPD circuit can develop a rich and complex dynamical behaviors such as hyperchaos, metastable chaos (or transient chaos), strange-nonchoatic attractors and quasiperiodic orbits. However, the external periodic signal required to generate chaotic behaviors are not always easy to obtain because frequency generator is expensive. Moreover, in some applications where chaotic behavior is required, the frequency generator used to provide the external periodic signal might not be available or, in some situations, the space to put all the devices can be very small. Therefore, having an autonomous, minor and less bulky system capable of achieving the tasks required is advantageous. A self-sustained system can provide wide benefits for the physical equipment while reducing its cost. On the dynamical point of view, the power spectrum of a chaotic system driven by a periodic signal presents some clear peaks at the multiple of the frequency of the periodic signal indicating that the chaos found in system is not fully developed (Kingni et al. 2012,

2014). And therefore this chaotic behavior found is not indicating for some relevant engineering applications such as secure communications. While the power spectrum of chaos found in an autonomous system has a randomly distributed harmonics peaks indicating the robustness of chaotic signal (Nana et al. 2009; Sprott 2010; Kingni et al. 2015).

The external periodical signal required to drive some systems in order to generate complex behaviors such as chaos can be overcome by using two techniques: (i) by replacing the external periodic drive source of the nonlinear damped systems by a direct positive feedback loop (Tamaševičius et al. 2009) or (ii) by using the jerk architecture (Benítez et al. 2006). The authors of reference (Tamaševičius et al. 2009) designed and built a three-dimensional autonomous Duffing-Holmes chaotic oscillator as an alternative for the nonautonomous Duffing-Holmes two dimensional chaotic oscillator. In comparison with the well-known nonautonomous Duffing-Holmes circuit, it lacks the external periodic drive, but includes two extra linear feedback subcircuits, namely a direct positive feedback loop and an inertial negative feedback loop. Inspired by reference (Tamaševičius et al. 2009), in this work, we introduce as an alternative for the VdPD oscillator driven by the external periodic signal, an autonomous three dimensional VdPD type oscillator with a direct positive feedback loop. Our objective in this work is to investigate analytical, numerically and analogically the proposed autonomous three dimensional VdPD type oscillator in order to shed more light on its dynamics and synchronization.

The paper is organized as follows. Section 2 is devoted to the analytical and numerical analysis of the proposed autonomous three dimensional VdPD type oscillator. In Sect. 3, an appropriate analog computer is proposed for the investigation of the dynamical behavior of the proposed autonomous three dimensional VdPD type oscillator. Section 4 focuses on synchronization of identical coupled proposed autonomous VdPD type oscillators in bistable and coexisting regimes is studied using the unidirectional linear feedback methods. The conclusion is given in Sect. 5.

2 Design and Analysis of the Proposed Three-Dimensional Autonomous Van der Pol–Duffing Type Oscillator

The nonautonomous VdPD oscillator is a classical example of a nonlinear dynamical system exhibiting complex behaviors such as chaos (Ueda and Akamatsu 1981; Rudowski and Szemplinska-Stupnicka 1997; Kozlov et al. 1999). It is described by the two dimensional differential equation with an external periodic drive term:

$$\ddot{x} - \varepsilon\left(1 - x^2\right)\dot{x} + \alpha x + \beta x^3 = f \sin(\omega t) \tag{1}$$

where the parameters ε, α and β are the dimensionless damping coefficient, linear and cubic nonlinearity parameters, respectively. ω is the external frequency of periodic signal and f stands for the amplitude of the external excitation. The potential associated to VdPD oscillator is given by $V(x) = \frac{\alpha}{2}x^2 + \frac{\beta}{4}x^4$ and it is a Φ^4 potential. Depending on the signs of α and β, the potential $V(x)$ can as three main shapes: single-well potential (for $\alpha > 0$ and $\beta > 0$), double-well potential (for $\alpha < 0$ and $\beta > 0$) and double hump potential (for $\alpha > 0$ and $\beta < 0$). These three shapes correspond to three physical situations of VdPD. In this work, we focus our attention on the study of VdPD oscillator with double-well potentials. For $f = 0$ and $\alpha = -1$, Eq. (1) has three fixed points $(x^*, y^* = dx^*/dt) = (0,0)$ and $(\pm 1/\sqrt{\beta}, 0)$. According to the Routh-Hurwitz criteria, the equilibrium point $(0,0)$ is unstable for any values of parameters ε and β. The equilibrium points $(\pm 1/\sqrt{\beta}, 0)$ are stable for $\varepsilon > 0$ and $0 < \beta < 1$. The trajectories of Eq. (1) for $f = 0$ and $\alpha = -1$ converge to a limit cycle. Inspired by reference (Tamaševičius et al. 2009) in this section, an autonomous version of the VdPD type oscillator is designed by replacing the external periodic drive source by a direct positive feedback loop. That is given by:

$$\ddot{x} - \varepsilon(1 - x^2)\dot{x} - x + \beta x^3 + kz = 0 \tag{2a}$$

$$\dot{z} = \dot{x} - z \tag{2b}$$

By setting $\dot{x} = y$, we obtain the following the three-dimensional autonomous VdPD type oscillator:

$$\dot{x} = y \tag{3a}$$

$$\dot{y} = x - \beta x^3 + \varepsilon(1 - x^2)y - kz \tag{3b}$$

$$\dot{z} = y - z \tag{3c}$$

Here z is the third independent dynamical variable and $k > 0$ is the feedback coefficient. It is apparent that the system (3a, 3b, 3c) has a natural symmetry under the transformation $S(x, y, z) \rightarrow (-x, -y, -z)$. The system (3a, 3b, 3c) has three equilibrium points $O = (0, 0, 0)$ and $E_{1,2} = (\pm 1/\sqrt{\beta}, 0, 0)$. The characteristic equation associated to the equilibrium point $E = (x^*, y^*, z^*)$ is

$$\lambda^3 + [1 - \varepsilon(1 - x^2)]\lambda^2 + [k^2 - \varepsilon(1 - x^2) - 1 + 3\beta x^2 + \varepsilon xy]\lambda$$
$$- 1 + 3\beta x^2 + \varepsilon xy = 0 \tag{4}$$

For the equilibrium point O, we have $\lambda^3 + (1 - \varepsilon)\lambda^2 + (k^2 - 1 - \varepsilon)\lambda - 1 = 0$. According to the Routh-Hurwitz criteria, the equilibrium point O is unstable for any values of ε and k. For the equilibrium points $E_{1,2}$, we have $\lambda^3 + (1 - \varepsilon - \varepsilon/\beta)\lambda^2 + (2 - \varepsilon + k^2)\lambda + 2 = 0$. According to the Routh-Hurwitz criteria, the equilibrium points $E_{1,2}$ is stable if

$$\varepsilon(1 + 1/\beta) < 1 \tag{5a}$$

$$(1 - \varepsilon - \varepsilon/\beta)(2 + k^2 - \varepsilon) - 2 > 0 \tag{5b}$$

Since $k > 0$ and $\beta > 0$, the set of inequality (5a, 5b) is not met therefore the equilibrium points $E_{1,2}$ are unstable.

The dynamical behavior of system (3a, 3b, 3c) is illustrated by bifurcation diagrams, Lyapunov exponent, basin of attraction and phase portraits. In Fig. 1, we present the two parameters (k, β) bifurcation diagram of the dynamical behavior of system (3a, 3b, 3c) for $\varepsilon = 1$.

The two parameters (k, β) bifurcation diagram of Fig. 1 is constructed by examining the Lyapunov exponents and time series for each cell. From Fig. 1, we see that system (3a, 3b, 3c) can display periodic (cyan regions) and chaotic behaviors (red regions). In order to know the route to chaotic behavior exhibited by system (3a, 3b, 3c), we plot in Figs. 2 and 4 the bifurcation diagrams depicting the local maxima of $x(t)$ as a function of the parameter k or parameter β for $\varepsilon = 1$ and for a specific value of parameter β or k.

The bifurcation diagram of Fig. 2a is obtained by plotting local maxima of the output $x(t)$ versus the parameter k which increased (or decreased) in tiny steps in the range $1.46 \leq k \leq 2$. The final state at each iteration of the control parameter

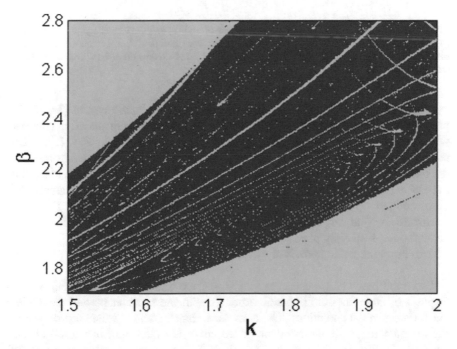

Fig. 1 Regions of dynamical behavior in the parameters k and β for $\varepsilon = 1$. Periodic oscillations are in light blue color and chaotic oscillations are in red color

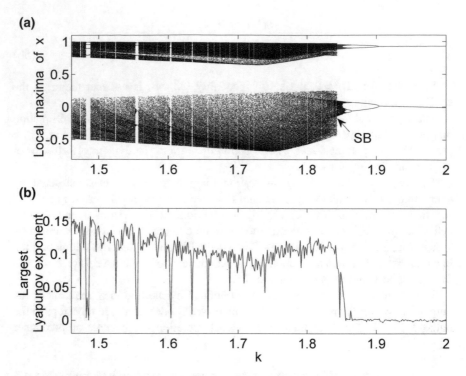

Fig. 2 The bifurcation diagrams depicting maxima of $x(t)$ (**a**) and the largest Lyapunov exponent (**b**) of system (3a, 3b, 3c) versus the parameter k for $\varepsilon = 1$ and $\beta = 2.0$. Bifurcation diagrams are obtained by scanning the parameter k upwards (black) and downwards (red). The acronym SB means symmetry-breaking

k serves as the initial state for the next iteration. In Fig. 2a, the system (3a, 3b, 3c) presents a reverse period-doubling to chaos interspersed with periodic windows when the parameter k varies from 1.46 to 2 (see black dot in Fig. 2a). When performing the same analysis by ramping the parameter k (see red dot in Fig. 2a), the output $x(t)$ displays the same dynamical behaviors as in Fig. 2a (see black dot) but at $k = 1.842$ where the symmetry breaking bifurcation appeared the amplitudes of the output $x(t)$ are not the same. The chaotic behavior found in Fig. 2a is confirmed by the largest Lyapunov exponent shown in Fig. 2b. The chaotic behavior and bistability phenomenon shown in Fig. 2 is further detailed in Fig. 3, which shows the phase portrait of system (3a, 3b, 3c) for specific values of β, ε and k.

From the black and red curves in Fig. 3a, we notice that system (3a, 3b, 3c) displays bistable one-scroll chaotic attractor with the same shape and parameters but different initial conditions. The attractor in black curve is called left one-scroll chaotic attractor while the attractor in red curve is called rigth one-scroll chaotic attractor. In Fig. 3b, the system (3a, 3b, 3c) presents monostable double-scroll chaotic attractor.

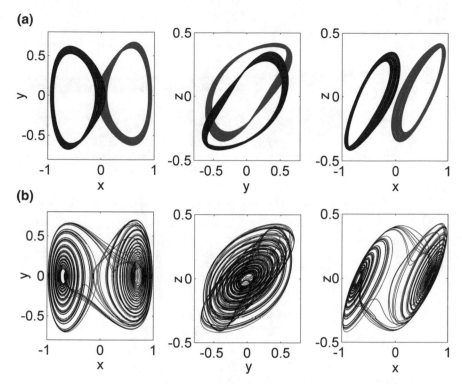

Fig. 3 The phase portrait of system (3a, 3b, 3c) in planes (x, y), (x, z) and (y, z) for specific value of parameter k: **a** $k = 1.85$ and **b** $k = 1.7$. The curve in black line is obtained by using the initial conditions $(x(0), y(0), z(0)) = (0.1, 0.1, 0.1)$ whereas the curve in red line is obtained by using the initial conditions $(x(0), y(0), z(0)) = (-0.1, -0.1, -0.1)$. The other parameter are $\beta = 2$ and $\varepsilon = 1$

For $\varepsilon = 1$ and $k = 1.7$, we plot the bifurcation diagrams depicting maxima of $x(t)$ and the largest Lyapunov exponent of system (3a, 3b, 3c) versus the parameter β as shown in Fig. 4.

The bifurcation diagram of Fig. 4a is obtained by plotting local maxima of the output $x(t)$ versus the parameter β which increased (or decreased) in tiny steps in the range $1.7 \leq \beta \leq 2.6$. The final state at each iteration of the control parameter β serves as the initial state for the next iteration. When the parameter β varies from 1.7 to 2.6 (see black dot in Fig. 4a), the bifurcation diagram of the output $x(t)$ shows a period-1-oscillations followed by a period-doubling bifurcation to chaos interspersed with periodic windows. When performing the same analysis by ramping the parameter β (see red dot in Fig. 4a), the output $x(t)$ displays the same dynamical behaviors as in Fig. 4 (a) (see black dot) in the range $1.7 \leq \beta \leq 1.882$ but the amplitudes of the output $x(t)$ are not the same from $\beta = 1.7$ up to $\beta = 1.76$ where the symmetry recovering appeared. Therefore one can notice that the system (3a, 3b, 3c) shows bistability in the range $1.7 \leq \beta \leq 1.76$. While in the range

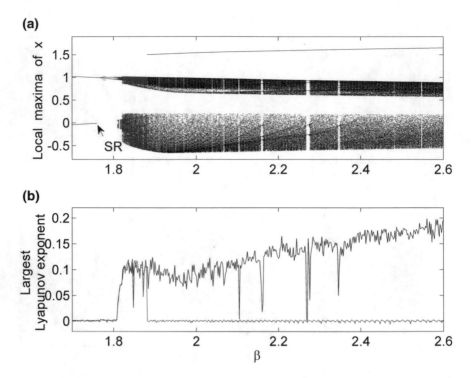

Fig. 4 The bifurcation diagrams depicting maxima of $x(t)$ (**a**) and the largest Lyapunov exponent (**b**) of system (3a, 3b, 3c) versus the parameter β for $\varepsilon = 1$ and $k = 1.7$. Bifurcation diagrams are obtained by scanning the parameter β upwards (black) and downwards (red). The acronym SR corresponds to symmetry-restoring

$1.882 < \beta \leq 2.6$, the output $x(t)$ exhibits period-1-oscillations. By comparing the two set of data [for increasing (black) and decreasing (red)] used to plot Fig. 4a, one can notice that period-1-oscillations coexist with chaotic oscillations and periodic oscillations in the range $1.882 < \beta \leq 2.6$. The chaotic behavior found in Fig. 4a is confirmed by the largest Lyapunov exponent shown in Fig. 4b. The coexistence of attractors found in Fig. 4 is presented in Fig. 5 which depicts the phase portraits of the resulting attractors of the system (3a, 3b, 3c) in the plane (x, y) for specific value of β and initial conditions.

At $\beta = 2$, the system (3a, 3b, 3c) displays period-1-oscillations and double-scroll chaotic attractor for two different initial conditions as shown in Fig. 5. The coexistence of attractors shown in Fig. 5 is further detailed in Fig. 6 which shows the basin of attraction of system (3a, 3b, 3c) in the plane $z = 0$ for $\beta = 2, \varepsilon = 1.0$ and $k = 1.7$.

One can see from Fig. 6 that the system (3a, 3b, 3c) can exhibit either chaotic attractor or periodic attractor depending of the initial conditions. Since basin of attraction of coexisting attractor is associated with an unstable equilibrium.

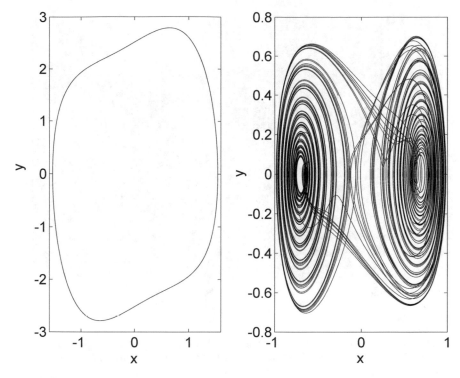

Fig. 5 Coexistence of attractors for specific value of $\beta = 2$ and initial conditions. In the left panel the initial conditions are $(x(0), y(0), z(0)) = (1.5, 0.1, 0.1)$ whereas in the right panel, the initial conditions are $(x(0), y(0), z(0)) = (0.1, 0.1, 0.1)$. The remaining parameters are $\varepsilon = 1$ and $k = 1.7$

The proposed autonomous VdPD type oscillator belongs to chaotic systems with self-excited attractor attractors.

3 Electronic Simulations of the Proposed Three-Dimensional Autonomous Van der Pol–Duffing Type Oscillator

Electronic circuit implementation of theoretical chaotic models (Chedjou et al. 2001; Kengne et al. 2018; Kingni et al. 2014; Pehlivan and Uyarogglu 2012; Pham et al. 2017a) is an interesting approach to investigate the dynamical behavior of such systems and has many practical technological applications including cryptography, image encryption, random bit generator (Nana and Woafo 2015; Yalcin et al. 2004; Volos et al. 2012, 2013; Akgul et al. 2016; Pham et al. 2017b) and so on. The aim of this section is to design and simulate a proposed electronic circuit able to emulate the dynamical behavior of system (3a, 3b, 3c) in order to validate the numerical results.

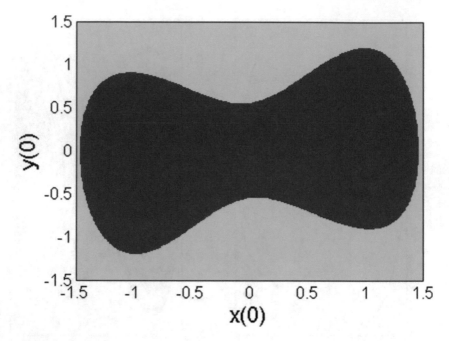

Fig. 6 Cross section of the basin of attraction of system (3a, 3b, 3c) in the xy-plane at z = 0 for $\beta = 2, \varepsilon = 1.0$ and $k = 1.7$. Periodic attractors are in light blue color and chaotic attractors are in red color

To this end, we propose the electronic circuit depicted in Fig. 7 which is designed according to the theoretical model of the proposed three-dimensional autonomous VdPD type oscillator described by system (3a, 3b, 3c).

The circuit consists of operational amplifiers configured as integrators and negative gain amplifiers, analog multipliers used to implement the different non-linearities, capacitors and resistors. Each state variable (x, y and z) of system (3a, 3b, 3c) is implemented as the output voltage of operational amplifiers U_2A, U_4A and U_6A, respectively. By applying the Kirchhoff's laws into the circuit of Fig. 7, we obtain its state equations given as follows

$$\frac{dV_x}{dt} = \frac{V_y}{RC} \tag{6a}$$

$$\frac{dV_y}{dt} = \frac{V_x}{RC} - \frac{V_x^3}{k_m R_b C} + \frac{V_y}{RC} - \frac{V_x^2 V_y}{R_e k_m} - \frac{V_z}{R_k C} \tag{6b}$$

$$\frac{dV_z}{dt} = \frac{V_y}{RC} - \frac{V_z}{RC} \tag{6c}$$

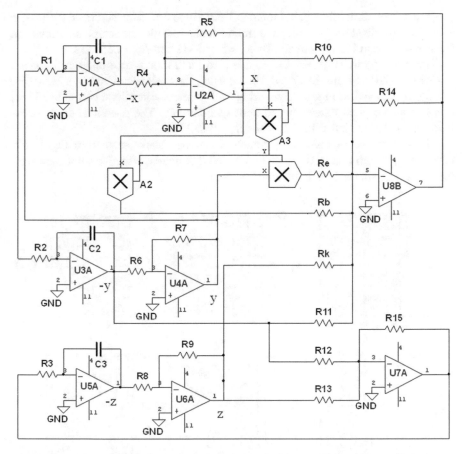

Fig. 7 The schematic diagram of the proposed three-dimensional autonomous VdPD type oscillator described by system (3a, 3b, 3c)

where $k_m = 100$ is a fixed constant introduced by the analog multipliers, $R = R_i$ ($i = 1, 2, 3, \ldots, 15$) and $C = C_i$ ($i = 1, 2, 3$). Using the following dimensionless states variables $x = V_x/1V$, $y = V_y/1V$, $z = V_z/1V$, $t = \tau RC$, system (7a, 7b, 7c) becomes

$$\dot{x} = y \tag{7a}$$

$$\dot{y} = x - \frac{R}{R_b k_m} x^3 + y - \frac{R}{R_e k_m} x^2 y - \frac{R}{R_k} z \tag{7b}$$

$$\dot{z} = y - z \tag{7c}$$

The system (6a, 6b, 6c) and (7a, 7b, 7c) are equivalent if and only if $\beta = R/k_m R_b$, $\varepsilon = R/k_m R_b$ and $k = R/R_k$. When $R = 10\,\text{k}\Omega$, $R_b = 5\,\text{k}\Omega$ and $R_e = 10\,\text{k}\Omega$, the proposed three-dimensional autonomous VdPD type oscillator described by system (3a, 3b, 3c) displays bistable and monostable chaotic attractors as shown in Figs. 8a1, a2 and b, respectively for $R_k = 5.263\,\mathrm{k\Omega}$ and $R_k = 5.882\,\mathrm{k\Omega}$.

The phase portraits of bistable double-scroll and monostable one-scroll chaotic attractors obtained from the Orcad-PSpice using the designed circuit of system (3a, 3b, 3c) is depicted in Fig. 8. Comparing with numerical results reported in Fig. 3, one can notice good agreement, at least qualitatively. The phenomenon of coexisting attractors is shown in Fig. 9 for $R_k = 5.882\,\mathrm{k\Omega}$.

The occurrence of coexisting attractors can be clearly seen from Fig. 9. By comparing, it with Fig. 5, it can be concluded that good qualitative agreement with the numerical simulations is obtained, as well.

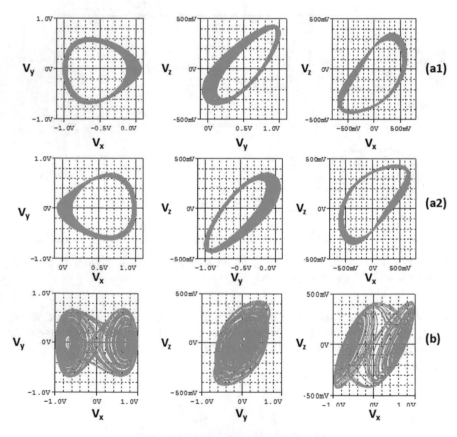

Fig. 8 The phase portraits of the chaotic attractors in planes (V_x, V_y), (V_y, V_z) and (V_x, V_z) for specific values of R_k (a) $R_k = 5.263\,\mathrm{k\Omega}$ and (b) $R_k = 5.882\,\mathrm{k\Omega}$. The initial conditions are setting as: **a1** $(V_x(0), V_y(0), V_z(0)) = (0.1, 0.1, 0.1)$, **a2** $(V_x(0), V_y(0), V_z(0)) = (-0.1, -0.1, -0.1)$ and **b** $(V_x(0), V_y(0), V_z(0)) = (0.1, 0.1, 0.1)$

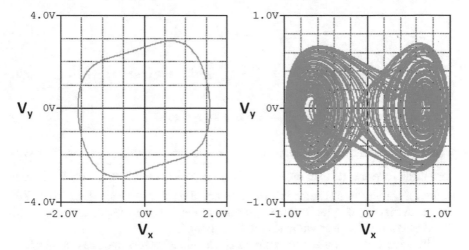

Fig. 9 The phase portraits in plane (V_x, V_y) of the coexisting attractors for specific values of $R_k = 5.882\,\text{k}\Omega$. The limit cycle is obtained with $(V_x(0), V_y(0), V_z(0)) = (1.5, 0, 0)$ while, the chaotic attractor one is obtained with $(V_x(0), V_y(0), V_z(0)) = (0.1, 0.1, 0.1)$

4   Synchronization of Identical Coupled Proposed Three-Dimensional Autonomous Van der Pol–Duffing Type Oscillators in Bistable Regime

With the intent of utilising the bistable one-scroll chaotic attractor property of the proposed three-dimensional autonomous VdPD type oscillator described by system (3a, 3b, 3c), synchronization between left one-scroll chaotic attractor as drive and right one-scroll chaotic attractor as response is introduced so as to enhance the safety factor of secure communication. In this section, the unidirectional linear error feedback coupling scheme is used to achieve chaos synchronization between two identical coupled proposed three-dimensional autonomous VdPD oscillators. The drive system is described by:

$$\frac{dx_1}{dt} = y_1 \tag{8a}$$

$$\frac{dy_1}{dt} = x_1 - \beta x_1^3 + \varepsilon\left(1 - x_1^2\right)y_1 - kz_1 \tag{8b}$$

$$\frac{dz_1}{dt} = y_1 - z_1 \tag{8c}$$

and the response system is given by:

$$\frac{dx_2}{dt} = y_2 + \kappa_1 (x_1 - x_2) \tag{9a}$$

$$\frac{dy_2}{dt} = x_1 - \beta x_2^3 + \varepsilon\left(1 - x_2^2\right) y_2 - k z_2 + \kappa_2 (y_1 - y_2) \tag{9b}$$

$$\frac{dz_2}{dt} = y_2 - z_2 + \kappa_3 (z_1 - z_2) \tag{9c}$$

where the constant parameters κ_1, κ_2 and κ_3 are the coupling strengths. The synchronization error is defined by $e(t) = \sqrt{(x_1 - x_2)^2 + (y_1 - y_2)^2 + (z_1 - z_2)^2}$, the aim of the synchronization scheme is to find the coupling strength such that $e(t) \to 0$ as $t \to \infty$. In Fig. 10, we plot the phase portraits of the drive and response systems and the synchronization error without coupling strength.

The drive and response systems display chaotic attractors and are not synchronized without coupling as shown in Fig. 10. It is interesting to explore the

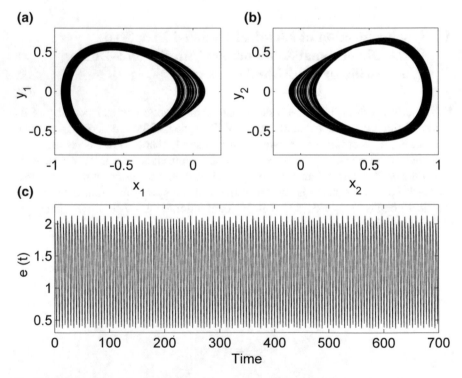

Fig. 10 The phase portraits of the drive (**a**) and response (**b**) systems and synchronization error (**c**) for $k = 1.85$, $\beta = 2$, $\varepsilon = 1$ and the coupling strength $\kappa_1 = \kappa_2 = \kappa_3 = 0$. The initial conditions of the drive and response systems are $(x_1(0), y_1(0), z_1(0)) = (0.1, 0.1, 0.1)$ and $(x_2(0), y_2(0), z_2(0)) = (-0.1, -0.1, -0.1)$ respectively

synchronization between the drive and response systems (8a, 8b, 8c) and (9a, 9b, 9c) by setting different coupling strengths and the results are plotted in Fig. 11 when the first and second variables are coupled respectively.

The drive and response systems (8a, 8b, 8c) and (9a, 9b, 9c) are synchronized, if the maximum synchronization errors of the three state variables become close to zero. In Fig. 11a, only the first variable is coupled ($\kappa_2 = \kappa_3 = 0$), one can observe that the drive and response systems (8a, 8b, 8c) and (9a, 9b, 9c) are synchronized for $\kappa_1 \geq 0.67$. While when only the second variable is coupled ($\kappa_1 = \kappa_3 = 0$), the drive and response systems (8a, 8b, 8c) and (9a, 9b, 9c) are synchronized for $\kappa_2 \geq 0.79$ as shown in Fig. 11b. The projection of the phase space trajectory of the response and drive systems (8a, 8b, 8c) and (9a, 9b, 9c) for a specific value of the coupling strengths are depicted in Fig. 12.

The drive and response systems (8a, 8b, 8c) and (9a, 9b, 9c) are synchronized for specific value of the coupling strengths as shown in Fig. 12. It is worth noting that when the third variable is set as linear coupling our investigation shown that the drive and response systems (8a, 8b, 8c) and (9a, 9b, 9c) are not synchronized.

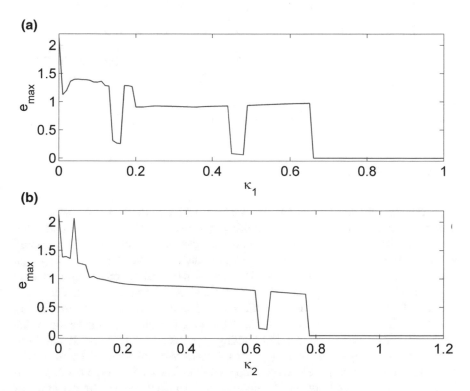

Fig. 11 Maximum synchronization error $e(t)$ versus the coupling strengths: κ_1 (**a**) and κ_2 (**b**) using the parameter values $k = 1.85$, $\beta = 2$, $\varepsilon = 1$

(a)

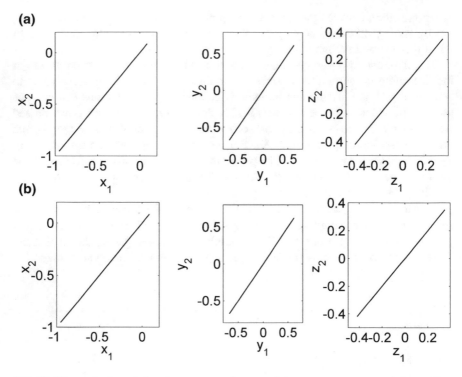

(b)

Fig. 12 The projection of the phase space trajectory of the response system and its auxiliary system for specific value of the coupling strengths: **a** $\kappa_1 = 0.7$, $\kappa_2 = \kappa_3 = 0$ and **b** $\kappa_2 = 0.9$, $\kappa_1 = \kappa_3 = 0$

5 Conclusion

This chapter reported results on the analysis and electronic implementation of proposed autonomous Van der Pol-Duffing type oscillator. This autonomous Van der Pol-Duffing type oscillator belongs to chaotic systems with self-excited attractors. For specific parameters, proposed autonomous Van der Pol-Duffing type oscillator exhibited interesting dynamics such as monostable double-scroll chaotic attractor, bistable one-scroll chaotic attractor and coexisting attractors. The coexistence of attractors, one- and double-scroll chaotic attractors obtained during numerical simulations have been confirmed using electronic implementation of the proposed autonomous Van der Pol-Duffing type oscillator. Finally, It has been found from the numerical simulations that chaos synchronization of identical unidirectional coupled proposed autonomous Van der Pol-Duffing type oscillators in bistable regime depends on the coupling coefficient as well as the selection of coupling variables.

References

Akgul A, Moroz I, Pehlivan I, Vaidyanathan S (2016) A new four–scroll chaotic attractor and its engineering applications. Optik 127:5491–5499

Benítez, M.S., Zuppa, L.A., and Guerra, R.J.R. (2006). Chaotification of the Van der Pol system using Jerk architecture. IEICE Trans. On Fundamentals E89-A:375–378

Chedjou J, Fotsin H, Woafo P, Domngang S (2001) Analog simulation of the dynamics of a Van der Pol oscillator coupled to a duffing oscillator. IEEE Trans Circuits Syst I Fundam Theory Appl 48:748–757

Cui J, Liang J, Lin Z (2016) Stability analysis for periodic solutions of the Van der Pol-Duffing forced oscillator. Phys Scr 91:015201–015208

Hilborn RC (2006) Chaos and nonlinear dynamics: an introduction for scientists and engineers. Oxford University Press

Kapitaniak T (1998) Chaos for engineers: theory, applications and control. Springer, New York

Kengne J, Nguomkam Negou A, Tchiotsop D, Kamdoum Tamba V, Kom GH (2018) On the dynamics of chaotic systems with multiple attractors: a case study, In: Kyamakya K, Mathis W, Stoop R, Chedjou J, Li Z (eds) Recent advances in nonlinear dynamics and synchronization. studies in systems, decision and control, vol 109. Springer, Cham, pp 17–32

Kingni ST (2014) Dynamics and Synchronization of VCSELs and SRLs subject to Current Modulation and Optical/Optoelectronic Feedback, Joint PhD thesis between University of Yaounde I, Cameroon and Vrije Universiteit Brussel, Belguim

Kingni ST, Talla Mbé JH, Woafo P (2012) Semiconductor lasers driven by self-sustained chaotic electronic oscillators and applications to optical chaos cryptography. Chaos 22:033108–033115

Kingni ST, Jafari S, Simo H, Woafo P (2014) Three-dimensional chaotic autonomous system with only one stable equilibrium: Analysis, circuit design, parameter estimation, control, synchronization and its fractional-order form. Eur Phys J Plus 129:76–91

Kingni ST, Nana B, Mbouna Ngueuteu GS, Woafo P, Danckaert J (2015) Bursting oscillations in a 3D system with asymmetrically distributed equilibria: mechanism, electronic implementation and fractional derivation effect. Chaos Solitons Fractals 71:29–40

Kozlov AK, Sushchik MM, Molkov YI, Kuznetsov AS (1999) Bistable phase synchronization and chaos in a system of coupled Van-der-Pol. Int J Bifurcat Chaos 9:2271–2277

Liu H, Wang X, Zhu Q (2011) Asynchronous anti-noise hyper chaotic secure communication system based on dynamic delay and state variables switching. Phys Lett A 375:2828–2835

Maccari A (2008) Vibration amplitude control for a Van der Pol-Duffing oscillator with time delay. J Sound Vib 317:20–29

Nana B, Woafo P (2015) Chaotic masking of communication in an emitter–relay–receiver electronic setup. Nonlinear Dyn 24:899–908

Nana B, Woafo P, Domngang S (2009) Chaotic synchronization with experimental application to secure communications. Commun Nonlinear Sci Numer Simul 14:2266–2276

Otto E (1993) Chaos in dynamical systems. Cambridge University Press

Pehlivan I, Uyarogglu Y (2012) A new 3D chaotic system with golden proportion equilibria: analysis and electronic circuit realization. Comput Electr Eng 38:1777–1784

Pham VT, Volos CK, Kapitaniak T (2017a) Circuitry realization. In: Systems with hidden attractors. springerbriefs in applied sciences and technology. Springer, Cham

Pham VT, Vaidyanathan S, Volos C, Tlelo-Cuautle E, Tahir FR (2017b). A memristive system with hidden attractors and its engineering application. In: Vaidyanathan S, Volos C (eds) Advances in memristors, memristive devices and systems. studies in computational intelligence, vol 701. Springer, Cham, pp 81–99

Rudowski W, Szemplinska-Stupnicka J (1997) The coexistence of periodic, almost-periodic and chaotic attractors in the Van der Pol-Duffing oscillator. J Sound Vib 199:165–175

Sprott JC (2010) Elegant chaos: algebraically simple Flow. World Scientific Publishing, Singapore

Tamaševičius A, Bumelienė S, Kirvaitis R, Mykolaitis G, Tamaševičiūtė E, Lindberg E (2009). Autonomous duffing-holmes type chaotic oscillator, electronics and electrical engineering, vol 93, Technologija, Kaunas, pp 43–46

Ueda Y, Akamatsu N (1981) Chaotically transitional phenomena in the forced negative-resistance oscillator. IEEE Trans CAS 28:217–224

Vincent UE, Odunaike RK, Laoye JA, Gbindinninuola AA (2011) Adaptive backstepping control and synchronization of a modified and chaotic Van der Pol-Duffing oscillator. J Control Theory Appl 9:141–145

Volos CK, Kyprianidis IM, Stouboulos IN (2012) A chaotic path planning generator for autonomous mobile robots. Robot Auto Syst 60:651–656

Volos CK, Kyprianidis IM, Stouboulos IN (2013) Image encryption process based on chaotic synchronization phenomena. Sig Process 93:1328–1340

Xiaofan W, Guanrong C (2000) Chaotification via arbitrarily small feedback controls: Theory, method, and applications. Int J Bifurcat Chaos 10:549–570

Yalcin ME, Suykens JAK, Vandewalle J (2004) True random bit generation from a double–scroll attractor. IEEE Trans Circuits Syst I 51:1395–1404

Dynamic Analysis, Electronic Circuit Realization of Mathieu-Duffing Oscillator and Its Synchronization with Unknown Parameters and External Disturbances

Victor Kamdoum Tamba, François Kapche Tagne,
Elie Bertrand Megam Ngouonkadi and Hilaire Bertrand Fotsin

Abstract This chapter deals with dynamic analysis, electronic circuit realization and adaptive function projective synchronization (AFPS) of two identical coupled Mathieu-Duffing oscillators with unknown parameters and external disturbances. The dynamics of the Mathieu-Duffing oscillator is investigated with the help of some classical nonlinear analysis techniques such as bifurcation diagrams, Lyapunov exponent plots, phase portraits as well as frequency spectrum. It is found that the oscillator experiences very rich and striking behaviors including periodicity, quasi-periodicity and chaos. An appropriate electronic circuit capable to mimic the dynamics of the Mathieu-Duffing oscillator is designed. The correspondences are established between the parameters of the system model and electronic components of the proposed circuit. A good agreement is obtained between the experimental measurements and numerical results. Furthermore, based on Lyapunov stability theory, adaptive controllers and sufficient parameter updating laws are designed to achieve the function projective synchronization between two identical drive-response structures of Mathieu-Duffing oscillators. The external disturbances are taken into account in the drive and response systems in order to verify the robustness of the proposed strategy. Analytical calculations and numerical simulations are performed to show the effectiveness and feasibility of the method.

Keywords Dynamic analysis · Mathieu-Duffing oscillator
Electronic circuit realization · Synchronization · Parameter identification
External disturbances

V. K. Tamba (✉) · F. K. Tagne
Department of Telecommunication and Network Engineering, IUT-Fotso,
Victor of Bandjoun, University of Dschang, P. O. Box: 134, Bandjoun, Cameroon
e-mail: vkamdoum@gmail.com

V. K. Tamba · E. B. M. Ngouonkadi · H. B. Fotsin
Laboratory of Electronics and Signal Processing (LETS), Department of Physics,
Faculty of Science, University of Dschang, P. O. Box: 67, Dschang, Cameroon

© Springer International Publishing AG 2018
V.-T. Pham et al. (eds.), *Nonlinear Dynamical Systems with Self-Excited and Hidden Attractors*, Studies in Systems, Decision and Control 133,
https://doi.org/10.1007/978-3-319-71243-7_8

181

1 Introduction

Several problems in physics, chemistry, biology, electronics, neurology and many other disciplines are related to nonlinear self-excited oscillators (Rajasekar et al. 1992). Examples include, the self-excited oscillations in bridges and airplane wings, the beating of a heart and the nonlinear model of a machine tool chatter, the vortex-or flow-induced oscillations in the cylinder of square cross-section and the galloping of transmission lines (Moon and Johnson 1998; Corless and Parkison, 1988, 1993; Yu et al. 1992, 1993a, b). Self-excited oscillators (e.g. Van der Pol, damped Duffing, Duffing-Van der Pol and Duffing-Rayleigh) have been intensively studied and demonstrated to exhibit complex and rich dynamical behaviors including harmonic, subharmonic and superharmonic frequency entrainment (Hayashi 1964), devil's staircase in the behavior of the winding number (Parlitz and Lauterborn 1987) and chaotic behavior with period-doubling cascades (Hayashi 1964; Parlitz and Lauterborn 1987; Guckenheimer and Holmes 1984; Steeb and Kunick 1987). A two-well Duffing oscillator with nonlinear damping term proportional to the power of velocity has been investigated in Anjali et al. (2012). The authors focused their attention on how the damping exponent affects the global dynamical behavior of the oscillator. Analytically, the threshold condition for the occurrence of homoclinic bifurcation using Melnikov technique is derived. The results were supported by numerical simulations. In Venkatesan and Lakshmanan (1997) the authors have demonstrated that a driven Duffing-van der Pol oscillator with a double well potential exhibits rich and striking bifurcation structures such as period-doubling phenomena, intermittencies, crises, transient chaos, and quasi-periodicity. Siewe Siewe and colleagues (Siewe Siewe et al. 2010) have investigated the dynamics of a Duffing-Rayleigh oscillator under harmonic external excitation. They used the Melnikov technique to derive the necessary conditions for chaotic motion of this deterministic system. The effect of damping parameter on phase portraits and Poincaré maps, in addition to the numerical simulations of bifurcation diagram and maximum Lyapunov exponents have been also examined. In Shen et al. (2008), the bifurcation and route to chaos of the Mathieu-Duffing oscillator have been reported using the incremental harmonic balance (IHB) procedure. The authors proposed a new scheme for selecting the initial conditions used for predicting the higher order periodic solutions. The phase portraits and bifurcation points obtained from the IHB method and numerical time-integration were compared yielding a very good agreement. Shen and Chen (2009) investigated the control of chaos in Mathieu-Duffing oscillator using open-plus-closed-loop (OPCL) method. A controller composed of an external excitation and a linear feedback has been designed to entrain chaotic trajectories of Mathieu-Duffing oscillator to its periodic and higher periodic orbits. The critical feedback coefficients under which the chaotic Mathieu-Duffing oscillator is globally and locally OPCL controllable respectively are obtained theoretically and demonstrated numerically. Many other interesting works (Yang et al. 2015; Wen et al. 2016, 2017) have been reported on the fractional-order form of the Mathieu-Duffing oscillator. The authors of these references studied the effects of the fractional-order on the dynamical behaviors of the integer-order Mathieu-Duffing oscillator. Numerical simulations are performed in their works to validate the theoretical investigations. Motivated by complex dynamical behaviors of self-excited oscillators and their potential applications in many fields, in this chapter, we investigate numerically and experimentally the dynamics and synchronization of a Mathieu-Duffing oscillator in presence of unknown parameters and external disturbances.

Since the idea of synchronization of chaotic systems was introduced by Pecora and Carroll in 1990 (Pecora and Carrol 1990), chaos synchronization has received an increasing attention due to its theoretical challenge and its potential applications in secure communications, chemical reactions, biological systems, information science, and plasma technologies (Zhan et al. 2003). Up to now, many types of synchronization phenomena have been reported. These include complete synchronization (Vincent et al. 2008), phase synchronization (Chitra and Kuriakose 2008), lag synchronization (Zhu and Wu 2004), anticipating synchronization (Zhu and Wu 2004), projective synchronization (Yang et al. 2010), modified projective synchronization (Zhu and Zhang 2009), function projective synchronization (FPS) (Li and Chen 2007), etc. In projective synchronization, the drive and the response systems synchronize up to a scaling factor whereas in modified projective synchronization, the response of the synchronized dynamical state variables synchronizes up to a constant matrix (Kareem et al. 2012). Recently, a more general form of projective synchronization called function projective synchronization (An and Chen 2009; Ping and Yu-Xia 2010) in which drive and response systems are synchronized up to a desired scaling function has attracted much attention of scientists and engineers as it provides more security in its applications to secure communication because the unpredictability of the scaling function matrix. Also, FPS of discrete chaotic systems has now been widely investigated for its great practical application (Fei et al. 2013). Therefore, the research on FPS is more valuable in practice. The majority of the mentioned works are carried out by using the known (certain) parameters of drive and response systems, and the controller is constructed from those known parameters. However, some system's parameters may not be exactly known in advance. In real physical systems, or experimental situations, chaotic systems may have some uncertain or time varying parameters (Mahmoud and Mansour 2011). Moreover, the influence of the uncertainties has been taken into account rarely. It is known that in the real world applications (e.g. secure communication), the systems are affected by various uncertainties including parameter perturbations and external disturbances, which can influence the accuracy of the communication. To our understanding, function projective synchronization of Mathieu-Duffing oscillators with the consideration of unknown parameters and external disturbances has not been explored. The objectives of this chapter are threefold: (a) to consider the dynamics of the Mathieu-Duffing oscillator and investigate its bifurcation structures with particular emphasis on the effects of the amplitude of the parametric excitation; (b) to carry out an experimental study of the dynamics of the system in order to validate the theoretical and numerical results;

and (c) to investigate the synchronization of such a coupled oscillators with unknown parameters and subjected to the external disturbances.

The layout of chapter is as follows. Section 2 deals with the analytical and numerical analysis of the system under study. Some basic properties and bifurcation structures of the system are investigated. The experimental study is carried out in Sect. 3. The laboratory experimental measurements show a qualitative agreement with numerical results. Section 4 deals with FPS between two identical Mathieu-Duffing oscillators in presence of unknown parameters and external disturbances. Numerical simulations are performed in order to illustrate and verify the effectiveness, feasibility and the robustness of the synchronization scheme. Finally, we summarize our results and draw the conclusions of this chapter in Sect. 5.

2 Theoretical Analysis of Mathieu-Duffing Oscillator

2.1 Description of the Model

In this chapter, we consider the Mathieu-Duffing oscillator (Shen et al. 2008) which is described by the following equation of motion:

$$\ddot{x} + 2\varepsilon\dot{x} - (\alpha + \beta \sin \omega t)x + \gamma x^3 = 0 \tag{1}$$

in which $\dot{x} = dx/dt$ represents the derivative with respect to time, ε is the damping coefficient, β and ω are respectively, the amplitude and the frequency of the parametric excitation, α and γ represent respectively, the linear and nonlinear stiffness coefficients. Many mechanical and engineering problems can be really described by Eq. (1). Indeed, it has been used to model the one-mode transverse vibration of the axially moving beam with harmonic fluctuated speed (Shen et al. 2008). The second-order differential Eq. (1) can be transformed into a set of first-order differential equations as follows:

$$\begin{cases} \dot{x} = y \\ \dot{y} = -2\varepsilon y + \alpha x - \gamma x^3 + \beta x \sin \omega t \end{cases} \tag{2}$$

System (2) involves five independent parameters. Due to the relatively large number of parameters, the detailed influence of each parameter on dynamics of the system (2) will be not presented here. The bifurcation structure will be carried out with respect to the amplitude of the parametric excitation because this parameter can be easily varied in the practical situation using a low frequency generator. For the numerical analysis, the following values of parameters will be employed: $\varepsilon = 0.125$, $\alpha = \gamma = 1$, $\omega = 2$ and β variable.

2.2 Dissipativity and Symmetry

To generate chaotic signal, it is necessary for the system to be dissipative. The divergence of system (2) in absence of the external force is evaluated as

$$\nabla V = \frac{\partial \dot{x}}{\partial x} + \frac{\partial \dot{y}}{\partial y} = -2\varepsilon \tag{3}$$

System (2) is dissipative since $\nabla V < 0$. This implies that any volume element $V_0 = V(t=0)$ will be continuously contracted by the flow (i.e. each volume element containing the trajectory shrinks to zero as time evolves to infinity). Then, all system orbits will be confined to a specific bounded subset of zero volume in state space and the asymptotic dynamics settles onto an attractor. The symmetry is one of the interesting characteristics of the dynamical system. This property commonly exists in many nonlinear systems. It is easy to check in absence of external force that system (2) has a natural symmetry since the transformation S: $(x,y) \leftrightarrow (-x,-y)$ is invariant for a specific set of the system parameters. The solution of system (2) that is invariant under the above transformation is called a symmetry solution; otherwise it is called an asymmetry solution.

2.3 Fixed Points Analysis

In absence of external force and by setting the right hand side of system (2) to zero, it is found that there are three equilibrium points $E_1(0,0)$ and $E_{2,3}(\pm\sqrt{\alpha/\gamma},0)$. The characteristic equation obtained at any equilibrium point $E(\bar{x},\bar{y})$ is defined as

$$\lambda^2 + 2\varepsilon\lambda - (\alpha - 3\gamma\bar{x}) = 0 \tag{4}$$

The characteristic equation for the equilibrium point $E_1(0,0)$ is

$$\lambda^2 + 2\varepsilon\lambda - \alpha = 0 \tag{5}$$

It is obvious that $E_1(0,0)$ is always unstable provided that the corresponding characteristic Eq. (5) has coefficients with different signs. For the analysis of stability of the equilibrium points $E_{2,3}(\pm\sqrt{\alpha/\gamma},0)$, one only needs to consider since the system is invariant under the transformation $(x,y) \leftrightarrow (-x,-y)$ as mentioned above. Thus, the characteristic equation associated to one of them is defined as follows

$$\lambda^2 + 2\varepsilon\lambda + 2\alpha = 0 \tag{6}$$

Since ε and α are positive, the equilibrium points $E_{2,3}(\pm\sqrt{\alpha/\gamma},0)$ are stable.

2.4   Bifurcation and Chaos

In the numerical results that follow, we investigate the dependence of the system behavior at given angular frequency, linear and nonlinear stiffness coefficients by varying the amplitude of the parametric excitation. The bifurcation diagram shows the projection of the attractors in the Poincaré section onto one of the system coordinates with respect to the chosen control parameter. In order to gain further insight about the dynamics of the oscillator under investigation, we compute the frequency spectrum as well as the largest Lyapunov exponent with the help of the algorithm proposed by Wolf et al. (1985). These results are obtained by solving system (2) with aid of the standard fourth-order Runge Kutta algorithm (Press et al. 1992). The system is integrated for sufficiently long time and the transient is cancelled. The bifurcation diagram showing the local maxima of the coordinate y and the corresponding graph of the largest Lyapunov exponent in terms of the control parameter β varying in the range $3.6 \leq \beta \leq 6$ are provided in Fig. 1 for $\varepsilon = 0.125$, $\alpha = \gamma = 1$ and $\omega = 2$.

In light of Fig. 1a, the extreme sensitivity of the oscillator with respect to small parameter changes is clearly observed. Some interesting dynamical behaviors such

Fig. 1 Bifurcation diagram **a** showing the local maxima of the coordinate y and the corresponding graph of the largest Lyapunov exponent **b** in terms of the control parameter β for $\varepsilon = 0.125$, $\alpha = \gamma = 1$ and $\omega = 2$

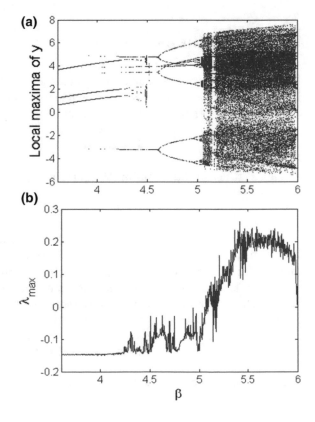

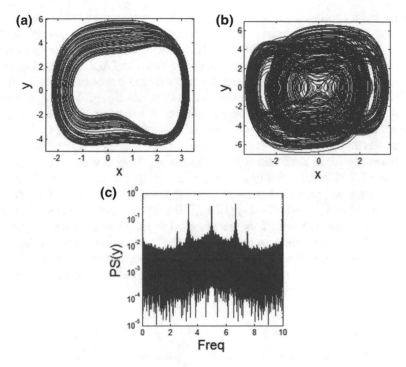

Fig. 2 Chaotic phase portraits of Mathieu-Duffing oscillator computed for $\varepsilon = 0.125$, $\alpha = \gamma = 1$ and $\omega = 2$. **a** Single band chaotic attractor for $\beta = 5.2$, **b** double band chaotic attractor and **c** frequency spectrum of coordinate y for $\beta = 5.8$

as periodicity, quasi-periodicity and chaos are visible. The bifurcation structure is perfectly traced by the largest Lyapunov exponent. Accordingly, some phase portraits showing chaotic states with corresponding frequency spectrum of the system are depicted in Fig. 2.

Asymmetric chaotic attractor is observed in Fig. 2a while a double band strange attractor is depicted in Fig. 2b. The broadband noise-like of frequency spectrum (see Fig. 2c) is signature of the chaotic steady state.

3 Electronic Circuit Realization of Mathieu-Duffing Oscillator

Implementing the theoretical chaotic models using electronic circuits is of great importance for various engineering applications such as robotics, chaos based communications, image encryption and random number generation (Banerjee 2010; Volos et al. 2012, 2013a, b). Moreover, the electronic circuit realization of theoretical chaotic models is an effective approach to investigate the dynamics of such systems via for instance the experimental bifurcation diagram obtained by varying the values of variable resistors associated to the control bifurcation parameters (Ma et al. 2014; Buscarino et al. 2009). The dynamics of the system under scrutiny has been investigated in preceding paragraphs using theoretical and numerical methods. It is predicted that the system can exhibit very rich and complex behaviors. In this section, in order to validate the numerical results, we design and implement an electronic circuit capable to mimic the dynamical behaviors of system (2). The schematic diagram of the proposed electronic circuit is depicted in Fig. 3.

The electronic circuit of Fig. 3 consists of some analog multipliers used to implement the cubic nonlinear term of the model. They operate over a dynamic range of ± 1 V with typical tolerance less than 1%. The output signal (W) is connected to those at inputs $(+X_1)$, $(-X_2)$, $(+Y_1)$, $(-Y_2)$, and $(+Z)$ by the following expression $W = (X_1 - X_2)(Y_1 - Y_2)/10 + Z$. The operational amplifiers accompanied with resistors and capacitors are exploited to implement the basic operations such as addition, subtraction and integration. The bias is provided by a 15 Volts DC symmetry source. Using the Kirchhoff's laws into the circuit of Fig. 3, we obtain its mathematical model given by two coupled first-order nonlinear differential equations

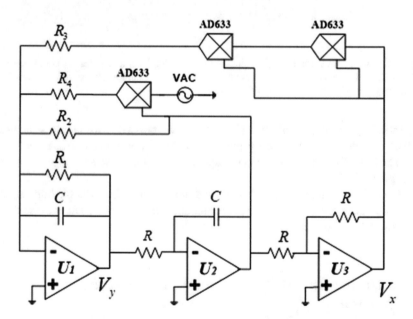

Fig. 3 Electronic circuit realization of the Mathieu-Duffing oscillator. The value of electronic components are fixed as $C = 10\,\text{nF}$, $R = 10\,\text{k}\Omega$, $R_1 = 40\,\text{k}\Omega$, $R_2 = 100\,\Omega$, $R_3 = 100\,\Omega$ and R_4 variable. The analog multipliers devices are AD633JN-type while operational amplifiers (U_1, U_2 and U_3) are TL084CN-type ones

$$\begin{cases} \dfrac{dV_x}{dt} = \dfrac{V_y}{RC} \\[3mm] \dfrac{dV_y}{dt} = -\dfrac{V_y}{R_1 C} + \dfrac{V_x}{R_2 C} - \dfrac{V_x^3}{10 k_m R_3 C} + \dfrac{V_x \sin \omega t}{k_m R_4 C} \end{cases} \tag{7}$$

where V_x and V_y are the output voltages of the operational amplifiers and $k_m = 10$ is a constant introduced by the analog multiplier. The values of components of electronic circuit in Fig. 3 are chosen in order to match system (2) and according to the following change of state variables and parameters: $t = \tau RC$; $x = V_x/1\ V$; $y = V_y/1\ V$; $2\varepsilon = R/R_1$; $\alpha = R/R_2$; $\gamma = R/10 k_m R_3$, $\beta = R/k_m R_4$ as follows: $C = 10\ \text{nF}$, $R = 10\ \text{k}\Omega$, $R_1 = 40\ \text{k}\Omega$, $R_2 = 10\ \text{k}\Omega$, $R_3 = 100\ \Omega$ and R_4 variable.

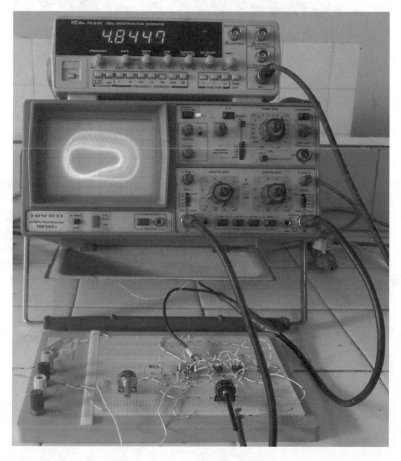

Fig. 4 Photograph of the analog oscilloscope displaying a single band chaotic attractor obtained from the electronic circuit of Fig. 3

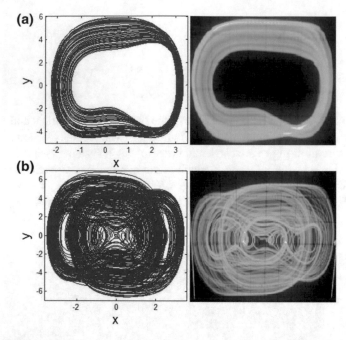

Fig. 5 Experimental phase portrait obtained from the Mathieu-Duffing oscillator using a dual-trace oscilloscope in XY mode. Corresponding numerical phase portraits are shown in the left. Output voltages V_x and V_y are fed to the X and Y input, respectively. **a** Single-band chaotic attractor for $R_4 = 1.92\,k\Omega$ and **b** double-band chaotic attractor for $R_4 = 1.72\,k\Omega$

A photograph of the analog oscilloscope displaying a single band chaotic attractor obtained from the electronic circuit of Fig. 3 is shown in Fig. 4.

The experimental results showing some dynamical behaviors of electronic circuit of Fig. 3 for some specific values of control parameter R_4 are shown in Fig. 5.

In light of the pictures in Fig. 5, one can note the good similarity of experimental portraits with those obtained numerically. This shows that the proposed electronic circuit is capable to reproduce the dynamics of the system under investigation. It should be stressed that system (2) can be also implemented using many other techniques such as integrated circuit technology (Trejo-Guerra et al. 2012), Field Programmable Analog Array (FPAA) technologies (Koyuncu et al. 2014) and Field

Programmable Gate Array (FPGA) (Fatma et al. 2016). The latter technology provides a fast prototype for investigating chaotic systems.

4 Adaptive Function Projective Synchronization of Two Identical Coupled Mathieu-Duffing Oscillators

In this section, we consider the problem of synchronization of Mathieu-Duffing oscillators with unknown parameters and external disturbances using adaptive function projective synchronization technique.

4.1 Problem Formulation

Let the drive and response systems be defined as

$$\dot{x} = F_1(t, x) + G_1(t, x)\varphi + D'(t) \tag{8}$$

$$\dot{y} = F_2(t, y) + G_2(t, y)\theta + D''(t) + u(t, x, y) \tag{9}$$

where $x, y \in R^n$, are the state variables of the drive and response systems, respectively, $F_1(t, x)$, $F_2(t, y)$; $R^n \to R^n$, $G_1(t, x) \in R^{n \times p}$, $G_2(t, y) \in R^{n \times q}$ are the continuous nonlinear functions, $\varphi \in R^p$, $\theta \in R^q$ are the unknown parameters of the drive and response system respectively, $D'(t) = [d_{11}, d_{12}, \ldots, d_{1n}]^T \in R^n$ and $D''(t) = [d_{21}, d_{22}, \ldots, d_{2n}]^T \in R^n$ represent the disturbance inputs with $|d_{1i}| \leq \lambda_i, (i = 1, 2, \ldots, n)$ and $|d_{2i}| \leq \lambda_i, (i = 1, 2, \ldots, n)$, and assume that $\lambda_i \geq 0$ are given, $\lambda_i = [\lambda_1, \lambda_2, \ldots, \lambda_i]^T$ and $u(t, x, y)$ is the control function to be determined. Let us define the error state between the drive (8) and response (9) systems as follows

$$e(t) = x(t) - m(t)y(t) \tag{10}$$

where $m(t)$ is a continuously differentiable bounded function with $m(t) \neq 0$ for all t. The objective is to synchronize both drive and response systems to a scaling function $m(t)$ in presence of unknown parameters and external disturbances such that the error system (10) can be asymptotically stable at the zero equilibrium, i.e. $\|e(t)\| \to 0$ as $t \to \infty$.

Remark 1 If the scaling function $m(t)$ is a constant different from 1, the problem of function projective synchronization becomes projective synchronization. In the cases that $m(t) = 1$ and $m(t) = -1$, it turns out to be complete synchronization and antisynchronization, respectively (Chen and Li. 2007).

4.2 Main Results

Here, we suppose that the parameters in the drive system or response system are unknown. From Eq. (10), we can obtain the error dynamical system as

$$\dot{e}(t) = \dot{y}(t) - m(t)\dot{x}(t) - \dot{m}(t)x(t) \tag{11}$$

By substituting systems (8) and (9), we obtain

$$\begin{aligned}
\dot{e}(t) &= F_2(t, y) + G_2(t, y)\theta - m(t)[F_1(t, x) + G_1(t, x)\varphi + D^{'}(t)] \\
&\quad - \dot{m}(t)x(t) + D^{''}(t) + u(t, x, y)
\end{aligned} \tag{12}$$

Theorem 1 *For the given scaling function $m(t)$, the adaptive FPS between drive system (8) and response system (9) can be achieved by the control function (13) and sufficient parameters update laws (14) and (15) as below*

$$\begin{aligned}
u(t, x, y) &= m(t)[F_1(t, x) + G_1(t, x)\hat{\varphi}] - F_2(t, y) \\
&\quad - G_2(t, y)\hat{\theta} + \dot{m}(t)x - ke - H(t, e)\rho
\end{aligned} \tag{13}$$

$$\dot{\hat{\varphi}} = -G_1^T(t, x)m(t)e \tag{14}$$

$$\dot{\hat{\theta}} = G_2^T(t, y)e \tag{15}$$

where $H(t, e) = \tanh[m(t)(e_1, e_2, \ldots, e_n)]$. In Eq. (15), $\hat{\varphi}$ and $\hat{\theta}$ are estimated values of unknown parameters φ and θ of the drive and the response system, respectively; $\tanh(.)$ denotes the tangent hyperbolic function; $\rho = [\rho_1, \rho_2, \ldots, \rho_n]^T$ is the boundaries of the uncertainties and $k = diag(k_1, k_2, \ldots, k_n)$ is a gain matrix for each controller. The desired convergence rate can be adjusted by the gain matrix k.

Remark 2 In the control function (13), $H(t, e)\rho$ is a compensation term which is introduced to eliminate the influence of the disturbance inputs.

It is interesting to note that the conventional control methods often use the sign function (Hongyue et al. 2011; Fu 2012; Srivastava et al. 2013), but the discontinuity of the sign function causes the chattering and undesirable oscillations. In order to avoid these problems, in this chapter the discontinuous sign function is replaced by the continuous tangent hyperbolic function.

Proof Let $\tilde{\varphi} = \varphi - \hat{\varphi}$ and $\tilde{\theta} = \theta - \hat{\theta}$ be the parameter estimation errors. Choose the storage Lyapunov function of system (11) as

$$V = \frac{1}{2}\left(e^T e + \tilde{\varphi}^T \tilde{\varphi} + \tilde{\theta}^T \tilde{\theta}\right) \tag{16}$$

Thus, the time derivation of the Lyapunov function V along the trajectory of the error dynamics system (11) with following notations ($\dot{\tilde{\varphi}} = -\dot{\hat{\varphi}}$ and $\dot{\tilde{\theta}} = -\dot{\hat{\theta}}$) is

$$
\begin{aligned}
\dot{V} &= e^T \dot{e} + \tilde{\varphi}^T \dot{\tilde{\varphi}} + \tilde{\theta}^T \dot{\tilde{\theta}} \\
&= e^T \left[G_2(t,y)(\theta - \hat{\theta}) - m(t)G_1(t,x)(\varphi - \hat{\varphi}) - H(t,e)\rho \right] \\
&\quad + \tilde{\varphi}^T \left[G_1^T(t,x)m(t)e \right] + \theta^T \left[-G_2^T(t,y)e \right] - e^T k e - m(t)D'(t)e^T \\
&\quad + D''(t)e^T = -e^T k e - e^T H(t,e)\rho - m(t)D'(t)e^T + D''(t)e^T
\end{aligned}
\tag{17}
$$

Let

$n_1 = e^T[D''(t) - m(t)D'(t)]$ and $n_2 = e^T H(t,e)\rho$ where n_1, $n_2 \in R$ and $n_2 \geq 0$. According to the definition and assumption of $D'(t)$, $D''(t)$, φ and θ, it is guaranteed that $n_1 \leq n_2$, i.e. $n_1 - n_2 \leq 0$, then $\dot{V}$ is written as

$$
\dot{V} = -e^T k e + n_1 - n_2 \leq -e^T k e \leq 0
\tag{18}
$$

Provided that $\dot{V}$ is negative semi-definite, and since V is positive definite, it follows that $e \in L_\infty$, φ, $\theta \in L_\infty$. Thus $\dot{e} \in L_\infty$, and according to Eq. (11), it can be obtained that

$$
\int_0^t \|e\|^2 dt = \int_0^t e^T e \, dt \leq -\frac{1}{l} \int_0^t \dot{V} \, dt = \frac{1}{l}[V(0) - V(t)] \leq \frac{1}{l} V(0)
\tag{19}
$$

Since $V(0) \leq \infty$ and $e \in L_2$, according to Barbalat's lemma, we have $\|e(t)\| \to 0$ as $t \to \infty$, i.e. the error dynamical system (11) will be stabilized at the zero equilibrium asymptotically. Thus, according to the Lyapunov stability theorem, the adaptive function projective synchronization between drive system (8) and response system (9) in presence of unknown parameters and external disturbances is achieved under the control function (13) and sufficient parameter update laws (14) and (15). However, we cannot conclude that the unknown parameters can be automatically estimated to their true values. The unknown parameters should be almost constant in some bounded interval (i.e. $\dot{\hat{\varphi}} = 0$ and $\dot{\hat{\theta}} = 0$). Another sufficient condition to guarantee the parameter identification based on the Linear Independence (LI) condition which is elaborated as follows. Using the Lasalle's Invariant Set Theorems (Zhiyong et al. 2012), the largest invariant set M can be obtained as

$$
M = \{ e \in R^n, \varphi \in R^p, \theta \in R^q \mid, e = 0, G_2(t,y)\tilde{\theta} - m(t)G_1(t,x)\tilde{\varphi} - ke = 0 \}
\tag{20}
$$

Thus one can get $G_2(t,y)\tilde{\theta} - m(t)G_1(t,x)\tilde{\varphi} = 0$. To ensure that this equation has the unique solution of $\tilde{\varphi} = 0$ and $\tilde{\theta} = 0$ (which implies that the unknown parameters are estimated to their true values as $\hat{\varphi} = \varphi$, $\hat{\theta} = \theta$), the following condition (Linear Independence condition) should be satisfied. To achieve synchronization based on parameter identification of systems (8) and (9) with unknown parameters, the function elements in the function vector groups $-G_1^T(t,x)m(t)$ and $G_2^T(t,y)$ should be linearly independent on the synchronization manifold. Interested readers would consult (Yu et al. 2007) for more discussions. This completes the proof.

4.3 Application to Mathieu-Duffing Oscillator

For application we consider that the parameters of the drive system are known and those of the response system are unknown. With these considerations, the drive system is given as

$$\begin{cases} \dot{x}_1 = x_2 + d_1'(t) \\ \dot{x}_2 = -2\varepsilon x_2 + \alpha x_1 - \gamma x_1^3 + \beta x_1 \sin(\omega t) + d_2'(t) \end{cases} \tag{21}$$

and the response system is given as

$$\begin{cases} \dot{y}_1 = y_2 + d_1''(t) + u_1(t,x,y) \\ \dot{y}_2 = -2\hat{\varepsilon} y_2 + \hat{\alpha} y_1 - \hat{\gamma} y_1^3 + \hat{\beta} y_1 \sin(\omega t) + d_2''(t) + u_2(t,x,y) \end{cases} \tag{22}$$

Based on Theorem 1, the control functions and and parameter update laws are determined by

$$u_1(t,x,y) = m(t)x_2 - y_2 + \dot{m}(t)x_1 - ke_1 - H(t,e_1) \tag{23a}$$

$$\begin{aligned} u_2(t,x,y) = m(t)[-2\varepsilon x_2 + \alpha x_1 - \gamma x_1^3 + \beta x_1 \sin \omega t] + 2\hat{\varepsilon} y_2 - \hat{\alpha} y_1 + \hat{\gamma} y_1^3 \\ - \hat{\beta} y_1 \sin \omega t + \dot{m}(t)x_2 - ke_2 - H(t,e_2) \end{aligned} \tag{23b}$$

and

$$\begin{cases} \dot{\hat{\varepsilon}} = -y_2 e_2 \\ \dot{\hat{\alpha}} = y_1 e_2 \\ \dot{\hat{\gamma}} = -y_2^3 e_2 \\ \dot{\hat{\beta}} = y_1 \sin(\omega t)e_2 \end{cases} \tag{24}$$

In what follows, numerical simulations are given to verify the feasibility and the robustness of the proposed methods. The standard fourth-order Runge-Kutta method is applied to solve the differential equations describing the drive (21) and the response (22) systems with time step size equal to 0.005. The parameters values of the drive system are selected as $\varepsilon = 0.125$, $\alpha = 1$, $\gamma = 1$, $\omega = 2$ and $\beta = 5.8$ so that it exhibits chaotic behaviors. The initial states are chosen as $x(0) = [0.1, 0.2]$ and $y(0) = [0.3, 0.4]$, respectively for the drive and response systems. The initial values of unknown parameters are set to be $\hat{\varepsilon}(0) = 0.006$, $\hat{\alpha}(0) = 0.015$, $\hat{\gamma}(0) = 0.07$ and $\hat{\beta}(0) = 0.08$. The control gains are set as $k_i = 20 (i = 1, 2)$. The scaling function is selected as $m(t) = 0.6 + 0.1 \sin(0.15\pi t)$. In order to verify the robustness of the method, we perform the numerical simulations in three cases (i) without external disturbances, (ii) with continuous time varying (sinusoidal type) and (iii) with white Gaussian noise. For the first case, the time response of the error system (11) and the synchronization quality which is defined by $e = \sqrt{e_1^2 + e_2^2}$ are shown in Fig. 6.

In Fig. 7, we show the time evolution of the parameter estimations in the response system.

Obviously, the synchronization errors converge to zero with exponentially asymptotical speed and two systems with different initial states achieve FPS very quickly. The unknown parameters of the response system are simultaneously successfully estimated to their true values.

For the second case, the external disturbances subjected to the drive and response systems are selected as $d_1' = 0.1 \cos(0.2\pi t)$, $d_2' = 0.2 \sin(0.3\pi t)$ and $d_1'' = 0.1 \sin(0.2\pi t)$, $d_2'' = 0.2 \cos(0.3\pi t)$. The boundaries of the uncertainties are chosen randomly as $\rho_1 = 0.8$ and $\rho_2 = 0.5$. The time response of the error system (11) and the synchronization quality are depicted in Fig. 8.

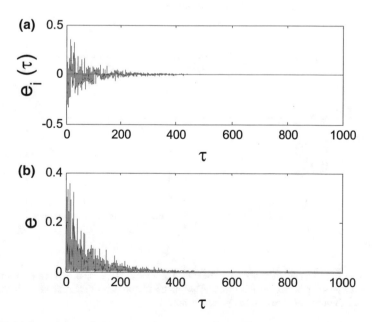

Fig. 6 Time dependence of the errors dynamics e_i $(i = 1, 2)$ (**a**) and synchronization quality (**b**) between two coupled identical Mathieu-Duffing oscillators with FPS scheme without external disturbances

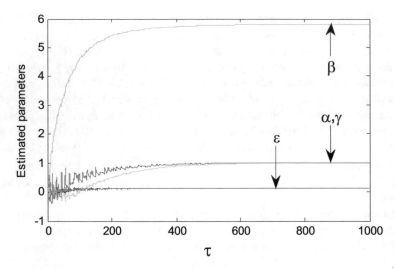

Fig. 7 Time evolution of the parameter estimations in the response system with FPS scheme without external disturbances

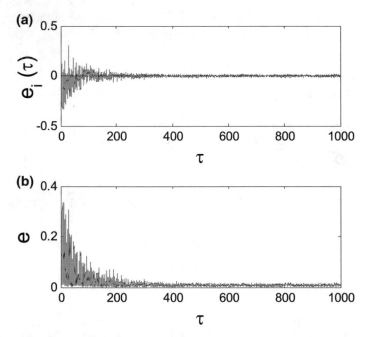

Fig. 8 Time dependence of the errors dynamics e_i $(i = 1, 2)$ (**a**) and synchronization quality (**b**) between two coupled identical Mathieu-Duffing oscillators with FPS scheme with continuous time varying (sinusoidal type) external disturbances

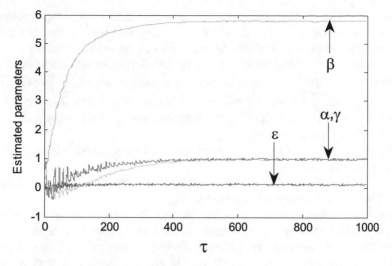

Fig. 9 Time evolution of the parameter estimations in the response system with FPS scheme with continuous time varying (sinusoidal type) external disturbances

The time evolution of the parameter estimations in the response system subjected to a continuous time varying (sinusoidal type) external disturbances are depicted in Fig. 9.

As can be seen from those figures, the synchronization errors and synchronization quality arrive at zero in finite time and the unknown parameters in the

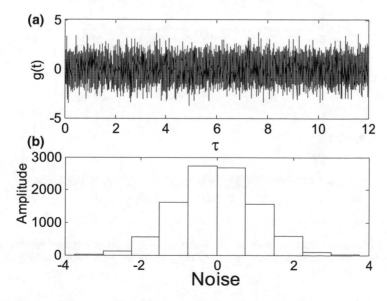

Fig. 10 Time evolution of a white Gaussian noise (**a**) and its histogram in the range [−4, 4] (**b**)

response system have been estimated to their true values in spite of the presence of the external disturbances. The effect of the external disturbances is clearly visible on the errors dynamics as well as on the estimation of the unknown parameters.

In the last case, external disturbances subjected to drive and response systems are the white Gaussian noises chosen as $d'_1 = d''_2 = \sqrt{(-2 \ln \lambda)} \sin(2\pi\lambda)$ and $d'_2 = d''_1 = \sqrt{(-2 \ln \lambda)} \cos(2\pi\lambda)$ where λ is the random function. The control gains are set as $k_i = 10$ $(i = 1, 2)$. In Fig. 10, we show the white Gaussian noise and its histogram in the range $[-4, 4]$.

The time response of the error system (11) and the synchronization quality are displayed in Fig. 11.

The time evolution of the parameter estimations in the response system subjected to a white Gaussian noise are depicted in Fig. 12.

One can see from those figures that the synchronization errors and synchronization quality converge to zero and the unknown parameters in the response system have been estimated approximatively to their true values. The effect of the external disturbances is more visible on the errors dynamics as well as on the estimation of the unknown parameters. All these results demonstrate that the FPS in the coupled identical Mathieu-Duffing oscillators via control functions (23) and

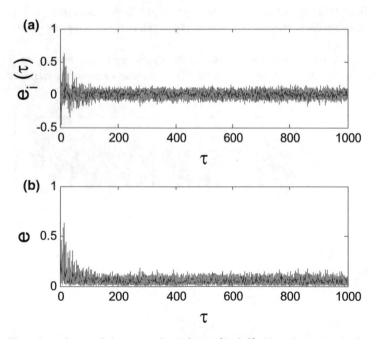

Fig. 11 Time dependence of the errors dynamics e_i $(i = 1, 2)$ (**a**) and synchronization quality (**b**) between two coupled identical Mathieu-Duffing oscillators with FPS scheme with white Gaussian external noise

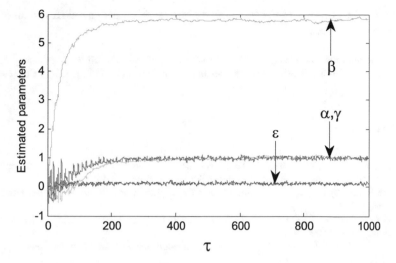

Fig. 12 Time evolution of the parameter estimations in the response system with FPS scheme with white Gaussian external noise

parameter update laws (24) is obtained with estimation of the unknown parameters of the response system and in presence of external uncertainties at the desired scaling function $m(t)$.

5 Concluding Remarks

The dynamics of a Mathieu-Duffing oscillator has been investigated numerically and experimentally in this chapter. The dynamical properties of the oscillator have been examined using classical nonlinear analysis techniques such as bifurcation diagram, plot of largest Lyapunov exponent and frequency spectrum. It was found from the bifurcation structure that the system experiences very rich and complex behaviors including periodicity, quasi-periodicity and chaos. An experimental study has been carried out and the laboratory experimental measurements were in a good qualitative agreement with numerical results. Furthermore, using the Lyapunov stability theory, we have designed adaptive controllers and sufficient parameter update laws able to achieve the function projective synchronization between two identical drive-response structures of Mathieu-duffing oscillators with unknown parameters. We have introduced the external disturbances in the drive and response systems in order to verify the robustness of our proposed strategy. It has been noted that the unpredictable properties of the scaling function $m(t)$ can additionally enhance the security of the communication. Theoretical results and numerical simulations were finally included to visualize the effectiveness and feasibility of the developed methods. We stress also that the approach followed in this chapter may be exploited rigorously to the study of any other nonlinear dynamical system driven by an external force.

Acknowledgements V. Kamdoum Tamba wishes to thank Dr. Sifeu Takougang Kingni (University of Maroua, Cameroon) for interesting discussions and careful reading of the chapter.

References

An H-L, Chen Y (2009) The function cascade synchronization scheme for discrete-time hyperchaotic systems. Commun Nonlinear Sci Numer Simul 14:1494–1501

Banerjee R (2010) Chaos Synchronization and Cryptography for Secure communications. IGI Global, USA

Buscarino A, Fortuna L, Frasca M (2009) Experimental robust synchronization of hyperchaotic circuits. Phys D 238:1917–1922

Chen Y, Li X (2007) Function projective synchronization between two identical chaotic systems. Int J Mod Phys C 18:2246–2255

Chitra RN, Kuriakose VC (2008) Phase synchronization in an array of driven Josephson junctions. Chaos 18:0131251

Corless RM, Parkinson GV (1988) A model of the combined effects of vortex-induced vibration and galloping. J Fluids Struct 2:203–220

Corless RM, Parkinson GV (1993) A model of the combined effects of vortex-induced vibration and galloping part II. J Fluids Struct 7:825–848

Dalkiran Fatma Yildirim, Sprott JC (2016) Simple chaotic hyperjerk system. Int J Bifurcat Chaos 26:1650189

Fei Y, Chunhua W, Qiuzhen W, Yan H (2013) Complete switched modified function projective synchronization of a five-tern chaotic system with uncertain parameters and disturbances. Pramana J Phys 80:223–235

Fu G (2012) Robust adaptive modified function projective synchronization of hyperchaotic systems subject to external disturbance. Commun Nonlinear Sci Numer Simul 17:2602–2608

Guckenheimer J, Holmes PJ (1984) Nonlinear oscillations, dynamical systems, and bifurcations of vector fields. Springer, Berlin

Hayashi C (1964) Nonlinear oscillations in physical systems. McGraw Hill, New York

Hongyue D, Feng L, Guangshi M (2011) Robust function projective synchronization of two different chaotic systems with unknown parameters. J Franklin Inst 348:2782–2794

Kareem SO, Ojo KS, Njah AN (2012) Function projective synchronization of identical and non-identical modified finance and Shimizu-Morioka systems. Pramana J Phys 79:71–79

Koyuncu I, Ozecerit AT, Pehlivan I (2014) Implementation of FPGA-based real time novel chaotic oscillator. Nonlinear Dyn 77:49–59

Li X, Chen Y (2007) Function projective synchronization of two identical new hyperchaotic systems. Commun Theoret Phys 48:864–870

Ma J, Wu X, Chu R et al (2014) Selection of multi-scroll attractors in Jerk circuits and their verification using Pspice. Nonlinear Dyn 76:1951–1962

Mahmoud GM, Mansour EA (2011) A hyperchaotic complex system generating two-, three-, and four-scroll attractors. J Vib Control 18:841–849

Moon FC, Johnson MA (1998) Nonlinear dynamics and chaos in manufacturing processes. Wiley, New York

Parlitz U, Lauterborn W (1987) Period-doubling cascades and devil's staircases of the driven van der Pol oscillator. Phys Rev A 36:1428–1434

Pecora LM, Carroll TL (1990) Synchronization in chaotic systems. Phys Rev Lett 64:821–824

Ping Z, Yu-xia C (2010) Function projective synchronization between fractional-order chaotic systems and integer-order chaotic systems. Chin Phys 19:100507–100511

Press WH, Teukolsky SA, Vetterling WT, Flannery BP (1992) Numerical recipes in Fortran, vol 77. Cambridge University Press

Rajasekar S, Parthasarathy S, Lakshmanan M (1992) Prediction of horseshoe chaos in BVP and DVP oscillators. Chaos, Solitons Fractals 2:208–271

Sharma Anjali, Patidar Vinod, Purohit G, Sud KK (2012) Effects on the bifurcation and chaos in forced Duffing oscillator due to nonlinear damping. Commun Nonlinear Sci Numer Simul 17:2254–2269

Shen J, Chen S (2009) An open-plus-closed-loop control for chaotic Mathieu-Duffing oscillator. Appl Math Mech Engl Ed 30:19–27

Shen JH, Lin KC, Chen SH, Sze KY (2008) Bifurcation and route-to-chaos analyses for Mathieu-Duffing oscillator by the incremental harmonic balance method. Nonlinear Dyn 52:403–414

Siewe Siewe M, Tchawoua C, Woafo P (2010) Melnikov chaos in a periodically driven Rayleigh-Duffing oscillator. Mech Res Commun 37:363–368

Srivastava M, Agrawal SK, Subir D (2013) Adaptive projective synchronization between different chaotic systems with parametric uncertainties and external disturbances. Pramana J Phys 81:417–437

Steeb WH, Kunick A (1987) Chaos in limit cycle systems with external periodic excitation. Int J Nonlinear Mech 1987(349):361–422

Trejo-Guerra R, Tlelo-Cuautle E, Jimenez-Fuentes JM, Sanchez-Lopez C, Munoz-Pacheco JM, Espinosa-Flores-Verdad G, Rocha-Perez JM (2012) Integrated circuit generating 3- and 5-scroll attractors. Commun Nonlinear Sci Numer Simul 17:4328–4335

Venkatesan A, Lakshmanan M (1997) Bifurcation and chaos in the double-well Duffing-van der Pol oscillator: numerical and analytical studies. Phys Rev E 56:6321–6330

Vincent UE, Laoye JA, Kareem SO (2008) Control and synchronization of chaos in RCL-shunted Josephson junction using backstepping design. Physica C 468:374–382

Volos CK, Kyprianidis IM, Stouboulus INA (2012) Chaotic path planning generator for autonomous mobile robots. Robot Auton Syst 60:651–656

Volos CK, Kyprianidis IM, Stouboulus IN (2013a) Image encryption process based on chaotic synchronization phenomena. Signal Process 93:1328–1340

Volos CK, Kyprianidis IM, Stouboulus IN (2013b) Experimental investigation on coverage performance of a chaotic autonomous mobile robot. Robot Auton Syst 61:1314–1322

Wen SF, Shen YJ, Wang XN, Yang SP, Xing HJ (2016) Dynamical analysis of strongly nonlinear fractional-order Mathieu-Duffing equation. Chaos 26:084309

Wen SF, Shen YJ, Yang SP, Wang J (2017) Dynamical response of Mathieu-Duffing oscillator with fractional-order delayed feedback. Chaos, Solitons Fractals 94:54–62

Wolf A, Swift JB, Swinney HL, Wastano JA (1985) Determining Lyapunov exponents from time series. Phys D 16:285–317

Yang M, Cai B, Cai G (2010) Projective synchronization of a modified three dimensional chaotic finance system. Int J Nonlinear Sci 10:32–38

Yang JH, Sanjuan MAF, Liu HG (2015) Bifurcation and resonance in a fractional Mathieu-Duffing oscillator. Eur Phys J B 88:310–318

Yu P, Shah AH, Popplewell N (1992) Inertially coupled galloping of iced conductors. ASME J Appl Mach 59:140–145

Yu P, Desai YM, Shah AH, Popplewell N (1993a) Three-degree-of-freedom model for galloping. Part I Formulation ASME J Eng Mech 119:2404–2425

Yu P, Desai YM, Popplewell N, Shah AH (1993b) Three-degree-of-freedom model for galloping. Part II Solutions ASME J Eng Mech 119:2426–2448

Yu W, Chen G, Cao J, Lu J, Parlitz U (2007). Parameter identification of dynamical systems from time series, Phys. Rev. E, 75: 067201–067204

Zhan M, Wang X, Gong X, Wei GW, Lai CH (2003) Complete synchronization and generalized synchronization of one-way coupled time-delay systems. Phys Rev E Stat Nonlinear Sot Matter Phys 68:036208

Zhiyong S, Gangauan S, Fuhong M, Yanbin Z (2012) Adaptive modified function projective synchronization and parameter identification of uncertain hyperchaotic (chaotic) systems with identical or non-identical structures. Nonlinear Dyn 68:471–486

Zhu S, Wu L (2004) Anticipating and lag synchronization in chaotic laser system. Int. J. Mod. Phys. 18:2547–2551

Zhu HL, Zhang XB (2009) Modified Projective Synchronization of different hyperchaotic systems. J Inf Comput Sci 4:33–40

An Autonomous Helmholtz Like-Jerk Oscillator: Analysis, Electronic Circuit Realization and Synchronization Issues

Victor Kamdoum Tamba, Gaetan Fautso Kuiate,
Sifeu Takougang Kingni and Pierre Kisito Talla

Abstract This chapter introduces an autonomous self-exited three-dimensional Helmholtz like oscillator which is built by converting the well know autonomous Helmholtz two-dimensional oscillator to a jerk oscillator. Basic properties of the proposed Helmholtz like-jerk oscillator such as dissipativity, equilibrium points and stability are examined. The dynamics of the proposed jerk oscillator is investigated by using bifurcation diagrams, Lyapunov exponent plots, phase portraits, frequency spectra and cross-sections of the basin of attraction. It is found that the proposed jerk oscillator exhibits some interesting phenomena including Hopf bifurcation, period-doubling bifurcation, reverse period-doubling bifurcation and hysteretic behaviors (responsible of the phenomenon of coexistence of multiple attractors). Moreover, the physical existence of the chaotic behavior and the coexistence of multiple attractors found in the proposed autonomous Helmholtz like-jerk oscillator are verified by some laboratory experimental measurements. A good qualitative agreement is shown between the numerical simulations and the experimental results. In addition, the synchronization of two identical coupled Helmholtz

V. K. Tamba (✉)
Department of Telecommunication and Network Engineering,
IUT-Fotso Victor of Bandjoun, University of Dschang, P. O. Box: 134
Bandjoun, Cameroon
e-mail: vkamdoum@gmail.com; victorkamdoum@yahoo.fr

G. F. Kuiate
Department of Physics, Higher Teacher Training College, University of Bamenda,
P. O. Box 39, Bamenda, Cameroon
e-mail: fautsokuiate@yahoo.co.uk

S. T. Kingni
Department of Mechanical and Electrical Engineering, Institute of Mines
and Petroleum Industries, University of Maroua, P. O. Box 46, Maroua, Cameroon
e-mail: stkingni@gmail.com

P. K. Talla
Laboratory of Mechanics and Modelling of Physical Systems (L2MPS),
Department of Physics, Faculty of Science, University of Dschang,
P. O. Box: 67, Dschang, Cameroon
e-mail: tpierrekisito@yahoo.com

© Springer International Publishing AG 2018
V.-T. Pham et al. (eds.), *Nonlinear Dynamical Systems with Self-Excited
and Hidden Attractors*, Studies in Systems, Decision and Control 133,
https://doi.org/10.1007/978-3-319-71243-7_9

like-jerk oscillators is carried out using an extended backstepping control method. Based on the considered approach, generalized weighted controllers are designed to achieve synchronization in chaotic Helmholtz like-jerk oscillators. Numerical simulations are performed to verify the feasibility of the synchronization method. The approach followed in this chapter shows that by combining both numerical and experimental techniques, one can gain deep insight about the dynamics of chaotic systems exhibiting hysteretic behavior.

Keywords Helmholtz like-jerk oscillator · Bifurcation analysis
Coexistence of attractors · Electronic circuit realization · Synchronization

1 Introduction

Chaos is an interesting phenomenon which has been extensively studied in the last three decades. Chaotic systems are characterized by their extreme sensitivity both to initial conditions as well as to parameters changes. The great interest allowed to chaotic systems is motivated by their important applications in various fields including for instance physics, chemistry, biology, ecology, engineering and economics just to name a few (Azar et al. 2015; Hilborn 2001; Lakshmanan and Rajasekhar 2003; Strogatz 1994). In the recent past, there is increasing interest in the study of robust chaotic systems with as simple as possible mathematical model and simple electronic circuit. Some typical examples of this class of chaotic systems have been investigated in Sprott (2000a, b), Vaidyanathan et al. (2016). In these references, the authors studied several new systems with many nonlinearities that show chaotic behavior with easy electronic circuit realization. These systems are modelled by the time evolution of a single scalar variable x given by $\dddot{x} = F(x, \dot{x}, \ddot{x})$ and namely jerk equation. In this equation, x, $\dot{x}$, $\ddot{x}$ and $\dddot{x}$ represent position, velocity, acceleration and jerk (the time derivative of the acceleration), respectively. According to the simplicity and elegance (in the mathematical model and electronic circuit) of this class of chaotic systems, development of new jerk systems is of great importance. In this point of view, in Benitez et al. (2006), the authors introduced and investigated theoretically and experimentally a new jerk system obtained by converting the well know Van der Pol architecture into a third order differential equation. The proposed mathematical model and electronic circuit are relatively simple. Also, using the same technique, an autonomous chaotic Duffing oscillator based on a jerk architecture is reported in Louodop et al. (2014). The finite-time synchronization of two identical proposed chaotic jerk systems via a simple linear feedback control is examined. The authors used theoretical proofs, numerical and PSpice simulations, as well as practical implementation to demonstrate the feasibility of their proposed scheme. Another interesting works on jerk systems are reported in Kengne et al. (2016, 2017), where the authors introduced and analyzed a new jerk oscillators with hyperbolic sine and smooth piecewise quadratic nonlinearities. They proved through theoretical analysis, numerical simulations and experimental measurements that both systems experience very rich and complex dynamical behaviors such as period-doubling, symmetry recovering crises events, antimonotonicity (i.e. the concurrent creation and annihilation of periodic orbits) and coexistence of multiple attractors.

Motivated by the above mentioned works reported in Sprott (2000a, b), Benitez et al. (2006), Louodop et al. (2014), Kengne et al. (2016, 2017) and many others, in this chapter, we consider an autonomous chaotic jerk oscillator which is obtained by converting the second-order well know Helmholtz oscillator (Del Rio et al. 1992) into a third order differential equations using the jerk architecture. The Helmholtz oscillator is a second order differential equation with a quadratic non-linearity (Del Rio et al. 1992; Thompson 1989; Gottwald et al. 1995). This oscillator known to naval-architects as the Helmholtz-Thompson equation, provides a simple archetype to describe ship stability to waves in windy situations and its potential and eventual capsize (Thompson et al. 1990). It plays an important role in a large number of developments. In Del Rio et al. (1992) the author provides an overview of the dynamic response of the system which has been studied experimentally by Gottwald et al. (1995). The inhibition of chaotic escape is considered in the context of Balibrea et al. (1998) and a more general approach to the problem is discussed in Lenci and Rega (2001). It is interesting to note that the escape of a dynamical system from a potential well is a common topic in physics and engineering and under periodic forcing it is known that the escape will often be triggered by chaotic motions. The Helmholtz equation finds direct application in the study of bubble dynamics (Kang and Leal 1990) and is much discussed in the naval architecture literature (Thompson 1997). The engineering integrity diagram (Soliman and Thompson 1989) and the use of Melnikov theory to predict parameter values for which erosion of basins of attraction takes place were developed in this context. These concepts continue to find fruitful applications in quantification of capsize resistance, see Spyrou et al. (2002).

Chaos synchronization is one the important issues in nonlinear dynamical science because of its various applications in physics, chemical reactors, control theory, biological networks, artificial neural networks, secure communication, etc. (Blekhman 1988; Pikovsky et al. 2001; Nagaev 2003; Pecora and Carrol 1990; Junde and Parlitz 2000). Various types of synchronization including complete synchronization, generalized synchronization, phase synchronization, lag synchronization, anticipated synchronization and measure synchronization (Vincent et al. 2005; Rullkov et al. 1995; Rosemblum et al. 1997; Voss 2000; Hampton and Zanette 1999) have been investigated in the literature. To achieve these different type of synchronization in dynamical systems, many several methods have been identified and studied. Among these include adaptive control, active control, nonlinear control, sliding mode control and backstepping design. The backstepping technique has been recognized as a powerful design technique for stabilization, tracking and synchronization of chaotic systems. It has been reported in Krstic et al. (1995) that backstepping can guarantee global stability, tracking and transient performance of a broad class of strict-feedback nonlinear systems. Due to the many advantages of backstepping design, in this chapter, we develop an extended backstepping technique to achieve the synchronization of two identical coupled autonomous Helmholtz like-jerk oscillators.

The goal of this chapter is fourfold: (i) to enrich the literature by proposing a relatively simple autonomous chaotic jerk system obtained by converting the second-order well know Helmholtz oscillator into a third order differential equations using the jerk architecture; (ii) to point out the stability and bifurcation analyses in order to reveal different dynamics of the system with respect to its parameter as well as highlighting some of its singularities; (iii) to carry out an experimental study of the system to validate the theoretical analyses and (iv) to investigate the synchronization of a coupled autonomous Helmholtz jerk oscillators via extended backstepping method in order to promote chaos-based synchronization designs of this type of oscillators. Such an approach is particularly useful as it provides important tools for the design of such types of oscillators for relevant engineering applications.

The layout of chapter is as follows. Section 2 describes the system under study and highlights some of its basic properties. The stability of the equilibrium points is also examined and conditions for the occurrence of the Hopf bifurcation are derived. Section 3 deals with numerical study. The bifurcation structures of the system are investigated in order to depict some interesting transitions such as period-doubling and reverse period-doubling scenarios to chaos. Some windows showing the hysteretic dynamics (responsible of the occurrence of the coexistence of multiple attractors) are depicted. The multistability is illustrated by plotting the cross-sections of the basins of attraction of various coexisting attractors. The experimental study is carried out in Sect. 4. The laboratory experimental measurements show a qualitative agreement with numerical results. Section 5 discusses the design of the extended backstepping controllers for synchronization of chaos in the jerk systems. Numerical simulations are given for the illustration and verification of the effectiveness and feasibility of the synchronization technique. Finally in Sect. 6, we summarize our results and draw the conclusions of this chapter.

2 Description and Analytical Analysis of the Proposed Autonomous Helmholtz Like-Jerk Oscillator

2.1 System Description and Basic Properties

In this chapter we consider an autonomous Helmholtz like-jerk oscillator derived from the standard nonautonomous Helmholtz oscillator (Del Rio et al. 1992) described by a two-dimensional differential equation as follows

$$\frac{d^2x}{dt^2} + \delta\frac{dx}{dt} + x + x^2 = f\sin(\omega t) \tag{1}$$

where x denotes the vibratory displacement, $\delta > 0$ is a dimensionless damping coefficient, f and ω are respectively, the amplitude and the pulsation of the harmonic external force. It is demonstrated in Thompson (1989), Gottwald et al. (1995) that Eq. (1) can oscillate chaotically for some specific parameters setting. Motivated by the fact that jerk systems are simple in the mathematical representation and easy to realize its corresponding electronic circuit, we propose in this section a three dimensional autonomous Helmholtz oscillator based on jerk architecture. The jerk systems are the third-order equation defined as in Sect. 1. The Helmholtz oscillator defined in Eq. (1) with $f = 0$ can be converted to a jerk oscillator, as follows

$$\frac{d^3x}{dt^3} = -\left(\frac{d^2x}{dt^2} + \delta\frac{dx}{dt} + x + x^2\right) \tag{2}$$

where the parameter δ has the same signification as in Eq. (1). Obviously, Eq. (2) can be converted to the following set of three coupled first-order nonlinear differential equations:

$$\begin{cases} \frac{dx}{dt} = y \\ \frac{dy}{dt} = \gamma z \\ \frac{dz}{dt} = -z - \delta y - x - x^2 \end{cases} \tag{3}$$

where $dx/dt = y$, $d^2x/dt^2 = z$ and γ a new parameter introduced in order to achieve chaotic behavior in Eq. (2). The simplicity of the model is remarkable. It can be implemented experimentally using an appropriate analog electronic circuit as well as integrated circuit technology.

The divergence of system (3) can be obtained as follows

$$\nabla V = \frac{\partial \dot{x}}{\partial x} + \frac{\partial \dot{y}}{\partial y} + \frac{\partial \dot{z}}{\partial z} = -1 \tag{4}$$

Obviously, ∇V is less than zero and therefore, system (3) is dissipative. This means that the system can support attractors.

2.2 Analytical Analysis of the Proposed Autonomous Helmholtz Jerk Oscillator

It is well known that the equilibrium points play an important role on the dynamics of nonlinear system (Hilborn 1994). By setting the right hand side of system (3) to zero, it is found that there are two equilibrium points $E_1(0, 0, 0)$ and $E_1(-1, 0, 0)$. The characteristic equation obtained at any equilibrium point $E^*(x^*, y^*, z^*)$ is defined as

$$\lambda^3 + \lambda^2 + \gamma\delta\lambda + \gamma(2x^* + 1) = 0 \qquad (5)$$

The characteristic equation for the equilibrium point $E_1(0, 0, 0)$ is

$$\lambda^3 + \lambda^2 + \gamma\delta\lambda + \gamma = 0 \qquad (6)$$

Based on Routh-Hurwitz conditions, Eq. 6 has all roots with negative real parts if and only if $\gamma(\delta - 1) > 0$. The equilibrium point $E_1(0,0,0)$ is stable if $\delta > 1$ and unstable if $\delta < 1$ provided that the parameters γ and δ are strictly positive. The two situations (stable for $\delta > 1$ and unstable for $\delta < 1$) suggest the existence of the Hopf bifurcation from the equilibrium point $E_1(0,0,0)$ when δ is selected as the control parameter.

Theorem *The system under scrutiny undergoes a Hopf bifurcation at the equilibrium point $E_1(0, 0, 0)$ when δ passes through the critical value $\delta_H = 1$.*

Proof Let a root of the characteristic Eq. (6) as $\lambda = i\omega_0(\omega_0 > 0)$. By substituting this root into Eq. (6) and after some manipulations, we have

$$\omega = \omega_0 = \sqrt{\gamma} \qquad (7a)$$

$$\delta_H = 1 \qquad (7b)$$

The differentiation of both sides of Eq. (6) with respect to bifurcation parameter δ leads to the following expression

$$\frac{d\lambda}{d\delta} = \frac{-\gamma\lambda}{3\lambda^2 + 2\lambda + \gamma\delta} \qquad (8)$$

By substituting λ and δ by their corresponding expressions defined above into Eq. (8), the following relation is obtained

$$\left. \text{Re}\left(\frac{d\lambda}{d\delta}\right) \right|_{\substack{\lambda = i\omega_0 \\ \delta = \delta_H}} = -\frac{\gamma}{2(\gamma + 1)} \neq 0 \qquad (9)$$

Provided that the characteristic equation of the system at the equilibrium point $E_1(0, 0, 0)$ has two purely imaginary eigenvalues and the real part satisfy Eq. (9), thus all the conditions for occurrence of Hopf bifurcation are satisfied. As consequence, system (3) undergoes Hopf bifurcation at critical value of control parameter $\delta_H = 1$ and periodic solutions will exist in a neighbourhood of this critical point. Thus the proof is completed.

The characteristic equation for the equilibrium point $E_2(-1, 0, 0)$ is

$$\lambda^3 + \lambda^2 + \gamma\delta\lambda - \gamma = 0 \qquad (10)$$

It is obvious that the equilibrium point $E_2(-1, 0, 0)$ is always unstable provided that the corresponding characteristic Eq. (10) has coefficients with different signs and the parameters γ and δ are positive. Thus, the system under consideration is a self-exited system since its basin of attraction is associated with an unstable equilibrium.

3 Complex Dynamics in the Autonomous Helmholtz Like-Jerk Oscillator

In order to investigate various bifurcation structures in the proposed jerk oscillator, system (3) is integrated numerically using the standard fourth-order Runge-Kutta integration algorithm (Press et al. 1992). Throughout this chapter, the time grid is always $\Delta t = 0.001$ and the calculations are pointed out using variables and constants parameters in extended mode. The system is integrated for a sufficiently long time and the transient is discarded. The transition to chaos is characterized by the bifurcation diagram and graph of largest Lyapunov exponent noted $(\lambda_{\max})$. The bifurcation diagram is computed by plotting the local maxima or local minima of the state variable with respect to the control parameter that is changed in tiny steps and the final state at each iteration of the control parameter is used as the initial conditions for the next iteration, while the largest Lyapunov exponent is obtained numerically using the algorithm of Wolf et al. (1985). It measures the exponential rates of divergence or convergence of nearby trajectories in phase space, which can also be used to measure the sensitive dependence of the initial conditions. In particular, the sign of the largest Lyapunov exponent is used to determine the rate of almost all small perturbations to the system's state, and consequently, the nature of the underlined dynamical attractor. For $\lambda_{\max} < 0$, all perturbations vanish and trajectories starting sufficiently close to each other converge to the same stable fixed point in state space; for $\lambda_{\max} = 0$, initially close orbits remains close but distinct, corresponding to oscillatory dynamics of a limit-cycle or torus; and finally for $\lambda_{\max} > 0$, small perturbations grow exponentially, and the system evolves chaotically within the folded space of a strange attractor. In addition, the complexity of system (3) is examined by using the Lyapunov dimension of the attractors which is computed with the help of the definition proposed by Kaplan and Yorke expressed (Frederickson et al. 1983) as

$$D_L = k + \frac{1}{|\lambda_{k+1}|} \sum_{j=1}^{k} \lambda_j \tag{11}$$

where k is the largest integer satisfying the following conditions $\sum_{j=1}^{k} \lambda_j \geq 0$ and $\sum_{j=1}^{k+1} \lambda_j < 0$. The Kaplan-Yorke dimension indicates the complexity of the attractor. In other words, it is a measure of the degree of disorder of the points on the attractor. The dynamics of system (3) is also investigated by using another numerical tools such as, the time series of state variables, the frequency spectra as well as the cross sections of the basin of attraction. The latter tool is computed by taking a grid 500 × 500 initial states, testing each initial state on the grid to determine which attractor is goes to, and then plotting those which lead to the chaotic attractors ($\lambda_{max} > 0$) and periodic solutions ($\lambda_{max} \leq 0$). The same strategy is used to compute the two-parameter phase diagram in which we take a grid 500 × 500 points of parameter γ and δ with fixed initial states.

3.1 Transition to Chaos

In order to select the values of the system parameters accordingly, we plot in Fig. 1, the two-parameter phase diagram showing the regions of periodic, chaotic and unbounded behaviors in the (γ, δ) plane with $4 \leq \gamma \leq 6$ and $0.5 \leq \delta \leq 0.6$.

In Fig. 1, one can see that different regions of chaotic, periodic and unbounded solutions intertwined intricately. This diagram is of great importance for a practical implementation of system (3). Indeed, it helps to choose the parameters of the system according to the desired bahavior.

We select δ as the control parameter in order to examine its sensitivity on the dynamics of the system. To this end, we fix $\gamma = 4$ and vary δ in the range $0.53 \leq \delta \leq 1.02$. The bifurcation diagram showing the local maxima (magenta dots) and local minima (blue dots) of coordinate x associated to the graph of largest Lyapunov exponent versus control parameter δ are provided in Fig. 2.

It found from Fig. 2 that the system under study can experience various and rich dynamical behaviors such as fixed point motion, Hopf bifurcation, period-doubling bifurcation, periodic and chaotic motions when the control parameter is monitored. For values of δ above the critical value $\delta_c = 1$, the system exhibits a fixed point motion (i.e. no oscillations) and the associated largest Lyapunov exponent is negative. By decreasing the control parameter δ past this critical value, the system

Fig. 1 Two-parameter phase diagram showing different dynamical behaviors of system (3) in the (γ, δ) plane. The chaotic regions are shown in red, the periodic regions are shown in cyan and unbounded regions are in green

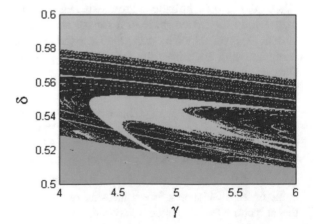

Fig. 2 Bifurcation diagram **a** depicting the local maxima (magenta dots) and local minima (blue dots) of the coordinate x and the corresponding graph of largest Lyapunov exponent (**b**) versus control parameter δ computed with $\gamma = 4$. Chaotic and regular behaviors are indicated respectively, by positive and zero largest Lyapunov exponents while fixed points are marked by negative largest Lyapunov exponent

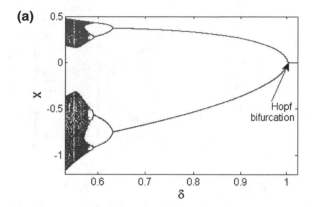

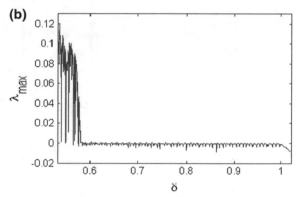

displays a stable period-1 limit cycle born from the Hopf bifurcation. Further decreasing δ, this period-1 limit cycle metamorphoses to chaos via a series of period-doubling bifurcations. It can be also seen that the bifurcation diagram well coincides with the graph of largest Lyapunov exponent.

To confirm the transition to chaos observed in Fig. 2, sample phase portraits in plane (x, y) accompanied with the corresponding frequency spectra along the control parameter δ for $\gamma = 4$ are shown in Fig. 3.

One can observe that the scenario to chaos predicted by the bifurcation diagram of Fig. 2 is confirmed by the phase portraits of Fig. 3.

Using $\gamma = 4$ and $\delta = 0.55$, the corresponding Lyapunov exponents are $\lambda_1 = 0.070$, $\lambda_2 = -0.001$ and $\lambda_3 = -1.069$. From this Lyapunov spectrum, we find that $\sum_{j=1}^{k+1} \lambda_j = -1 < 0$, which confirms that system (3) is dissipative. The calculated fractional dimension with the same system parameters setting is

$$D_L = 2 + \frac{\lambda_1 + \lambda_2}{|\lambda_3|} = 2.064 \tag{12}$$

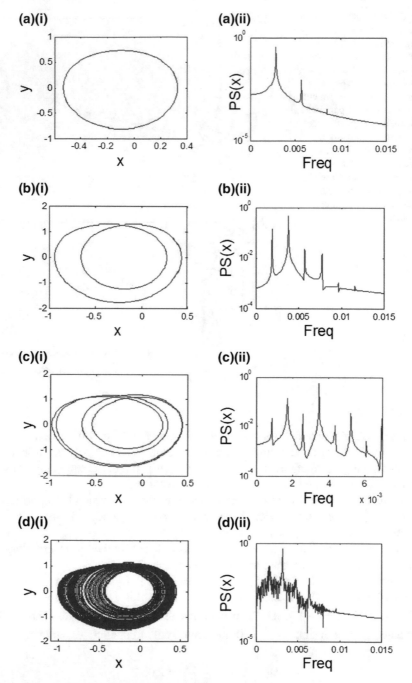

Fig. 3 Phase portraits **a(i)**–**d(i)** and corresponding frequency spectra **a(ii)**–**d(ii)** depicting routes to chaos in the system with respect to the control parameter δ for $\gamma = 4$. **a** Period-1 for $\delta = 0.8$, **b** period-2 for $\delta = 0.6$, **c** period-4 for $\delta = 0.585$ and **d** chaos for $\delta = 0.56$

Equation (12) clearly indicates that the dissipative system under investigation displays chaotic behavior.

Now, we select also γ as the control parameter in order to investigate its effect on the dynamics of the system. The bifurcation diagram depicting the local maxima of the coordinate x and corresponding graph of largest Lyapunov exponent with γ varying in the range $3 \leq \gamma \leq 14$ are shown in Fig. 4 for $\delta = 0.55$.

The bifurcation diagram (Fig. 4a) and the corresponding graph of largest Lyapunov exponent (Fig. 4b) indicate clearly that there are some windows of periodic and chaotic behaviors. In Fig. 4a, one can observe that two set of data corresponding, respectively, for increasing (blue) and decreasing (magenta) values of control parameter γ are superimposed. This method is a simple way to localize the window in which the system experiences hysteretic phenomenon which is at the origin of multistability (i.e. coexistence of attractors). This striking and exciting phenomenon is examined in the next subsection.

Fig. 4 Bifurcation diagram **a** depicting the local maxima of the coordinate x with respect to the control parameter γ and the corresponding graph of largest Lyapunov exponent **b** computed in the range $3 \leq \gamma \leq 14$ for $\delta = 0.55$

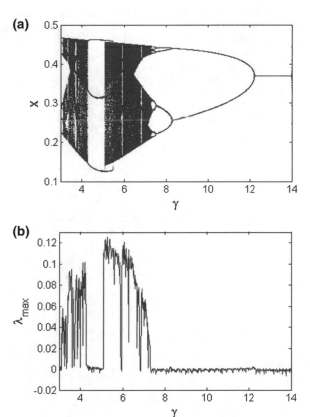

3.2 Coexistence of Attractors in Autonomous Helmholtz Like-Jerk Oscillator

Multistability (i.e. coexistence of multiple attractors) is one of the most striking and exciting phenomenon commonly encountered in dynamical systems. This phenomenon has been reported in almost all natural sciences, including electronics (Maurer and Libchaber 1980; Kamdoum et al. 2016; Kengne et al. 2017; Kengne et al. 2018), optics (Brun et al. 1985), mechanics (Thompson and Stewart 1986), and biology (Foss et al. 1996). During the numerical investigations of the system under consideration, the effects of the initial states on the dynamics of the model were observed. In fact, when we made an enlargement of the bifurcation diagram (Fig. 4a) and the graph of largest Lyapunov exponent (Fig. 4b) in the range $4 \leq \gamma \leq 5.8$, the region in which the system experiences hysteretic dynamics (coexisting bifurcation) is clearly visible as shown in Fig. 5.

Fig. 5 Enlargement of the bifurcation diagram (**a**) and corresponding graph of largest Lyapunov exponent (**b**) of Fig. 4 in the range $4 \leq \gamma \leq 5.8$ in order to make more visible the region in which the system exhibits coexisting attractors

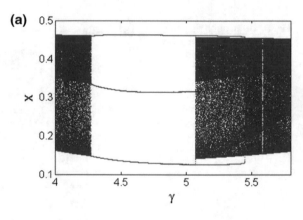

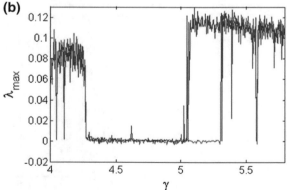

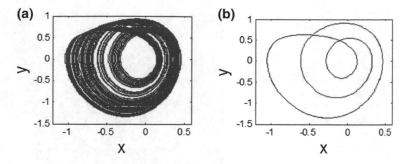

Fig. 6 Coexistence of two different attractors obtained with $\delta = 0.55$ and $\gamma = 5.2$. **a** Chaotic attractor and **b** period-3 limit cycle. The initial conditions are respectively, $x(0), y(0), z(0) = (0.1, 0.1, 0.1)$ and $x(0), y(0), z(0) = (0.2, 0.3, 0.1)$

For the values of γ selected in this window, the dynamics of the system depends on the initial states. For instance, the coexistence of chaotic attractor with period-3 limit cycle obtained respectively, for $x(0), y(0), z(0) = (0.1, 0.1, 0.1)$ and $x(0), y(0), z(0) = (0.2, 0.3, 0.1)$ is shown in Fig. 6 for $\delta = 0.55$ and $\gamma = 5.2$.

In order to further characterize the phenomenon of coexisting attractors observed in the system, we provide in Fig. 7 the cross-sections of the basin of attraction, respectively for $z(0) = 0$, $y(0) = 0$ and $x(0) = 0$ for $\delta = 0.55$ and $\gamma = 5.2$.

In Fig. 7, the complexity of the basin boundaries is clearly highlighted. It can also be noted that the regions of initial conditions leading to chaotic attractors (red dots) are more abundant than those leading to period-3 limit cycle (cyan dots). This implies that the chaotic behavior in the coexisting windows dominates the period-3 limit cycle. It is known that the occurrence of multiple attractors represents an additional source of randomness (Luo and Small 2007) and system which experience this phenomenon can be used for many applications such as chaos based communication, image encryption and generation of random numbers. However, in many practical situations, this singular phenomenon is not desirable and requires control. Interested readers can see the interesting work of Pisarshik and collaborators (2014) about the review on control of multistability. This direction is an important challenge in the continuation of this work.

4 Electronic Circuit Realization of Proposed Autonomous Helmholtz Like-Jerk Oscillator

The physical realization of theoretical chaotic models is of great importance in various engineering applications such as robotics, chaos based communications, image encryption and random number generation (Banerjee 2010; Volos et al. 2012, 2013a, b). Moreover, the electronic circuit realization of theoretical chaotic models is an effective approach to investigate the dynamics of such systems via for

Fig. 7 (Color online)
Cross-sections of the basin of
attraction respectively, for
$z(0) = 0$, $y(0) = 0$ and
$x(0) = 0$ for $\delta = 0.55$ and
$\gamma = 5.2$ showing the regions of
initial conditions leading to
chaotic attractors (red) and
period-3 limit cycle (cyan).
The green dots regions
correspond to the unbounded
solutions

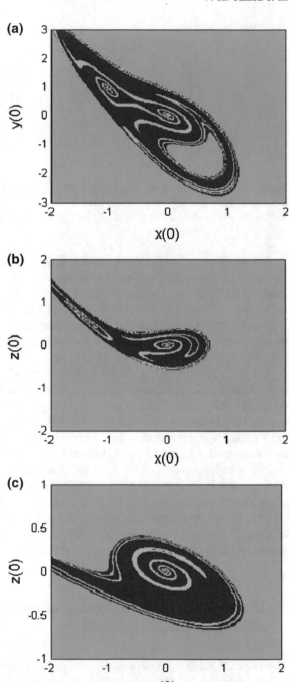

instance the experimental bifurcation diagram obtained by varying the values of variable resistors associated to the control bifurcation parameters (Ma et al. 2014; Buscarino et al. 2009). The dynamics of the system under scrutiny has been examined in preceding paragraphs using theoretical and numerical methods. It is predicted that the system can exhibit very rich and complex behaviors. In this section, to validate the numerical results, we design and realize an electronic circuit capable to emulate the dynamics of system (3). The schematic diagram of the proposed electronic circuit is depicted in Fig. 8.

The electronic circuit of Fig. 8 comprises the analog multipliers used to implement the nonlinear term of the model. They operate over a dynamic range of ± 1 V with typical tolerance less than 1%. The output signal (W) is connected to those at inputs $(+X_1)$, $(-X_2)$, $(+Y_1)$, $(-Y_2)$, and $(+Z)$ by the following expression $W = (X_1 - X_2)(Y_1 - Y_2)/10 + Z$. The operational amplifiers accompanied with resistors and capacitors which are exploited to implement the basic operations such as addition, subtraction and integration. The bias is provided by a 15 V DC symmetry source. By applying the Kirchhoff's laws into the circuit of Fig. 8, we obtain its mathematical model given by three coupled first-order nonlinear differential equations

$$\begin{cases} \frac{dV_x}{dt} = \frac{V_y}{RC} \\ \frac{dV_y}{dt} = \frac{V_z}{R_1 C} \\ \frac{dV_z}{dt} = -\frac{V_z}{RC} - \frac{V_y}{R_2 C} - \frac{V_x}{RC} - \frac{V_x^2}{k_m RC} \end{cases} \tag{13}$$

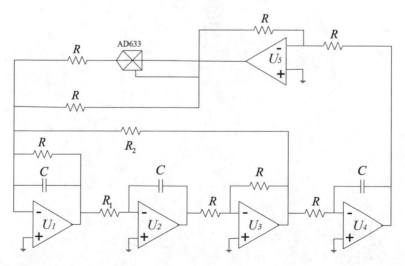

Fig. 8 Electronic circuit realization of the autonomous Helmholtz like-jerk oscillator. The value of electronic components are fixed as $C = 10$ nF, $R = 10$ kΩ, R_1 and R_2 variable. The analog multipliers devices are AD633J N-type while operational amplifiers are TL084C N-type ones

where V_x, V_y and V_z are the output voltages of the operational amplifiers. Systems (13) and (3) are equivalent according to the following change of state variables and parameters: $t = \tau RC; x = V_x/1V; y = V_y/1V; z = V_z/1V; \gamma = R/R_1; \delta = R/R_2$. The electronic circuit components are selected as $C = 10\,nF$, $R = 10\,k\Omega$, $R_1 = 2.5\,k\Omega$ and R_2 variable. When monitoring the control resistor R_2, it is found that the electronic circuit under study displays a rich and striking behaviors including period-doubling route to chaos and coexistence of attractors. A photograph of the experimental hardware on breadboard in operation is shown in Fig. 9. The analog oscilloscope presents the single-band chaotic attractor.

The comparison between numerical (left) and experimental (right) phase portraits is provided in Fig. 10 and a very good similarity can be noted.

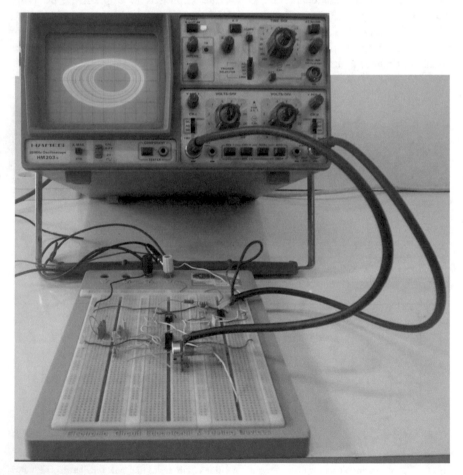

Fig. 9 Photograph of the experimental hardware on breadboard in operation. The analog oscilloscope displays the single-band chaotic attractor captured from the experimental circuit

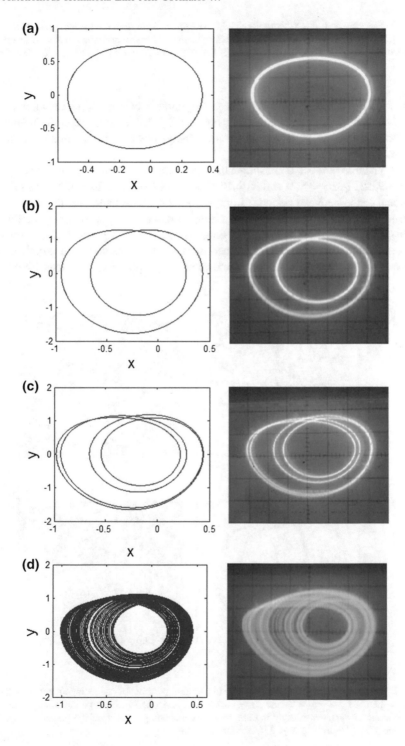

◀Fig. 10 Experimental phase portraits (right) obtained from the circuit using a dual trace oscilloscope in the XY mode; the corresponding numerical phase portraits are shown in the left obtained by a direct integration of system (3). The scales are $X = 1$ V/div and $X = 2$ V/div for all pictures

In Fig. 10, it clearly appears that the dynamics of the proposed autonomous Helmholtz like-jerk system is well reproduced by the electronic circuit. Moreover, one can note that the experimental circuit displays the same bifurcation scenarios as those obtained numerically.

The phenomenon of coexisting attractors is also validated experimentally for $R_1 = 1.92$ kΩ (i.e. $\gamma = 5.2$) and $R_2 = 18.18$ kΩ (i.e. $\delta = 0.55$). In Fig. 11, we provide the coexistence between period-3 limit cycle with single band chaotic attractor for different initial states obtained by switching on and off the power supply randomly.

From Fig. 11, one can note the good similarity of experimental portraits of coexisting attractors with those obtained numerically. This serves to proof that the phenomenon of coexisting attractors exists in the proposed autonomous Helmholtz

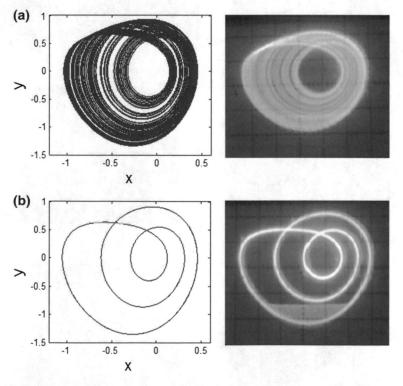

Fig. 11 Numerical (left) and experimental (right) phase portraits of coexisting attractors obtained for $R_1 = 1.92$ kΩ (i.e. $\gamma = 5.2$) and $R_2 = 18.18$ kΩ (i.e. $\delta = 0.55$). Both period-3 limit cycle and chaotic attractor appear randomly in experiment by switching on and off the power supply. The scales are $X = 1$ V/div and $X = 2$ V/div for all pictures

like-jerk oscillator. It should be stressed that system (3) can be also implemented using many other ways such as integrated circuit technology (Trejo Guerra et al. 2012), Field Programmable Gate Array (FPGA) (Koyuncu et al. 2014) and Field Programmable Analog Array (FPAA) technologies (Fatma and Sprott 2016).

5 Synchronization of Two Identical Coupled Autonomous Helmholtz Like-Jerk Oscillators via Extended Backstepping Method

In this section, we synchronize two identical coupled autonomous Helmholtz like-jerk oscillators via extended backstepping method. Based on the proposed approach, generalized weighted controllers are designed to achieve synchronization in chaotic systems. The effectiveness and feasibility of the proposed weighted controllers are verified numerically.

5.1 Design of the Extended Backstepping Controllers for Synchronization in Chaotic Autonomous Helmholtz Like-Jerk Oscillators

Brief recall that the classical backstepping technique has been widely exploited to achieve synchronization in chaotic and hyperchaotic systems. It has the advantage to achieve global stability tracking transient performance for a large class of strict-feedback nonlinear systems (Njah 2010) and the references therein. Moreover, the scheme requires less control effort in comparison to the differential geometric approach (Mascolo 1997). However, the control function designed via this method has been demonstrated to be difficult for practical implementation because of its complexity and high signal strength (Olusola et al. 2011). To overcome these limitations of classical backstepping technique, the improved version of this method namely extended backstepping technique is proposed in Onma et al. (2014). The proposed approach in latter reference is suitable for practical implementation.

To derive the controllers for synchronization in chaotic autonomous Helmholtz like jerk oscillators, we rewrite system (3) in the form

$$\begin{cases} x_1 = x_2 \\ x_2 = x_3 \\ x_3 = -x_3 - x_2 - x_1 - x_1^2 \end{cases} \tag{14}$$

Let system (14) be the drive system and the following be the response

$$\begin{cases} y_1 = y_2 + u_1(t) \\ y_2 = y_3 + u_2(t) \\ y_3 = -y_3 - y_2 - y_1 - y_1^2 + u_3(t) \end{cases} \tag{15}$$

where $u_i(t) i = 1, 2, 3$ are the control functions.

Let the error state between (14) and (15) defined as

$$e_i = y_i - x_i \tag{16}$$

From (16), we can obtain the error dynamical system as

$$\begin{cases} \dot{e}_1 = e_2 + u_1(t) \\ \dot{e}_2 = e_3 + u_2(t) \\ \dot{e}_3 = -e_3 - \delta e_2 - e_1(1 + e_1 + 2x_1) + u_3(t) \end{cases} \tag{17}$$

The challenge is to determine the control function $u_i(t)$ that can stabilize the error states in (17) at the origin. To this end, we design the controllers in three steps. In the first step, we stabilize the first equation in (17) by considering e_2 as controller. Choosing the storage Lyapunov function as $V_1(e_1) = e_1^2/2$, the time derivative of V_1 along the trajectory of error dynamical subsystem $\dot{e}_1$ is

$$\dot{V}_1 = e_1 \dot{e}_1 = e_1(e_2 + u_1) \tag{18}$$

We suppose that the controller e_2 has the following form $e_2 = \alpha_1(e_1)$, then (18) can be written as $\dot{V}_1 = e_1(\alpha_1 e_1 + u_1)$. The time derivative of Lyapunov function $\dot{V}_1$ is negative definite if the estimate function $\alpha_1(e_1) = -e_1$ and $u_1 = 0$. Thus, the subsystem e_1 is stabilized. In the second step, we choose the error between e_2 and $\alpha_1(e_1)$ as

$$w_2 = e_2 - \alpha_1 e_1 = e_2 + e_1 \tag{19}$$

The time derivative of (19) is

$$\dot{w}_2 = e_3 + w_2 - e_1 + u_2 \tag{20}$$

We now stabilize the (e_1, w_2) subsystem defined by (20) as follows. Let a Lyapunov function $V_2(e_1, w_2) = V_1(e_1) + w_2^2/2$ and its time derivative is

$$\dot{V}_2 = \dot{V}_1(e_1) + w_2 \dot{w}_2 = -e_1^2 + w_2(-w_2 + e_3 + 2w_2 - e_1 + u_2) \tag{21}$$

Estimating that the controller $e_3 = \alpha_2(e_1, w_2)$, then (21) can be written as

$$\dot{V}_2 = -e_1^2 + w_2(-w_2 + \alpha_2 + 2w_2 - e_1 + u_2) \tag{22}$$

If the estimative function $\alpha_2(e_1, w_2) = 0$ and $u_2 = e_1 - 2w_2$, then $\dot{V}_2 = -e_1^2 - w_2^2$ is negative definite and hence the (e_1, w_2) subsystem is stabilized. For the last step, we define the error between e_3 and $\alpha_2(e_1, w_2)$ as

$$w_3 = e_3 - \alpha_2(e_1, w_2) = e_3 \tag{23}$$

The time derivative of (23) is

$$\dot{w}_3 = \dot{e}_3 = -e_3 - \delta e_2 - e_1(1 + e_1 + 2x_1) + u_3 \tag{24}$$

We now stabilize the (e_1, w_2, w_3) complete system defined by (17) as follows. Let a Lyapunov function $V_3(e_1, w_2, w_3) = V_2(e_1, w_2) + w_3^2/2$ and its time derivative is

$$
\begin{aligned}
\dot{V}_3 &= \dot{V}_2(e_1, w_2) + w_3\dot{w}_3 = -e_1^2 - w_2^2 + w_3\dot{w}_3 \\
&= -e_1^2 - w_2^2 + w_3[-w_3 - \delta e_2 - e_1(1 + e_1 + 2x_1) + u_3]
\end{aligned}
\tag{25}
$$

Estimating that $u_3 = \delta e_2 + e_1(1 + e_1 + 2x_1)$, (25) becomes

$$\dot{V}_3 = -e_1^2 - w_2^2 - w_3^2 < 0 \tag{26}$$

Thus, the synchronization goal is realized with the weight added to the control functions as follows

$$
\begin{cases}
u_1(t) = [0]\varepsilon_1 \\
u_2(t) = [e_1 - 2w_2]\varepsilon_1 \\
u_3(t) = [\delta e_2 + e_1(1 + e_1 + 2x_1)]\varepsilon_3
\end{cases}
\tag{27}
$$

5.2 Numerical Simulations

In this subsection, numerical simulations are given in order to verify and demonstrate the effectiveness and feasibility of the proposed method. The fourth-order Runge-Kutta method is applied to integrate the drive (14) and response (15) system with time step size equal to 0.001. The initial conditions are selected randomly as $x_1(0), x_2(0), x_3(0) = (0.1, 0.1, 0.1)$ and $y_1(0), y_2(0), y_3(0) = (-1, 1, -0.1)$, respectively for the drive and the response system, while the parameter values are chosen to be $\gamma = 4$ and $\delta = 0.56$ so that the systems exhibited chaotic behaviors if no control functions are applied. The time response of the error system and the synchronization quality which is defined by $e = \sqrt{e_1^2 + e_2^2 + e_3^2}$ are shown in Fig. 12 for the control function applied after approximately 100 units of time.

From Fig. 12, it is found that the error dynamics moves chaotically with time when the controllers are switched off in the interval $50 < t < 100$. After this

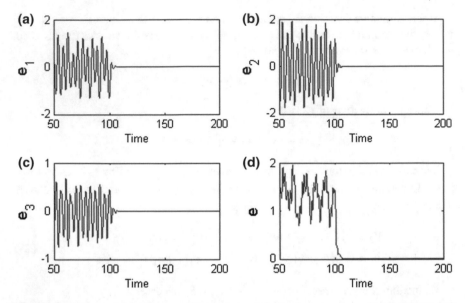

Fig. 12 Error dynamics between two identical coupled chaotic autonomous Helmholtz like-jerk oscillators (**a**)–(**c**) and the synchronization quality (**d**) with the controller deactivated $50 < t < 100$. The values of parameters and initial conditions are indicated in the text and $\varepsilon_i = 0.3$ where $i = 1, 2, 3$

interval, it is very clear that the synchronization is achieved since the error dynamics between the two identical coupled chaotic autonomous Helmholtz like-jerk oscillators approaches zero as $t \to \infty$. The synchronization between the drive and the response systems is also confirmed by the synchronization quantity e which approaches also zero as $t \to \infty$. It is important to notice that, the control strength and its complexity are reduced by about 70% when the extended back-stepping method is used compared with the classical backstepping approach (Mascolo 1997; Njah 2010; Olusola et al. 2011). Thus, the new approach investigated in this chapter produces economic controllers with low energy consumption which may be of vital importance for practical applications (Onma et al. 2014). The effectiveness and feasibility of study of synchronization of chaotic systems is verified and is found to be good to be used in secure communication field.

6 Conclusions

This chapter has focused on the dynamical analysis, electronic circuit realization and synchronization of an autonomous oscillator obtained by converting the two-dimensional well know Helmholtz oscillator into a third-order differential equations using the jerk architecture. The stability of the equilibrium points has been examined and conditions for the occurrence of the Hopf bifurcation have been derived. Some basic properties of the model have been studied using standard nonlinear analysis tools. The bifurcation structures of the proposed jerk oscillator were revealed some interesting transitions and phenomena such as period-doubling and reverse period-doubling scenarios to chaos, periodic windows and hysteretic behavior. The latter phenomena has been further illustrated by computing some cross-sections of the basin of attraction showing the domains of initial conditions where coexisting attractors are found. An experimental study has been carried out and the laboratory experimental measurements were in a good qualitative agreement with numerical results. Finally, appropriate controllers have been designed via extended backstepping technique to synchronize the proposed jerk oscillators. Numerical simulations are given to illustrate and verify the effectiveness and feasibility of the synchronization technique. It is worth pointing out that the extended backstepping technique has several advantages over other methods of synchronization as mentioned in Sect. 5. Thus, synchronization of the autonomous proposed jerk oscillators via extended backstepping technique is of practical interest. We stress also that the approach followed in this chapter may be exploited rigorously to the study of any other nonlinear dynamical system exhibiting coexisting bifurcations.

References

Azar AT, Vaidyanathan S (2015) Chaos modeling and control systems design. Springer, Berlin

Balibrea F, Chacon R, Lopez MA (1998) Inhibition of chaotic escape by an additional driven term. Int J Bifurcat Chaos 8:1719–1724

Banerjee R (2010) Chaos Synchronization and Cryptography for Secure communications. IGI Global, USA

Benitez MS, Zuppa LA, Guerra RJR (2006) Chaotification of the Van der Pol system using Jerk architecture. IEICE Trans. Fundam. 89-A, 375–378

Blekhman II (1988) Synchronization in science and technology. AMSE Press, New York

Brun E, Derighetti B, Meier D, Holzner R, Ravani M (1985) Observation of order and chaos in a nuclear spin-flip laser. J. Opt. Soc. Am. B 2:156–167

Buscarino A, Fortuna L, Frasca M (2009) Experimental robust synchronization of hyperchaotic circuits. Physica D 238:1917–1922

Del Río E, Rodriguez Lozano A, Velarde MG (1992) A prototype Helmholtz-Thompson nonlinear oscillator. AIP Rev Sci Instrum 63:4208–4212

Dalkiran FY, Sprott JC (2016) Simple chaotic hyperjerk system. Int J Bifurcat Chaos 26:1650189

Foss J, Longtin A, Mensour B, Milton J (1996) Multistability and delayed recurrent loops. Phys Rev Lett 76:708–711

Frederickson P, Kaplan JL, Yorke HL et al (1983) The Lyapunov dimension of strange attractor. J Differ Equ 49:185–207

Gottwald JA, Virgin LN, Dowell EH (1995) Routes to escape from an energy well. J Sound Vib 187:133–144

Hampton A, Zanette HD (1999) Measure synchronization in coupled Hamiltonian systems. Phys Rev Lett 83:2179–2182

Hilborn RC (1994) Chaos and nonlinear dynamics: an introduction for scientists and engineers. Oxford University Press, Oxford, UK

Hilborn RC (2001) Chaos and nonlinear dynamics: an introduction for scientists and engineers, 2nd edn. Oxford University Press, Oxford, UK

Junge L, Parlitz U (2000) Synchronization of coupled Ginzburg-Landau equations using local potential. Phys Rev E 61:3736–3742

Kamdoum Tamba V, Fotsin HB, Kengne J, Megam Ngouonkadi EB, Talla PK (2016) Emergence of complex dynamical behaviors in improved Colpitts oscillators: antimonotonicity, coexisting attractors, and metastable chaos. Int J Dyn Control. https://doi.org/10.1007/s40435-016-0223-4

Kang IS, Leal LG (1990) Bubble dynamics in time-periodic straining flows. J Fluid Mech 218:41–69

Kengne J, Njitacke ZT, Nguomkam Negou A, Fouodji Tsostop M, Fotsin HB (2016) Coexistence of multiple attractors and crisis route to chaos in a novel chaotic jerk circuit. Int J Bifurcat Chaos 26:1650081

Kengne J, Folifack Signing VR, Chedjou JC, Leutcho GD (2017) Nonlinear behavior of a novel chaotic jerk system: antimonotonicity, crises, and multiple coexisting attractors. Int J Dyn Control 1–18

Kengne J, Nguomkam Negou A, Tchiotsop D, Kamdoum Tamba V, Kom GH (2018) On the dynamics of chaotic systems with multiple attractors: a case study. In: Recent advances in nonlinear dynamics and synchronization, studies in systems, decision and control, vol. 109. Springer

Koyuncu I, Ozecerit AT, Pehlivan I (2014) Implementation of FPGA-based real time novel chaotic oscillator. Nonlinear Dyn 77:49–59

Krstic M, Kanellakopoulos I, Kokotovic PO (1995) Nonlinear and adaptive control design. Wiley, New York

Lakshmanan M, Rajasekhar S (2003) Nonlinear dynamics: integrability, chaos, and patterns. Springer, Berlin

Lenci S, Rega G (2001) Optimal control of homoclinic bifurcation in a periodically driven Helmholtz oscillator. In: Proceedings of the ASME design engineering technical conference, Pittsburgh, Pennsylvania, USA

Louodop P, Kountchou M, Fotsin H, Bowong S (2014) Practical finite-time synchronization of jerk systems: theory and experiment. Nonlinear Dyn 78:597–607

Luo X, Small M (2007) On a dynamical system with multiple chaotic attractors. Int J Bifurcat Chaos 17(9):3235–3251

Ma J, Wu X, Chu R et al (2014) Selection of multi-scroll attractors in Jerk circuits and their verification using Pspice. Nonlinear Dyn 76:1951–1962

Mascolo S (1997) Backstepping design for controlling Lorenz chaos. In: Proceedings of the 36th IEEE conference on decision and control, San Diego, California, USA, pp 1500–1501

Maurer J, Libchaber A (1980) Effect of the Prandtl number on the onset of turbulence in liquid-He-4. J Phys Lett 41:515–518

Nagaev RF (2003) Dynamics of synchronizing systems. Springer, Berlin-Heidelberg

Njah AN (2010) Tracking control and synchronization of the new hyperchaotic Liu system via backstepping techniques. Nonlinear Dyn 61:1–9

Olusola OI, Vincent UE, Njah AN, Ali E (2011) Control and synchronization of chaos in biological systems via backstepping design. Int J Nonlinear Sci 11:121–128

Onma OS, Olusola OI, Njah AN (2014) Control and synchronization of chaotic and hyperchaotic Lorenz systems via extended backstepping techniques. J Nonlinear Dyn, ID 861727

Pecora LM, Carrol TL (1990) Synchronization in chaotic systems. Phys Rev Lett 64:821–824

Pikovsky A, Rosemblum M, Kurths J (2001) Synchronization: a universal concept in nonlinear science. Cambridge University Press, New York

Pisarchik AN, Feudel U (2014) Control of multistability. Phys Rep 540(4):167–218

Press WH, Teukolsky SA, Vetterling WT, Flannery BP (1992) Numerical recipes in Fortran 77. Press, Cambridge U

Rosemblum MG, Pikovsky AS, Kurths J (1997) From phase to lag synchronization in coupled chaotic oscillators. Phys Rev Lett 78:4193–4196

Rullkov NF, Sushchik MM, Tsimring LS, Abarbanel HDI (1995) Generalized synchronization of chaos in directionally coupled chaotic systems. Phys Rev E 51:980–994

Soliman MS, Thompson JMT (1989) Integrity measures quantifying the erosion of smooth and fractal basins of attraction. J Sound Vib 135:453–475

Sprott JC (2000a) A new class of chaotic circuit. Phys Lett A 266:19–23

Sprott JC (2000b) Simple chaotic systems and circuits. Am J Phys 68:758–763

Spyrou KJ, Cotton B, Gurd B (2002) Analytical expressions of capsize boundary for a ship with roll bias in beam waves. J Ship Res 46:167–174

Strogatz SH (1994) Nonlinear dynamics and chaos: with applications to physics, biology, chemistry, and engineering. Perseus Books, Massachussetts

Thompson JMT, Stewart HB (1986) Nonlinear dynamics and chaos. Wiley, Chichester

Thompson JMT (1989) Chaotic phenomena triggering the escape from a potential well. Proc R Soc Lond A 421:195–225

Thompson JMT, Rainey RCT, Soliman MS (1990) Ship stability criteria based on chaotic transients from incursive fractals. Philos Trans R Soc Lond A 332:149–167

Thompson JMT (1997) Designing against capsize in beam seas: recent advances and new insights. Appl Mech Rev 50:307–325

Trejo-Guerra R, Tlelo-Cuautle E, Jimenez-Fuentes JM, Sanchez-Lopez C, Munoz-Pacheco JM, Espinosa-Flores-Verdad G, Rocha-Perez JM (2012) Integrated circuit generating 3- and 5-scroll attractors. Commun Nonlinear Sci Numer Simul 17:4328–4335

Vincent UE, Njah AN, Akinlade O, Solarin ART (2005) Synchronization of cross-well chaos in coupled duffing oscillators. Int J Mod Phys B 19:3205–3216

Vaidyanathan S, Pham VT, Volos CK (2016) Adaptive backstepping control, synchronization and circuit simulation of a novel jerk chaotic system with a quartic nonlinearity. In: Advances and applications in chaotic systems, studies in computational intelligence. Springer

Volos CK, Kyprianidis IM, Stouboulus INA (2012) Chaotic path planning generator for autonomous mobile robots. Robot Auton Syst 60:651–656

Volos CK, Kyprianidis IM, Stouboulus IN (2013a) Image encryption process based onchaotic synchronization phenomena. Signal Process 93:1328–1340

Volos CK, Kyprianidis IM, Stouboulus IN (2013b) Experimental investigation on coverage performance of a chaotic autonomous mobile robot. Robot Auton Syst 61:1314–1322

Voss HU (2000) Anticipating chaotic synchronization. Phys Rev E 61:5115–5119

Wolf A, Swift JB, Swinney HL, Wastano JA (1985) Determining Lyapunov exponents from time series. Phys D 16:285–317

Synchronization in Kuramoto Oscillators Under Single External Oscillator

Gokul P. M., V. K. Chandrasekar and Tomasz Kapitaniak

Abstract In this chapter we study the influence of a single strongly attractively coupled external oscillator on a system of coupled Kuramoto oscillators. First we go
through the original method used by Kuramoto to solve this system of coupled oscillators. Then we use a later approach developed by Ott and Antonsen. We will use this
approach first to solve the original system and show that the results match. Next we
will solve a variations of the this system using Ott-Antonsen method, after which we
will use it to solve our particular system. We consider a variation of the Kuramoto
system which shows multiple regions of synchronization. First we observe the effects
of attractive and repulsive couplings. Next we qualitatively study the effect of the initial frequency distribution of the internal oscillators, both the mean and the standard
deviation of different distributions like the Gaussian and Lorentzian distributions,
on these synchronization regions.

1 Introduction

Synchronization has always been an interesting topic of study since its discovery
by Huygen's (1673) in coupled pendulum. In a field as diverse and encompassing
as non-linear dynamics, coupled systems and their synchronization behaviour has

Gokul P. M. (✉) · T. Kapitaniak
Division of Dynamics, Lodz University of Technology, ul. Stefanowskiego 1/15,
90-924 Lodz, Poland
e-mail: gokulnappu@gmail.com

T. Kapitaniak
e-mail: tomasz.kapitaniak@p.lodz.pl

V. K. Chandrasekar
Center for Nonlinear Science and Engineering, School of Electrical
and Electronics Engineering, SASTRA University, Thanjavur 613401, Tamil Nadu, India
e-mail: chandrasekar@eee.sastra.edu

© Springer International Publishing AG 2018

229

V.-T. Pham et al. (eds.), *Nonlinear Dynamical Systems with Self-Excited
and Hidden Attractors*, Studies in Systems, Decision and Control 133,
https://doi.org/10.1007/978-3-319-71243-7_10

been used to model a wide variety of systems from those in Biology and Physics to Economics. Many studies on the mathematical aspects of collective synchronization have been done in the past decades. These systems have a lot of applications in physical systems like Josephson junction, electrochemical array, etc. (Yamaguchi et al. 2003; Kiss et al. 2002; Wiesenfeld and Swift 1995; Pantaleone 1998; Hubler et al. 1997).

Many types of synchronization have been observed. Two types of synchronization are of immediate interest in our case. They are spontaneaous synchronization and froced synchronization. In many coupled systems, there can be spontaneous synchronization (Pikovsky et al. 2001; Strogatz 2004; Boccaletti et al. 2002). That is, for a critical value of a parameter, the system shows some collective behavior without any external influence. In case of forced synchronization, the system shows collective behavior due to an external forcing term. There are many more different types of synchronization that may be induced due to many factors, including noise (Flandoli et al. 2017).

One interesting coupled system that shows a variety of different synchronization behaviour was proposed by Kuramoto (1975). The Kuramoto system is a system of N-coupled phase oscillators. These can be thought of as a collection of limit cycles. Under certain conditions, these coupled phase oscillators were seen to undergo the phenomenon of synchronization. Many different variations of the system, including even second-order differential forms were studied (Bountis et al. 2014; Olmi et al. 2014; Jaros et al. 2015), many of whom showed partial synchronizations, chimera (Maistrenko et al. 2017) and even solitary states (Jaros et al. 2017).

A system similar to the one that will be studied in this chapter was analyzed by Childs and Strogatz in 2008 (Strogatz 2008). These systems show different types of synchronization. An interesting fact about this specific system is that even repulsive coupling of oscillators lead to synchronization. Not only that, they show behavior very similar to that shown when the oscillators have attractive coupling. This is true for different distributions of the initial frequencies making it a very general phenomenon.

To understand the system better, we can consider the Kuramoto system to be a set of points moving around in a unit circle with θ position and angular velocity ω. We can see that some transformations, like $\theta \longrightarrow \theta + c$, where c is a constant does not change the system. A commonly used transformation of the system during calculations is $\omega \longrightarrow \omega + c$. This is called the rotating frame transformation. This can be seen as the unit circle with all the point oscillators itself rotating at a frequency. Also, now addition of external force could be seen as the unit circle itself experiencing the force rather than the same force being applied to every single oscillator, since both scenarios give the same equation.

There are many ways of solving the basic system. First we will see the method that was used by Kuramoto (1975, 1984) to solve this system, after which we will use a method suggested by Ott and Antonsen (2008).

2 Solving Kuramoto System

First we will see the method used by Kuramoto to solve this system (Kuramoto 1975, 1984). The Kuramoto model is a simple model of N-coupled oscillators with different frequencies. The simple Kuramoto system is given by

$$\frac{d\theta_i}{dt} = \omega_i + \frac{K}{N} \sum_{j=1}^{N} sin(\theta_j - \theta_i) \tag{1}$$

where $i = 1, \ldots, N$ denote the N oscillators. Here ω_i gives the frequencies of the ith oscillator. That is the frequency at which the oscillator would move had it not been coupled to other oscillators. θ_i denotes the phase of the ith oscillator. The parameter K gives the coupling strength. We have taken the coupling to be a sine function as was done in the original Kuramoto article, although there have been many generalizations done in later years.

We can see from the system equations that in the case of $K > 0$, which is the one we are working with, the coupling is making the system come closer. That is, if the phase of an individual oscillator is smaller than the average phase, then the coupling will increase it, while if the individual oscillator has a larger than average phase, then it will be decreased by the coupling. The system has a natural tendency to synchronize towards the average phase.

It should also be noted that N^{-1} is also an important term in the coupling. If not for this term, the coupling would not be N-independent in the thermodynamic limit of $N \longrightarrow \infty$, which is the case we consider for analysis.

We define ψ as $R = re^{i\psi} = \frac{1}{N} \sum_{i=1}^{n} e^{i\theta_i}$ where R is called an order parameter. It can be seen from this expression, that ψ gives the average phase, and is hence called the 'mean field', while r gives the variation of the individual oscillators phase from ψ. This will be used as the measure for synchronization as when all sysntems are synchronized, $r = 1$.

We now write the system equations as a function of r and ψ

$$\frac{d\theta_i}{dt} = \omega_i + \frac{K}{N} Im \left[\sum_{i=1}^{n} e^{i(\theta_j - \theta_i)} \right]$$

$$= \omega_i + \frac{K}{N} Im \left[e^{-i\theta_i} \sum_{i=1}^{n} e^{i\theta_j} \right]$$

$$= \omega_i + Im[Ke^{-i\theta_i} re^{i\psi}]$$

$$= \omega_i + Krsin(\psi - \theta_i)$$

From this form of the system equations we see that Kuramoto system can be thought of as a system of oscillators being forced by a mean field. This also explains neatly the tendency of the system to synchronize into the mean field.

Now we do the transformation $\theta \longrightarrow \theta - \psi$ and Ω is assumed to be the steady state frequency of the oscillators. One thing to be noted is that, for solving his system, Kuramoto used a few assumption, like the existence of a steady state. To understand this, we can think of it as the system of N-oscillators being thought of as a single oscillator moving in the unit circle at ψ angle and Ω frequency. This assumption was later verified by the self-consistency condition. Coming back to the calculation, we do one more transformation in which we go to a rotating frame with Ω frequency. This eliminated the ψ and Ω terms leaving us with the equation

$$\Rightarrow \frac{d\theta_i}{dt} = \omega_i - Krsin(\theta_i)$$

Since we are taking steady state solutions, we have $r = constant$.

There are two types of solutions for the oscillators described above depending on the two terms on the right-hand side.

If $|\omega_i| \leqslant Kr$, then the oscillators converge to a steady state and reach synchronization. This implies that at steady state $|\omega_i| = Krsin\theta_i$

$$\Rightarrow \theta_i = sin^{-1}(\frac{Kr}{\omega_i}) \leqslant \frac{pi}{2}$$

This set of oscillators are called phase locked, as undoing the transformations would mean that these are moving at the same frequency Ω.

Now let us consider the other case where $|\omega_i| \geq Kr$. These oscillators do not synchronize but move freely around the unit circle.

Now the problem which arises is: Is r and ψ constant? This was solved by Kuramoto by assuming that the mean of the drifting oscillators form a stationary distribution on the circle. If $\rho(\theta, \omega)d\theta$ denote the fraction of oscillators with frequency ω that lie between θ and $\theta + d\theta$, then this ρ, to satisfy the stationary distribution condition, should be inversely proportional to the speed.

$$\Rightarrow \rho(\theta, \omega) = \frac{C}{\omega - Krsin\theta}$$

Here $C = \frac{1}{2\pi}\sqrt{\omega^2 - (kr)^2}$ is the normalization constant.

Now to solve the system and to justify our assumptions, we will invoke the self-consistency condition. Since the order parameter has to be a constant as assumed,

$$< e^{i\theta} >=< e^{i\theta} >_{lock} + < e^{i\theta} >_{drift}$$

where $<>$ denote the population averages. Continuing from here, since $\psi = 0$, $< e^{i\theta} >= r$.

$$\Rightarrow r = <e^{i\theta}>_{lock} + <e^{i\theta}>_{drift}$$

Due to symmetry of the system as $N \longrightarrow \infty$,

$$<e^{i\theta}>_{lock} = <cos\theta>_{lock} = \int_{-Kr}^{Kr} cos\theta(\omega)g(\omega)d\omega$$

For the drifting oscillators

$$<e^{i\theta}>_{drift} = \int_{-\pi}^{\pi}\int_{|\omega|>Kr} e^{i\theta}\rho(\theta,\omega)g(\omega)d\omega d\theta$$

This integral vanishes due to the symmetry of ρ. Therefore now the whole self-consistency is given just by the locked terms, which is written in terms of θ as

$$r = Kr \int_{-\pi/2}^{\pi/2} cos^2\theta g(Krsin\theta)d\theta$$

which has two solutions, the trivial $r = 0$ and the other given by

$$1 = K \int_{-\pi/2}^{\pi/2} cos^2\theta g(Krsin\theta)d\theta$$

This shows the increase in r beyond a critical K_c given by

$$K_c = \frac{2}{\pi g(0)}$$

For Lorentzian distribution, the integral can be calculated exactly and gives $r = \sqrt{1 - \frac{K_c}{K}}$.

3 Ott-Antonsen Method for Solving Kuramoto-Like Systems

3.1 Original Kuramoto System

As of now, we have defined and solved the simple Kuramoto system. We have shown the existence of a synchronous state in the $N \longrightarrow \infty$ limit and derived the expression for the the order parameter r. But as intuitive as the proof and calculations for the system have been, it is not a general method. As in, even though this works for a simple Kuramoto system, the same method cannot be used for systems derived from it, like the Kuramoto system with an external forcing term.

There are many methods used to solve this system, most of which are similar to the method shown above. This process is done by separating the system into different sub-populations and then using self-consistency to arrive at a solution for r. We will be using a different approach proposed by Ott and Antonsen (2008). The crux of this method is the use of an anzatz given by them. The reason and proof for the use of this anzatz is given in has been studied in their publication and further used in many others. Here we will just be showing the process involved with the solving.

The simple Kuramoto system is again given by

$$\frac{d\theta_i}{dt} = \omega_i + \frac{K}{N} \sum_{j=1}^{N} sin\left(\theta_j - \theta_i\right) \tag{2}$$

where $i = 1 \ldots N$ denote the N oscillators.

We define ψ and r the same way as before

$$re^{i\psi} = \frac{1}{N} \sum_{i=1}^{n} e^{i\theta_i}$$

Re-writing the equation again in terms of ψ and r

$$\frac{d\theta_i}{dt} = \omega_i + \frac{K}{N} Im \left[\sum_{i=1}^{n} e^{i(\theta_j - \theta_i)} \right]$$

$$= \omega_i + \frac{K}{N} Im \left[e^{-i\theta_i} \sum_{i=1}^{n} e^{i\theta_i} \right]$$

$$= \omega_i + Im \left[Ke^{-i\theta_i} re^{i\psi} \right]$$

$$\frac{d\theta_i}{dt} = \omega_i + Krsin\left(\psi - \theta_i\right) \tag{3}$$

It is easier to use Eq. (3) for simulations since as is shown, it is the same system in a different form.

For our calculations

$$\frac{d\theta_i}{dt} = \omega_i + \frac{K}{N}\left(\sum_{i=1}^{n} \frac{e^{i(\theta_j-\theta_i)} - e^{-i(\theta_j-\theta_i)}}{2i}\right)$$

$$= \omega_i + \frac{K}{2i}\left(Re^{-i\theta_i} - R^*e^{i\theta_i}\right)$$

In the continuous limit of $N \to \infty$, let f be the probability distribution of θ and

$$g(\omega) = \int_0^{2\pi} f(\omega, \theta, t)\, d\theta \tag{4}$$

be the time-independent oscillator frequency distribution.

The continuity equation is given by

$$\frac{\partial f}{\partial t} + \frac{\partial}{\partial \theta}(f\dot{\theta}) = 0$$

$$\Rightarrow \frac{\partial f}{\partial t} + \frac{\partial}{\partial \theta}\left[\left(\omega_i + \frac{K}{2i}(Re^{-i\theta} - R^*e^{i\theta})\right)f\right] \tag{5}$$

where

$$R = \int_0^{2\pi} d\theta \int_{-\infty}^{\infty} d\omega f e^{i\theta} \tag{6}$$

Expanding f using Fourier series gives us

$$f = \frac{g(\omega)}{2\pi}\left[1 + \left[\sum_{n=1}^{\infty} f_n(\omega, t)e^{in\theta} + f_n^*(\omega, t)e^{-in\theta}\right]\right]$$

We will now use the anzatz provided by Ott and Antonsen

$$f_n(\omega, t) = [\alpha(\omega, t)]^n$$

where $|\alpha(\omega, t)| \leqslant 1$ so that the system is convergent.

Putting this back in Eq. (5) and taking only the coefficients of $e^{i\theta}$, we get

$$\frac{\partial \alpha}{\partial t} + \left[\alpha\omega i + \frac{k}{2i}(R\alpha^2 - R^*)i\right] = 0$$

$$\Rightarrow \frac{\partial \alpha}{\partial t} + \frac{k}{2}\left(R\alpha^2 - R^*\right) + i\omega\alpha = 0 \tag{7}$$

and into Eq. (6)

$$R^* = \int_\infty^\infty d\omega \alpha(\omega, t) g(\omega) \tag{8}$$

As of now, this converted an infinite dimensional system in θ to an infinite dimensional system in ω. To solve the system, we need to make this into a finite dimensional set of differential equations, which is what this method will do in the following steps.

Now we will solve the equation for an example of ω—distribution. Here we take it to be Lorentzian.

$$g(\omega) = \frac{\Delta}{\pi} \frac{1}{\left[(\omega - \omega_0)^2 + \Delta^2\right]}$$

$$= \frac{1}{2\pi i} \left[\frac{1}{\omega - \omega_0 - i\Delta} - \frac{1}{\omega - \omega_0 + i\Delta} \right]$$

By going into a rotating frame using the transformation $\omega \longrightarrow \frac{\omega - \omega_0}{\Delta}$ and $\theta \longrightarrow \theta - \omega_0 t$, we can see that it is possible to put $\omega_0 = 0$ and $\Delta = 1$.

Putting this in Eq. (8) and solving using the residue method, we get

$$R = \alpha^*(-i, t) \tag{9}$$

Using this and $R = re^{i\psi}$ into Eq. (7), we get

$$\frac{\partial}{\partial t}\left(re^{-i\psi}\right) + \frac{k}{2}\left(r^3 e^{-i\psi} - re^{-i\psi}\right) + re^{-i\psi} = 0$$

$$\Rightarrow \frac{\partial r}{\partial t} e^{-i\psi} - ire^{-i\psi}\frac{\partial \psi}{\partial t} + \frac{k}{2}(r^3 e^{-i\psi} - re^{-i\psi}) + re^{-i\psi} = 0$$

Separating the real and imaginary parts, we have

$$\frac{\partial r}{\partial t} + \frac{k}{2}\left(r^3 - r\right) + r = 0 \tag{10}$$

$$\frac{\partial \psi}{\partial t} = 0 \tag{11}$$

On solving Eqs. (10) and (11), we get

$$r(t) = \sqrt{\frac{(1 - \frac{2}{K})}{\left|1 + \left[\frac{1 - \frac{2}{K}}{r(0)^2} - 1\right]e^{(1 - \frac{K}{2})t}\right|}}$$

From this we can see that for $K < 2$, which is the critical k value, the order parameter r goes to zero, signifying that the system is not synchronized. For $K > 2$, the order parameter is non-zero asymptotically. This is the same as was seen in the previous section, with $K_c = 2$.

3.2 Kuramoto System with External Force

The Kuramoto system with an external driving force (Sakaguchi 1988) is given by

$$\frac{d\theta_i}{dt} = \omega_i + \frac{K}{N} \sum_{j=1}^{N} sin(\theta_j - \theta_i) + Fsin(\Omega t - \theta_i) \tag{12}$$

where $i = 1 \ldots N$ denote the N oscillators.

R and ψ are defined exactly the same as in the last section. $R = re^{i\psi} = \frac{1}{N} \sum_{i=1}^{n} e^{i\theta_i}$.

Following the same method (Antonsen et al. 2008; Ott and Antonsen 2008; Childs and Strogatz 2008), we can get a different form of the same equation which is easier to work with in simulations. Therefore Eq. (12) is rewritten as

$$\frac{d\theta_i}{dt} = \omega_i + Krsin\left(\psi - \theta_i\right) + Fsin\left(\Omega t - \theta_i\right)$$

Now to reduce this infinite dimensional equation, we follow the same procedure as in the previous section.

For our calculations

$$\frac{d\theta_i}{dt} = \omega_i + \frac{K}{N} \left(\sum_{i=1}^{n} \frac{e^{i(\theta_j-\theta_i)} - e^{-i(\theta_j-\theta_i)}}{2i} \right) + F\frac{e^{i(\Omega t-\theta_i)} - e^{-i(\Omega t-\theta_i)}}{2i}$$

Now we put $\theta_i \longrightarrow \theta_i + \Omega t$

$$\frac{d\theta_i}{dt} = \omega_i - \Omega + \frac{K}{N} \left(\sum_{i=1}^{n} \frac{e^{i(\theta_j-\theta_i)} - e^{-i(\theta_j-\theta_i)}}{2i} \right) + F\frac{e^{-i\theta_i} - e^{i\theta_i}}{2i}$$

$$= \omega_i - \Omega + \frac{1}{2i}[(KR + F)e^{-i\theta_i} - (KR^* + F)e^{i\theta_i}]$$

Since K and R are real

$$\frac{d\theta_i}{dt} = \omega_i - \Omega + \frac{1}{2i} \left[(KR + F)e^{-i\theta_i} - (KR + F)^* e^{i\theta_i} \right]$$

For the continuous system as $N \longrightarrow \infty$, first we write the continuity equation which gives us

$$\Rightarrow \frac{\partial f}{\partial t} + \frac{\partial}{\partial \theta} \left[\omega - \Omega + \frac{1}{2i} \left[(KR + F)e^{-i\theta} - (KR + F)^* e^{i\theta} \right] f \right] \tag{13}$$

and R being the same as before. Now we do the Fourier expansion, take the same ansatz and check for the coefficients of $e^{i\theta}$, which gives us

$$\frac{\partial \alpha}{\partial t} + \frac{1}{2} \left[(KR + F)^* \alpha^2 - (KR + F) \right] + \left[1 + i(\Omega - \omega) \right] \alpha = 0$$

We again consider the Lorentzian distribution for ω where we put $\Delta = 1$ and $\omega = \omega_0$ and repeat the same process, after which we get

$$\frac{\partial R}{\partial t} + \frac{1}{2} \left[(KR + F)^* R^2 - (KR + F) \right] + \left[1 + i(\Omega - \omega_0) \right] R = 0$$

Considering $R = re^{i\psi}$, we can simplify this into two equations

$$\dot{r} = -\Delta r + \frac{(1 - r^2)}{2} (F \cos(\psi) + Kr) \tag{14}$$

$$\dot{\psi} = -(\Omega - \omega_0) - \frac{F}{2r}(1 + r^2) \sin(\psi) \tag{15}$$

This can be solved with the system being depended on the parameters $K, F, (\Omega - \omega_0)$ (Childs and Strogatz 2008).

4   Model of the System

In this section we study a synchronization of Kuramoto-like phase oscillators with a time-dependent external force. The equations of motion are as follows:

$$\left. \begin{array}{l} \dot{\theta}_i = \omega_i + \frac{K}{N} \sum_{j=1}^{N} \sin(\theta_i - \theta_j) - F \sin(\theta_i - \xi) \\ \dot{\xi} = \sigma - \frac{F}{N} \sum_{j=1}^{N} \sin(\xi - \theta_j) \end{array} \right\} \tag{16}$$

where $i = 1 \ldots N$ and N is number of single Kuramoto-like oscillators. Parameter K refers to strength of internal oscillators, ω_i is frequency of single node, F is external force and σ refers to external frequency. The phase of each Kuramoto-like system is given by θ_i and phase of external oscillator is ξ. In our study we vary parameters F, K and ω_i while external frequency is always fixed to $\sigma = 1.5$. The frequencies

ω_i are given by mean frequency ω_0, standard derivation and distribution (Gaussian, uniform or Lorenzian).

There are two ways to visualize this system. One way is to consider this as a set of Kuramoto oscillators experiencing a force. But unlike the normal force, in our system the force is dependent to an extent on the internal oscillators. That is this force is influence by the oscillators themselves.

The other way is to think of this system as a set of kuramoto oscillators being influence by a lone strongly coupled external kuramoto oscillator. We have mainly used this approach in explaining the results that we observed.

5 Numerical Study

Fig. 1 shows how the order parameter r changes with the strength of the external oscillator for both repulsive and attractive coupling (Figs. 1a and 1b respectively). The synchronization state appears for both type of coupling, however for a repulsive coupling, it occurs for smaller values of F then for the attractively coupled oscillators. Additionally, we observe two ranges of synchronous motion with desynchronization between them.

To see if these two regions are the same or distinct we calculate the ratio of mean frequency $\overline{\omega}$ of the oscillators to the frequency of the external oscillator σ (see Fig. 2) was plotted as shown in Fig. 2a. As can be seen here, these regions are qualitatively different. This implies that the first synchronization region refers to phase locked solution with ratio 2:1 between synchronized Kuramoto-like oscillators and external frequency. The second region is typical 1:1 locked state.

To study this phenomenon in detail, we increased the external frequency to 30 times higher then internal frequency. The ratios are plotted in Fig. 2b. Transition to synchronization state occurs every time the internal frequency reaches an integer multiple value of the external frequency. That is to say that the repulsively coupled oscillators synchronize separately into natural frequencies which are integer multiples of the external frequency and that this is a more general phenomenon with not two but many different internal synchronizations depending on the initial difference between the external and internal frequencies. Nevertheless it is clear that for higher ratios the plateau of synchronous motion becomes narrower.

Aforementioned results are for Lorentzian internal frequency distribution. To see how general this phenomenon is, we repeat the process with different initial frequency distribution for the internal oscillators, each time varying the standard deviation. Figure 3 shows the plot of the change of order parameter for varying strength of external oscillator for the Lorentzian distribution for varying standard deviation, that is 0.1, 0.01, 0.001. Figure 4 shows the same plot but with the internal oscillators now in an initial distribution that is Gaussian and Fig. 5 shows again the same plot, but now for a Uniform initial internal frequency distribution. As it is easy to see, the observed phenomenon is independent of the frequency distribution, although we observe that it is sensitive to the standard deviation. The region

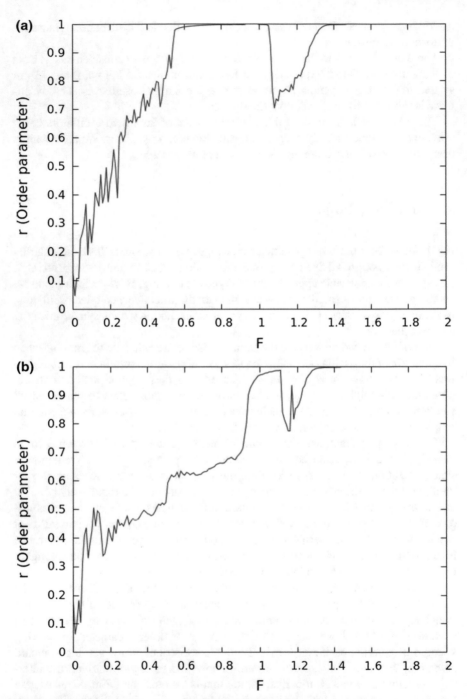

Fig. 1 Order parameter in repulsive and attractive coupling. The parameters are $\omega_0 = 0.5$, standard deviation $= 0.01$, $\sigma = 1.5$: **a** $K = 1.5$ for repulsive and **b** $K = -1.5$ for attractive coupling

Fig. 2 Plot of the change in the ratio of mean frequency $\overline{\omega}$ of the oscillators to frequency of the the external oscillator σ with respect to the parameter F for repulsive coupling for three values of $\sigma = 1.5$ (panel (a)), $\sigma = 2.5$ (panel (b)) and $\sigma = 15$ (panel (c)). The other parameters are $\omega_0 = 0.5$, standard deviation $= 0.01$ and $K = 1.5$

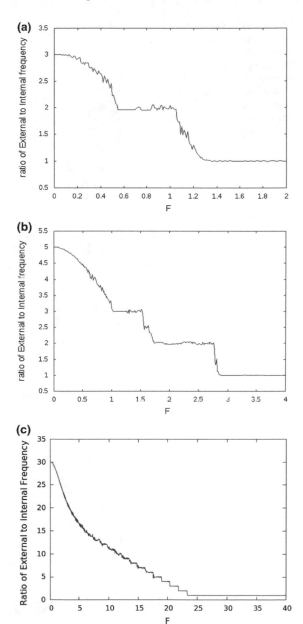

Fig. 3 Panels (a), (b) and (c) show the variation of order parameter r with increasing external oscillator interaction(F) for an initial ω_i distribution as Lorentzian with varying standard deviation of 0.1, 0.01, 0.001 respectively

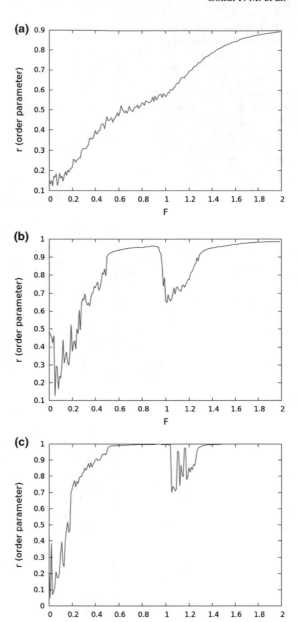

Fig. 4 Panels (a), (b) and (c) show the variation of order parameter r with increasing external oscillator interaction(F) for an initial ω_i distribution as Gaussian with varying standard deviation of 0.1, 0.01, 0.001 respectively

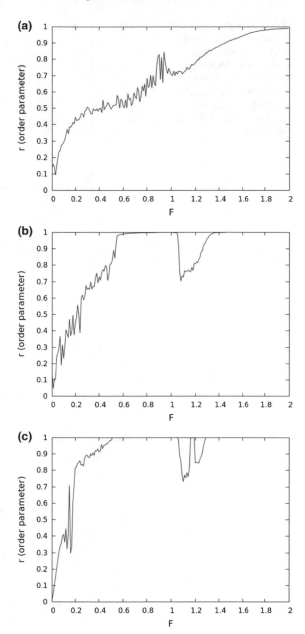

Fig. 5 Panels (a), (b) and
(c) show the variation of
order parameter r with
increasing external oscillator
interaction (F) for an initial
ω_i distribution as Uniform
with varying standard
deviation of 0.1, 0.01, 0.001
respectively

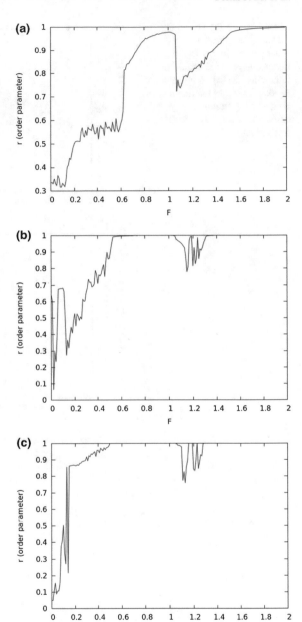

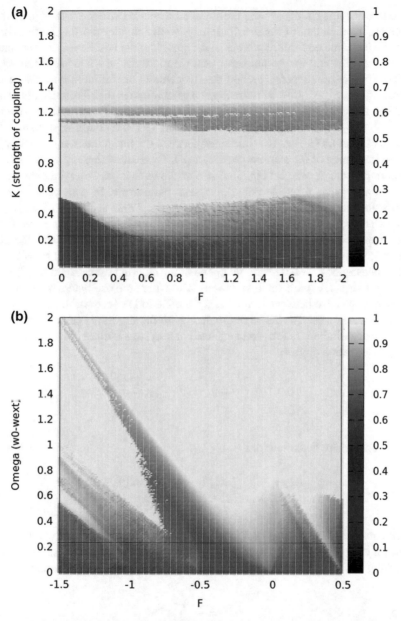

Fig. 6 **a** Shows the variation of the order parameter (in color) with respect to external oscillator interaction (F) and the strength of internal coupling (K) for a given $\Omega = -(\sigma - \omega_0)$, where $\omega_0 = 0.5, \sigma = 1.5$, for Lorentzian distribution as the initial oscillator frequency distribution. **b** Shows the variation of the order parameter (in false color) with respect to the external force/oscillator interaction (F) and $\Omega = -(\sigma - \omega_0)$, where $\omega_0 = 0.5, \sigma$ increases from -1 to 1, for Lorentzian distribution as the initial oscillator frequency distribution for a fixed strength of internal coupling (K) = 1.5

of the transition is changing with the standard deviation whereas the regions themselves are independent of the distribution. Also the smaller the standard deviation, the more pronounced the effect. That is, for more identical oscillator, we see that the different regions and their boundaries become more distinct. This could mean that the transition region depends on how nearly identical the initial frequencies are.

Now we base the initial $\omega - distribution$ as Lorentzian distribution again. As can be noted, there are three main parameters of consideration in the system equations. They are, the strength of the external oscillators (F), the coupling strength of the internal oscillators (K) and the difference between the mean internal frequency (ω_0) and the frequency of the external oscillator (σ). To see how theses parameters affect the order parameter, we plot Fig. 6. This plot shows that the observed phenomenon is robust and exists for wide range of system parameters. In both panels we vary external force F in range $F \in [0, 2]$, however vertical axis is different, i.e., in panel (a) we change strength of external oscillators K from 0 to 2.0 and in panel (b) the difference $\Omega = -(\sigma - \omega_0)$ between the external frequency and the mean internal frequency. In color we show the order parameter r. In Fig. 6a synchronous range (yellow color) appears for $K > 0.6$ nearly independ of the external oscillator strength F. It is present till $K \approx 1.1$ (with some variations with where the oscillators desynchronize a change in the ratio between frequencies from 2:1 to 1:1 occurs. Above $K \approx 1.3$ the synchronous state persists and is stable. In Fig. 6b the structure is more complex due to change of ratio between the frequencies. For most of Ω values two or more ranges of synchronization appear.

6 Analysis

The system we are dealing with is

$$\left.\begin{array}{c} \dot{\theta}_i = \omega_i + \frac{K}{N} \sum_{j=1}^{N} \sin(\theta_i - \theta_j) - F \sin(\theta_i - \xi) \\ \dot{\xi} = \sigma - \frac{F}{N} \sum_{j=1}^{N} \sin(\xi - \theta_j) \end{array}\right\} \tag{17}$$

which consists of N coupled oscillators which are acting under a force, whose frequency is coupled with the system using an attractive coupling. It is easier to see this by writing Eq. (17) using the mean field ψ and the order parameter R defined as

$$Re^{i\psi} = \frac{1}{N} \sum_{j=1}^{N} e^{i\theta_j} \tag{18}$$

using (18), we can rewrite (17) as

$$\left.\begin{array}{c} \dot{\theta}_i = \omega_i + KR \sin(\theta_i - \psi) - F \sin(\theta_i - \xi) \\ \dot{\xi} = \sigma - FR \sin(\xi - \psi) \end{array}\right\} \tag{19}$$

This is an easier form to understand. This system has atleast two different ways of attaining synchronisation: with either the mean field ψ or the forcing field ϕ. In fact both these types of synchronisation has been oberved in this system as can be seen from the Fig. 1.

Now to analyze this system we go into a rotating frame. That is we do the transformation $\theta_i \rightarrow \theta_i - \xi$. After which we follow the procedure given by Ott and Antonsen and taking the $\omega - distribution$ to be Lorentzian to finally arrive at

$$\dot{R} = \frac{1}{2}\left[(KR + F)^* R^2 - (KR + F)\right] - \frac{F}{2}(R - R^*) - \left[\Delta + i(\sigma - \omega_0)\right] R$$

where ω_0 is the mean and Δ the width of the $\omega - distribution$.

Now by putting $R = re^{i\psi}$, we can simplify the whole system into the following two equations:

$$\dot{r} = -\Delta r + \frac{(1 - r^2)}{2}(F \cos(\psi) - Kr) \tag{20}$$

$$\dot{\psi} = \Omega - \frac{F}{2r}(1 + 3r^2)\sin(\psi) \tag{21}$$

where $\Omega = -(\sigma - \omega_0)$, ψ is the mean field of the full system.

Figure 7 shows the order parameter as calculated from the equation with increasing value of F for different Δ values along with the respective numerical result.

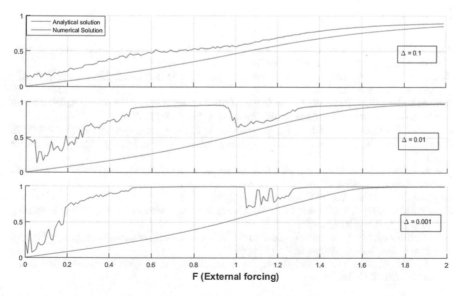

Fig. 7 Analytical and numerical plots for order parameter with F for different Δ values

7 Conclusion

We observed that in the presence of a strong attractively coupled oscillator, a system of repulsively coupled kuramoto oscillators reach synchronization faster than a system of attractively coupled oscillators. There are different regions of synchronization of the internal oscillators, characterized by their frequencies. They are such that the mean of the internal oscillators are integer multiples of the external oscillator frequency.

Next found out that this phenomenon is independent of the initial frequency distribution by repeating the numerical simulations for a total of three different initial internal frequency distributions: namely the Lorentzian, Gaussian and uniform distributions.

We also observed that even though this phenomenon occurs independent of the initial internal frequency distribution, they are dependent on the width of the distribution, which characterizes how close these initial values are and the smaller the width, the more clearer these separate regions became. In other words the more identical the initial distribution, the more clear the observed phenomenon.

Finally, we solved this system of equations using the method given by Ott-Antonsen. After having done the analysis, it was discovered that this phenomenon was not seen analytically.

Acknowledgements This work has been supported by the Polish National Centre, Maestro program-Project No. 2013/08/ST8/00/780.

References

Antonsen TM, Faghih R Jr, Girvan M, Ott E, Platig J (2008) External periodic driving of large systems of globally coupled phase oscillators. Chaos 18:037112
Boccaletti S, Kurths J, Osipov G, Valladares DL, Zhou CS (2002) The synchronization of chaotic systems. Phys Rep 366:1–101
Bountis T, Kanas VG, Hizanidis J, Bezerianos A (2014) Chimera states in a twopopulation network of coupled pendulumlike elements. Eur Phys J Spec Top 223:721–728
Childs LM, Strogatz SH (2008) Stability diagram for the forced Kuramoto model. Chaos 18:043128
Flandoli F, Gess B, Scheutzow M (2017) Synchronization by noise. Probab Theory Relat Fields 168:511–556
Hubler R, Bartussek R, Hnggi P (1997) Coupled Brownian rectifiers. In: AIP conference proceedings vol 411, pp 243–248
Huygens C (1673) Horologium oscillatorium, Muget, Paris
Jaros P, Maistrenko Y, Kapitaniak T (2015) Chimera states on the route from coherence to rotating waves. Phys Rev E 91:022907
Jaros P, Brezetsky S, Levchenko R, Dudkowski D, Kapitaniak T, Maistrenko Y (2017) Solitary states for coupled oscillators. arXiv:1703.06950
Kiss IZ, Zhai Y, Hudson JL (2002) Emerging coherence in a population of chemical oscillators. Science 296:1676–1678

Kuramoto Y (1975) Self-entrainment of a population of coupled non-linear oscillators In: Araki H, (ed) Lecture notes in physics, international symposium on mathematical problems in theoretical physics, vol 34, Springer, New York, pp 420–422

Kuramoto Y (1984) Chemical oscillations, waves, and turbulence. Springer, Berlin

Maistrenko Y, Brezetsky S, Jaros P, Levchenko R, Kapitaniak T (2017) Smallest chimera states. Phys Rev E 95:010203

Olmi S, Navas A, Boccaletti S, Torcini A (2014) Hysteretic transitions in the Kuramoto model with inertia. Phys Rev E 90:042905

Ott E, Antonsen T (2008) Low dimensional behavior of large systems of globally coupled oscillators. Chaos 18:037113

Pantaleone J (1998) Stability of incoherence in an isotropic gas of oscillating neutrinos. Phys Rev D 58:073002

Pikovsky A, Rosenblum M, Kurths J (2001) Synchronization: a universal concept in nonlinear science. Cambridge University Press, Cambridge

Sakaguchi H (1988) Cooperative phenomena in coupled oscillator systems under external fields. Prog Theor Phys 79(1):3946

Strogatz SH (2004) Sync: the emerging science of spontaneous order. Penguin Science, New York

Strogatz SH (2008) From kuramoto to crawford: exploring the onset of synchronization in populations of coupled oscillators. Physica D 143:1–4

Wiesenfeld K, Swift JW (1995) Averaged equations for Josephson junction series arrays. Phys Rev E 51:1020–1025

Yamaguchi S, Isejima H, Matsuo T, Okura R, Yagita K, Kobayashi M, Okamura H (2003) Synchronization of cellular clocks in the suprachiasmatic nucleus. Science 302:1408–1412

Analysis, Circuit Design and Synchronization of a New Hyperchaotic System with Three Quadratic Nonlinearities

A. A. Oumate, S. Vaidyanathan, K. Zourmba, B. Gambo
and A. Mohamadou

Abstract Hyperchaos has important applications in many branches of science and engineering. In this work, we propose a new 4-D hyperchaotic system with three quadratic nonlinearities by modifying the dynamics of hyperchaotic Wang system (Wang et al. 2010). The proposed new hyperchaotic system has a unique equilibrium at the origin, which is a saddle point and unstable. Thus, the new hyperchaotic system exhibits self-excited hyperchaotic attractor. We describe qualitative properties of the new hyperchaotic system such as symmetry, Lyapunov exponents, Kaplan-Yorke dimension, etc. Furthermore, an active control method is derived for the synchronization of two identical new hyperchaotic systems. The circuit experimental results of the new hyperchaotic system show agreement with the numerical simulations.

Keywords Hyperchaos · Hyperchaotic systems · Nonlinear systems · Active control · Synchronization · Circuit design

A. A. Oumate (✉) · K. Zourmba · B. Gambo · A. Mohamadou
Faculty of Science, Department of Physics, University of Maroua,
P.O. Box 814, Maroua, Cameroon
e-mail: oumat_oaa@yahoo.fr

K. Zourmba
e-mail: zourmba@yahoo.com

B. Gambo
e-mail: gambobetch@yahoo.fr

A. Mohamadou
e-mail: mohdoufr@yahoo.fr

S. Vaidyanathan
Research and Development Centre,
Vel Tech University, Avadi, Chennai 600062, Tamil Nadu, India
e-mail: sundarvtu@gmail.com

© Springer International Publishing AG 2018   251
V.-T. Pham et al. (eds.), *Nonlinear Dynamical Systems with Self-Excited and Hidden Attractors*, Studies in Systems, Decision and Control 133,
https://doi.org/10.1007/978-3-319-71243-7_11

1 Introduction

The subject of chaos synchronization has received a great attention since 1990 (Pecora and Carroll 1990) and grown rapidly, both theoretically and experimentally (Azar and Vaidyanathan 2015, 2016, 2017; Vaidyanathan and Volos 2016a, b, 2017). Chaos control and synchronization have applications in several areas such as memristors (Azar and Vaidyanathan 2017; Pham et al. 2015, 2016b), chemical reactors (Vaidyanathan 2015b, c, e), oscillators (Yu et al. 2016; Vaidyanathan 2015f, i), secure communication (Xu et al. 2017; Wang et al. 2016), cryptosystems (Ahmad et al. 2015), robotics (Jafarov et al. 2016), neural networks (Sadeghpour et al. 2012; Vaidyanathan 2015a, j), etc.

Synchronization of chaotic systems considers a pair of chaotic systems called *master* and *slave* systems, and it aims to achieve asymptotic tracking of the states of the slave system to the states of the master system. Because of the butterfly effect of chaotic systems, the synchronization of chaotic systems is a challenging problem in the literature (Azar and Vaidyanathan 2015, 2016, 2017; Vaidyanathan and Volos 2016a, b). In the literature, there are also other type of synchronization problems such as generalized synchronization (Kocarev and Parlitz 1996; Yang and Duan 1998), phase synchronization (Mg et al. 1996), anti-synchronization (Hammami et al. 2010; Vaidyanathan 2015d; Vaidyanathan et al. 2015b), hybrid synchronization (Li 2005; Vaidyanathan 2015g, h), lag synchronization (Taherion and Lai 1999), generalized projective synchronization (Sarasu and Sundarapandian 2012, 2011b; Vaidyanathan and Pakiriswamy 2016), etc.

Many control techniques are developed to synchronize chaotic systems such as active control (Sundarapandian 2013; Sarasu and Sundarapandian 2011a; Karthikeyan and Sundarapandian 2014), adaptive control (Tirandaz and Hajipour 2017; Fotsin and Bowong 2006; Sundarapandian and Karthikeyan 2012), backstepping control (Vaidyanathan and Rasappan 2014; Vaidyanathan 2017), sliding mode control (Vaidyanathan and Sampath 2012, 2017; Vaidyanathan et al. 2015a; Lakhekar et al. 2016; Vaidyanathan and Rhif 2017; Vaidyanathan 2014), etc.

There is good interest in exploiting chaotic dynamics in engineering applications, where some attention has been focused on effectively creating chaos via simple physical systems, such as electronic circuits (Sundarapandian and Pehlivan 2012; Pehlivan et al. 2014; Akgul et al. 2016; Pham et al. 2016a; Volos et al. 2017). The pursuit of designing circuits to produce chaotic attractors has become a focal point for engineers, not only because of their theoretical interest, but also due to their potential real-world applications in various chaos-based technologies and information systems (Vaidyanathan and Volos 2016a, b, 2017; Azar and Vaidyanathan 2017).

Recently, chaotic systems are classified into two types of attractors, viz. *self-excited* and *hidden attractors* (Leonov et al. 2011, 2012, 2015). An attractor is called a self-excited attractor if its basin of attraction intersects an arbitrarily small open neighborhood of equilibrium. Otherwise, the attractor is called a hidden attractor. Thus, hidden attractor has basin of attraction which does not overlap with an arbitrarily small vicinity of equilibria. For example, hidden attractors are attractors in systems without equilibria or with only one stable equilibrium.

In this chapter, we first discuss the qualitative properties of the hyperchaotic Wang system (Wang et al. 2010) with three quadratic nonlinearities. We show that the hyperchaotic Wang system exhibits self-excited hyperchaotic attractor. Next, by modifying the dynamics of the hyperchaotic Wang system (Wang et al. 2010), we propose a new 4-D hyperchaotic system with three quadratic nonlinearities.

The proposed new hyperchaotic system has a unique equilibrium at the origin, which is a saddle point and unstable. Thus, the new hyperchaotic system exhibits self-excited hyperchaotic attractor. Indeed, the new hyperchaotic system has a unique equilibrium at the origin, which is unstable. We describe the qualitative properties of the new hyperchaotic system such as symmetry, Lyapunov exponents, Kaplan-Yorke dimension, etc.

Furthermore, we discuss the numerical simulation and circuit realization of synchronization of the new hyperchaotic systems (Wang et al. 2010) via active control. The stability of the complete synchronization error system is assured by Lyapunov criterion to prove that the error vector approaches zero as time approaches infinity.

The rest of this chapter is organized as follows. In Sect. 2, we describe the dynamics and analysis of the new hyperchaotic system. In Sect. 3, we derive new results for the active control of identical new hyperchaotic systems. In Sect. 4, we discuss the numerical simulations of the synchronization scheme developed in Sect. 3 for the identical new hyperchaotic systems. In Sect. 5, we provide a circuit design of the new hyperchaotic system and the active controller designed in Sect. 3. The circuital design results confirm the feasibility of the theoretical model. Section 6 contains the main conclusions.

2 A New Hyperchaotic System

In this section, we first consider the hyperchaotic Wang system (Wang et al. 2010) given by

$$\begin{cases} \dot{x} = a(y - x) \\ \dot{y} = bx + cy - xz + w \\ \dot{z} = y^2 - hz \\ \dot{w} = -dx \end{cases} \tag{1}$$

where x, y, z, w are the states and a, b, c, d, h are real parameters.

The hyperchaotic Wang system (1) is a nonlinear autonomous system with three quadratic nonlinearities.

In Wang et al. (2010), it was shown that the system (1) undergoes hyperchaotic behavior when the parameters take the values

$$a = 27.5, \quad b = 3, \quad c = 19.3, \quad d = 3.3, \quad h = 2.9 \tag{2}$$

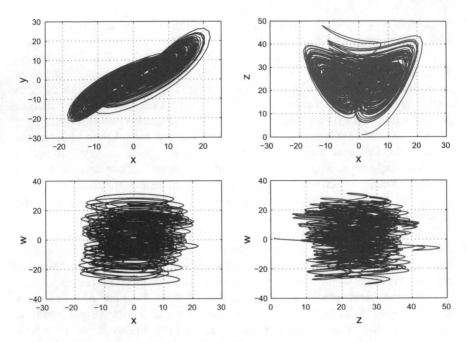

Fig. 1 2-D phase portraits of the hyperchaotic Wang system

For numerical simulations, we take the initial values

$$x(0) = 1, \quad y(0) = 1, \quad z(0) = 1, \quad w(0) = 1 \tag{3}$$

Figure 1 shows the phase portraits of the hyperchaotic Wang system (1) for the parameter values (2) and the initial state (3).

The Lyapunov exponents of the hyperchaotic Wang system (1) for the parameter values (2) and the initial state (3) are calculated using Wolf's algorithm (Wolf et al. 1985) as

$$L_1 = 1.67, \quad L_2 = 0.1, \quad L_3 = 0, \quad L_4 = -12.87 \tag{4}$$

Since there are two positive Lyapunov exponents in (4), we find that the 4-D Wang system (1) is hyperchaotic.

Also, the Kaplan-Yorke dimension of the hyperchaotic Wang system (1) is found as

$$D_{KY} = 3 + \frac{L_1 + L_2 + L_3}{|L_4|} = 3.1375, \tag{5}$$

which shows the complexity of the system.

Since the sum of the Lyapunov exponents in (4) is negative, the hyperchaotic Wang system (1) is dissipative.

Figure 2 shows the Lyapunov exponents of the hyperchaotic Wang system (1).

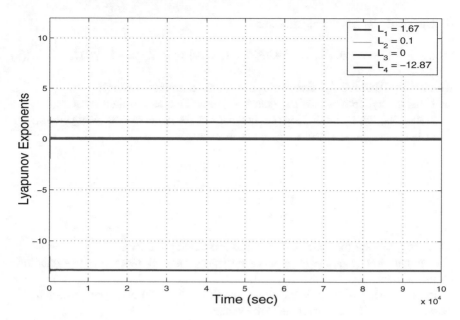

Fig. 2 Lyapunov exponents of the hyperchaotic Wang system

The equilibrium points of the hyperchaotic Wang system (1) are obtained by solving the following system of equations:

$$a(y - x) = 0 \tag{6a}$$
$$bx + cy - xz + w = 0 \tag{6b}$$
$$y^2 - hz = 0 \tag{6c}$$
$$-dx = 0 \tag{6d}$$

From (6d), we get $x = 0$. Substituting $x = 0$ in (6a), we get $y = 0$.
Substituting $y = 0$ in (6c), we get $z = 0$.
Substituting $x = y = z = 0$ in (6b), we get $w = 0$.
Thus, we conclude that the hyperchaotic Wang system (1) has a unique equilibrium point at $E_0 = \mathbf{0}$.
To test the stability type of the equilibrium E_0, we take the parameter values of the hyperchaotic Wang system (1) as in the hyperchaotic case (2).
Then we obtain the Jacobian matrix

$$J_0 = J(E_0) = \begin{bmatrix} -27.5 & 27.5 & 0 & 0 \\ 3 & 19.3 & 0 & 1 \\ 0 & 0 & -2.9 & 0 \\ -3.3 & 0 & 0 & 0 \end{bmatrix} \tag{7}$$

The matrix J_0 has the eigenvalues

$$\lambda_1 = -2.9, \quad \lambda_2 = -29.2627, \quad \lambda_3 = 0.1483, \quad L_4 = -29.2627, \tag{8}$$

which shows that the equilibrium E_0 is a saddle-point and unstable.

Thus, the hyperchaotic Wang system (1) exhibits a self-excited attractor.

In this chapter, by modifying the dynamics of the hyperchaotic Wang system (1), we obtain a new hyperchaotic system as follows:

$$\begin{cases} \dot{x} = a(y - x) \\ \dot{y} = bx + cy - xz + w \\ \dot{z} = y^2 - hz \\ \dot{w} = -dy \end{cases} \tag{9}$$

where x, y, z, w are the states and a, b, c, d, h are real parameters.

Our new 4-D system (9) is a nonlinear autonomous system with three quadratic nonlinearities.

In this work, we shall demonstrate that the system (9) undergoes hyperchaotic behavior when the parameters take the values

$$a = 27.5, \quad b = 3.5, \quad c = 19.5, \quad d = 3.3, \quad h = 3 \tag{10}$$

For numerical simulations, we take the initial values

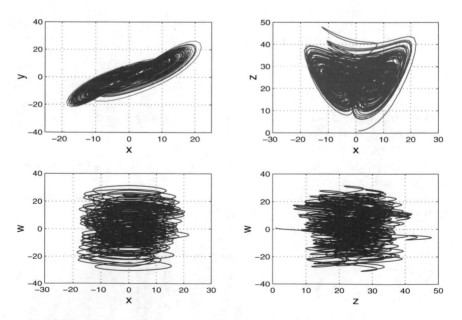

Fig. 3 2-D phase portraits of the new hyperchaotic system

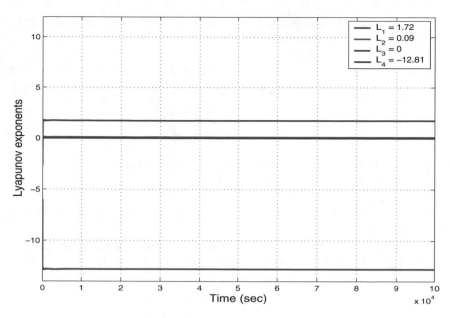

Fig. 4 Lyapunov exponents of the new hyperchaotic system

$$x(0) = 1, \quad y(0) = 1, \quad z(0) = 1, \quad w(0) = 1 \tag{11}$$

Figure 3 shows the phase portraits of the new hyperchaotic system (9) for the parameter values (10) and the initial state (11).

The Lyapunov exponents of the new hyperchaotic system (9) for the parameter values (10) and the initial state (11) are calculated using Wolf's algorithm (Wolf et al. 1985) as

$$L_1 = 1.67, \quad L_2 = 0.1, \quad L_3 = 0, \quad L_4 = -12.87 \tag{12}$$

Since there are two positive Lyapunov exponents in (12), we find that our 4-D system (9) is hyperchaotic.

Also, the Kaplan-Yorke dimension of the new hyperchaotic system (9) is found as

$$D_{KY} = 3 + \frac{L_1 + L_2 + L_3}{|L_4|} = 3.1413, \tag{13}$$

which is greater than the Kaplan-Yorke dimension of the hyperchaotic Wang system (1). This shows that our hyperchaotic system (9) exhibits more complexity than the hyperchaotic Wang system (9).

Since the sum of the Lyapunov exponents in (12) is negative, the hyperchaotic Wang system is dissipative.

Figure 4 shows the Lyapunov exponents of the new hyperchaotic system (9).

The equilibrium points of the new hyperchaotic system (9) are obtained by solving the following system of equations:

$$a(y - x) = 0 \tag{14a}$$
$$bx + cy - xz + w = 0 \tag{14b}$$
$$y^2 - hz = 0 \tag{14c}$$
$$-dy = 0 \tag{14d}$$

From (14d), we get $y = 0$. Substituting $y = 0$ in (14a), we get $x = 0$.
Substituting $y = 0$ in (14c), we get $z = 0$.
Substituting $x = y = z = 0$ in (14b), we get $w = 0$.
Thus, we conclude that the new hyperchaotic system (9) has a unique equilibrium point at $E_0 = \mathbf{0}$.

To test the stability type of the equilibrium E_0, we take the parameter values of the new hyperchaotic system (9) as in the hyperchaotic case (10).

Then we obtain the Jacobian matrix

$$J_0 = J(E_0) = \begin{bmatrix} -27.5 & 27.5 & 0 & 0 \\ 3.5 & 19.5 & 0 & 1 \\ 0 & 0 & -3 & 0 \\ 0 & -3 & 0 & 0 \end{bmatrix} \tag{15}$$

The matrix J_0 has the eigenvalues

$$\lambda_1 = -3, \quad \lambda_2 = -29.4617, \quad \lambda_3 = 0.1313, \quad L_4 = -29.4617, \tag{16}$$

which shows that the equilibrium E_0 is a saddle-point and unstable.

Thus, the new hyperchaotic system (9) exhibits a self-excited attractor.

We also note that the new hyperchaotic system (9) stays invariant under the change of coordinates given by

$$(x, y, z, w) \mapsto (-x, -y, z, -w) \tag{17}$$

for all values of the parameters. This shows that the new hyperchaotic system (9) has rotation symmetry about the z-axis and that every non-trivial trajectory of the new hyperchaotic system (9) must have a twin trajectory.

We also observe that the z-axis is invariant under the flow of the new hyperchaotic system (9), and the invariant flow on the z-axis is characterized by the 1-D dynamics

$$\dot{z} = -hz \tag{18}$$

which is globally exponentially stable since $h > 0$.

3 Synchronization of the Identical New Hyperchaotic Systems

3.1 Problem Description

As the master system, we consider the new hyperchaotic system given by

$$
\begin{cases}
\dot{x}_m(t) = a(y_m(t) - x_m(t)) \\
\dot{y}_m(t) = bx_m(t) + cy_m(t) - x_m(t)z_m(t) + w_m(t) \\
\dot{z}_m(t) = y_m^2(t) - hz_m(t) \\
\dot{w}_m(t) = -dy_m(t)
\end{cases}
\tag{19}
$$

As its slave system, we consider the following hyperchaotic system given by

$$
\begin{cases}
\dot{x}_s(t) = a(y_s(t) - x_s(t)) + u_1(t) \\
\dot{y}_s(t) = bx_s(t) + cy_s(t) - x_s(t)z_s(t) + w_s(t) + u_2(t) \\
\dot{z}_s(t) = y_s^2(t) - hz_s(t) + u_3(t) \\
\dot{w}_s(t) = -dy_s(t)
\end{cases}
\tag{20}
$$

where $u(t) = \begin{bmatrix} u_1(t) & u_2(t) & u_3(t) \end{bmatrix}^T$ is the active control to be determined to ensure the complete synchronization of the new identical hyperchaotic systems.

We define the synchronization error between the new hyperchaotic systems (19) and (20) as

$$
\begin{cases}
e_{1s}(t) = x_s(t) - x_m(t) \\
e_{2s}(t) = y_s(t) - y_m(t) \\
e_{3s}(t) = z_s(t) - z_m(t) \\
e_{4s}(t) = w_s(t) - w_m(t)
\end{cases}
\tag{21}
$$

The error dynamics is obtained as

$$
\begin{cases}
\dot{e}_{1s} = a(e_{2s} - e_{1s}) + u_1 \\
\dot{e}_{2s} = be_{1s} + ce_{2s} + e_{4s} - x_s z_s + x_m z_m + u_2 \\
\dot{e}_{3s} = -he_{3s} + y_s^2 - y_m^2 + u_1 \\
\dot{e}_{4s} = -de_{2s}
\end{cases}
\tag{22}
$$

We can express the error dynamics (22) in matrix form as

$$
\dot{e}_s = A_e e_s + Bu
\tag{23}
$$

where

$$
e_s(t) = \begin{bmatrix} e_{1s}(t) & e_{2s}(t) & e_{3s}(t) & e_{4s}(t) \end{bmatrix}^T
\tag{24}
$$

and

$$
A_e = \begin{bmatrix} -a & a & 0 & 0 \\ b - z_m & c & -x_s & 1 \\ 0 & y_s + y_m & -h & 0 \\ 0 & -d & 0 & 0 \end{bmatrix}, \quad B = \begin{bmatrix} 1 & 0 & 0 \\ 0 & 1 & 0 \\ 0 & 0 & 1 \\ 0 & 0 & 0 \end{bmatrix} \tag{25}
$$

3.2 Main Results

In this subsection, we derive a new active control law for achieving complete synchronization of the new hyperchaotic systems (19) and (20). Our active controller design makes use of the practical stability criterion of Borne and Gentina (Borne and Benjerab 2008).

We consider an active control law of the form

$$
u = -K e_s \tag{26}
$$

Substituting (26) into the error dynamics (23) leads to the closed-loop error system

$$
\dot{e}_s = A_{es} e_s \tag{27}
$$

where

$$
A_{es} = A_e - BK = \begin{bmatrix} -a - k_{11} & a - k_{12} & -k_{13} & -k_{14} \\ b - z_m - k_{21} & c - k_{22} & -x_s - k_{23} & 1 - k_{24} \\ -k_{31} & y_s + y_m - k_{32} & -h - k_{33} & -k_{34} \\ 0 & -d & 0 & 0 \end{bmatrix} \tag{28}
$$

We establish stability of the closed-loop error system (27) by applying the practical stability criterion of Borne and Gentina (Borne and Benjerab 2008), which is associated with the Benrejeb arrow form matrix.

To satisfy this aim, the parameters of the gain matrix K can be chosen as follows.

$$
\begin{cases} a - k_{12} = 0 \\ -k_{13} = 0 \\ b - z_m - k_{21} = 0 \\ -k_{31} = 0 \\ y_s + y_m - k_{32} = 0 \\ -x_s - k_{23} = 0 \end{cases} \implies \begin{cases} k_{12} = a \\ k_{13} = 0 \\ k_{21} = b - z_m \\ k_{31} = 0 \\ k_{32} = y_s + y_m \\ k_{23} = -x_s \end{cases} \tag{29}
$$

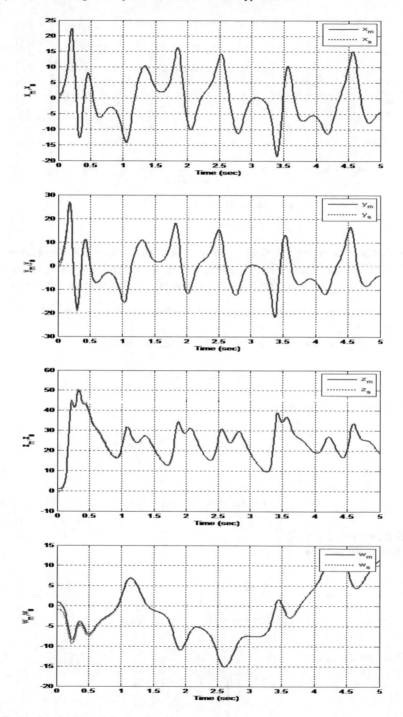

Fig. 5 Synchronization of the new hyperchaotic systems

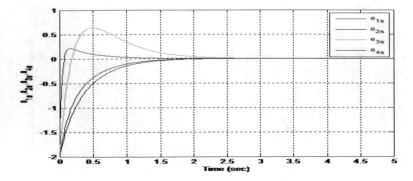

Fig. 6 Time-history of the synchronization errors

By application of the classical Borne and Gentina stability criterion (Borne and Benjerab 2008), associated to the particular canonical Benrejeb arrow matrix, the instantaneous characteristic matrix A_{es} must satisfy the following three conditions.

(C1) The nonlinear elements are isolated in either one row or one column of the matrix A_{es}.
(C2) The diagonal elements, a_{ii}, of the matrix A_{es} are such that:

$$a_{ii} < 0, \quad \forall i = 1, 2, 3 \tag{30}$$

(C3) There exist $\varepsilon > 0$ such that:

$$(c - k_{22})[-(k_{13}k_{34} - dk_{14}(h + k_{33}))] < -\varepsilon \tag{31}$$

The condition (C1) leads to

$$1 - k_{24} = 0 \quad \text{or} \quad k_{24} = 1 \tag{32}$$

Thus, we take $k_{24} = 1$.
The condition (C2) leads to

$$\begin{cases} -a - k_{11} < 0 \\ c - k_{22} < 0 \\ -h - k_{33} < 0 \end{cases} \implies \begin{cases} -a < k_{11} \\ c < k_{22} \\ -h < k_{33} \end{cases} \tag{33}$$

Thus, we choose

$$\begin{cases} k_{11} = 1 \\ k_{22} = 30 \\ k_{33} = 1 \end{cases} \tag{34}$$

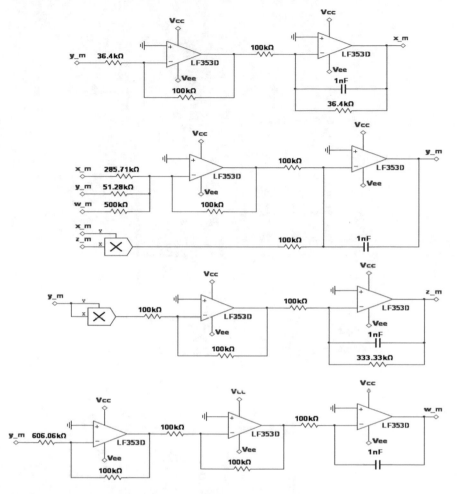

Fig. 7 Circuit diagram of the drive system (19)

For the gains k_{12}, k_{13}, k_{21}, k_{31}, k_{32}, k_{23}, k_{24}, k_{11}, k_{22} and k_{33} defined in (29), (32) and (34), the condition (C3) leads to

$$\begin{cases} k_{14} = 5 \\ k_{34} = 5 \end{cases}$$

Among the various choices of the gain matrix K, one possible choice is the following matrix.

$$K = \begin{bmatrix} 1 & a & 0 & 5 \\ b - z_m & 30 & -x_s & 1 \\ 0 & y_m + y_s & 1 & 5 \end{bmatrix} \tag{35}$$

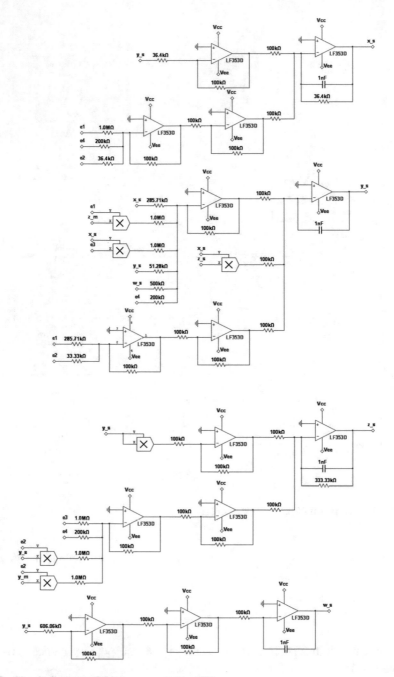

Fig. 8 Circuit diagram of the response system (20)

Fig. 9 Circuit diagram of the synchronization error dynamics (22)

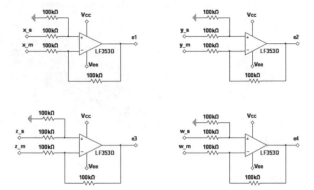

4 Numerical Simulation

For numerical simulation, we use classical fourth-order Runge-Kutta method in MATLAB.

We take parameters of the new hyperchaotic systems (19) and (20) as in the hyperchaotic case (10).

The gain matrix K is chosen as in (35).

The initial conditions of the master system (19) are taken as

$$x_m(0) = 1, \quad y_m(0) = 1, \quad z_m(0) = 1, \quad w_m(0) = 1 \tag{36}$$

The initial conditions of the slave system (20) are taken as

$$x_s(0) = -1, \quad y_s(0) = -1, \quad z_s(0) = -1, \quad w_s(0) = -1 \tag{37}$$

Figure 5 shows the synchronization between the new hyperchaotic systems (19) and (20).

Figure 6 shows the time-history of the synchronization errors.

5 Circuit Design of the New Hyperchaotic System

In this section, we realize the circuit of the synchronized new hyperchaotic systems with Electronic Work Bench (EWB). We use electronic components: operational amplifiers, resistors and capacitors to realize the system equations. We select LF353D as the amplifier and AD633JN as the multiplier to design the synchronous circuit. The input supply was $V_{cc} = +15$ V and $V_{ee} = -15$ V. In order to restrict the change of state variables to the operating voltage of the analog circuit, the state variables are reduced by 10 and 20 times, namely let $(x_m, y_m, z_m, w_m) \longrightarrow (10x_m, 10y_m, 10z_m, 20w_m)$ and $(x_s, y_s, z_s, w_s) \longrightarrow (10x_s, 10y_s, 10z_s, 20w_s)$.

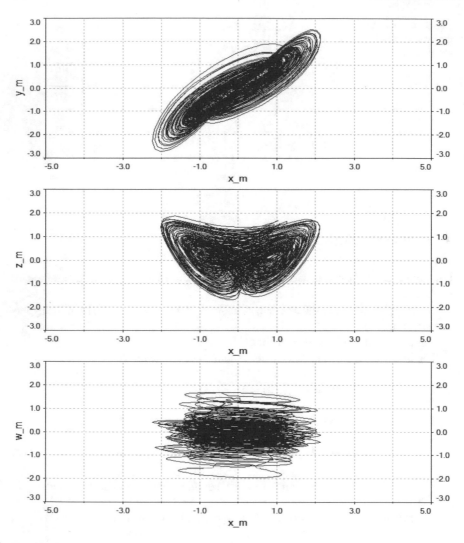

Fig. 10 Circuit simulation of phase portraits of the new hyperchaotic system (19)

The circuit schematics of the drive system (19) and the response system (20) are given respectively in Figs. 7 and 8, where the values of resistors and capacitors are indicated. Figure 9 depicts the circuit design of the error dynamics. Figure 10 presents the experimental simulation of the phase portraits of the drive system (19). Figure 11 shows the circuit simulations of the synchronization of drive and response systems which indicates that complete synchronization of new hyperchaotic systems is achieved.

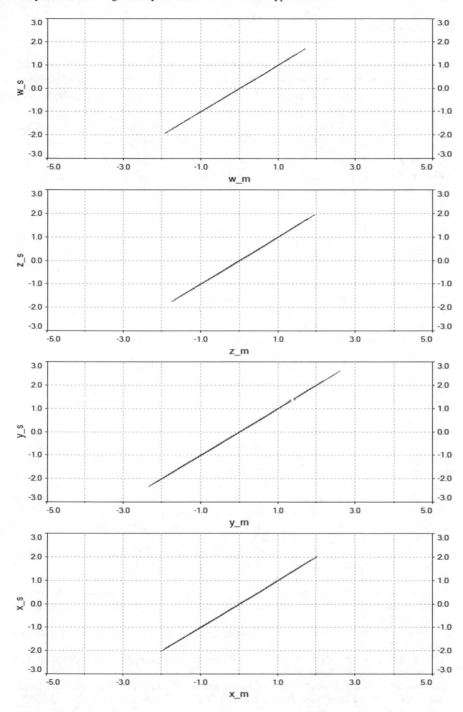

Fig. 11 Circuit simulation of synchronization between new hyperchaotic systems

6 Conclusion

In this paper, we introduced a new hyperchaotic system and investigated the qualitative properties of the system such as Lyapunov exponents, Kaplan-Yorke dimension, equilibria, dissipativity, etc. We noted that the new system with three quadratic non-linearities exhibits a self-excited hyperchaotic attractor. Next, we derived new results for the complete synchronization of new hyperchaotic system via active control law. To verify the feasibility of the theoretical model, the electronic circuits of the new hyperchaotic systems have been designed and the circuital simulation results are in good accordance with the theoretical model for the synchronization of the new hyperchaotic systems via active control.

References

Ahmad M, Shamsi U, Khan IR (2015) An enhanced image encryption algorithm using fractional chaotic systems. Procedia Comput Sci 57:852–859

Akgul A, Moroz I, Pehlivan I, Vaidyanathan S (2016) A new four-scroll chaotic attractor and its engineering applications. Optik 127(13):5491–5499

Azar AT, Vaidyanathan S (2015) Chaos modeling and control systems design. Springer, Berlin, Germany

Azar AT, Vaidyanathan S (2016) Advances in chaos theory and intelligent control. Springer, Berlin, Germany

Azar AT, Vaidyanathan S (2017) Fractional order control and synchronization of chaotic systems. Springer, Berlin, Germany

Borne B, Benjerab M (2008) On the representation and the stability study of the large scale systems. Int J Comput Commun Syst 3:55–66

Fotsin H, Bowong S (2006) Adaptive control and synchronization of chaotic systems consisting of Van der Pol oscillators coupled to linear oscillators. Chaos Solitons Fractals 27:822–835

Hammami S, Benrejeb M, Feki M, Borne P (2010) Feedback control design for Rössler and Chen chaotic systems antisynchronization. Phys Lett A 374:2835–2840

Jafarov SM, Zeynalov ER, Mustafayeva AM (2016) Synthesis of the optimal fuzzy T-S controller for the mobile robot using the chaos theory. Procedia Comput Sci 102:302–308

Karthikeyan R, Sundarapandian V (2014) Hybrid chaos synchronization of four-scroll systems via active control. J Electr Eng 65(2):97–103

Kocarev L, Parlitz U (1996) Generalized synchronization, predictability, and equivalence of unidirectionally coupled dynamical systems. Phys Rev Lett 76:1816–1819

Lakhekar GV, Waghmare LM, Vaidyanathan S (2016) Diving autopilot design for underwater vehicles using an adaptive neuro-fuzzy sliding mode controller. In: Vaidyanathan S, Volos C (eds) Advances and applications in nonlinear control systems. Springer, Berlin, Germany, pp 477–503

Leonov G, Kuznetsov N, Vagaitsev V (2011) Localization of hidden Chua's attractors. Phys Lett A 375:2230–2233

Leonov G, Kuznetsov N, Vagaitsev V (2012) Hidden attractor in smooth Chua systems. Phys D: Nonlinear Phenom 241:1482–1486

Leonov G, Kuznetsov N, Mokaev T (2015) Homoclinic orbits, and self-excited and hidden attractors in a Lorenz-like system describing convective fluid motion. Eur Phys J: Spec Top 224:1421–1458

Li GH (2005) Synchronization and anti-synchronization of Colpitts oscillators using active control. Chaos Solitons Fractals 26:87–93

Pecora LM, Carroll TL (1990) Synchronization in chaotic systems. Phys Rev Lett 64:821–824

Pehlivan I, Moroz IM, Vaidyanathan S (2014) Analysis, synchronization and circuit design of a novel butterfly attractor. J Sound Vib 333(20):5077–5096

Pham VT, Volos CK, Vaidyanathan S, Le T, Vu VY (2015) A memristor-based hyperchaotic system with hidden attractors: dynamics, synchronization and circuital emulating. J Eng Sci Technol Rev 8(2):205–214

Pham VT, Jafari S, Volos C, Giakoumis A, Vaidyanathan S, Kapitaniak T (2016a) A chaotic system with equilibria located on the rounded square loop and its circuit implementation. IEEE Trans Circuits Syst-II: Express Br 63(9):878–882

Pham VT, Vaidyanathan S, Volos CK, Jafari S, Kuznetsov NV, Hoang TM (2016b) A novel memristive time-delay chaotic system with equilibrium points. Eur Phys J: Spec Top 225(1):127–136

Mg R, Arkady SP, Jürgen K (1996) Phase synchronization of chaotic oscillators. Phys Rev Lett 76:1804–1807

Sadeghpour M, Khodabakhsh M, Salarieh H (2012) Intelligent control of chaos using linear feedback controller and neural network identifier. Commun Nonlinear Sci Numer Simul 17(12):4731–4739

Sarasu P, Sundarapandian V (2011a) Active controller design for generalized projective synchronization of four-scroll chaotic systems. Int J Syst Signal Control Eng Appl 4(2):26–33

Sarasu P, Sundarapandian V (2011b) The generalized projective synchronization of hyperchaotic Lorenz and hyperchaotic Qi systems via active control. Int J Soft Comput 6(5):216–223

Sarasu P, Sundarapandian V (2012) Generalized projective synchronization of three-scroll chaotic systems via adaptive control. Eur J Sci Res 72(4):504–522

Sundarapandian V (2013) Analysis and anti-synchronization of a novel chaotic system via active and adaptive controllers. J Eng Sci Technol Rev 6(4):45–52

Sundarapandian V, Karthikeyan R (2012) Adaptive anti-synchronization of uncertain Tigan and Li systems. J Eng Appl Sci 7(1):45–52

Sundarapandian V, Pehlivan I (2012) Analysis, control, synchronization and circuit design of a novel chaotic system. Math Comput Model 55:1904–1915

Taherion IS, Lai YC (1999) Observability of lag synchronization of coupled chaotic oscillators. Phys Rev E 59:6247–6250

Tirandaz H, Hajipour A (2017) Adaptive synchronization and anti-synchronization of TSUCS and Lü unified chaotic systems with unknown parameters. Optik 130:543–549

Vaidyanathan S (2014) Global chaos synchronisation of identical Li-Wu chaotic systems via sliding mode control. Int J Model Identif Control 22(2):170–177

Vaidyanathan S (2015a) 3-cells cellular neural network (CNN) attractor and its adaptive biological control. Int J PharmTech Res 8(4):632–640

Vaidyanathan S (2015b) A novel chemical chaotic reactor system and its adaptive control. Int J ChemTech Res 8(7):146–158

Vaidyanathan S (2015c) Adaptive control of a chemical chaotic reactor. Int J PharmTech Res 8(3):377–382

Vaidyanathan S (2015d) Anti-synchronization of brusselator chemical reaction systems via integral sliding mode control. Int J ChemTech Res 8(11):700–713

Vaidyanathan S (2015e) Anti-synchronization of chemical chaotic reactors via adaptive control method. Int J ChemTech Res 8(8):73–85

Vaidyanathan S (2015f) Global chaos control of Mathieu-Van der Pol system via adaptive control method. Int J ChemTech Res 8(9):406–417

Vaidyanathan S (2015g) Hybrid chaos synchronization of FitzHugh-Nagumo neuron models via adaptive control method. Int J PharmTech Res 8(8):48–60

Vaidyanathan S (2015h) Hybrid chaos synchronization of Rikitake two-disk dynamo chaotic systems via adaptive control method. Int J ChemTech Res 8(11):12–25

Vaidyanathan S (2015i) Output regulation of the forced Van der Pol chaotic oscillator via adaptive control method. Int J PharmTech Res 8(6):106–116

Vaidyanathan S (2015j) Synchronization of 3-cells cellular neural network (CNN) attractors via adaptive control method. Int J PharmTech Res 8(5):946–955

Vaidyanathan S (2017) A new 3-D jerk chaotic system with two cubic nonlinearities and its adaptive backstepping control. Arch Control Sci 27(3):365–395

Vaidyanathan S, Pakiriswamy S (2016) A five-term 3-D novel conservative chaotic system and its generalized projective synchronization via adaptive control method. Int J Control Theory Appl 9(1):61–78

Vaidyanathan S, Rasappan S (2014) Global chaos synchronization of n-scroll Chua circuit and Lur'e system using backstepping control design with recursive feedback. Arab J Sci Eng 39(4):3351–3364

Vaidyanathan S, Rhif A (2017) A novel four-leaf chaotic system, its control and synchronisation via integral sliding mode control. Int J Model Identif Control 28(1):28–39

Vaidyanathan S, Sampath S (2012) Anti-synchronization of four-wing chaotic systems via sliding mode control. Int J Autom Comput 9(3):274–279

Vaidyanathan S, Sampath S (2017) Anti-synchronisation of identical chaotic systems via novel sliding control and its application to a novel chaotic system. Int J Model Identif Control 27(1):3–13

Vaidyanathan S, Volos C (2016a) Advances and applications in chaotic systems. Springer, Berlin, Germany

Vaidyanathan S, Volos C (2016b) Advances and applications in nonlinear control systems. Springer, Berlin, Germany

Vaidyanathan S, Volos C (2017) Advances in memristors, memristive devices and systems. Springer, Berlin, Germany

Vaidyanathan S, Sampath S, Azar AT (2015a) Global chaos synchronisation of identical chaotic systems via novel sliding mode control method and its application to Zhu system. Int J Model Identif Control 23(1):92–100

Vaidyanathan S, Volos CK, Kyprianidis IM, Stouboulos IN, Pham VT (2015b) Analysis, adaptive control and anti-synchronization of a six-term novel jerk chaotic system with two exponential nonlinearities and its circuit simulation. J Eng Sci Technol Rev 8(2):24–36

Volos C, Maaita JO, Vaidyanathan S, Pham VT, Stouboulos I, Kyprianidis I (2017) A novel four-dimensional hyperchaotic four-wing system with a saddle-focus equilibrium. IEEE Trans Circuits Syst-II: Express Br 64(3):339–343

Wang B, Zhong SM, Dong XC (2016) On the novel chaotic secure communication scheme design. Commun Nonlinear Sci Numer Simul 39:108–117

Wang HX, Cai GL, Sheng M, Tian LX (2010) Nonlinear feedback control of a novel hyperchaotic system and its circuit implementation. Chin Phys B 19(3):030,509. http://stacks.iop.org/1674-1056/19/i=3/a=030509

Wolf A, Swift JB, Swinney HL, Vastano JA (1985) Determining Lyapunov exponents from a time series. Phys D 16:285–317

Xu G, Xiu C, Liu F, Zang Y (2017) Secure communication based on the synchronous control of hysteretic chaotic neuron. Neurocomputing 227:108–112

Yang SS, Duan K (1998) Generalized synchronization in chaotic systems. Chaos Solitons Fractals 10:1703–1707

Yu F, Li P, Gu K, Yin B (2016) Research progress of multi-scroll chaotic oscillators based on current-mode devices. Optik 127(13):5486–5490

A New Chaotic Finance System: Its Analysis, Control, Synchronization and Circuit Design

Babatunde A. Idowu, Sundarapandian Vaidyanathan, Aceng Sambas, Olasunkanmi I. Olusola and O. S. Onma

Abstract This chapter announces a new chaotic finance system and show that it is a self-excited chaotic attractor. The phase portraits and qualitative properties of the new chaotic system are described in detail. An electronic circuit realization of the new chaotic finance system is carried out to verify the feasibility of the theoretical model. Next, this chapter examines the control and synchronization of the new chaotic financial system with uncertain parameters as well as known parameters using adaptive control and backstepping control techniques. The designed adaptive controller control and globally synchronizes two identical chaotic financial systems evolving from different initial conditions. The designed controller is capable of stabilizing the financial system at any position as well as controlling it to track any trajectory that is a smooth function of time. Numerical simulations are presented to demonstrate the feasibility of the proposed schemes.

Keywords Chaos · Chaotic systems · Finance system · Lyapunov exponents · Adaptive control · Backstepping control · Circuit design

B. A. Idowu
Deparment of Physics, Lagos State University, Ojo, Lagos, Nigeria
e-mail: babatunde.idowu@lasu.edu.ng

S. Vaidyanathan (✉)
Research and Development Centre, Vel Tech University, Avadi,
Chennai 600062, India
e-mail: sundarvtu@gmail.com

A. Sambas
Department of Mechanical Engineering, Universitas Muhammadiyah,
Tasikmalaya, Indonesia
e-mail: acenx.bts@gmail.com

O. I. Olusola
Department of Physics, University of Lagos, Akola, Lagos, Nigeria
e-mail: Olasunkanmi2000@gmail.com

O. S. Onma
Nonlinear Dynamics Research Group, Department of Physics,
Federal University of Agriculture, P.M.B. 2240, Abeokuta, Nigeria
e-mail: onmaokpabisunday@yahoo.com

© Springer International Publishing AG 2018
V.-T. Pham et al. (eds.), *Nonlinear Dynamical Systems with Self-Excited and Hidden Attractors*, Studies in Systems, Decision and Control 133,
https://doi.org/10.1007/978-3-319-71243-7_12

271

1 Introduction

Due to complexity and diversity of systems in real-world applications, there has been increasing interest in presenting new systems to fit into it and also describe the phenomenon observed. In order to achieve this, new chaotic systems are been developed, studied and modelled. Thus, Chaos modelling has applications in several areas of science, engineering (Vaidyanathan and Volos 2016a, b, 2017), social sciences and other areas of human endeavor. Chaos control (which is stabilizing a desired unstable periodic solution or one of the systems equilibrium points) and synchronization (two systems are required to co-operate with each other, of which many potential applications abounds) which looks impracticable several decades ago now have applications in several areas such as memristors (Pham et al. 2015, 2016a), chemical reactors (Vaidyanathan 2015a, b, c), neural networks (Sadeghpour et al. 2012; Vaidyanathan 2015d, e), robotics (Jafarov et al. 2016), oscillators (Yu et al. 2016; Vaidyanathan 2015f, g), secure communications (Xu et al. 2017), financial economics etc.

In order to achieve chaos control and synchronization, many control techniques have been developed over time, such as active control (Sundarapandian 2013; Sarasu and Sundarapandian 2011a; Karthikeyan and Sundarapandian 2014; Idowu et al. 2009), adaptive control (Tirandaz and Hajipour 2017; Sarasu and Sundarapandian 2011b, 2012; Fotsin and Bowong 2006; Sundarapandian and Karthikeyan 2012; Vaidyanathan and Idowu 2016; Idowu et al. 2013; Guo et al. 2009) backstepping control (Vaidyanathan 2017; Vaidyanathan and Rasappan 2014; Vaidyanathan et al. 2015a, 2016; Idowu et al. 2009), sliding mode control (Vaidyanathan 2014; Vaidyanathan and Sampath 2012, 2017; Vaidyanathan and Rhif 2017; Lakhekar et al. 2016), etc. The techniques can be used for systems with either known or unknown parameters, although, most applications are based on systems with known parameters. In real-world situations, many systems are nonlinear with unknown parameters and are desirable and this showcases the butterfly effect.

In engineering applications, some attention has been focused on effectively creating chaos via simple physical systems, such as electronic circuits (Sundarapandian and Pehlivan 2012; Pehlivan et al. 2014; Akgul et al. 2016; Pham et al. 2016b; Volos et al. 2017). The pursuit of designing circuits to produce chaotic attractors has become a focal point for engineers, not only because of their theoretical interest, but also due to their potential real-world applications in various chaos-based technologies and information systems (Vaidyanathan and Volos 2016a, b, 2017; Azar and Vaidyanathan 2015, 2016, 2017). As a result we will be presenting the circuit design of our new system.

Recently, chaotic systems are classified into two types of attractors, viz. self-excited and hidden attractors (Leonov et al. 2011, 2015). An attractor is called a *self-excited attractor* if its basin of attraction intersects an arbitrarily small open neighborhood of equilibrium. Otherwise, the attractor is called a *hidden attractor*. Thus, hidden attractor has a basin of attraction which does not overlap with an arbitrarily small neighborhood of equilibria. For example, hidden attractors are attractors in systems without equilibria or with only one stable equilibrium. In this chapter, we present a novel new chaotic finance system that is a self-excited attractor.

According to Cai and Huang (2007), since the chaotic phenomenon in economics was first found in 1985, great impact has been imposed on the prominent economics at present, because the occurrence of the chaotic phenomenon in the economic system means that the macroeconomic operation has in itself the inherent indefiniteness. As a result utilizing the nonlinear dynamical theory to study the complexity of economy and finance system has wide foreground, important theoretical and practical meaning (Wang et al. 2010). Similarly, Zhao et al. (2014) stated that it is well known that the economic activity is a complex human behavior; it has many uncertainties, which is reflected in the nonlinear model for economic dynamics such as Goodwin's nonlinear accelerator model (Godwin 1951), forced van der Pol model on business cycle (Chian et al. 2006), the dynamic IS-LM model (Fanti and Manfredi 2007), and nonlinear dynamical model on finance system (Ma and Chen 2001a, b; Gao and Ma 2009). In these models, chaotic phenomena are common and have showed the importance of the chaotic finance system as well as the need for new models to be developed to fit into evolving scenarios. This we have considered by bringing forth a new chaotic finance model, with two quadratic nonlinearities and a quartic nonlinearity. To the best of our knowledge, this is the only chaotic finance system with quartic nonlinearity designed so far.

In 2007, Cai and Huang investigated the complicated dynamical behaviour and slow manifold of a new finance chaotic attractor and also presented the adaptive control of the system amongst other things. Cai et al. (2009), presented the projective synchronization of the finance attractor using the active sliding mode technique, whilst, Zheng and Du (2014), presented two feedback control schemes to control the system to any equilibrium points and N identical chaotic systems to achieve synchronization. In Abd-Elouahab et al. (2010), Yu et al. (2012) the chaos control of a fractional-order financial system and control of a new hyperchaotic finance system were investigated respectively and Xie et al. (2015), investigated chaos synchronization of financial chaotic system with external perturbation, while Chen (2006) investigated the dynamics and control of this system with multiple delayed feed-backs.

Most investigations on the chaotic finance system has been with known parameters, but due to nonlinearities involved in the system, it is ideal to consider situations with unknown parameters, because in real-world situations, some of the systems parameters are unknown. It has been established that the derivative of adaptive controller for control and synchronization of chaotic system in the presence of unknown parameters is an important issue (Vaidyanathan et al. 2015b).

It is a known fact that in economic activities, chaos is undesired sometimes, so there is need to control the chaotic orbits to a stable state or a periodic orbit and we are going to achieve this in this chapter. Similarly, the synchronization of the chaotic system with unknown parameters will be achieved, using adaptive control technique.

In this chapter, we describe a nine-term 3-D nonlinear finance chaotic system consisting of two quadratic nonlinearities and a quartic nonlinearity. Our novel finance chaotic system is obtained by adding a quartic nonlinearity to the finance chaotic system presented by Gao and Ma (2009).

This chapter is organized as follows. Section 2 describes the dynamic equations, phase portraits and qualitative properties of the novel 3-D finance chaotic system. We show that the novel 3-D finance chaotic systems exhibits a self-excited attractor. Section 3 describes the electronic circuit realization of the new finance chaotic system. Section 4 discusses the adaptive controller to control the new chaotic finance system with unknown parameters to equilibrium, while Sect. 5 is for synchronization of the system with unknown parameters via adaptive controller. In Sect. 6 we presented the chaos control and tracking while Sect. 7 discusses the synchronization of the new system using backstepping technique. Section 8 contains the conclusions.

2 A New Chaotic Finance System and Its Dynamic Properties

Gao and Ma (2009) studied the nonlinear chaotic finance system described by the 3-D dynamics

$$
\begin{aligned}
\dot{x}_1 &= x_3 + (x_2 - a)x_1 \\
\dot{x}_2 &= 1 - bx_2 - x_1^2 \\
\dot{x}_3 &= -x_1 - cx_3
\end{aligned}
\tag{1}
$$

where x_1, x_2, x_3 are the states of the nonlinear system (1) with the following economic interpretation.

In (1), x_1 represents the interest rate, x_2 represents the investment demand and x_3 denotes the price index. Also, the parameter a represents the savings, b represents the investment cost and c represents the commodities demand elasticity.

In the work (Gao and Ma 2009), it was shown that the system (1) exhibits a strange chaotic attractor for the parameter values

$$
a = 0.6, \quad b = 0.1, \quad c = 1
\tag{2}
$$

For numerical simulations, we take the initial values as

$$
x_1(0) = 0.2, \quad x_2(0) = 0.2, \quad x_3(0) = 0.2
\tag{3}
$$

Then the Lyapunov exponents of the finance system (1) are obtained for the parameter values (2) and initial values (3) using Wolf's algorithm (Wolf et al. 1985) as

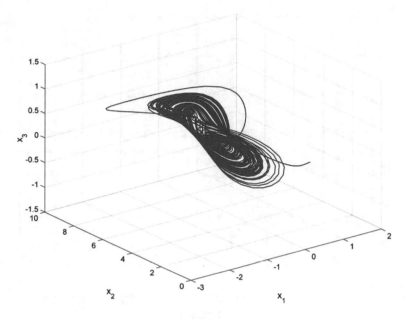

Fig. 1 Strange chaotic attractor of the nonlinear finance system (1)

$$L_1 = 0.0903, \quad L_2 = 0, \quad L_3 = -0.3933 \tag{4}$$

This shows that the finance system (1) is chaotic. Since the sum of the Lyapunov chaos exponents in (4) is negative, the nonlinear finance system (1) is dissipative. Also, the Maximal Lyapunov Exponent (MLE) of the chaotic finance system (1) is $L_1 = 0.0903$.

The Kaplan Yorke dimension of the chaotic finance system (1) is calculated as

$$D_{KY} = 2 + \frac{L_1 + L_2}{|L_3|} = 2.2296 \tag{5}$$

which shows the complexity of the nonlinear finance system (1).

Figure 1 shows the strange chaotic attractor of the nonlinear finance system (1) for the parameter values (2) and the initial state (3). We note that the strange attractor of the finance system (1) is a two-scroll attractor. Figure 2 shows the Lyapunov exponents of the nonlinear finance system (1).

In this chapter, we announce a new chaotic system by adding a quartic non-linearity in the second differential equation of the nonlinear finance system (1). Thus, we obtain a new 3-D nonlinear finance system given by

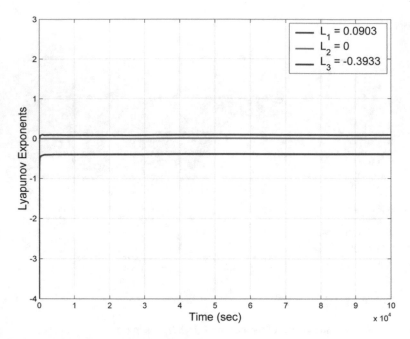

Fig. 2 Lyapunov exponents of the nonlinear finance system (1)

$$\dot{x}_1 = x_3 + (x_2 - a)x_1$$
$$\dot{x}_2 = 1 - bx_2 - x_1^2 - dx_1^4 \qquad (6)$$
$$\dot{x}_3 = -x_1 - cx_3$$

where x_1, x_2, x_3 have the same economic interpretation as in the nonlinear finance system (1). It is also remarked that the parameters a, b, c, d are all constants and positive. Here, the parameter a represents the savings, b represents the investment cost and c represents the commodities demand elasticity. The parameter d is a positive scaling parameter.

In this work, we shall show that the new nonlinear finance system (6) is chaotic when the parameter values are chosen as

$$a = 0.6, \quad b = 0.1, \quad c = 1, d = 0.1 \qquad (7)$$

For numerical simulations, we take the initial values as

$$x_1(0) = 0.2, \quad x_2(0) = 0.2, \quad x_3(0) = 0.2 \qquad (8)$$

Then the Lyapunov exponents of the new finance system (6) are obtained for the parameter values (7) and initial values (8) using Wolf's algorithm (Wolf et al. 1985) as

$$L_1 = 0.0964, \quad L_2 = 0, \quad L_3 = -0.4078 \tag{9}$$

This shows that the new finance system (6) is chaotic. Since the sum of the Lyapunov chaos exponents in (9) is negative, the new nonlinear finance system (6) is dissipative. Also, the Maximal Lyapunov Exponent (MLE) of the new chaotic finance system (6) is $L_1 = 0.0964$.

The Kaplan Yorke dimension of the new chaotic finance system (6) is calculated as

$$D_{KY} = 2 + \frac{L_1 + L_2}{|L_3|} = 2.2364, \tag{10}$$

which shows the complexity of the nonlinear finance system (6).

Since the Maximal Lyapunov Exponent (MLE) and Kaplan-Yorke dimension of the new finance chaotic system (6) are greater than the Maximal Lyapunov Exponent (MLE) and Kaplan-Yorke dimension of the nonlinear finance chaotic system (1) respectively, it is clear that the new finance chaotic system (6) is more complex and chaotic than the finance chaotic system (1).

Figure 3 shows the strange chaotic attractor of the new chaotic finance system (6) for the parameter values (7) and the initial state (8). We note that the strange attractor of the finance system (6) is a two-scroll attractor. Figure 4 shows the Lyapunov exponents of the nonlinear finance system (6).

The equilibrium points of the new chaotic finance system are obtained by solving the system of equations

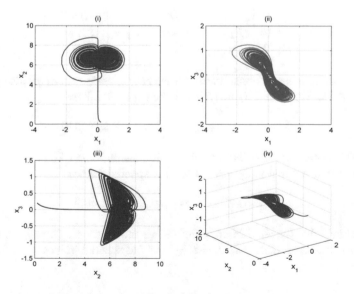

Fig. 3 Strange chaotic attractor of the new chaotic finance system (6)

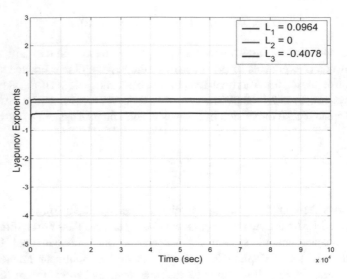

Fig. 4 Lyapunov exponents of the new chaotic finance system (6)

$$
\begin{aligned}
x_3 + (x_2 - a)x_1 &= 0 \\
1 - bx_2 - x_1^2 - dx_1^4 &= 0 \\
-x_1 - cx_3 &= 0
\end{aligned}
\tag{11}
$$

We take the parameter values as in the chaotic case (7). It is easy to check that the new chaotic finance system has three equilibrium points given by

$$
E_1 = \begin{bmatrix} 0 \\ 10 \\ 0 \end{bmatrix}, E_2 = \begin{bmatrix} 0.8828 \\ 1.6000 \\ -0.8828 \end{bmatrix}, E_3 = \begin{bmatrix} -0.8828 \\ 1.6000 \\ 0.8828 \end{bmatrix}
\tag{12}
$$

Let $J(x)$ denote the Jacobian matrix for the new chaotic system (6) at any $x \in R^3$. We find that the matrix $J_1 = J(E_1)$ has the spectral values

$$
\lambda_1 = -0.1, \quad \lambda_2 = -0.9029, \quad \lambda_3 = 9.3029
\tag{13}
$$

This shows that the equilibrium point E_1 is a saddle-point. Thus, E_1 is unstable. We find that the matrix $J_2 = J(E_2)$ has the spectral values

$$
\lambda_1 = -0.7750, \quad \lambda_{2,3} = 0.3375 \pm 1.4869i
\tag{14}
$$

This shows that the equilibrium point E_2 is a saddle-focus. Thus, E_2 is unstable. We also find that the matrix $J_3 = J(E_3)$ has the spectral values

$$\lambda_1 = -0.7750, \quad \lambda_{2,3} = 0.3375 \pm 1.4869i \tag{14}$$

This shows that the equilibrium point E_3 is a saddle-focus. Thus, E_3 is unstable.

Since all the equilibrium points of the new chaotic finance system are unstable, this chaotic system exhibits a self-excited strange attractor.

Furthermore, we see that the new chaotic finance system (6) is invariant under the change of coordinates

$$(x_1, x_2, x_3) \mapsto (-x_1, x_2, -x_3) \tag{15}$$

Thus, it follows that the new chaotic finance system (6) has a rotation symmetry about the x_2-axis and any non-trivial trajectory of the system (6) must have a twin trajectory.

3 Circuit Simulation Results

In this section, circuit design of the new chaotic finance system (6) is presented. The state variable x_1, x_2, x_3 of system (6) are the scaled up to display in a larger range. Therefore the system (6) will be changed to:

$$\begin{cases} \dot{x}_1 = x_3 + (4x_2 - a)x_1 \\ \dot{x}_2 = \dfrac{1}{4} - bx_2 - 4x_1^2 - 64dx_1^4 \\ \dot{x}_3 = -x_1 - cx_3 \end{cases} \tag{16}$$

By applying Kirchhoff's circuit laws, the corresponding circuital equations of designed circuit can be written as

$$\begin{cases} \dot{x}_1 = \dfrac{1}{C_1R_1}x_3 + \dfrac{1}{C_1R_2}x_1x_2 - \dfrac{1}{C_1R_3}x_1 \\ \dot{x}_2 = \dfrac{1}{C_2R_7}V_1 - \dfrac{1}{C_2R_4}x_2 - \dfrac{1}{C_2R_5}x_1^2 - \dfrac{1}{C_2R_6}x_1^4 \\ \dot{x}_3 = -\dfrac{1}{C_3R_8}x_1 - \dfrac{1}{C_3R_9}x_3 \end{cases} \tag{17}$$

where x_1, x_2, x_3 are the voltages in the outputs of the operational amplifiers U1A, U2A and U3A. The TL082CD operational amplifiers are used in this work. The supplies of all active devices are ± 15 V. We choose the values of the circuital elements as:

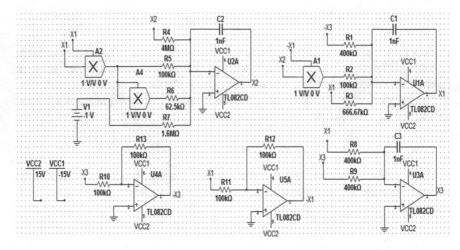

Fig. 5 The schematic of the circuit that emulates the proposed new finance chaotic system (6)

$$\begin{cases} R_1 = R_8 = R_9 = 400\,\text{K}\Omega, R_3 = 666.67\,\text{K}\Omega, \\ R_4 = 4\,\text{M}\Omega, R_6 = 62.5\,\text{K}\Omega, R_7 = 1.6\,\text{M}\Omega, V_1 = -1V_{DC} \\ R_2 = R_5 = R_{10} = R_{11} = R_{12} = R_{13} = 100\,\text{K}\Omega \\ C_1 = C_2 = C_3 = 1\,\text{nF} \end{cases} \tag{18}$$

Using the design approach based on the operational amplifiers, we have the electronic circuit as shown in Fig. 5. Oscilloscope results are displayed in Fig. 6 where we show various phase portraits of the new chaotic finance system (6) obtained in MultiSIM. It is observed that the obtained oscilloscope results (See Fig. 6) confirm the feasibility of the theoretical model (See Fig. 3).

4 Design of Adaptive Controllers for Controlling a New Chaotic Financial System with Unknown Parameters to Equilibrium

The new chaotic fiancé attractor is describable by the following set of differential equations:

$$\begin{aligned} \dot{x}_1 &= x_3 + (x_2 - a)x_1 \\ \dot{x}_2 &= 1 - bx_2 - x_1^2 - dx_1^4 \\ \dot{x}_3 &= -x_1 - cx_3 \end{aligned} \tag{19}$$

Fig. 6 MultiSIM chaotic
attractors of the new finance
chaotic system (6), **a** x_1-x_2
plane, **b** x_1-x_3 plane and **c** x_2-
x_3 plane

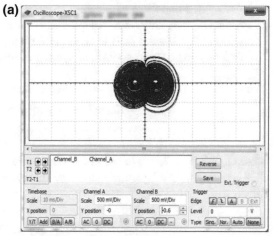

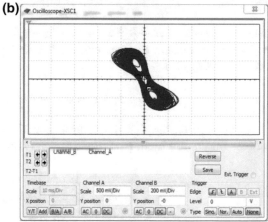

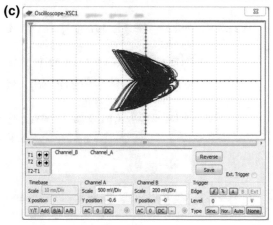

We assumed that the parameters a, b, c and d are as defined earlier in Sect. 2 and in system (19) are unknown. Adaptive control method will be used to control the financial chaotic system (19) to the equilibrium point.

We then consider the controlled system as follows:

$$\begin{aligned}
\dot{x}_1 &= x_3 + (x_2 - a)x_1 + u_1 \\
\dot{x}_2 &= 1 - bx_2 - x_1^2 - dx_1^4 + u_2 \\
\dot{x}_3 &= -x_1 - cx_3 + u_3
\end{aligned}$$ (20)

where $u_i(t)$ $(i = 1, 2, 3)$ are the controllers to be determined appropriately.

According to the Lyapunov stability theory, we select a Lyapunov function as follows:

$$V = \frac{1}{2}(x_1^2 + x_2^2 + x_3^2 + \tilde{a}^2 + \tilde{b}^2 + \tilde{c}^2 + \tilde{d}^2)$$ (21)

where $\tilde{a} = a - \bar{a}$, $\tilde{b} = b - \bar{b}$, $\tilde{c} = c - \bar{c}$ and $\tilde{d} = d - \bar{d}$ are the estimated values of these unknown parameters respectively. Differentiating Eq. (21) with respect to time, result into Eq. (22).

$$\dot{V} = x_1\dot{x}_1 + x_2\dot{x}_2 + x_3\dot{x}_3 + \tilde{a}\dot{\tilde{a}} + \tilde{b}\dot{\tilde{b}} + \tilde{c}\dot{\tilde{c}} + \tilde{d}\dot{\tilde{d}}$$ (22)

In order to ensure that the controller in Eq. (20) converges to the origin, we choose the control input from Eq. (20) as follows

$$\begin{aligned}
u_1 &= -(x_2 - a)x_1 - x_3 - x_1 \\
u_2 &= bx_2 + x_1^2 + dx_1^4 - 1 - x_2 \\
u_3 &= x_1 + cx_3 - x_3
\end{aligned}$$ (23)

We substituted Eq. (20) in Eq. (22) as follows;

$$\dot{V} = x_1[x_3 + (x_2 - a)x_1 + u_1] + x_2[1 - bx_2 - x_1^2 - dx_1^4 + u_2]$$
$$\quad + x_3[-x_1 - cx_3 + u_3] + \tilde{a}(-\dot{\tilde{a}}) + \tilde{b}(-\dot{\tilde{b}}) + \tilde{c}(-\dot{\tilde{c}}) + \tilde{d}(-\dot{\tilde{d}})$$

$$\dot{V} = x_1[x_3 + (x_2 - a)x_1 + u_1] + x_2[1 - bx_2 - x_1^2 - dx_1^4 + u_2]$$
$$\quad + x_3[-x_1 - cx_3 + u_3] + \tilde{a}(-\dot{\tilde{a}} - x_1^2) + \tilde{b}(-\dot{\tilde{b}} - x_2^2) + \tilde{c}(-\dot{\tilde{c}} - x_3^2) \quad (24)$$
$$\quad + \tilde{d}(-\dot{\tilde{d}} - x_2 x_1^4)$$

We choose the following parameter estimation update laws from Eq. (24).

$$\dot{a} = -x_1^2$$
$$\dot{b} = -x_2^2$$
$$\dot{c} = -x_3^2$$
$$\dot{d} = -x_2 x_1^4$$

(25)

Substituting Eqs. (25) and (23) respectively in Eq. (24) we have,

$$\dot{V} = -x_1^2 - x_2^2 - x_3^2 < 0$$

According to Lyapunov stability theory, the condition above ensures that the controlled system (20) converges to the equilibrium point with the controllers in Eq. (23) and the parameter update laws in Eq. (25).

Furthermore, we verify the effectiveness and feasibility of the derived controllers in (25) above by simulating the dynamics of drive system and response system using the fourth-order Runge-Kutta algorithm with initial conditions $(x_1, x_2, x_3) = (1.0, 5.0, -5.0)$, a time step of 0.001 and fixing the parameter values of the system to ensure a chaotic dynamics of the state variables, we solved system (20) with the control function as defined in (25). The results shows that the error state variable moves chaotically with time when the controllers are deactivated and when the controllers are switched on at $t = 20$ as shown in Fig. 7. However, the initial values of the parameter update laws (25) are chosen as $a_1(0) = -0.5$,

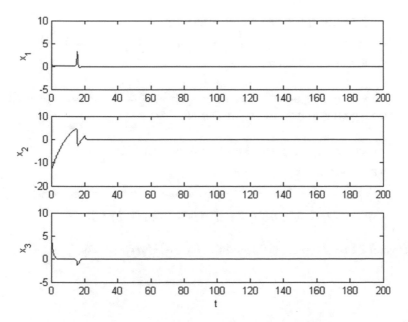

Fig. 7 Controlled states of the new chaotic finance system to equilibrium

$b_1(0) = 0.4$, $c_1(0) = 2.0$ and $d_1(0) = 8.0$. The system is controlled to a stable equilibrium when the controllers (23) are activated as shown in Fig. 7.

5 Design of Adaptive Controllers for Synchronization of a New Chaotic Financial System with Unknown Parameters to Equilibrium

In this section, by using adaptive control technique, we designed controllers that can globally synchronize two identical financial chaotic systems evolving from different initial conditions.

We regard the system (19) as the drive (transmitter) while the following is the response (receiver).

$$\begin{aligned}
\dot{y}_1 &= y_3 + (y_2 - a)y_1 + u_1 \\
\dot{y}_2 &= 1 - by_2 - y_1^2 - dy_1^4 + u_2 \\
\dot{y}_3 &= -y_1 - cy_3 + u_3
\end{aligned} \tag{26}$$

where $u_i(t)$ $(i = 1, 2, 3)$ are the control functions to be designed.

We assume that the parameters a, b, c and d are unknown parameters.

Using this notation; $e_i = y_i - x_i$ or $y_i = e_i + x_i$, we obtain the error vector given in Eq. (27) below:

$$\begin{aligned}
\dot{e}_1 &= e_3 + x_1 e_2 + x_2 e_1 + e_1 e_2 - ae_1 + u_1 \\
\dot{e}_2 &= -be_2 - (2x_1 e_1 + e_1^2) - d(e_1^4 + 4x_1 e_1^3 + 2x_1^2 e_1^2 + 4x_1^3 e_1) + u_2 \\
\dot{e}_3 &= -e_1 - ce_3 + u_3
\end{aligned} \tag{27}$$

We choose a Lyapunov function; $V = \frac{1}{2}(e_1^2 + e_2^2 + e_3^2 + \tilde{a}^2 + \tilde{b}^2 + \tilde{c}^2 + \tilde{d}^2)$ and differentiating with respect to time, to have;

$$\begin{aligned}
\dot{V} &= e_1 \dot{e}_1 + e_2 \dot{e}_2 + e_3 \dot{e}_3 + \tilde{a}\dot{\tilde{a}} + \tilde{b}\dot{\tilde{b}} + \tilde{c}\dot{\tilde{c}} + \tilde{d}\dot{\tilde{d}} \\
\dot{V} &= e_1[e_3 + x_1 e_2 + x_2 e_1 + e_1 e_2 + u_1] \\
&\quad + e_2\left[-be_2 - (2x_1 e_1 + e_1^2) - d(e_1^4 + 4x_1 e_1^3 + 2x_1^2 e_1^2 + 4x_1^3 e_1) + u_2\right] \\
&\quad + e_3[-e_1 - ce_3 + u_3] + \tilde{a}(-\dot{a}) + \tilde{b}(-\dot{b}) + \tilde{c}(-\dot{c}) + \tilde{d}(-\dot{d})
\end{aligned} \tag{28}$$

where $\bar{a}, \bar{b}, \bar{c}$ and $\bar{d}$ are the estimate of a, b, c and d respectively.

Hence,

$$
\begin{aligned}
\dot{V} = {} & e_1[e_3 + x_1 e_2 + x_2 e_1 + e_1 e_2 - a e_1 + u_1] \\
& + e_2\left[-b e_2 - (2x_1 e_1 + e_1^2) - d(e_1^4 + 4x_1 e_1^3 + 2x_1^2 e_1^2 + 4x_1^3 e_1) + u_2\right] \\
& + e_3[-e_1 - c e_3 + u_3] + \tilde{a}(-\dot{a} - e_1^2) + \tilde{b}\left(-\dot{b} - e_2^2\right) + \tilde{c}(-\dot{c} - e_3^2) \\
& + \tilde{d}\left(-\dot{d} - e_2(e_1^4 + 4x_1 e_1^3 + 2x_1^2 e_1^2 + 4x_1^3 e_1)\right)
\end{aligned}
$$

From Eq. (27), the controller function is chosen as follows:

$$
\begin{aligned}
u_1 &= a e_1 - (x_1 e_2 + x_2 e_1 + e_1 e_2) - e_3 - e_1 \\
u_2 &= b e_2 + 2x_1 e_2 + e_1^2 + d(e_1^4 + 4x_1 e_1^3 + 2x_1^2 e_1^2 + 4x_1^3 e_1) - e_2 \qquad (29)\\
u_3 &= e_1 + c e_3 - e_3
\end{aligned}
$$

The parameter updates estimation law is chosen as follows:

$$
\begin{aligned}
\dot{\tilde{a}} &= -e_1^2 - a \\
\dot{\tilde{b}} &= -e_2^2 - b \\
\dot{\tilde{c}} &= -e_3^2 - c \qquad (30)\\
\dot{\tilde{d}} &= -e_2(e_1^4 + 4x_1 e_1^3 + 2x_1^2 e_1^2 + 4x_1^3 e_1) - d
\end{aligned}
$$

Substituting Eq. (30) into Eq. (28) yields,

$$
\dot{V} = -e_1^2 - e_2^2 - e_3^2 - \tilde{a}^2 - \tilde{b}^2 - \tilde{c}^2 - \tilde{d}^2 < 0 \qquad (31)
$$

With condition (31), the error dynamical system converges to the origin asymptotically in line with the Lyapunov stability theory. Also the drive system (19) is synchronized with the response system (26) with controller (29) and the parameter update law (30).

Using fourth-order Runge-Kutta routine with initial conditions $(x, y, z) = (5.0, 10.0, 7.0)$, a time step of 0.001 and fixing the parameter values as in Fig. 7 to ensure chaotic dynamics of the state variables, we solved system (20) with the controllers $u_i(t)$, $i = 1, 2, 3$ as defined in (29). The results obtained show that the state variables move chaotically with time when the controllers are deactivated and when the controllers are switched on at time $t = 20$, the state variables converges to the equilibrium point. The results are shown in Figs. 8 and 9.

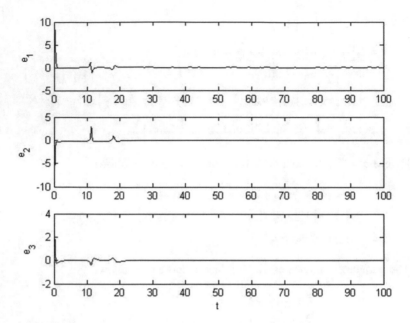

Fig. 8 Error dynamics between the two chaotic financial system with the controller deactivated for $0 < t < 20$ and activated for $t \geq 20$

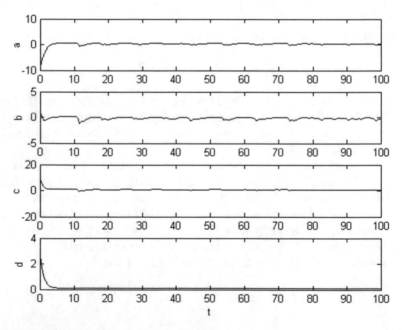

Fig. 9 Time responses of the parameter estimation errors

6 Chaos Control and Tracking of a New Finance Chaotic Attractor Based on Backstepping Approach

Here, we present a recursive backstepping tracking control for the new chaotic financial system, to track a smooth function of time $f(t)$.

From Eq. (19), we have

$$
\begin{aligned}
\dot{x}_1 &= x_3 + (x_2 - a)x_1 + u_1 \\
\dot{x}_2 &= 1 - bx_2 - x_1^2 - dx_1^4 + u_2 \\
\dot{x}_3 &= -x_1 - cx_3 + u_3
\end{aligned}
\tag{32}
$$

The state variables x_1, x_2 and x_3 of the system (20) have the desired values x_{1d}, x_{2d} and x_{3d} respectively.

The error states between the states variable and the desired value are;

$$
\begin{aligned}
e_{x1} &= x_1 - x_{1d} \\
e_{x2} &= x_2 - x_{2d} \\
e_{x3} &= x_3 - x_3 d
\end{aligned}
\tag{33}
$$

To determine a general control function, $u_i(t)$, $(i = 1, 2, 3)$ that can control system (20), to track any trajectory $f(t)$ that is a smooth function of time, we let;

$$
\begin{aligned}
x_{1d} &- f(t) \\
x_{2d} &= c_1 e_{x1} \\
x_{3d} &= c_2 e_{x1} + c_3 e_{x2}
\end{aligned}
\tag{34}
$$

where c_i $(i = 1, 2, 3)$ are the arbitrary control parameters to be chosen appropriately. By substituting Eq. (34) into Eq. (33) and differentiating the resulting equation with respect to time, we have the following system of error vector.

$$
\begin{aligned}
\dot{e}_{x1} &= e_{x3} + c_2 e_{x1} + c_3 e_{x2} + (e_{x2} + c_1 e_{x1} - a)(e_{x1} + f(t)) - \dot{f}(t) + u_1 \\
\dot{e}_{x2} &= 1 - b(e_{x2} + c_1 e_{x1}) - (e_{x1} + f(t))^2 - d(e_{x1} + f(t))^4 - c_1 \dot{e}_{x1} + u_2 \\
\dot{e}_{x3} &= -(e_{x1} + f(t)) - c(e_{x3} + c_2 e_{x1} + c_3 e_{x2}) - c_2 \dot{e}_{x1} - c_3 \dot{e}_{x2} + u_3
\end{aligned}
\tag{35}
$$

To stabilize the error system (33), we consider a Lyapunov function of the form;

$$
V = \frac{1}{2}(k_{x1}e_{x1}^2 + k_{x2}e_{x2}^2 + k_{x3}e_{x3}^2)
\tag{36}
$$

where k_{x1}, k_{x2} and k_{x3} are positive constant coefficient and the derivative of Eq. (36) are as follows;

$$\dot{V} = k_{x1}e_{x1}\dot{e}_{x1} + k_{x2}e_{x2}\dot{e}_{x2} + k_{x3}e_{x3}\dot{e}_{x3} \tag{37}$$

To satisfy the condition for asymptotic stability of the error system (35), necessary for tracking, i.e. $\dot{V} = -\sum k_{xi}^2 e_{xi}^2 < 0; i = 1, 2$ and 3, we substitute Eq. (35) into Eq. (37) with the choice of control input function $u_i(t)$ ($i = 1, 2, 3$) as follows;

$$
\begin{aligned}
u_1 &= e_{x3} - c_2 e_{x1} - c_3 e_{x2} - (e_{x2} + c_1 e_{x1} - a)(e_{x1} + f(t)) - \dot{f}(t) - e_{x1} \\
u_2 &= -1 + b(e_{x2} + c_1 e_{x1}) + (e_{x1} + f(t))^2 + d(e_{x1} + f(t))^4 - e_{x2} \\
u_3 &= (e_{x1} + f(t)) + c(e_{x3} + c_2 e_{x1} + c_3 e_{x2}) - e_{x3}
\end{aligned}
\tag{38}
$$

We observed from the numerical simulation that for system (19) to be effectively controlled to follow a smooth function of time, we chose $c_1 = c_2 = c_3 = 1$, which reduces the controller in Eq. (38) to;

$$
\begin{aligned}
u_1 &= e_{x3} - e_{x1} - e_{x2} - (e_{x2} + e_{x1} - a)(e_{x1} + f(t) - \dot{f}(t)) - e_{x1} \\
u_2 &= -1 + b(e_{x2} + e_{x1}) + (e_{x1} + f(t))^2 + d(e_{x1} + f(t))^4 - e_{x2} \\
u_3 &= (e_{x1} + f(t)) + c(e_{x3} + e_{x1} + e_{x2}) - e_{x3}
\end{aligned}
\tag{39}
$$

In order to verify the effectiveness of the proposed scheme, the fourth-order Runge-Kutta routine is applied with the initial conditions $(x, y, z) = (5.0, 10.0, 7.0)$, a time step of 0.001 and fixing the parameter values as in Fig. 7 to ensure chaotic

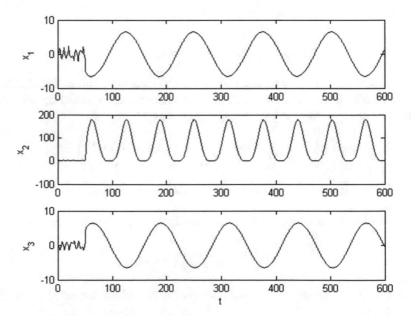

Fig. 10 Time responses of the state variables (x, y, z) for chaotic financial system with the control $u_i(t)$ activated at $t = 20$ to track $f(t) = 27 \cos 0.05t$

dynamics of the state variables, we solve system (32) with the controllers $u_i(t)$, $i = 1, 2, 3$ as defined in (39). The result obtained show that the state variables move chaotically with time when the controllers are deactivated and when the controllers are switched on at $t \geq 20$ the state variables are controlled to track any desired smooth functions of time such as $f(t) = 27 \cos 0.05t$ as displayed in Fig. 10. The results showed that the recursive backstepping controllers (39) are effective in the stabilization and to track any desired smooth function $f(t)$ of the new chaotic financial system.

7 Synchronization of a New Financial Attractor Using Active Backstepping Method

Our aim here is to design an appropriate active backstepping nonlinear control vector $u = [u_1, u_2, u_3]^T$ that can stabilize the error states at the origin also to make the state variables of the response (receiver) system to track the ones of drive (transmitter) system.

From Eqs. (19) and (26) the drive and response systems are given in Eqs. (34) and (35) respectively:

$$
\begin{aligned}
\dot{x}_1 &= x_3 + (x_2 - a)x_1 \\
\dot{x}_2 &= 1 - bx_2 - x_1^2 - dx_1^4 \\
\dot{x}_3 &= -x_1 - cx_3
\end{aligned}
\tag{40}
$$

$$
\begin{aligned}
\dot{y}_1 &= y_3 + (y_2 - a)y_1 + u_1 \\
\dot{y}_2 &= 1 - by_2 - y_1^2 - dy_1^4 + u_2 \\
\dot{y}_3 &= -y_1 - cy_3 + u_3
\end{aligned}
\tag{41}
$$

Using this notation; $e_i = y_i - x_i$ or $y_i = e_i + x_i$, we obtained the error vector given in Eq. (42).

$$
\begin{aligned}
\dot{e}_1 &= e_3 + x_1 e_2 + x_2 e_1 + e_1 e_2 - ae_1 + u_1 \\
\dot{e}_2 &= -be_2 - (2x_1 e_1 + e_1^2) - d(e_1^4 + 4x_1 e_1^3 + 2x_1^2 e_1^2 + 4x_1^3 e_1) + u_2 \\
\dot{e}_3 &= -e_1 - ce_3 + u_3
\end{aligned}
\tag{42}
$$

We stabilize the first equation in system (42) by regarding e_2 as a virtual controller. Selecting a Lyapunov function $V_1(e_1) = \frac{1}{2}e_1^2$, and differentiating it with respect to time, we have;

$$\dot{V}_1(e_1) = e_1(e_3 + x_2e_1 - ae_1 + (x_1 + e_1)e_2 + u_1) \tag{43}$$

We estimate that the virtual controller is $e_2 = \alpha_1(e_1)$. Then Eq. (43) becomes;

$$\dot{V}_1(e_1) = e_1(e_3 + x_2e_1 - ae_1 + (x_1 + e_1)\alpha_1(e_1) + u_1).$$

We choose $u_1 = -(e_3 + x_2e_1)$ and estimating that $\alpha_1(e_1) = 0$, $\dot{V}_1(e_1) = -ae_1^2$(negative definite since $a > 0$). This means that the e_1 subsystem is stabilized since the virtual controller $\alpha_1(e_1)$ is measurable. The error ω_2 between e_2 and $\alpha_1(e_1)$ is defined as;

$$\omega_2 = e_2 - \alpha_1(e_1) = e_2 \tag{44}$$

Substituting for $\dot{e}_2$ and e_2 from Eqs. (42) and (44) respectively into the time derivative of Eq. (38) yields;

$$\dot{\omega}_2 = -b\omega_2 - (2x_1e_1 + e_1^2) - d(e_1^4 + 4x_1e_1^3 + 2x_1^2e_1^2 + 4x_1^3e_1) + u_2 \tag{45}$$

We now stabilize (e_1, ω_2) subsystem given by Eq. (45) as follows. We select another Lyapunov function, $V_2(e_1, \omega_2) = v_1(e_1) + \frac{1}{2}\omega_2^2$ and its time derivative yields;

$$\dot{V}_2(e_1, \omega_2) = \dot{v}_1 + \omega_2\dot{\omega}_2 \tag{46}$$

From Eqs. (41) and (42),

$$\dot{V}_1(e_1) = -ae_1^2 + (x_1 + e_1)e_1\omega_2 \tag{47}$$

Thus,

$$\dot{V}_2(e_1, \omega_2) = -ae_1^2 + [-b\omega_2 - x_1e_1 \\ - d(e_1^3 + 4x_1e_1^2 + 2x_1^2e_1 + 4x_1^3)e_1 + u_2] \tag{48}$$

If $u_2 = x_1e_1$, and $e_1 = \alpha_2(e_1, \omega_2)$ then Eq. (48) reduces to;

$$\dot{V}_2(e_1, \omega_2) = -ae_1^2 + \omega_2[-b\omega_2 - d(e_1^3 + 4x_1e_1^2 + 2x_1^2e_1 + 4x_1^3)\alpha_2].$$

If the estimative function $\alpha_2(e_1, \omega_2) = 0$, then $\dot{V}_2 = -ae_1^2 - b\omega_2^2 < 0$ (negative definite since $a, b > 0$). Thus, we can conclude that the (e_1, ω_2) subsystem is stable since $\alpha_2(e_1, \omega_2)$ is estimated.

The error ω_3 between e_3 and $\alpha_2(e_1, \omega_2)$ is:

$$\omega_3 = e_3 - \alpha_2(e_1, \omega_2) = e_3 \tag{49}$$

Substituting for $\dot{e}_3$ and e_3 from Eqs. (40) and (49) into the time derivative of Eq. (49) gives;

$$\dot{\omega}_3 = -e_1 - c\omega_3 + u_3 \tag{50}$$

In order to stabilize $(e_1, \omega_2, \omega_3)$ complete system given by Eqs. (40), (42) and (50), we select a Lyapunov function $V_3(e_1, \omega_2, \omega_3)$ as follows;

$$V_3(e_1, \omega_2, \omega_3) = v_2(e_1, \omega_2) + \omega_3\dot{\omega}_3 \tag{51}$$

We differentiate Eq. (51) with respect to time as;

$$\dot{V}_3(e_1, \omega_2, \omega_3) = \dot{v}_2(e_1, \omega_2) + \omega_3\dot{\omega}_3 \tag{52}$$

Hence,

$$\dot{V}_3(e_1, \omega_2, \omega_3) = -ae_1^2 - b\omega_2^2 + \omega_3[-e_1 - c\omega_3 + u_3].$$

But $e_1 = \alpha_2(e_1, \omega_2) = 0$.
Thus, we have

$$\dot{V}_3(e_1, \omega_2, \omega_2) = -ae_1^2 - b\omega_2^2 + \omega_3[-c\omega_3 + u_3].$$

If $u_3 = 0$, $\dot{V}_3 = -ae_1^2 - b\omega_2^2 - c\omega_3^2 < 0$ (i.e. the time derivative of Lyapunov is negative definite).

According to Lassalle-Yoshizawa theorem, it follows that all the solution of Eq. (36) converges to the manifold $e_i = 0$ ($i = 1, 2, 3$) as $t \to \infty$. Hence, the systems (40) and (41) are globally synchronized.

The synchronization aim is achieved with the control input;

$$\begin{aligned} u_1 &= -(e_3 + x_2e_1) \\ u_2 &= x_1e_1 \\ u_3 &= 0 \end{aligned} \tag{53}$$

To verify the feasibility and effectiveness of the designed backstepping control function (53), we simulate the dynamics of drive system and response system using the fourth-order Runge-Kutta algorithm with initial conditions $(x_1, x_2, x_3) = (1.0, 5.0, -5.0), (y_1, y_2, y_3) = (2.0, 8.0, -1.0)$, with a time grid of 0.001 and fixing the parameter values as in Fig. 7 to ensure a chaotic dynamics of the state variables. We solve systems (40) and (41) with the control function as defined in (53). The results shows that the error state variable moves chaotically

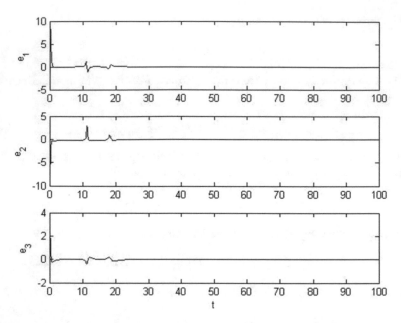

Fig. 11 Error dynamics between the two chaotic financial with the controller deactivated for $0 < t < 20$ and activated for $t \geq 20$

with time when the controllers are deactivated and when the controllers are switched on at t = 20 as depicted in Fig. 11 the error state variables converges to zero and thereby guaranteeing the synchronization of systems (40) and (41).

8   Conclusions

In this chapter, we presented a new chaotic finance system that is a self-excited chaotic attractor with two quadratic nonlinearities and a quartic nonlinearity and also describe its phase portrait and qualitative properties. The Maximal Lyapunov Exponent (MLE) and Kaplan-Yorke dimension of the new finance chaotic system (6) are greater than the Maximal Lyapunov Exponent (MLE) and Kaplan-Yorke dimension of the nonlinear finance chaotic system (1) respectively, thus, it is clear that the new finance chaotic system (6) presented here is more complex and chaotic than the finance chaotic system (1) by Gao and Ma (2009). The electronic circuit realization of the new chaotic finance system was also carried out to verify the feasibility of the theoretical model.

We utilized the adaptive and backstepping control techniques to synchronize the new chaotic financial system with fully unknown and known parameters respectively. Since in economic activities, chaos is sometimes not desired, we designed control functions to control and track the new chaotic financial system. The designed controllers control and globally synchronize two identical chaotic financial systems evolving from different initial conditions.

References

Abd-Elouahab MS, Hamri N-E, Wang J (2010) Chaos control of a fractional-order financial system. Math Prob Eng Article ID 270646, 18 pp

Azar AT, Vaidyanathan S (2015) Chaos modeling and control systems design. Springer, Berlin, Germany

Azar AT, Vaidyanathan S (2016) Advances in chaos theory and intelligent control. Springer, Berlin, Germany

Azar AT, Vaidyanathan S (2017) Fractional order control and synchronization of chaotic systems. Springer, Berlin, Germany

Akgul A, Moroz I, Pehlivan I, Vaidyanathan S (2016) A new four-scroll chaotic attractor and its engineering applications. Optik 127(13):5491–5499

Cai G, Huang J (2007) A new finance chaotic attractor. Int J Nonlinear Sci 3:213–220

Cai N, Jing Y, Zhang S (2009) Modified projective synchronization of chaotic systems with disturbances via active sliding mode control. Commun Nonlinear Sci Numer Simul 15:1613–1620

Chian AC-L, Rempel EL, Rogers C (2006) Complex economic dynamics: chaotic saddle, crisis and intermittency. Chaos, Solitons Fractals 29(5):1194–1218

Fanti L, Manfredi P (2007) Chaotic business cycles and fiscal policy: an IS-LM model with distributed tax collection lags. Chaos, Solitons Fractals 32(2):736–744

Fotsin H, Bowong S (2006) Adaptive control and synchronization of chaotic systems consisting of Van der Pol oscillators coupled to linear oscillators. Chaos, Solitons Fractals 27:822–835

Gao Q, Ma J (2009) Chaos and Hopf bifurcation of a finance system. Nonlinear Dyn 58(1–2):209–216

Goodwin RM (1951) The nonlinear accelerator and the persistence of business cycles. Econometrica 19(1):1–17

Guo R, Vincent UE, Idowu BA (2009) Synchronization of chaos in RCL-shunted Josephson junction using a simple adaptive controller. Physica Scripta 79:035801

Idowu BA, Guo R, Vincent UE (2013) Adaptive control for the stabilization and synchronization of nonlinear gyroscopes. Int J Chaos Control Model Simul 2(2):27–44

Idowu BA, Vincent UE, Njah AN (2009) Synchronization of chaos in non-identical parametrically excited systems. Chaos Solitons Fractals 39:2322–2331

Jafarov SM, Zeynalov ER, Mustafayeva AM (2016) Synthesis of the optimal fuzzy T-S controller for the mobile robot using the chaos theory. Procedia Comput Sci 102:302–308

Karthikeyan R, Sundarapandian V (2014) Hybrid chaos synchronization of four-scroll systems via active control. J Electr Eng 65(2):97–103

Lakhekar GV, Waghmarem LM, Vaidyanathan S (2016) Diving autopilot design for underwater vehicles using an adaptive neuro-fuzzy sliding mode controller. In: Vaidyanathan S, Volos C (eds) Advances and applications in nonlinear control systems. Springer, Berlin, Germany, pp 477–503

Leonov G, Kuznetsov N, Vagaitsev V (2011) Localization of hidden Chua's attractors. Phys Lett A 375:2230–2233

Leonov G, Kuznetsov N, Mokaev T (2015) Homoclinic orbits, and self-excited and hidden attractors in a Lorenz-like system describing convective fluid motion. Eur Phys J: Spec Top 224:1421–1458

Ma J, Chen Y (2001a) Study for the bifurcation topological structure and the global complicated character of a kind of nonlinear finance system (I). Appl Math Mech 22(11):1119–1128

Ma J, Chen Y (2001b) Study for the bifurcation topological structure and the global complicated character of a kind of nonlinear finance system (II). Appl Math Mech 22(12):1236–1242

Pehlivan I, Moroz IM, Vaidyanathan S (2014) Analysis, synchronization and circuit design of a novel butterfly attractor. J Sound Vib 333(20):5077–5096

Pham VT, Volos CK, Vaidyanathan S, Le T, Vu VY (2015) A memristor-based hyperchaotic system with hidden attractors: dynamics, synchronization and circuital emulating. J Eng Sci Technol Rev 8(2):205–214

Pham VT, Jafari S, Volos C, Giakoumis A, Vaidyanathan S, Kapitaniak T (2016a) A chaotic system with equilibria located on the rounded square loop and its circuit implementation. IEEE Trans Circuits Syst-II: Express Br 63(9):878–882

Pham VT, Vaidyanathan S, Volos CK, Jafari S, Kuznetsov NV, Hoang TM (2016b) A novel memristive time-delay chaotic system with equilibrium points. Eur Phys J: Spec Top 225(1):127–136

Sadeghpour M, Khodabakhsh M, Salarieh H (2012) Intelligent control of chaos using linear feedback controller and neural network identifier. Commun Nonlinear Sci Numer Simul 17(12):4731–4739

Sarasu P, Sundarapandian V (2011a) Active controller design for generalized projective synchronization of four-scroll chaotic systems. Int J Syst Signal Control Eng Appl 4(2):26–33

Sarasu P, Sundarapandian V (2011b) The generalized projective synchronization of hyperchaotic Lorenz and hyperchaotic Qi systems via active control. Int J Syst Signal Control Eng Appl 4(2):26–33

Sarasu P, Sundarapandian V (2012) Generalized projective synchronization of three-scroll chaotic systems via adaptive control. Eur J Sci Res 72(4):504–522

Sundarapandian V (2013) Analysis and anti-synchronization of a novel chaotic system via active and adaptive controllers. J Eng Sci Technol Rev 6(4):45–52

Sundarapandian V, Karthikeyan R (2012) Adaptive anti-synchronization of uncertain Tigan and Li systems. J Eng Appl Sci 7(1):45–52

Sundarapandian V, Pehlivan I (2012) Analysis, control, synchronization and circuit design of a novel chaotic system. Math Comput Model 55:1904–1915

Tirandaz H, Hajipour A (2017) Adaptive synchronization and antisynchronization of TSUCS and Lü unified chaotic systems with unknown parameters. Optik 130:543–549

Vaidyanathan S (2014) Global chaos synchronisation of identical Li-Wu chaotic systems via sliding mode control. Int J Model Identif Control 22(2):170–177

Vaidyanathan S (2015a) A novel chemical chaotic reactor system and its adaptive control. Int J ChemTech Res 8(7):146–158

Vaidyanathan S (2015b) Adaptive control of a chemical chaotic reactor. Int J PharmTech Res 8 (3):377–382

Vaidyanathan S (2015c) Anti-synchronization of chemical chaotic reactors via adaptive control method. Int J ChemTech Res 8(8):73–85

Vaidyanathan S (2015d) 3-cells cellular neural network (CNN) attractor and its adaptive biological control. Int J PharmTech Res 8(4):632–640

Vaidyanathan S (2015e) Synchronization of 3-cells cellular neural network (CNN) attractors via adaptive control method. Int J PharmTech Res 8(5):946–955

Vaidyanathan S (2015f) Output regulation of the forced Van der Pol chaotic oscillator via adaptive control method. Int J PharmTech Res 8(6):106–116

Vaidyanathan S (2015g) Global chaos control of Mathieu-Van der Pol system via adaptive control method. Int J ChemTech Res 8(9):406–417

Vaidyanathan S (2017) A new 3-D jerk chaotic system with two cubic nonlinearities and its adaptive backstepping control. Arch Control Sci 27(3):365–395

Vaidyanathan S, Rasappan S (2014) Global chaos synchronization of n-scroll Chua circuit and Lur'e system using backstepping control design with recursive feedback. Arab J Sci Eng 39(4):3351–3364

Vaidyanathan S, Rhif A (2017) A novel four-leaf chaotic system, its control and synchronisation via integral sliding mode control. Int J Model Identif Control 28(1):28–39

Vaidyanathan S, Sampath S (2012) Anti-synchronization of four-wing chaotic systems via sliding mode control. Int J Autom Comput 9(3):274–279

Vaidyanathan S, Sampath S (2017) Anti-synchronisation of identical chaotic systems via novel sliding control and its application to a novel chaotic system. Int J Model Identif Control 27 (1):3–13

Vaidyanathan S, Volos C (2016a) Advances and applications in nonlinear control systems. Springer, Berlin, Germany

Vaidyanathan S, Volos C (2016b) Advances and applications in chaotic systems. Springer, Berlin, Germany

Vaidyanathan S, Volos C (2017) Advances in memristors, memristive devices and systems. Springer, Berlin, Germany

Vaidyanathan S, Madhavan K, Idowu BA (2016) Backstepping control design for the adaptive stabilization and synchronization of the Pandey Jerk chaotic system with unknown parameters. Int J Control Theory Appl 9(1):299–319

Vaidyanathan S, Idowu BA (2016) Adaptive control and synchronization of Chlouverakis-Spott hyperjerk system via backstepping control. In: Vaidyanathan S, Volos C (eds) Advances and applications in nonlinear control systems. Studies in computation intelligence 635. Springer International Publishing, Switzerland, pp 117–142

Vaidyanathan S, Idowu BA, Azar AT (2015a) Backstepping controller design for the global chaos synchronization of Sprott's jerk systems. In: Chaos modeling and control systems design. Studies in computational intelligence 581. Springer International Publishing, pp 39–58

Vaidyanathan S, Volos CK, Rajagopal K, Kyprianidis IM, Stouboulos IN (2015b) Adaptive backstepping controller for the anti-synchronization of identical WINDMI chaotic systems with unknown parameters and its SPICE implementation. J Eng Sci Tech Rev 8(2):74–82

Volos C, Maaita JO, Vaidyanathan S, Pham VT, Stouboulos I, Kyprianidis I (2017) A novel four-dimensional hyperchaotic four-wing system with a saddle-focus equilibrium. IEEE Trans Circuits Syst-II: Express Br 64(3):339–343

Wang Y, Zhai YH, Wang J (2010) Chaos and Hopf bifurcation of a finance system with distributed time delay. Int J Appl Math Mech 6(20):1–13

Chen Wei-Ching (2006) Dynamics and control of a financial system with time-delayed feedbacks. Chaos, Solitons Fractals 37:1198–1207

Wolf A, Swift JB, Swinney HL, Vastano JA (1985) Determining Lyapunov exponents from a time series. Phys D 16:285–317

Xie C, Xu Y, Tong D (2015) Chaos synchronization of financial chaotic system with external perturbation. Discrete Dyn Nat Soc Article ID 731376, 7 pp

Xu G, Xiu C, Liu F, Zang Y (2017) Secure communication based on the synchronous control of hysteretic chaotic neuron. Neurocomputing 227:108–112

Yu F, Li P, Gu K, Yin B (2016) Research progress of multi-scroll chaotic oscillators based on current-mode devices. Optik 127(13):5486–5490

Yu H, Cai G, Li Y (2012) Dynamic analysis and control of a new hyperchaotic finance system. Nonlinear Dyn 67:2171–2182

Zhao H, Lu M, Zuo J (2014) Anticontrol Hopf bifurcation and control of chaos for a finance system through washout filters with time delay. Volume 2014, Article ID 983034, 11 pp

Zheng J, Du B (2014). Research on feedback control for a kind of chaotic finance system. J Math Inform 2:1–11

Part II
Nonlinear Dynamical Systems with Hidden Attractors

Periodic Orbits, Invariant Tori and Chaotic Behavior in Certain Nonequilibrium Quadratic Three-Dimensional Differential Systems

Alisson C. Reinol and Marcelo Messias

Abstract In (Jafari et al, Phys Lett A 377(9):699-702, 2013) the authors gave the expressions of seventeen classes of quadratic differential systems defined in $\mathbb{R}^3$, depending on one real parameter a, which present chaotic behavior even without having any equilibrium point, for suitable choices of the parameter $a > 0$. In that paper, such systems are denoted by NE_1 to NE_{17}. As these systems have no equilibrium points, a natural question arises: how chaotic motion is generated in their nonequilibrium phase spaces? In this note we combine analytical and numerical results in order to study the integrability and dynamics of systems NE_1, NE_6, NE_8 and NE_9 among those listed in Jafari et al. (2013). We show that they exhibit a quite similar dynamical behavior and, consequently, the mechanisms for birth of chaos in these systems are similar. In this way, we intend to give at least a partial answer to the above question and contribute to better understand the complicated dynamics of the considered systems, in particular concerning the existence of periodic orbits and invariant tori and the emergence of chaotic behavior. The periodic orbits are studied using the Averaging Theory while the invariant tori are proved to exist via KAM Theorem. The chaotic dynamics arises from the broken of some of these invariant tori.

Keywords Periodic orbits · Invariant tori · Chaotic dynamics · Invariant algebraic surfaces · Averaging theory · KAM theorem

A. C. Reinol
Departamento de Matemática, Universidade Estadual Paulista (UNESP),
Instituto de Biociências, Letras e Ciências Exatas, São José do Rio Preto, Brazil
e-mail: alissoncarv@gmail.com

M. Messias (✉)
Faculdade de Ciências e Tecnologia, Departamento de Matemática e Computação,
Universidade Estadual Paulista (UNESP), Presidente Prudente, Brazil
e-mail: marcelo@fct.unesp.br

© Springer International Publishing AG 2018
V.-T. Pham et al. (eds.), *Nonlinear Dynamical Systems with Self-Excited and Hidden Attractors*, Studies in Systems, Decision and Control 133,
https://doi.org/10.1007/978-3-319-71243-7_13

1 Introduction

Let $\mathbb{K}[x, y, z]$ be the ring of polynomials in the variables x, y, z with coefficients in $\mathbb{K}$, where $\mathbb{K} = \mathbb{R}$ or $\mathbb{C}$, and

$$X = P \frac{\partial}{\partial x} + Q \frac{\partial}{\partial y} + R \frac{\partial}{\partial z}$$

the vector field associated to the differential system

$$\dot{x} = P(x, y, z), \qquad \dot{y} = Q(x, y, z), \qquad \dot{z} = R(x, y, z), \tag{1}$$

where P, Q, R are relatively prime polynomials in $\mathbb{R}[x, y, z]$ and the dot denotes derivative with respect to the independent variable t, usually called the *time*. We say that $d = \max\{\deg P, \deg Q, \deg R\}$ is the degree of system (1) (or the degree of the vector field X).

Besides their theoretical importance, polynomial differential systems like (1) are used as mathematical models of many natural phenomena arising in Physics, Chemistry, Biology, Engineering, and other sciences, see for instance Strogatz (2001), Wiggins (2003) and references therein. Hence understanding the dynamical behavior of the solutions of these systems is a very important matter, since it enables to better understand the natural phenomena modeled by them. In this way hundreds of books and papers have been published in the last 30 years aiming to describe the dynamics of system (1), which is far from being completely understood, even in the quadratic case, that is when it has degree $d = 2$. Indeed the dynamics generated by the flow of system (1) with degree $d \geq 2$ is, in general, very complex and difficult to be studied. Beyond equilibrium points, periodic, homoclinic and heteroclinic orbits, which are commonly encountered in the phase space of systems like (1), they may present *chaotic behavior*, which indicates the occurrence of complicated dynamical phenomena (Guckenheimer and Holmes 2002; Lorenz 1963; Wiggins 1988).

In general, it is possible to find a *chaotic attractor* in differential systems presenting chaotic behavior. An attractor of system (1) is a compact and connected set in its phase space for which all solutions in an open neighborhood of this set tend to as $t \to +\infty$. It can be a stable equilibrium point or periodic orbit, or, in some cases, a more complicated set, called chaotic attractor. The first chaotic attractor in a quadratic polynomial differential system of the form (1) was reported by Edward Lorenz in 1963 while he was studying the thermal convection of fluids in the atmosphere (Lorenz 1963; Sparrow 1982). Since then, several differential systems defined in $\mathbb{R}^n$, with $n \geq 3$, having this kind of attractors have been found and intensively studied, from theoretical and physical points of view, as for instance in the Chua system (Chua 1994), Chen system (Chen and Ueta 1999), Lü system (Lü and Chen 2002), Rabinovich system (Pikovskii et al. 1978), Rössler system (Rössler 1976), among others. For an introduction and some recent applications of chaotic dynamics in differential systems like (1) see for instance (Alligood et al. 1996; Broer and Takens 2011; Cencini and Vulpiani 2010; Chen and Yu 2003; Ott 2002; Wiggins 2003).

Recently, chaotic attractors of differential system (1) have been categorized as either *self-excited* or *hidden attractors* (Dudkowski et al. 2016; Leonov and Kuznetsov 2013a, b). Whereas a self-excited attractor has a basin of attraction that overlaps with the neighborhood of an unstable equilibrium point, a hidden attractor has a basin of attraction which does not intersect with small neighborhoods of any equilibrium point. The concept of hidden attractor was introduced in connection with the discovery of hidden oscillations in the classical and generalized Chua's circuit (Kuznetsov et al. 2010; Leonov et al. 2011, 2010). Chaotic attractors numerically observed in differential systems with no equilibrium points (Jafari et al. 2013; Li and Sprott 2014; Wang et al. 2012), with only one stable equilibrium point (Kingni et al. 2014; Lao et al. 2014; Molaie et al. 2013; Wang and Chen 2012; Wei and Pehlivan 2012; Wei and Yang 2011; Wei and Zhang 2014) or with an infinite number of equilibrium points (Gotthans and Petržela 2015; Jafari and Sprott 2013; Jafari et al. 2016) are examples of hidden attractors.

Differential systems with hidden attractors are rarely encountered, because there is no standard way of predicting the existence of this kind of attractors. In this way, only a few examples of such systems have been reported in the literature, as the ones mentioned above. Moreover, there is little knowledge about the formation of hidden attractors. Indeed, self-excited attractors in classical chaotic differential systems, as in the Lorenz and in the Chua systems, have some known routes to their formation (known as "routes to chaos"), as the bifurcation of homoclinic orbits (Shil'nikov-like theorems) or cascade of period doubling bifurcations, for more information about these bifurcations see Chaps. 4 and 6 of Kuznetsov (1998). On the other hand, very little is reported in the literature about the formation of hidden attractors.

In Jafari et al. (2013), the authors gave the expressions of seventeen classes of quadratic differential systems defined in $\mathbb{R}^3$, depending on one real parameter a, which present chaotic behavior for certain values of a, even without having any equilibrium point. In that paper such systems are denoted by NE_1 to NE_{17}. This kind of differential systems appear naturally in the study of various electromechanical models with rotation and electrical circuits with cylindrical phase space (Kuznetsov 2016). An analytical proof about the existence of chaotic attractors in these systems is yet needed. In this case, the attractors would be called hidden attractors because the systems have no equilibrium points. In this note we combine analytical and numerical results in order to study the integrability and dynamics of systems NE_1, NE_6, NE_8 and NE_9 shown in Table 1, which are among those provided in Jafari et al. (2013). We shall see in the course of this chapter that they present quite similar dynamical behavior. In this way we intend to give at least a partial answer to the question about how chaotic motion is generated in these nonequilibrium differential systems and contribute to better understand the complicated dynamics of them, in particular concerning the existence of periodic orbits and invariant tori and the emergence of chaotic behavior.

In the third column of Table 1 are shown the values of the real parameter a for which the corresponding systems present chaotic behavior, according to Jafari et al. (2013).

Table 1 Differential systems NE_1, NE_6, NE_8 and NE_9 given in Jafari et al. (2013)

Model	Equations	a
NE_1 (Sprott A)	$\dot{x} = y,$ $\dot{y} = -x - yz,$ $\dot{z} = y^2 - a$	1.0
NE_6	$\dot{x} = y,$ $\dot{y} = z,$ $\dot{z} = -y - xz - yz - a$	0.75
NE_8	$\dot{x} = y,$ $\dot{y} = -x - yz,$ $\dot{z} = xy + 0.5x^2 - a$	1.3
NE_9	$\dot{x} = y,$ $\dot{y} = -x - yz,$ $\dot{z} = -xz + 7x^2 - a$	0.55

System NE_1 in Table 1 is also called *Sprott A system* because it appeared in Sprott (1994) as the Case A system in a list of nineteen distinct differential systems in $\mathbb{R}^3$ with quadratic nonlinearities and presenting chaotic dynamics. In Messias and Reinol (2017b) we studied the integrability and global dynamics of NE_1 system, proving that for $a = 0$ it has a line of equilibria in the z–axis, its phase space is foliated by concentric invariant spheres with two equilibrium points located at their south and north poles and each one of these spheres is filled by heteroclinic orbits of south pole—north pole type. For $a \neq 0$, the spheres are no longer invariant algebraic surfaces and the heteroclinic orbits are destroyed. Then, from a detailed numerical study in the case $a > 0$ small, we observed that small nested invariant tori and a *limit set*, which encompasses these tori and is the α– and ω–limit set of almost all orbits in the phase space, are formed in a neighborhood of the origin. As the parameter a increases, this *limit set* expands and chaotic dynamics were detected in Sprott A system, for certain positive values of the parameter a. For details see Messias and Reinol (2017b).

In Messias and Reinol (2017a), through a further study of the line of equilibria at the z–axis, which exists in NE_1 system for $a = 0$, we showed that all its equilibrium points are normally hyperbolic, except the origin, which is a *non-isolated zero-Hopf equilibrium point*. We recall that an (isolated) equilibrium point of a differential system in $\mathbb{R}^3$ is called a *zero-Hopf equilibrium* if the Jacobian matrix of the system at this point has a zero and a pair of purely imaginary eigenvalues. As for $a = 0$ the origin of NE_1 system is a non-isolated equilibrium point with eigenvalues 0 and $\pm i$, then it is called a non-isolated zero-Hopf equilibrium point. This type of equilibrium was studied for instance in Llibre and Xiao (2014), where the authors proved the existence of one or two limit cycles bifurcating from it by using analysis techniques and the averaging theory of second order. In Messias and Reinol (2017a), by using the averaging theory and the KAM (Kolmogorov-Arnold-Moser) Theorem we proved that there exists a linearly stable periodic orbit which bifurcates from the non-isolated zero-Hopf equilibrium point located at the origin of system NE_1 and nested invariant tori are created around this periodic orbit. These dynamical elements play an important role in the emergence of chaotic behavior in NE_1 system. Indeed, chaotic seas are created through the destruction of some of these invariant tori, when the parameter a is varied. Furthermore, it was also proved in Messias and Reinol (2017a) that for $a \neq 0$ the NE_1 system has neither invariant algebraic surfaces nor polynomial first integrals. We observe that NE_1 system is a particular case of the well-known and widely studied Nosé-Hoover oscillator (Hoover 1985; Nosé 1984; Posch et al. 1986); some of the analytical results obtained in Messias and Reinol (2017a, b) were yet numerically described in Posch et al. (1986).

Based on the results obtained for NE_1 system in Messias and Reinol (2017a, b), we investigate the dynamics of the other sixteen differential systems provided in Jafari et al. (2013), that is NE_2–NE_{17}, trying to better understand the dynamics of these systems. After an extensive theoretical and numerical study, we obtained that NE_6, NE_8 and NE_9 systems shown in Table 1 have similar dynamical behavior than NE_1 system, in particular they seem to present similar mechanisms of birth of chaos. More precisely, although having no invariant algebraic surfaces for $a = 0$, NE_8 and NE_9 systems have a line of equilibria at the z–axis and in both cases, as well as in NE_1 system, all the equilibria at this line are normally hyperbolic, except the origin, which is a non-isolated zero-Hopf equilibrium point. The same is true for NE_6 system, except that in this case the line of equilibria existing for $a = 0$ is given by the x–axis. In this note, by using the averaging theory we show that for $a > 0$ small enough a linearly stable periodic orbit bifurcates from the non-isolated zero-Hopf equilibrium at the origin and nested invariant tori are formed around this periodic orbit, in NE_6, NE_8 and NE_9 systems, similarly to what happens in NE_1 system. The existence of these invariant tori are proved using the KAM Theorem. Finally, a detailed numerical investigation of the dynamics of NE_6, NE_8 and NE_9 systems suggests that the existence of such a periodic orbit bifurcating from the origin and nested invariant tori around it play an important role in the emergence of chaotic behavior in their nonequilibrium phase spaces, which is also determined by the broken of some of these invariant tori.

In the subsequent part of this chapter we give the proofs of the statements above, and it is organized as follows. In Sect. 2 we study the existence of invariant algebraic surfaces and polynomial first integrals for NE_6, NE_8 and NE_9 systems, while in Sect. 3 we prove the existence of periodic orbits and nested invariant tori for these systems. The numerical study of the dynamics of such three systems, including the creation of chaotic dynamics, are presented in Sect. 4. Some concluding remarks are given in Sect. 5.

2   Invariant Algebraic Surfaces and Polynomial First Integrals

As we observed in the Introduction, the dynamics generated by the flow of differential systems like (1) with degree $d \geq 2$ are, in general, very difficult to be studied. One of the tools used to study the dynamics of them is the determination of two-dimensional algebraic surfaces embedded in $\mathbb{R}^3$ which are invariant under the flow of these systems, called *invariant algebraic surfaces*, whose the precise definition is given below.

Definition 1 Let $f = f(x, y, z)$ be a non-constant polynomial in $\mathbb{C}[x, y, z]$. The algebraic surface $f = 0$ is an *invariant algebraic surface* of differential system (1) if for some polynomial $K = K(x, y, z)$ in $\mathbb{C}[x, y, z]$ we have

$$X(f) = P \frac{\partial f}{\partial x} + Q \frac{\partial f}{\partial y} + R \frac{\partial f}{\partial z} = Kf, \tag{2}$$

where X is the vector field associated to the polynomial differential system (1). The polynomial K is said the *cofactor* of the invariant algebraic surface $f = 0$ and f is called *Darboux polynomial*. If d is the degree of system (1), then K has degree at most $d - 1$.

Note that on the points of the algebraic surface $f = 0$ the gradient vector of f, that is $(\partial f / \partial x, \partial f / \partial y, \partial f / \partial z)$, is orthogonal to the vector field X associated to system (1). Hence at every point of $f = 0$ the vector field X is tangent to the surface $f = 0$, then this surface is formed by orbits of X. In this way the existence of invariant algebraic surfaces helps strongly the study of the dynamics of differential systems with complicated behavior, as it was shown, for example, in the study of Sprott A system (Messias and Reinol 2017b), Rikitake system (Llibre and Messias 2009), Rabinovich system (Llibre et al. 2008), Lorenz system (Llibre et al. 2010) and Chen system (Llibre et al. 2012). Hence it is important to investigate if system (1) has invariant algebraic surfaces.

If the cofactor K is identically zero in Eq. (2), then f is a *polynomial first integral* of differential system (1). A *first integral* of system (1) is a non-constant analytic function which is constant on all solution curves $(x(t), y(t), z(t))$ of the system. The knowledge of a first integral of system (1) in $\mathbb{R}^3$ allows to reduce the study of this system in one dimension. Moreover, if system (1) has two functionally independent first integrals f_1 and f_2, then we say that the system is *integrable*, which means that its phase space can be completely determined, since its orbits are on the intersection of the level surfaces $f_1 = c_1$ and $f_2 = c_2$, for $c_1, c_2 \in \mathbb{R}$.

In Definition 1 we allowed the invariant algebraic surface $f = 0$ to be complex, that is $f \in \mathbb{C}[x, y, z]$, even in the case that the polynomial differential system (1) is real, because even for real polynomial differential systems the existence of a real first integral can be forced by the existence of complex invariant algebraic surfaces. For more details about this fact see Chap. 8 of Dumortier et al. (2006).

In Messias and Reinol (2017b) it was proved that, for $a = 0$, NE_1 system has a polynomial first integral given by $f(x, y, z) = x^2 + y^2 + z^2$. Consequently, its phase space is foliated by the invariant spheres $x^2 + y^2 + z^2 = r^2$, with $r > 0$, which are formed by an infinite set of heteroclinic orbits of south pole—north pole type, as shown in Fig. 1. Note that the existence of concentric invariant spheres determines a compact structure of the orbits in the phase space of NE_1 system for $a = 0$. In Messias and Reinol (2017a) it was proved that, for $a \neq 0$, NE_1 system has neither invariant algebraic surfaces nor polynomial first integrals.

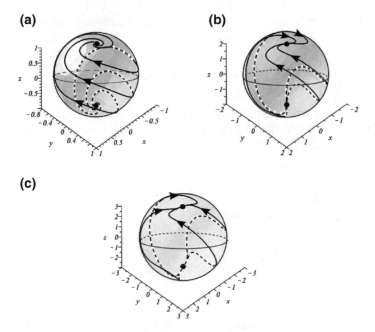

Fig. 1 Flow of NE_1 system with $a = 0$ restricted to the invariant spheres $x^2 + y^2 + z^2 = r^2$ for **a** $0 < r < 2$, **b** $r = 2$ and **c** $r > 2$. Observe that the heteroclinic orbits on the invariant spheres connect a pair of foci in **a**, improper nodes in **b** and nodes in **c**

Here we prove that NE_6 system has neither invariant algebraic surfaces nor polynomial first integrals for all $a \in \mathbb{R}$. In order to do it, the following lemma will be useful.

Lemma 1 *If $f(x, y, z) = 0$ is an invariant algebraic surface of NE_6 system with cofactor $K_f(x, y, z)$, then $F(X, Y, Z) = \mu^n f(X, Y, \mu^{-1}Z)$ is an invariant algebraic surface of the rescaled differential system*

$$
\begin{aligned}
\dot{X} &= \mu Y, \\
\dot{Y} &= Z, \\
\dot{Z} &= -\mu^2 Y - \mu XZ - \mu YZ - \mu^2 a.
\end{aligned}
\tag{3}
$$

with cofactor $K_F(X, Y, Z) = K_f(X, Y, \mu^{-1}Z)$.

Proof Let $f(x, y, z) = 0$ be an invariant algebraic surface of NE_6 system with cofactor $K_f(x, y, z)$. From the definition of invariant algebraic surface, f satisfies

$$
y \frac{\partial f}{\partial x} + z \frac{\partial f}{\partial y} + (-y - xz - yz - a) \frac{\partial f}{\partial z} = K_f f.
\tag{4}
$$

After the rescaling of time $T = \mu^{-1}t$ in NE_6 system, where T is the new time and $\mu \in \mathbb{R} \setminus \{0\}$, from Eq. (4) we obtain

$$y\frac{\partial f}{\partial x} + z\frac{\partial f}{\partial y} + (-y - xz - yz - a)\frac{\partial f}{\partial z} = \mu^{-1}K_f f.$$

Note that we obtain differential system (3) from NE_6 system after the rescaling of time $T = \mu^{-1}t$ and doing the change of variables $x = X$, $y = Y$, $z = \mu^{-1}Z$. Consider $F(X, Y, Z) = \mu^n f(X, Y, \mu^{-1}Z)$. Then,

$$\mu Y\frac{\partial F}{\partial X} + Z\frac{\partial F}{\partial Y} + (-\mu^2 Y - \mu XZ - \mu YZ - \mu^2 a)\frac{\partial F}{\partial Z} =$$
$$= \mu y\frac{\partial F}{\partial x}\frac{dx}{dX} + \mu z\frac{\partial F}{\partial y}\frac{dy}{dY} + (-\mu^2 y - \mu^2 xz - \mu^2 yz - \mu^2 a)\frac{\partial F}{\partial z}\frac{dz}{dZ} =$$
$$= \mu y\frac{\partial F}{\partial x} + \mu z\frac{\partial F}{\partial y} + \mu(-y - xz - yz - a)\frac{\partial F}{\partial z} =$$
$$= \mu^{n+1}\left(y\frac{\partial f}{\partial x} + z\frac{\partial f}{\partial y} + (-y - xz - yz - a)\frac{\partial f}{\partial z}\right) =$$
$$= \mu^{n+1}(\mu^{-1}K_f f) = K_F F.$$

Therefore, $F = 0$ is an invariant algebraic surface of system (3) with cofactor $K_F(X, Y, Z) = K_f(X, Y, \mu^{-1}Z)$. $\square$

Theorem 1 NE_6 *system has neither invariant algebraic surfaces nor polynomial first integrals for all $a \in \mathbb{R}$.*

Proof Suppose that $f = 0$ is an invariant algebraic surface of degree $n \geq 1$ of NE_6 system with cofactor $K = k_0 + k_1 x + k_2 y + k_3 z$, with $k_0, k_1, k_2, k_3 \in \mathbb{C}$, since NE_6 system has degree 2. Take f as the sum of its homogeneous parts, that is $f = \sum_{i=0}^{n} f_i$, where each f_i is a homogeneous polynomial of degree i, for $i = 0, ..., n$. At first assume that K is not identically zero. From the definition of invariant algebraic surface, f must satisfy equality (2), that is

$$y\frac{\partial f}{\partial x} + z\frac{\partial f}{\partial y} + (-y - xz - yz - a)\frac{\partial f}{\partial z} = (k_0 + k_1 x + k_2 y + k_3 z)f. \qquad (5)$$

From equality of polynomials, the terms of degree $n + 1$ in (5) satisfy

$$(-xz - yz)\frac{\partial f_n}{\partial z} = (k_1 x + k_2 y + k_3 z)f_n. \qquad (6)$$

Solving the partial differential equation above we get

$$f_n(x, y, z) = c_n(x, y)\, z^{g(x,y)} \exp\left(-\frac{k_3 z}{x + y}\right)$$

where c_n is an arbitrary function in the variables x and y, and

$$g(x, y) = -(k_1 x + k_2 y)/(x + y).$$

As f_n is a homogeneous polynomial of degree n, we must have $k_1 = k_2 = -m$ and $k_3 = 0$, with $0 \leq m \leq n$ an integer. Hence $f_n(x, y, z) = z^m c_n(x, y)$ and c_n is a polynomial of degree $n - m$.

Now, note that the terms of degree n in (5) satisfy

$$(-xz - yz)\frac{\partial f_{n-1}}{\partial z} + z\frac{\partial f_n}{\partial y} + y\left(\frac{\partial f_n}{\partial x} - \frac{\partial f_n}{\partial z}\right) = k_0 f_n - m(x + y)f_{n-1}. \qquad (7)$$

Consider first $m = 0$ in (7). Solving the partial differential equation for f_{n-1}, we get

$$f_{n-1}(x, y, z) = \frac{z}{x + y}\frac{\partial f_n}{\partial y} + \frac{\ln(z)}{x + y}\left[y\left(\frac{\partial f_n}{\partial x} - \frac{\partial f_n}{\partial z}\right) - k_0 f_n\right] + c_{n-1}(x, y),$$

where c_{n-1} is an arbitrary function in the variables x and y. As f_{n-1} is a homogeneous polynomial, f_n must satisfy the partial differential equation

$$y\left(\frac{\partial f_n}{\partial x} - \frac{\partial f_n}{\partial z}\right) - k_0 f_n = 0,$$

whose solution is

$$f_n(x, y, z) = c_n(y, x + z)\, \exp\left(\frac{k_0 x}{y}\right),$$

where c_n is an arbitrary function in the variables y and $x + z$. As f_n is a polynomial, then $k_0 = 0$. Hence $K = 0$, what is a contradiction, since we are considering K not identically zero. Therefore, $m > 0$.

Now, consider $m > 0$ and $f_n(x, y, z) = z^m c_n(x, y)$ (polynomial solution of the partial differential equation (6)) in the partial differential equation (7), that is

$$(-xz - yz)\frac{\partial f_{n-1}}{\partial z} + z^{m+1}\frac{\partial c_n}{\partial y} + y z^{m-1}\left(z\frac{\partial c_n}{\partial x} - m c_n\right) = k_0 z^m c_n - m(x + y)f_{n-1}.$$

Solving this equation for f_{n-1}, we get

$$f_{n-1}(x, y, z) = \frac{z^m \ln(z)}{x + y}\left(-y\frac{\partial c_n}{\partial x} + k_0 c_n\right) + \frac{z^{m-1}}{x + y}\left(z^2\frac{\partial c_n}{\partial y} + m y c_n\right) + z^m c_{n-1}(x, y),$$

where c_{n-1} is an arbitrary function in the variables x and y. As f_{n-1} is a polynomial, c_n must satisfy the partial differential equation

$$-y \frac{\partial c_n}{\partial x} + k_0 \, c_n = 0,$$

whose solution is

$$c_n(x, y) = \tilde{c}_n(y) + \exp\left(\frac{k_0 \, x}{y}\right),$$

where $\tilde{c}_n$ is an arbitrary function in the variable y. As f_n is a polynomial, then $k_0 = 0$.

In order to simplify the computations, consider the rescaling of time $T = \mu^{-1} t$, where T is the new time, and the change of variables $(x, y, z) \rightarrow (X, Y, Z)$, where $x = X$, $y = Y$, $z = \mu^{-1} Z$, with $\mu \in \mathbb{R} \setminus \{0\}$. Then, NE_6 system can be written as

$$\begin{aligned}
\dot{X} &= \mu X, \\
\dot{Y} &= Z, \\
\dot{Z} &= -\mu^2 Y - \mu XZ - \mu YZ - \mu^2 a.
\end{aligned} \tag{8}$$

Let

$$F(X, Y, Z) = \mu^n f(X, Y, \mu^{-1} Z) = \sum_{i=0}^{n} \mu^{n-i} F_i(X, Y, Z),$$

where F_i is the weight homogeneous part with degree i of F and n is the weight degree of F with weight exponents $s = (0, 0, -1)$. By Lemma 1, $F = 0$ is an invariant algebraic surface of system (8) with cofactor $K = -m(X + Y)$. From the definition of invariant algebraic surface, we have

$$\mu Y \sum_{i=0}^{n} \mu^{n-i} \frac{\partial F_i}{\partial X} + Z \sum_{i=0}^{n} \mu^{n-i} \frac{\partial F_i}{\partial Y} + (-\mu^2 Y - \mu XZ - \mu YZ - \mu^2 a) \sum_{i=0}^{n} \mu^{n-i} \frac{\partial F_i}{\partial Z}$$

$$= -m(X + Y) \sum_{i=0}^{n} \mu^{n-i} F_i(X, Y, Z). \tag{9}$$

The terms with μ^0 in (9) satisfy

$$Z \frac{\partial F_n}{\partial Y} = -m(X + Y) F_n(X, Y, Z).$$

Solving this partial differential equation, we readily obtain

$$F_n(X, Y, Z) = C_n(X, Z) \exp\left(-\frac{mY(2X + Y)}{2Z}\right),$$

where C_n is an arbitrary function in the variables X and Z. As F_n is a polynomial and $m > 0$, we must have $F_n = 0$, what is a contradiction, since F is a polynomial of degree n as well as f. Hence system NE_6 has no invariant algebraic surfaces.

Now, suppose that the cofactor K is identically zero, that is f is a polynomial first integral of NE_6 system. Then f satisfies equality (5) with $k_0 = k_1 = k_2 = k_3 = 0$. In this case, the terms of degree n in Eq. (5) satisfy

$$(-xz - yz)\frac{\partial f_n}{\partial z} = 0.$$

Then $\partial f_n / \partial z = 0$. Computing the terms of degree n in (5) with $k_0 = k_1 = k_2 = k_3 = 0$ and considering $\partial f_n / \partial z = 0$, we obtain

$$(-xz - yz)\frac{\partial f_{n-1}}{\partial z} + y\frac{\partial f_n}{\partial x} + z\frac{\partial f_n}{\partial y} = 0.$$

Solving this partial differential equation for f_{n-1}, we get

$$f_{n-1}(x, y, z) = \frac{y \ln(z)}{x + y}\frac{\partial f_n}{\partial x} + \frac{z}{x + y}\frac{\partial f_n}{\partial y} + c_{n-1}(x, y),$$

where c_{n-1} is an arbitrary function in the variables x and y. As f_{n-1} is a polynomial, we must have $\partial f_n / \partial x = 0$ and $\partial f_n / \partial y = (x + y)\, g$, with $g = g(x, y, z)$ an arbitrary polynomial. As $\partial f_n / \partial x = 0$, hence $g = 0$ and, consequently, $\partial f_n / \partial y = 0$ and $f_n = 0$, what is a contradiction, since f is a polynomial of degree n.

Therefore, NE_6 system has neither invariant algebraic surfaces nor polynomial first integrals for all $a \in \mathbb{R}$.   □

The next result is concerned with the existence of polynomial first integrals of NE_8 and NE_9 systems.

Theorem 2 NE_8 *and* NE_9 *systems have no polynomial first integrals for all* $a \in \mathbb{R}$.

Proof We start with NE_8 system. Suppose that $f = 0$ is a polynomial first integral of degree $n \geq 1$ of NE_8 system. Then, f satisfies the partial differential equation

$$y\frac{\partial f}{\partial x} + (-x - yz)\frac{\partial f}{\partial y} + (xy + 0.5x^2 - a)\frac{\partial f}{\partial z} = 0. \tag{10}$$

Take f as the sum of its homogeneous parts, that is $f = \sum_{i=0}^{n} f_i$, where each f_i is a homogeneous polynomial of degree i, for $i = 0, ..., n$. The terms of degree n in Eq. (10) satisfy

$$-yz\frac{\partial f_{n-1}}{\partial y} + (xy + 0.5x^2)\frac{\partial f_{n-1}}{\partial z} + y\frac{\partial f_n}{\partial x} - x\frac{\partial f_n}{\partial y} = 0. \tag{11}$$

Solving the partial differential equation above for f_n, we get

$$f_n(x, y, z) = -\frac{1}{4}(x^2 + y^2)\frac{\partial f_{n-1}}{\partial z} \arctan\left(\frac{x}{y}\right) - \frac{1}{2}x^2\frac{\partial f_{n-1}}{\partial z}$$

$$+ \frac{1}{4}x\left(y\frac{\partial f_{n-1}}{\partial z} + z\frac{\partial f_{n-1}}{\partial y}\right) + c_n(x^2 + y^2, z),$$

where c_n is an arbitrary function in the variables $x^2 + y^2$ and z. As f_n is a homogeneous polynomial, we must have $\partial f_{n-1}/\partial z = 0$. Considering $\partial f_{n-1}/\partial z = 0$ in partial differential equation (11) and solving it for f_{n-1}, we obtain

$$f_{n-1}(x, y, z) = \frac{y}{z}\frac{\partial f_n}{\partial x} - \frac{x}{z}\frac{\partial f_n}{\partial y} \ln(y) + c_{n-1}(x, z),$$

where c_{n-1} is an arbitrary function in the variables x and z. As f_{n-1} is a polynomial, then $\partial f_n/\partial x = z\,g$, with $g = g(x, y, z)$ an arbitrary polynomial, and $\partial f_n/\partial y = 0$.

Now, note that the terms of degree $n + 1$ in (10) satisfy

$$-yz\frac{\partial f_n}{\partial y} + (xy + 0.5x^2)\frac{\partial f_n}{\partial z} = 0.$$

As $\partial f_n/\partial y = 0$, we have that $\partial f_n/\partial z = 0$, which implies that $\partial f_n/\partial x = z\,g = 0$. Hence, $f_n = 0$, what is a contradiction, since f is a polynomial of degree n.

Using similar arguments, we can prove that NE_9 system also has no polynomial first integrals for all $a \in \mathbb{R}$. $\square$

Note that it remains an open problem to determine if NE_8 and NE_9 systems have or no invariant algebraic surfaces. On the other hand, it follows from Theorems 1 and 2 that, differently from NE_1 system, NE_6, NE_8 and NE_9 systems do not have their phase spaces foliated by invariant algebraic surfaces, even for $a = 0$. However other facts are common in the dynamical behavior of all the considered systems: by varying the parameter a from $a = 0$ to $a > 0$ small enough, a linearly stable periodic orbit bifurcates from the non-isolated zero-Hopf equilibrium at the origin and nested invariant tori are formed around it. This is the subject of the next section.

3   Periodic Orbits and Invariant Tori

The study of periodic and quasiperiodic solutions in differential systems is an important issue in qualitative theory of dynamical systems. Some known routes to chaos, for example, are characterized by the existence of this kind of solutions, as the cascade of period doubling bifurcations and quasiperiodic or Ruelle-Takens-Newhouse scenario, for more details about this and other known routes to chaos see for instance Cencini and Vulpiani (2010), Lakshmanan and Rajaseekar (2003), Ott (2002).

A *periodic solution* $\phi(t) = (x(t), y(t), z(t))$ of differential system (1) is a solution which is periodic in time, that is $\phi(t) = \phi(t + T)$ for some fixed positive constant T called the *period of* $\phi(t)$. A *periodic orbit* of differential system (1) is the orbit of any point through which a periodic solution passes. Periodic solutions determine *periodic motions*. The oscillatory motion of a pendulum, periodic oscillations of an electronic circuit and the bounded orbits of a Kepler particle are examples of periodic motions.

A function $h : \mathbb{R} \to \mathbb{R}^3$ is called *quasiperiodic* if it can be represented in the form $h(t) = H(\omega_1 t, \omega_2 t, \omega_3 t)$, where $H(x, y, z)$ is a continuous function of period 2π in x, y, z. The real numbers ω_1, ω_2, ω_3 are called the *basic frequencies*. A *quasiperiodic solution* $\phi(t) = (x(t), y(y), z(t))$ of differential system (1) is a solution defined by a quasiperiodic function with respect to the time. A *quasiperiodic orbit* is the orbit of any point through which a quasiperiodic solution passes.

In Messias and Reinol (2017a) it was proved the existence of periodic and quasiperiodic orbits in NE_1 system. Indeed, for $a = 0$, the z–axis is a line of equilibria of this system and the origin is a non-isolated zero-Hopf equilibrium. By using the averaging theory of first order, the authors proved that, for $a > 0$ sufficiently small, a linearly stable periodic orbit bifurcates from the origin of NE_1 system. Moreover, around this periodic orbit there exist nested invariant tori, whose orbits are dense and move quasiperiodically on them. The existence of such tori was also proved in Messias and Reinol (2017a) using the classical KAM Theorem, which provides a starting point for an explanation of the transition from regular or quasiperiodic to chaotic motion in Hamiltonian systems.

By using the same techniques, in this section we prove that there also exist periodic and quasiperiodic orbits in NE_6, NE_8 and NE_9 systems. In order to do it we start with an overview about the averaging theory of first order and the classical KAM Theorem. A general introduction about the averaging theory can be found in Sanders et al. (2007). See also Buică et al. (2012), Llibre et al. (2014), Llibre and Novaes (2015), Cândido et al. (2017) for recent works which extend and improve this theory. For more details about the KAM theory see, for instance, Pöschel (2001), Chap. 15 of Verhulst (1996), Chap. 7 of Lakshmanan and Rajaseekar (2003), Chap. 14 of Wiggins (2003) and references therein. KAM theory is a huge area and there are hundreds of books and papers about it. For the sake of completeness, in what follows we present some of the main results about the averaging and KAM theories, which will be used ahead in this section.

Averaging theory of first order. Consider the initial value problems

$$\dot{\mathbf{x}} = \varepsilon F_1(t, \mathbf{x}) + \varepsilon^2 F_2(t, \mathbf{x}, \varepsilon), \qquad \mathbf{x}(0) = \mathbf{x}_0, \tag{12}$$

and

$$\dot{\mathbf{y}} = \varepsilon g(\mathbf{y}), \qquad \mathbf{y}(0) = \mathbf{x}_0, \tag{13}$$

with $\mathbf{x}$, $\mathbf{y}$ and $\mathbf{x}_0$ in some open subset Ω of $\mathbb{R}^n$, $t \in [0, \infty)$ and $\varepsilon \in (0, \varepsilon_0]$, for some fixed $\varepsilon_0 > 0$ sufficiently small. Assume that F_1 and F_2 are periodic functions of period T in the variable t, and set

$$g(\mathbf{y}) = \frac{1}{T} \int_0^T F_1(t, \mathbf{y})dt.$$

Denote by $D_{\mathbf{x}}g$ all the first derivatives of g and by $D_{\mathbf{xx}}g$ all the second derivatives of g. Then, we have the following result, which is proved in Guckenheimer and Holmes (2002), Verhulst (1996).

Theorem 3 *Assume that F_1, $D_{\mathbf{x}}F_1$, $D_{\mathbf{xx}}F_1$ and $D_{\mathbf{x}}F_2$ are continuous and bounded by a constant independent of ε in $[0, \infty) \times \Omega \times (0, \varepsilon_0]$, and that $\mathbf{y}(t) \in \Omega$ for $t \in [0, 1/\varepsilon]$. Then, the following statements hold.*

1. *For $t \in [0, 1/\varepsilon]$, we have $\mathbf{x}(t) - \mathbf{y}(t) = O(\varepsilon)$ as $\varepsilon \to 0$.*
2. *If $p \neq 0$ is an equilibrium point of system (13) such that $\det[D_{\mathbf{y}}g(p)] \neq 0$, then there exists a periodic solution $\phi(t, \varepsilon)$ of period T for system (12) which is close to p and such that $\phi(0, \varepsilon) - p = O(\varepsilon)$ as $\varepsilon \to 0$.*
3. *The stability of the periodic solution $\phi(t, \varepsilon)$ is given by the stability of the equilibrium point p.*

Classical KAM Theorem. A differential system in $\mathbb{R}^{2n}$ is called *Hamiltonian* if there exists a non-constant analytic function $H : \Omega \to \mathbb{R}$, where Ω is an open subset of $\mathbb{R}^n \times \mathbb{R}^n$, such that

$$\dot{\mathbf{x}} = -\frac{\partial H}{\partial \mathbf{y}}(\mathbf{x}, \mathbf{y}), \qquad \dot{\mathbf{y}} = \frac{\partial H}{\partial \mathbf{x}}(\mathbf{x}, \mathbf{y}), \tag{14}$$

with $(\mathbf{x}, \mathbf{y}) \in \mathbb{R}^n \times \mathbb{R}^n$. In this case, we say that H is the *Hamiltonian function* of system (14).

If $n = 1$, then the Hamiltonian system (14) is *integrable*, because it is planar and has a first integral given by the Hamiltonian function H. By the Liouville-Arnold's Theorem (for more details about this theorem, see Chap. 14 of Wiggins (2003)), there exists a transformation $(\mathbf{x}, \mathbf{y}) \to (I, \varphi)$, with $(I, \varphi) \in U \times [0, 2\pi]$, where U is an open interval of $\mathbb{R}$, such that, in the variables I and φ, the obtained system is again Hamiltonian. This transformation leads to the differential system

$$\begin{aligned} \dot{I} &= 0, \\ \dot{\varphi} &= \omega(I), \end{aligned} \tag{15}$$

where I and φ are called *action-angle coordinates*. Let $H_0(I)$ be the Hamiltonian function of system (15). Hence $\omega(I) = \partial H_0/\partial I$. The advantage in using action-angle variables is that system (15) is easily integrable and its general solution is

$$\begin{aligned} I(t) &= I_0, \\ \varphi(t) &= \omega(I_0)t + \varphi_0, \end{aligned}$$

where $I(0) = I_0$ and $\varphi(0) = \varphi_0$ are the initial condition. Then the solutions of system (15) are straight lines which, due to the identification of the angular coordinate φ modulo 2π, are winding around the invariant tori

$$\mathbb{T}^1 = \{I_0\} \times [0, 2\pi]$$

with constant frequency $\omega(I_0)$. Thus, the whole phase space of system (15) is foliated by invariant tori with linear flow, which are also called *Kronecker tori*.

The frequency (or angular velocity) ω with which an orbit winds around an invariant torus is classified as resonant or nonresonant in the following way. Let $\mathbb{T}^n$ be the n-dimensional torus whose orbits wind around it with frequency $\omega = (\omega_1, ..., \omega_n)$. The frequency vector ω is said to be *resonant* (or *rationally dependent*) if there exists $k = (k_1, ..., k_n) \in \mathbb{Z}^n \setminus \{0\}$ such that $k \cdot \omega = \langle (k_1, ..., k_n), (\omega_1, ..., \omega_n) \rangle = 0$. Otherwise we say that ω is *nonresonant* (or *rationally independent*).

Note that, in the case $n = 1$, the torus $\mathbb{T}^1$ is a closed curve and resonance implies $\omega = 0$. A torus whose orbits wind around it with nonresonant frequency has the property that each orbit is dense on this torus. More precisely, given a point p on a nonresonant torus and a neighborhood V_p of that point, the orbit ϕ_p through the point p will re-intersect V_p in a future time, after leaving it. Furthermore, given any other point q and a neighborhood V_q of that point, the orbit ϕ_p will also intersect V_q. This is a classical result that goes back to Kronecker (the flow on a nonresonant torus is often referred to as Kronecker flow).

The introduction of action-angle variables is usually carried out by employing a *generating function* $S = S(I, \mathbf{y})$ (see Chap. 10 of Arnold (1989)). In general, we will not be able to find such a generating function S. However, these transformations are specially useful if system (14) is *nearly-integrable* in the following sense. Assume that the Hamiltonian function $H(\mathbf{x}, \mathbf{y})$ of system (14) contains a small parameter $\varepsilon > 0$ and the introduction of action-angle coordinates produces the system

$$\begin{aligned} \dot{I} &= \varepsilon\, f(I, \varphi), \\ \dot{\varphi} &= \omega(I) + \varepsilon\, g(I, \varphi). \end{aligned} \tag{16}$$

If $\varepsilon = 0$, system (16) is integrable; if $\varepsilon > 0$ sufficiently small, system (16) is called nearly-integrable. Let

$$H_0(I) + \varepsilon\, H_1(I, \varphi) \tag{17}$$

be the Hamiltonian function of system (16). The KAM theorem ensures that, under certain hypothesis, almost all tori which exist in the integrable case $\varepsilon = 0$, survive under small perturbations (that is, for $\varepsilon \neq 0$ small), although possibly slightly deformed. More precisely, the following result holds.

Theorem 4 (KAM) *Consider differential system (16) induced by the Hamiltonian function (17). If H_0 is non-degenerate, i.e.,*

$$\det \left(\frac{\partial^2 H_0}{\partial I^2} \right) \neq 0,$$

then almost all invariant tori which exist for the unperturbed system ($\varepsilon = 0$) will persist for $\varepsilon > 0$ sufficiently small, although slightly deformed. Furthermore, the Lebesgue-measure of the complement of the set of tori tends to zero as $\varepsilon \to 0$.

By using Theorems 3 and 4, we prove the following result about the existence of a periodic orbit and invariant tori around it in NE_6, NE_8 and NE_9 systems.

Theorem 5 *For $a > 0$ small enough, NE_6, NE_8 and NE_9 systems have a linearly stable periodic orbit, which tends to the origin as $a \to 0$. Around the periodic orbit of each one of these systems, there exist nested invariant tori whose orbits are dense and move quasiperiodically on them.*

Proof We start with NE_6 system. Before applying Theorem 3 in order to study the existence of periodic orbits of NE_6 system, we need to write its linear part at the origin into the real Jordan normal form. In order to do this, we consider the linear change of coordinates $(x, y, z) \to (u, v, w)$, where $x = -u + w$, $y = v$, $z = u$. In the new variables (u, v, w), NE_6 system becomes

$$
\begin{aligned}
\dot{u} &= -v - (-u + w)\, u - uv - a, \\
\dot{v} &= u, \\
\dot{w} &= -(-u + w)\, u - uv - a,
\end{aligned}
\tag{18}
$$

whose Jacobin matrix at the origin has eigenvalues 0 and $\pm i$. Now, writing system (18) in cylindrical coordinates (r, θ, w), where $u = r\cos\theta$, $v = r\sin\theta$, it becomes

$$
\begin{aligned}
\dot{r} &= -\cos\theta\, [r^2\cos\theta\, (\sin\theta - \cos\theta) + rw\cos\theta + a], \\
\dot{\theta} &= (-\cos^2\theta - \sin\theta\cos\theta + 1)\, r\cos\theta + w\cos\theta\sin\theta + 1 + \tfrac{a}{r}\sin\theta, \\
\dot{w} &= r^2\cos\theta\, (\cos\theta - \sin\theta) - rw\cos\theta - a.
\end{aligned}
\tag{19}
$$

We introduce the variable $\varepsilon > 0$ into system (19) considering $a = \varepsilon^2$ and doing the change of coordinates $(r, \theta, w) \to (R, \theta, W)$, where $r = \varepsilon R$, $w = \varepsilon W$. Then system (19) can be written as

$$
\begin{aligned}
\dot{R} &= -\varepsilon \cos\theta\, [R^2\cos\theta\, (\sin\theta - \cos\theta) + R\, W\cos\theta + 1], \\
\dot{\theta} &= [-\cos^2\theta - \sin\theta\cos\theta + 1]\, \varepsilon R\cos\theta + \varepsilon W \cos\theta\sin\theta + 1 + \tfrac{\varepsilon}{R}\sin\theta, \\
\dot{W} &= \varepsilon R^2\cos\theta\, (\cos\theta - \sin\theta) - \varepsilon RW\cos\theta - \varepsilon.
\end{aligned}
$$

Taking θ as the independent variable and doing the Taylor expansion of order 2 of the obtained equations at $\varepsilon = 0$, we get

$$
\begin{aligned}
\frac{dR}{d\theta} &= \varepsilon\, h(R, \theta, W)\cos\theta + O(\varepsilon^2), \\
\frac{dW}{d\theta} &= \varepsilon\, h(R, \theta, W) + O(\varepsilon^2),
\end{aligned}
\tag{20}
$$

where $h(R, \theta, W) = R^2\cos^2\theta - R\, W\cos\theta - R^2\cos\theta\sin\theta - 1$. Using the notation of Theorem 3, consider in system (20)

$$\mathbf{x} = \begin{pmatrix} R \\ W \end{pmatrix}, \qquad t = \theta, \qquad T = 2\pi, \qquad F_1(\theta, \mathbf{x}) = \begin{pmatrix} h(R, \theta, W)\cos\theta \\ h(R, \theta, W) \end{pmatrix}.$$

In this way,

$$g(\mathbf{y}) = \frac{1}{2\pi} \int_0^{2\pi} F_1(\theta, \mathbf{y}) \, d\theta = \begin{pmatrix} -\frac{1}{2} R W \\ \frac{1}{2} R^2 - 1 \end{pmatrix}.$$

Hence, $g(\mathbf{y}) = 0$ has the unique real solution $p = (R, W) = (\sqrt{2}, 0)$ (remember that $R > 0$), which satisfies $\det[D_{\mathbf{y}} g(p)] = 1 \neq 0$. Then, by Theorem 3, it follows that, for $\varepsilon > 0$ sufficiently small, system (20) has a periodic solution $\phi(\theta, \varepsilon) = (R(\theta, \varepsilon), W(\theta, \varepsilon))$ such that $\phi(0, \varepsilon) \to (\sqrt{2}, 0)$ as $\varepsilon \to 0$. Moreover, the eigenvalues of the matrix $[D_{\mathbf{y}} g(p)]$ are $\pm i$. Thus the obtained periodic solution is linearly stable, that is, any solution close enough to this periodic solution remains close enough forever, without tending to it.

Changing back the coordinates to NE_6 system, we have that, for $a > 0$ sufficiently small, such system has a periodic solution of period approximately 2π given by

$$\begin{aligned} x(t) &= -\sqrt{2a}\cos t + O(a), \\ y(t) &= \sqrt{2a}\sin t + O(a), \\ z(t) &= \sqrt{2a}\cos t + O(a). \end{aligned}$$

Note that this solution tends to the origin as $a \to 0$. Therefore, for $a > 0$ small enough, NE_6 system has a linearly stable periodic orbit which emerges from the origin.

Now we shall prove that around this periodic orbit there exist nested invariant tori. In order to use Theorem 4, we consider the changes of coordinates $(x, y, z) \to (u, v, w)$ and $(u, v, w) \to (r, \theta, w)$ as before and write NE_6 system in the form (19). This time around we introduce the variable $\varepsilon > 0$ into system (19) considering $a = \varepsilon$ and doing the change of coordinates $(r, \theta, w) \to (r, \Theta, W)$, where $\theta = \varepsilon \Theta, w = \varepsilon W$. Then system (19) can be written as

$$\begin{aligned} \dot{R} &= \cos(\varepsilon \Theta) \left[r^2 \cos(\varepsilon \Theta)(\cos(\varepsilon \Theta) - \sin(\varepsilon \Theta)) + \varepsilon \, r W \cos(\varepsilon \Theta) - \varepsilon \right], \\ \dot{\Theta} &= W \cos(\varepsilon \Theta) \sin(\varepsilon \Theta) - \frac{1}{\varepsilon} \left[r \cos(\varepsilon \Theta)(\sin(\varepsilon \Theta) \cos(\varepsilon \Theta) + \cos^2(\varepsilon \Theta) - 1) + 1 \right] \\ &\quad + \frac{1}{r} \sin(\varepsilon \Theta), \\ \dot{W} &= r \, W \cos(\varepsilon \, \Theta) - 1 + \frac{1}{\varepsilon} r^2 \cos(\varepsilon \Theta)[\cos(\varepsilon \Theta) - \sin(\varepsilon \Theta)]. \end{aligned}$$

Taking W as the independent variable and doing the Taylor expansion of order 2 of the obtained equations at $\varepsilon = 0$, we get

$$\frac{dr}{dW} = O(\varepsilon), \qquad \frac{d\Theta}{dW} = \frac{1}{r^2} + O(\varepsilon). \tag{21}$$

Consider $\varepsilon = 0$ in system (21). In this case, system (21) is Hamiltonian and its general solution is

$$r(W) = r_0, \qquad \Theta(W) = \frac{W}{r_0^2} + \Theta_0,$$

with $r(0) = r_0$ and $\Theta(0) = \Theta_0$. Hence, the solutions are straight lines which are winding around the invariant tori

$$\mathbb{T}^1 = \{r_0\} \times [0, 2\pi]$$

with constant frequency $\omega(r_0) = 1/r_0^2$ and the whole phase space of system (21) for $\varepsilon = 0$ is foliated by Kronecker tori. Furthermore the frequencies of the orbits which wind around these tori are nonresonant and hence each orbit is dense on them.

Using the notation of Theorem 4 we have that $I = r$ is the action variable and $\varphi = \Theta$ is the angle variable. Moreover

$$H_0(r) = -\frac{1}{r} \qquad \text{and} \qquad \frac{\partial^2 H_0}{\partial r^2} = -\frac{2}{r^3} \neq 0,$$

for all $r > 0$. By Theorem 4, almost all the invariant tori, which exist for the unperturbed system ($\varepsilon = 0$), persist, although slightly deformed, in the phase space of differential system (21) with $\varepsilon > 0$ sufficiently small. Changing back appropriately the coordinates, we have that NE_6 system has invariant tori whose orbits are dense and move quasiperiodically on them for $a > 0$ sufficiently small.

In a similar way, we prove the theorem for NE_8 and NE_9 systems. Indeed, for NE_8 system, consider the rescaling of time $T = -t$, where T is the new time, in order to write its linear part at the origin into the real Jordan normal form, and the changes of coordinates $(x, y, z) \rightarrow (r, \theta, z)$, where $x = r\cos\theta$, $y = r\sin\theta$, and $(r, \theta, z) \rightarrow (R, \theta, Z)$, where $r = \varepsilon R$, $z = \varepsilon Z$, with $a = \varepsilon^2$. After that, NE_8 system can be written as

$$
\begin{aligned}
\dot{R} &= \varepsilon \, R Z \sin^2\theta, \\
\dot{\theta} &= \varepsilon \, Z\cos\theta\sin\theta + 1, \\
\dot{Z} &= -\frac{1}{2} \varepsilon \, [R^2\cos\theta \, (2\sin\theta + \cos\theta) - 2].
\end{aligned}
\tag{22}
$$

Taking θ as the independent variable in system (22) and doing the Taylor expansion of order 2 of the obtained equations at $\varepsilon = 0$, we obtain

$$
\begin{aligned}
\frac{dR}{d\theta} &= \varepsilon \, R Z \sin^2\theta + O(\varepsilon^2), \\
\frac{dZ}{d\theta} &- \frac{1}{2} \varepsilon \, [R^2\cos\theta \, (2\sin\theta + \cos\theta) - 2].
\end{aligned}
\tag{23}
$$

Applying Theorem 3 for system (23), we obtain that, for $a > 0$ sufficiently small, NE_8 system has a periodic solution of period approximately 2π given by

$$x(t) = +2\sqrt{a}\cos t + O(a),$$
$$y(t) = -2\sqrt{a}\sin t + O(a),$$
$$z(t) = +O(a),$$

which tends to the origin as $a \to 0$. Therefore, for $a > 0$ small enough, NE_8 system has a linearly stable periodic orbit which emerges from the origin. Considering the same changes of coordinates, including the rescaling of time $T = -t$, for NE_9 system and using Theorem 3, we conclude that, for $a > 0$ sufficiently small, NE_9 system also has a periodic solution of period approximately 2π given by

$$x(t) = +\frac{1}{7}\sqrt{14\,a}\cos t + O(a),$$
$$y(t) = -\frac{1}{7}\sqrt{14\,a}\sin t + O(a),$$
$$z(t) = +O(a),$$

which tends to the origin as $a \to 0$ and, hence, NE_9 system also has a linearly stable periodic orbit which emerges from the origin.

Now we shall prove that around the periodic orbits of NE_8 and NE_9 systems there exist nested invariant tori. In order to do that for NE_8 system, we consider the rescaling of time $T = -t$ and the changes of variables $(x, y, z) \to (r, \theta, z)$, where $x = r\cos\theta$, $y = r\sin\theta$, and $(r, \theta, z) \to (r, \Theta, Z)$, where $\theta = \varepsilon\Theta$, $z = \varepsilon Z$, with $a = \varepsilon$. After that, NE_8 system can be written as

$$\dot{r} = \varepsilon r Z\,(1 - \cos^2(\varepsilon\Theta)),$$
$$\dot{\Theta} = Z\cos(\varepsilon\Theta)\sin(\varepsilon\Theta) + \frac{1}{\varepsilon}, \qquad (24)$$
$$\dot{Z} = 1 - \frac{1}{2\varepsilon}R^2\cos(\varepsilon\Theta)\,[2\sin(\varepsilon\Theta) + \cos(\varepsilon\Theta)].$$

Taking Z as the independent variable in system (24) and doing the Taylor expansion of order 2 of the obtained equations at $\varepsilon = 0$, we get

$$\frac{dr}{dZ} = O(\varepsilon^4), \qquad \frac{d\Theta}{dZ} = -\frac{2}{R^2} + O(\varepsilon). \qquad (25)$$

Applying Theorem 4 for system (25), we obtain that for $a > 0$ sufficiently small NE_8 system has invariant tori whose orbits are dense and move quasiperiodically on them. Considering the same changes of coordinates, including the rescaling of time, and using Theorem 4, we get the same result for NE_9 system. This concludes the proof of the theorem.   □

The existence of periodic orbits in nonequilibrium quadratic three-dimensional differential systems was also studied in Carvalho et al. (2016), Llibre et al. (2015).

In the second column of Table 2 are drawn the phase space of NE_1, NE_6, NE_8 and NE_9 systems for $a = 10^{-5}$, in which we can observe the (small) periodic orbit and nested invariant tori around it, for each one of these systems. Furthermore, in the third column of Table 2 are drawn the respective Poincaré section of these systems. The periodic orbit of the considered systems is represented in the Poincaré section by the

Table 2 Phase space and Poincaré section of NE_1, NE_6, NE_8 and NE_9 systems for $a = 10^{-5}$

Model	Phase space	Poincaré section
NE_1		
NE_6		
NE_8		
NE_9		

fixed point in the center of the closed curves, which in your turn represent the nested invariant tori. In this way, the numerical simulations developed by us corroborate the results stated in Theorem 5

4 On the Emergence of Chaotic Behavior

As we said in the Introduction, in Jafari et al. (2013) the authors proved that, for suitable choices of the value of parameter a, NE_1, NE_6, NE_8 and NE_9 systems present chaotic behavior even without having any equilibrium point. The values of a for which these systems present this kind of behavior can be found in Table 1. In this section, taking to account the results stated and proved in Sect. 3, we perform a numerical investigation of the dynamical consequences in the phase space of the considered systems as the parameter value a increases from $a = 0$, in order to better understand the mechanisms of emergence of chaotic behavior on them.

It was showed in Messias and Reinol (2017a) that although the periodic orbit of NE_1 system survives under small variations of the parameter $a > 0$, an increasing number of invariant tori are destroyed as the parameter value a increases, being those closest to the periodic orbit more persistent to perturbations. Consequently, due to the destruction of some of the invariant tori, a "turbulent" region of attraction/repulsion is formed around some remaining torus and this region leads to the creation of a homoclinic structure in the phase space of NE_1 system for $a > 0$ small, as we can see in Fig. 2. For suitable values of the parameter $a > 0$, as for example $a = 0.4$ (Messias and Reinol 2017b) and $a = 1$ (Sprott 1994), it is possible to detect chaotic behavior in NE_1 system.

This numerical study was corroborated by the results stated in Messias and Reinol (2017a), where we also studied the Poincaré section of NE_1 system for different values of the parameter a. See Fig. 3 where we can observe the formation of a new kind of behavior in the dynamics of NE_1 system: as the parameter a increases, small "loops", as suggested by the black points in Fig. 3b, arise around the closed curves. Regions determined by such loops are called *islands*, because they are surrounded by a sea of orbits that move randomly. This confirms the complicated dynamics of NE_1 system: around the remaining invariant tori there are regions with regular orbits (represented

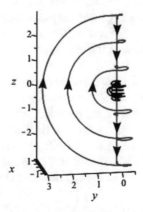

Fig. 2 Homoclinic structure in the phase space of NE_1 system for $a = 10^{-4}$

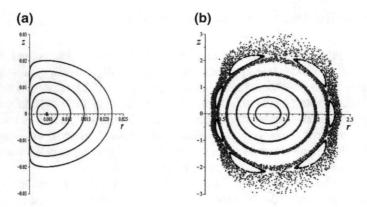

Fig. 3 Poincaré section of NE_1 system for **a** $a = 10^{-5}$ and **b** $a = 1$

by these islands) and regions densely filled by orbits without evidence of regularity. In Fig. 4 is drawn the phase space of NE_1 system for $a = 1$, where we can see that invariant tori are surrounded by a "chaotic sea".

As commented on before, these results about NE_1 system (or Sprott A system) motivated our studies of the other NE_i systems, $i = 2 \ldots 17$, from which we observed that, among them, NE_6, NE_8 and NE_9 have similar dynamical behavior than NE_1 system and consequently similar routes to the emergence of chaotic motion.

Indeed, let us consider one of the closed curves in the Poincaré section of NE_6 system for $a = 10^{-5}$ (Fig. 5a), defined by the same quasiperiodic orbit on the corresponding invariant torus, and let us observe what happens with this closed curve in the Poincaré section as the parameter value a increases. The closed curve becomes thicker as a increases and, for $a = 10^{-3}$, it determines the region drawn in Fig. 5b. In this region can be observed strings of Hénon attractors. In the phase space of NE_6 system for $a = 10^{-3}$, the toric structure is preserved and the orbits jump through the invariant tori, as we can see in Fig. 6a, where we plot the same orbit for different future intervals of times. Increasing more the parameter value a, the toric structure is deformed, as we can see in Fig. 6b for $a = 10^{-2}$, until that the invariant tori are apparently destroyed for larger values of the parameter a. A similar behavior is observed for NE_8 system.

Now we consider one of the closed curves in the Poincaré section of NE_9 system for $a = 10^{-5}$ (Fig. 7a), defined by the same quasiperiodic orbit on an invariant torus. As the parameter value a increases, the closed curve evolves into a chain of islands in the Poincaré section of NE_9 system for $a = 10^{-2}$, as we can see in Fig. 7b. A more detailed picture of the Poincaré section of NE_9 system in this case can be seen in Fig. 8. Observe that chains of islands appears between closed curves showing that layers of orbits without evidence of regular motion are sandwiched between regular orbits.

The study of the Poincaré section of NE_1, NE_6, NE_8 and NE_9 systems corroborates the complicated dynamics that these systems present as the parameter value a

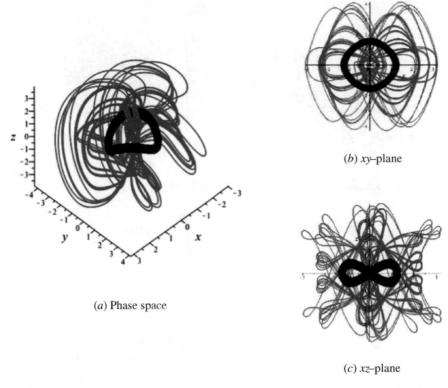

(b) xy–plane

(a) Phase space

(c) xz–plane

Fig. 4 (a) Phase space of NE_1 system for $a = 1$ with an invariant torus and an orbit in the "chaotic sea" and its projection in the (b) xy–plane and (c) xz–plane

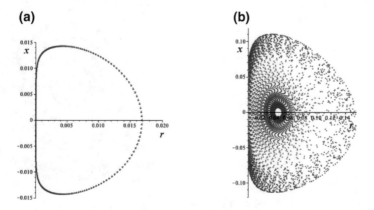

Fig. 5 Poincaré section of NE_6 system associated to the same orbit, for **a** $a = 10^{-5}$ and **b** $a = 10^{-3}$

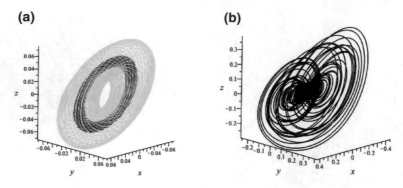

Fig. 6 Phase space of NE$_6$ system for **a** $a = 10^{-3}$ and **b** $a = 10^{-2}$

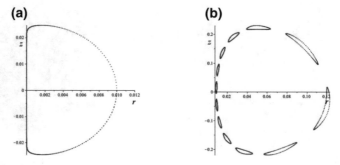

Fig. 7 Poincaré section of the same orbit of NE$_9$ system for **a** $a = 10^{-5}$ and **b** $a = 10^{-2}$

Fig. 8 Poincaré section of several orbits of NE$_9$ system for $a = 10^{-2}$: small regular islands between closed curves

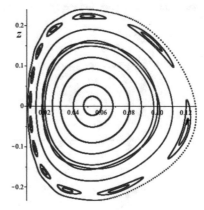

increases and the nested invariant tori are destroyed, evidencing that, in each system, the existence of a periodic orbit and invariant tori around it play an important role in the emergence of chaotic behavior in these systems.

5 Concluding Remarks

In this note we observed that the classes of differential systems NE_1, NE_6, NE_8 and NE_9 provided by Jafari et al. (2013) have a quite similar dynamical behavior as the parameter value a varies. Indeed, we proved that for $a = 0$ all the considered systems have a line of equilibria given by one of the coordinate axis in which the origin is a non-isolated zero-Hopf equilibrium point. By using the averaging theory we show that for $a > 0$ small enough a linearly stable periodic orbit bifurcates from the non-isolated zero-Hopf equilibrium at the origin and nested invariant tori are formed around it. The existence of such tori are proved using the classical KAM Theorem. Finally, a numerical investigation of the dynamics of these systems as the parameter value a increases from $a = 0$ suggests that the existence of such a periodic orbit bifurcating from the origin and nested invariant tori around it are important dynamical elements which leads to the emergence of chaotic motion in these systems. In particular, the mechanisms of birth of chaos presented here are different from cascade of period doubling bifurcations, which is suggested by the authors in Jafari et al. (2013) for the formation of hidden attractors in the nonequilibrium differential systems considered there.

About the integrability of NE_1, NE_6, NE_8 and NE_9 systems, it was proved in Messias and Reinol (2017a) that, for $a = 0$, NE_1 system has a polynomial first integral given by $f(x, y, z) = x^2 + y^2 + z^2$ and, consequently, its phase space is foliated by concentric invariant spheres centered at origin. For $a \neq 0$ the authors proved that NE_1 system has neither invariant algebraic surfaces nor polynomial first integrals. Here we proved that NE_6 system has neither invariant algebraic surfaces nor polynomial first integrals for all $a \in \mathbb{R}$, while NE_8 and NE_9 systems have no polynomial first integrals for all $a \in \mathbb{R}$. It remains an open question to determine if NE_1, NE_6, NE_8 and NE_9 systems have or not other kinds of first integrals, as rational or Darbouxian ones, and if NE_8 and NE_9 systems have or not invariant algebraic surfaces.

From the studies developed by us until now we can say that the NE_i systems, for $i \neq 1, 6, 8, 9$, have different mechanisms for the emergence of chaotic behavior from the one described here. We are studying these mechanisms and intend to present the results in a forthcoming paper.

Finally, we would like to thank Professor Julien Clinton Sprott for his valuable comments through an interesting e-mail correspondence, which helped us to improve the presentation of the results in this chapter.

References

Alligood KT, Sauer T, Yorke J (1996) Chaos: an introduction to dynamical systems. Springer, New York

Arnold VI (1989) Mathematical methods of classical mechanics. Springer, New York

Broer H, Takens F (2011) Dynamical systems and chaos. Springer, New York

Buică A, Giné J, Llibre J (2012) A second order analysis of the periodic solutions for nonlinear periodic differential systems with a small parameter. Phys D 241(5):528–533

Cândido MR, Llibre J, Novaes DD (2017) Persistence of periodic solutions for higher order perturbed differential systems via Lyapunov-Schmidt reduction. Nonlinearity 30(9):3560–3586

Carvalho T, Euzébio RD, Llibre J, Tonon DJ (2016) Detecting periodic orbits in some 3D chaotic quadratic polynomial differential systems. Discret Contin Dyn Sys 21(1):1–11

Cencini M, Cecconi F, Vulpiani A (2010) Chaos: from simple models to complex systems. World Scientific

Chen G, Ueta T (1999) Yet another chaotic attractor. Int J Bifurcat Chaos 9(7):1465–1466

Chen G, Yu X (2003) Chaos control: theory and applications. Springer, Berlin, Heidelberg

Chua LO (1994) Chua's circuit. An overview ten years later. J Circuits Syst Comput 4:117–159

Dudkowski D, Jafari S, Kapitaniak T, Kuznetsov NV, Leonov GA, Prasad A (2016) Hidden attractors in dynamical systems. Phys Rep 637:1–50

Dumortier F, Llibre J, Artés JC (2006) Qualitative theory of planar differential systems. Springer, Berlin

Gotthans T, Petržela J (2015) New class of chaotic systems with circular equilibrium. Nonlinear Dyn 81(3):1143–1149

Guckenheimer J, Holmes PJ (2002) Nonlinear oscillations, dynamical systems, and bifurcations of vector fields. Springer, New York

Hoover WG (1985) Canonical dynamics: equilibrium phase-space distributions. Phys Rev A 31:1695–1697

Jafari S, Sprott J, Golpayegani SMRH (2013) Elementary quadratic chaotic flows with no equilibria. Phys Lett A 377(9):699–702

Jafari S, Sprott JC (2013) Simple chaotic flows with a line equilibrium. Chaos Solitons Fractals 57:79–84

Jafari S, Sprott JC, Pham V-T, Volos C, Li C (2016) Simple chaotic 3D flows with surfaces of equilibria. Nonlinear Dyn 86(2):1349–1358

Kingni ST, Jafari S, Simo H, Woafo P (2014) Three-dimensional chaotic autonomous system with only one stable equilibrium: analysis, circuit design, parameter estimation, control, synchronization and its fractional-order form. Eur Phys J Plus 129(5):76

Kuznetsov NV (2016) Hidden attractors in fundamental problems and engineering models: a short survey. In: Duy V, Dao T, Zelinka I, Choi HS, Chadli M (eds) AETA 2015: recent advances in electrical engineering and related sciences. Lecture notes in electrical engineering, vol 371. Springer, Cham

Kuznetsov NV, Leonov GA, Vagaitsev VI (2010) Analytical-numerical method for attractor localization of generalized Chua's system. IFAC Proc 43(11):29–33

Kuznetsov YA (1998) Elements of applied bifurcation theory. Springer, Berlin

Lakshmanan M, Rajasekar S (2003) Nonlinear dynamics: integrability chaos and patterns. Springer, Berlin, Heidelberg

Lao S-K, Shekofteh Y, Jafari S, Sprott JC (2014) Cost function based on Gaussian mixture model for parameter estimation of a chaotic circuit with a hidden attractor. Int J Bifurcat Chaos 24(1):1450010 (11 pages)

Leonov GA, Kuznetsov NV (2013a) Analytical-numerical methods for hidden attractors' localization: the 16th Hilbert problem, Aizerman and Kalman conjectures, and Chua circuits. In: Repin S, Tiihonen T, Tuovinen T (eds) Numerical methods for differential equations, optimization, and technological problems. Computational methods in applied sciences, vol 27. Springer, Dordrecht

Leonov GA, Kuznetsov NV (2013b) Hidden attractors in dynamical systems. From hidden oscillations in Hilbert-Kolmogorov, Aizerman, and Kalman problems to hidden chaotic attractor in Chua circuits. Int J Bifurcat Chaos 23(1):1330002 (69 pages)

Leonov GA, Kuznetsov NV, Vagaitsev VI (2011) Localization of hidden Chua's attractors. Phys Lett A 375(23):2230–2233

Leonov GA, Vagaitsev VI, Kuznetsov NV (2010) Algorithm for localizing Chua attractors based on the harmonic linearization method. Dokl Math 82:663–666

Li C, Sprott JC (2014) Coexisting hidden attractors in a 4-D simplified Lorenz system. Int J Bifurcat Chaos 24(3):1450034 (12 pages)

Llibre J, Messias M (2009) Global dynamics of the Rikitake system. Phys D 238(3):241–252

Llibre J, Messias M, da Silva PR (2008) On the global dynamics of the Rabinovich system. J Phys A-Math Theor 41(27):275210 (21 pages)

Llibre J, Messias M, da Silva PR (2010) Global dynamics of the Lorenz system with invariant algebraic surfaces. Int J Bifurcat Chaos 20(10):3137–3155

Llibre J, Messias M, da Silva PR (2012) Global dynamics in the Poincaré ball of the Chen system having invariant algebraic surfaces. Int J Bifurcat Chaos 22(6):1250154 (17 pages)

Llibre J, Novaes DD (2015) Improving the averaging theory for computing periodic solutions of the differential equations. Z Angew Math Phys 66(4):1401–1412

Llibre J, Novaes DD, Teixeira MA (2014) Higher order averaging theory for finding periodic solutions via brouwer degree. Nonlinearity 27(3):563–583

Llibre J, Oliveira RDS, Valls C (2015) On the integrability and the zero-Hopf bifurcation of a Chen-Wang differential system. Nonlinear Dyn 80(1):353–361

Llibre J, Xiao D (2014) Limit cycles bifurcating from a non-isolated zero-Hopf equilibrium of three-dimensional differential systems. Proc Am Math Soc 142(6):2047–2062

Lorenz EN (1963) Deterministic nonperiodic flow. J Atmos Sci 20:130–148

Lü J, Chen G (2002) A new chaotic attractor coined. Int J Bifurcat Chaos 12(3):659–661

Messias M, Reinol AC (2017a) On the existence of periodic orbits and KAM tori in the Sprott A system—a special case of the Nosé-Hoover oscillator. Preprint

Messias M, Reinol AC (2017b) On the formation of hidden chaotic attractors and nested invariant tori in the Sprott A system. Nonlinear Dyn 88(2):807–821

Molaie M, Jafari S, Sprott JC, Golpayegani SMRH (2013) Simple chaotic flows with one stable equilibrium. Int J Bifurcat Chaos 23(11):1350188 (7 pages)

Nosé S (1984) A unified formulation of the constant temperature molecular-dynamics methods. J Chem Phys 81:511–519

Ott E (2002) Chaos in dynamical systems. Cambridge University Press

Pikovskii AS, Rabinovich MI, Trakhtengerts VY (1978) Onset of stochasticity in decay confinement of parametric instability. Sov Phys JETP 47:715–719

Posch HA, Hoover WG, Vesely FJ (1986) Canonical dynamics of the Nosé oscillator: stability, order, and chaos. Phys Rev A 33(6):4253–4265

Pöschel J (2001) A lecture on the classical KAM theorem. Proc Symp Pure Math 69:707–732

Rössler OE (1976) An equation for continuous chaos. Phys Lett A 57(5):397–398

Sanders JA, Verhulst F, Murdock J (2007) Averaging methods in nonlinear dynamical systems. Springer, New York

Sparrow C (1982) The Lorenz equations: bifurcations, chaos, and strange attractors. Springer, New York

Sprott JC (1994) Some simple chaotic flows. Phys Rev E 50:R647–R650

Strogatz SH (2001) Nonlinear dynamics and chaos: with applications to physics, biology, chemistry, and engineering. Westview Press

Verhulst F (1996) Nonlinear differential equations and dynamical systems. Springer, Berlin

Wang X, Chen G (2012) A chaotic system with only one stable equilibrium. Commun Nonlinear Sci 17:1264–1272

Wang Z, Cang S, Ochola EO, Sun Y (2012) A hyperchaotic system without equilibrium. Nonlinear Dyn 69(1):531–537

Wei Z, Pehlivan I (2012) Chaos, coexisting attractors, and circuit design of the generalized Sprott C system with only two stable equilibria. Optoelectron Adv Mat 6:742–745

Wei Z, Yang Q (2011) Dynamical analysis of a new autonomous 3-D chaotic system only with stable equilibria. Nonlinear Anal Real World Appl 12(1):106–118

Wei Z, Zhang W (2014) Hidden hyperchaotic attractors in a modified Lorenz-Stenflo system with only one stable equilibrium. *Int J Bifurcat Chaos* 24(10):1450127 (14 pages)

Wiggins S (1988) Global bifurcations and chaos. Springer, New York

Wiggins S (2003) Introduction to applied nonlinear dynamical systems and chaos. Springer, New York

Existence and Control of Hidden Oscillations in a Memristive Autonomous Duffing Oscillator

Vaibhav Varshney, S. Sabarathinam, K. Thamilmaran, M. D. Shrimali
and Awadhesh Prasad

Abstract Studying the memristor based chaotic circuit and their dynamical analysis has been an increasing interest in recent years because of its nonvolatile memory. It is very important in dynamic memory elements and neural synapses. In this chapter, the recent and emerging phenomenon such as hidden oscillation is studied by the new implemented memristor based autonomous Duffing oscillator. The stability of the proposed system is studied thoroughly using basin plots and eigenvalues. We have observed a different type of hidden attractors in a wide range of the system parameters. We have shown that hidden oscillations can exist not only in piecewise linear but also in smooth nonlinear circuits and systems. In addition, to control the hidden oscillation, the linear augmentation technique is used by stabilizing a steady state of augmented system.

Keywords Memristor system · Hidden oscillations · Linear augmentation
Memristor stability

V. Varshney · S. Sabarathinam (✉) · A. Prasad
Department of Physics and Astrophysics, University of Delhi, Delhi 110007, India
e-mail: saba.cnld@gmail.com

V. Varshney
e-mail: vaibhav.varshney1991@gmail.com

A. Prasad
e-mail: awadhesh.prasad@gmail.com

M. D. Shrimali
Department of Physics, Central University of Rajasthan, Ajmer 305817, India
e-mail: shrimali@curaj.ac.in

K. Thamilmaran
School of Physics, Centre for Nonlinear Dynamics, Bharathidasan University,
Tiruchirappalli 620024, India
e-mail: maran.cnld@gmail.com

© Springer International Publishing AG 2018
V.-T. Pham et al. (eds.), *Nonlinear Dynamical Systems with Self-Excited
and Hidden Attractors*, Studies in Systems, Decision and Control 133,
https://doi.org/10.1007/978-3-319-71243-7_14

1 Introduction

The new emerging fourth passive element named, memristor whose resistance depends on its internal state variables of the system was proposed by Chua in 1971 (Chua 1971). The concept of memristor is explained by state-dependent Ohm's law. The dependence is entirely on its past signals (applied voltage/current) across the memristor. The memristors have great paradigmatic usefulness for a circuit functionality because, it is not established with resistors, capacitors and inductors. The nonvolatile memory effect of the memristor is very important for potential applications in dynamics memory, neural synapses, spintronic devices, ultra-dense information storage, neuromorphic circuits, and programmable electronics (Xu et al. 2011; Mouttet 2008; Rajendran et al. 2010; Kim et al. 2011; Thomas 2013; Strukov 2008). This applications leads to a new method of high performance computing. Researchers have been motivated to investigate such memristor based oscillators from the dynamical system theory point of view. In recent years, there has been increasing interest to study the memristor based nonlinear circuits and their dynamics (Strukov 2008; Pershin et al. 2009; Pershin and Ventra 2008, 2009). The dynamics of the memristor based oscillator circuit systems is extraordinarily complex (Tour and He 2008). The basic idea of a memristor is that it is a two terminal resistive device, in which when current passes through it in one direction, the resistance increases, while when current flows in the opposite direction the resistance decreases. This gives the concept that a memristor maintains memory of its resistance, hence its name.

Numerous studies are available for understanding the conceptual background of memristor. The memristor relates the functional relationship of charge (q) and flux (ϕ) (Chua 1971). Memristor was considered to be the missing fourth circuit element, before it was postulated (the other known three being resistors, capacitors and inductors). Memristor was realized by Stan Williams group of HP Laboratory in 2008, is a passive two-terminal electronic device, described by nonlinear constitutive relation of charge and flux (Wang et al. 2009). The $v - i$ characteristic of the memristor is inherently nonlinear (pinched hysteresis) and is unique in the sense that no combination of nonlinear resistive, capacitive and inductive components can duplicate their circuit properties. Memristors have generated considerable excitement among circuit theorists. Memristor based chaotic circuit can be constructed using memristor as a nonlinear element and can easily generate chaotic dynamics and some novel features can be observed. Recently, the fabrication of single memristor element, memristive system, memristor circuits, designs and analysis of memristor based application circuit systems, etc. have attracted attention in engineering and biological sciences. Since memristors are commercially unavailable, it would be very useful to have a specific circuit that emulates a memristor. For the simulation of memristor devices there are different nonlinearities available in literature to mimick the feature of the memristor, such as HP memristor model (Radwan et al. 2010; Prodromakis et al. 2011), non-smooth piecewise linearity (Ahamed and Lakshmanan 2013; Chen et al. 2014), smooth cubic nonlinearity (Cheng et al. 2011; Talukdar 2011; Muthuswamy and Kokate 2009), smooth piecewise-quadratic nonlinearity (Bao et al. 2010), and so on. For the first time Itoh and Chua introduced the memristor instead of Chua diode as a nonlinear element based canonical Chua's oscillator (Itoh and Chua 2008). After that many studies on memristor based nonlinear circuit systems have been done.

Now a days numerous studies are available in the literature about the hidden attractors with no equilibrium point. Leonov et al. studied the hidden attractors in a Chua's system and suggested special procedure for localization of hidden attractors (Leonov et al. 2011). After that they considered the example of a Lorenz-like system derived from Glukhovsky-Dolghansky and Rabinovich systems, to demonstrate the analysis of self-excited and hidden attractors, and their characteristics. They also demonstrated the existence of a homoclinic orbit, proved the dissipativity and completeness of the system, and found absorbing and positively invariant sets (Li et al. 2014; Leonov and Kuznetsov 2013; Leonov et al. 2015). Recently, they investigated the hidden oscillations in dynamical systems, based on the development of numerical methods, computers, and applied bifurcation theory (Leonov et al. 2011). The simple four dimensional equilibrium free autonomous ODE system showed all the attractors are hidden reported by Li et al. (Li and Sprott 2014). Zhouchao Wei et al. found a four-dimensional (4D) non-Sil'nikov autonomous system with three quadratic nonlinearities, which exhibits some behavior previously unobserved: hidden hyperchaotic attractors with only one stable equilibrium (Wei and Zhang 2014). Jafari et al. reviewed several type of new rare chaotic flows with hidden attractors in many dynamical systems. They also explained the flows with no equilibrium, with a line of equilibrium points, and with a stable equilibrium (Jafari et al. 2015). Dawid Dudkowski et al. reviewed the most representative examples of hidden attractors and discussed their theoretical properties and experimental observations. They described numerical methods which allowed the identification of the hidden attractors (Dudkowski et al. 2016). They also discussed the use of perpetual points for tracing the hidden and the rare attractors of dynamical systems (Dudkowski 2015).

Kuznetsov et al. gave some rigorous nonlinear analysis and special numerical methods which should be used for the investigation of nonlinear control systems (Kuznetsov and Leonov 2014). He also described the formation of several different coexisting sets of hidden attractors, including the simultaneous presence of a pair of coinciding quasiperiodic attractors and of two mutually symmetric chaotic attractors (Kuznetsov et al. 2015). Chaudhuri et al. proposed a riddled-like complicated basins of coexisting hidden attractors both in coupled and uncoupled systems and a new route to amplitude death is observed in time-delay coupled hidden attractors (Chaudhuri and Prasad 2014). Brezetskyi et al. presented different types of dynamics for both the single and coupled multistable Vander Pol-Duffing oscillators that have very small basins of attraction considered as hidden or rare (Brezetskyi et al. 2015; Zhao 2014). The control of multistability in the hidden attractor through the scheme of linear augmentation, that can drive the multistable system to a monostable state was proposed by Sharma et al. (2015a, b). Kapitaniak tries to focus on different questions of present day interest in theory and applications of systems with multiple attractors. The particular attention is paid to uncovering and characterizing hidden attractors (Kapitaniak and Leonov 2015). For the real time application point of view, Kiseleva et al. studied the hidden oscillations appearing in electromechanical systems with and without equilibria (Kiseleva et al. 2016).

The above literature is devoted to the brief understanding of hidden attractors. But only limited studies are available for finding the hidden attractor in a memristor based dynamical systems. Chen et al. reported the hidden attractor in memristive Chua's circuit and also presented the coexisting hidden attractors (Chen et al. 2015). Saha et al. studied memristor based Lorenz system with no equilibrium (Saha et al. 2015). Mo Chen et al. proposed improved memristive Chua's circuit, from which some hidden attractors are found (Chen et al. 2015). Pham et al. proposed memristor based networks to elucidate the hidden attractors (Pham et al. 2015, 2014; Zhang et al. 2015). Duffing oscillator is well known and is widely used by the researchers for its enigmatic and simplicity. Recently the dynamics of the memristor based Duffing oscillator was studied by Sabarathinam et al. (2017). With the best of our knowledge none of above works studied the dynamics of memristor based autonomous form of the Duffing oscillator. The existence of hidden attractor in the proposed system has been found for wide range of parameters and with different initial condition.

Many physical, biological, and chemical phenomena are well modeled by coupled nonlinear equations (Pikovsky et al. 2001; Fujisaka and Yamada 1983). In most cases these systems are capable of displaying several types of dynamical behaviors such as limit cycle, bistability, birhythmicity, and chaos. In many real-world situations, it is often the case that stable output is required in spite of the nonlinear effects present in the system. Thus, the scope of control or self-regulation in systems with complex dynamics is of considerable interest. Generally, the stabilization of unstable fixed points of an oscillatory system is considered to be an important problem for many practical applications. Over the last two decades, chaos control in dynamical systems and stabilization of unstable dynamical states of the systems have been a topic of intense research from both the theoretical and experimental point of view (Rosa et al. 1998; Ira 1997; Triandaf and Schwartz 2000; Ott et al. 1990; Sinha 1990). Control of chaotic dynamics or stabilization of fixed points is important in many experimental studies; for example, removal of power fluctuation is highly desirable in coupled laser systems (Kim et al. 2005; Kumar et al. 2008, 2009; Prasad 2003). In all previous existing methods (Rosa et al. 1998; Ira 1997; Triandaf and Schwartz 2000; Ott et al. 1990; Sinha 1990) the stabilization of the fixed points can be obtained by changing the accessible internal parameters of the system. However, in many real situations, where the internal parameters of the systems are not accessible, the stabilization of fixed points can be done by using the phenomenon of amplitude death (AD) (Prasad 2003) using interactions between the coupled oscillators.

Recently, linear augmentation has been suggested as an another practical alternative method leading to oscillator suppression, which is achieved by coupling nonlinear systems to a linear system which simply consists of an exponentially decaying function (Bar-Eli 1985) in an uncoupled state. Interestingly, the coupling structure of linear augmentation is quite reminiscent of indirect or environmental coupling procedures (Sharma et al. 2011; Resmi et al. 2010; Sharma et al. 2012) which are motivated by the observations of collective behaviors in several real world systems, namely, behavior of chemical relaxation oscillators globally coupled through the concentration of chemicals in a common solution (Resmi et al. 2012), dynamics of multicell systems where the cells interact through common complex proteins (Toth et al. 2006), and collective behavior of cold atoms in the presence of a coherent electromagnetic field and atomic recoil (Zhang and Zou 2012; Kruse et al. 2003) for instance. These instances therefore also serve as good examples of systems where linear augmentation can exist naturally. Lately, studies have also effectively used linear augmentation in controlling bistability (Javaloyes et al. 2008), the dynamics of drive response systems (Sharma et al. 2013) and in controlling hidden attractors (Sharma et al. 2014). Motivated by the above studies in this chapter, we also applied the linear augmentation techniques in our memristor based autonomous Duffing oscillator to control the hidden oscillations.

The chapter is structured as follows. Section 2 explains the mathematical modeling of the memristor based autonomous Duffing oscillator. Section 3, investigates the systems stability with eigenvalue analysis. Section 4, explores the hidden dynamics with numerical results. Section 5 explains the linear augmentation technique used to stabilize the system. Finally, the chapter concludes with the summary in Sect. 6.

2 Mathematical Model of Memristor Based Duffing Oscillator

We all know about the six mathematical relations connecting the pairs of four fundamental entities namely, charge (q), current (i), flux (ϕ) and voltage (v) (Mohanty 2013). Three relations can be understood by the axiomatic definitions of the three classical two terminal circuit elements, namely, resistor (relationship between v and i), inductor (relationship between ϕ and i) and capacitor (relationship between q and v). One more relationship between ϕ and q remained undefined. From logical and axiomatic point of view, as well as for the sake of completeness, the necessity for the existence of fourth basic two terminal circuit element was postulated. Chua in 1971, suggested the fourth passive element namely '*memristor*' (Chua 1971). The memristor is a new passive two terminal element in which there is a functional relationship between the magnetic flux (ϕ) and electric charge (q). The memristor is governed by the relations $i = W(\phi)v$, and $v = M(q)i$, where, $W(\phi)$, $M(q)$ are called *memductance* and *memristance* respectively. The memristor used in this work is a charge controlled memristor that is characterized by its incremental memristance function $M(q)$ describing the charge-dependent rate of flux. We assume that the memristance $M(q)$ is characterized by a monotonically increasing and smooth cubic nonlinearity which is defined by, $\phi(q) = \omega_0^2 q + \beta q^3$, where $M(q)$ is defined as, $M(q) = \frac{d\phi(q)}{dq} = \omega_0^2 + 3\beta q^2$ is the memristance function. Figure 1, shows

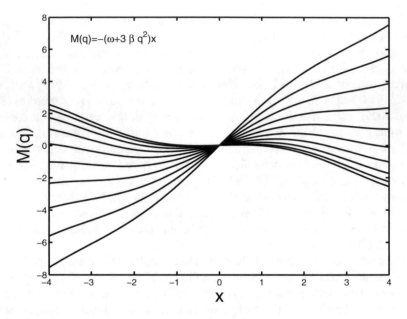

$$M(q)=-(\omega+3\,\beta\,q^2)x$$

Fig. 1 The variation of memristor nonlinearity $M(q)$ versus x of Eq. (1) for $q \in (0.1, 1.0)$ shown different slopes. The parameters of the memristor emulator consider as $\omega_0^2 = 0.35$ and $\beta = 0.85$

the nonlinearity curve $M(q)$ as a function of charge q of the memristor emulator. With increasing q, the slope changes clockwise (slope value changes drastically) which indicates that the proposed system is very sensitive to q. The effect of the memristor emulator (co-efficient's of memristor ω, β) in a single memristive system studied by Sabarathinam and Thamilmaran (2017). They found that the dynamics of the system controlled by the memductance profile of the memristor emulator. Figure 2a, shows that the schematic of the memristor emulator and its characteristic curve in Fig. 2b which is obtained from PSpice simulation. In our case, the memristor based Duffing oscillator is taken with the cubic nonlinearity simply replaced by the charge-controlled memristor characteristic nonlinear system (Fig. 2). Here, we have considered the autonomous form of the Duffing oscillator, so the external force is considered as $fsin(\omega t) = 0$ from the original Duffing equation (Sabarathinam et al. 2017). Based on the memristance concept (Chua 1971), the state equation for the memristor based autonomous Duffing oscillator is,

$$\ddot{x} + \alpha \dot{x} + M(q)x = 0, \tag{1}$$

For our convenience, we considered, $\alpha, \omega_0^2, \beta = a, b, c$. For the stability analysis as well as in the numerical study, the above equation is splitted into the following system of three first order coupled equations as:

(a)

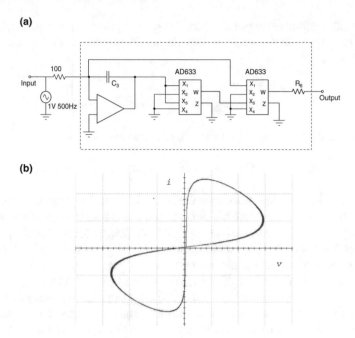

(b)

Fig. 2 PSpice analysis: **a** Schematic of the memristor nonlinearity with AC sweep analysis (1 V, 500 Hz) and **b** ($v - i$) characteristic curve of memristor. The input and output feed in to the oscilloscope

$$\dot{x}_1 = x_2$$
$$\dot{x}_2 = x_3$$
$$\dot{x}_3 = -ax_3 - bx_2 - 3cx_1^2 x_2. \tag{2}$$

Thus by replacing the cubic nonlinearity in the classical Duffing oscillator, a new memristor based autonomous form of Duffing oscillator is designed and its dynamical behavior is investigated in detail.

3 Stability Analysis

To study the stability of the above mentioned system (Eq. 2) we have used eigenvalue analysis. The system (2), is symmetric with respect to the origin and hence is invariant under the transformation,

$$(x_1, x_2, x_3) \rightarrow (-x_1, -x_2, -x_3).$$

Therefore, the equilibrium point calculated using system Eq. (2) is, $(x_1^*, x_2^*, x_3^*) = (0, 0, 0)$. In order, to find the eigenvalues, the stability matrix or the Jacobean matrix J is written as,

$$J = \begin{bmatrix} 0 & 1 & 0 \\ 0 & 0 & 1 \\ 6cx_{10}x_{20} & -b - 3cx_{10}^2 & -a \end{bmatrix} \quad (3)$$

In general the characteristic eigenvalue equation written as,

$$Det|J - \lambda I| = 0 \Rightarrow -\lambda^3 - \lambda^2 \alpha + \lambda(\omega + 3\beta x_1^{*2}) + 6\beta x_1^* x_2^* \quad (4)$$

We get eigenvalues for the above equation as, $\lambda_1 = -0.001 + 0.5916i$, $\lambda_2 = 0$, $\lambda_3 = -0.001 - 0.5916i$ for $\alpha = 0.0001$, $\omega = 0.35$, and $\beta = 0.85$. From the eigenvalues the system have *stable* equilibrium state. The potential (M(q)) depends on the parameter b. For positive b, we get *stable* equilibrium points, $\lambda_{1-3} = -0.001 \pm 0.5916i, 0$, and negative b, we get *saddle* equilibrium pointS, $\lambda_{1-3} = 0, 0.5916, -0.5917$. Figure 3, shows the changes in stability of the system with respect of $\pm b$ and $\pm x_1$ (detailed stability study were made Vaibhav et al. (2017)). For positive regions of b, we get *stable fixed points* (SFP) everywhere in the basin and for negative b the system have *saddle* equilibrium (UFP). In this chapter we have taken, $a > 0$ and $b > 0$ case for finding the hidden attractor. In that case, the system fall on *stable fixed point*. A vector field in the plane (for instance), can be visualized as a collection of different length of arrows which indicates the different magnitude and direction. Vector fields are often used to model, for example, the speed and direction throughout space, or the strength and direction of some force, as it changes from point to point. Vector fields can usefully be thought of as representing the velocity of a moving flow in space, and this physical intuition leads to notions such as the divergence (which represents the rate of change of volume of a flow) and curl (which represents the rotation of a flow) of the system. Figure 4 shown the vector field of $(y - z)$ plane of the proposed system in the range of ± 20. The different length of arrows which tells that the system stability changes by its basin. In the zero line magnitude of eigenvalues is high enough so,

Fig. 3 Basin of (i) $\pm b$ versus (x_{10}) based on the real part of the eigenvalues (λ_{1-3}) with fixed parameters $a, c = 0.0001, 0.85$ of system (2)

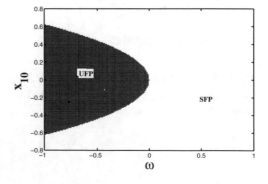

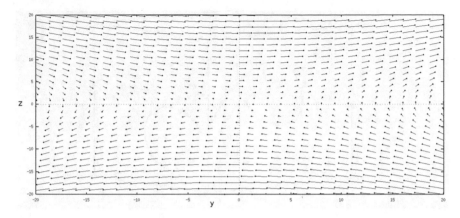

Fig. 4 Vector fields of system (2) in the $(x_{20} - x_{30})$ plane shows the arrows depict the field at discrete points, however, the field exists everywhere in the systems basin

if we start the trajectory very near to zero the system drastically fall into the fixed point in a shorter time.

4 Hidden Attractor

From the computational point of view, attractors are classified as self excited and hidden attractors. Self-excited attractors can be localized numerically by a standard computational procedure, in which after a transient process a trajectory, starting from a point of unstable manifold in the neighborhood of an equilibrium, reaches a state of oscillation, therefore one can easily identify it (Example: Lorenz, Rösseler, Chua oscillators, etc.) (Lakshmanan and Rajaseekar 2012). In contrast, for a hidden attractor, a basin of attraction does not intersect with any small neighborhoods of equilibria (Dudkowski et al. 2016). Hidden attractor can be chaotic as well as periodic e.g. the case of coexistence of the only stationary point which is stable and a stable limit cycle. It can easily predict the existence of self-excited attractor, while for hidden attractor the main problem is how to predict its existence in the phase space. Thus, for localization of hidden attractors it is important to develop special procedures, since there are no similar transient processes leading to such attractors. If the hidden attractor is present in the system dynamics and if coincidentally reached, then device (airplane, electronic circuit, etc.) starts to show the quasi-cyclic behavior that can, based on kind of device cause real disasters. In particular, to the extent that they have been known to exist, dynamical systems with no equilibrium have mostly been considered as nonphysical or mathematically incomplete. However, as experience shows, a system that presents hidden dynamical behaviour doesn't need to also display an unstable equilibrium state. In this section, the hidden attractor have

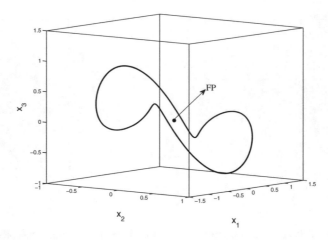

Fig. 5 Three dimensional phase plot for hidden oscillation at $(x_{10}, x_{20}, x_{30})=(0.8, 0.6, 0.0)$ with the parameters fixed at $a = 0.0001$, $b = 0.35$, and $c = 0.85$ of system (2). FP indicates fixed point of the system

been revealed in our proposed system (2). The parameters are fixed as $a = 0.0001$, $b = 0.35$, and $c = 0.85$ with the initial conditions $(x_{10}, x_{20}, x_{30}) = (0, 0, 0)$. From the stability, we start the system anywhere in the basin the trajectories approach to fixed point because of the nature of its stability. For instance in a particular range of initial conditions the system exhibit stable oscillations which does not intersect the neighbourhood of the equilibrium of the system named *hidden oscillations*. Figure 5 shows three dimensional plot for the hidden oscillation for particular set of initial conditions, $(x_{10}, x_{20}, x_{30}) = (0.8, 0.6, 0.0)$ and the parameter are fixed as $a = 0.0001$, $b = 0.35$, and $c = 0.85$ of system (2). The fixed point of the system are also replotted and its indicatcd as FP in Fig. 5. The different projection of the phase portraits shown in the Fig. 6. Figure 7 shows the time series plot of (a) $(x_1(t))$, (b) $(x_2(t))$, (c) $(x_3(t))$ variables for the corresponding phase plot in Fig. 5 which indicate the periodicity of the system. We got periodic hidden attractor in the wider range of initial conditions, We also intend to study the basin of the system. Figure 8 shows the various type of coexisting hidden attractor for different set of initial conditions indicates as $H_1 - H_6$. This attractors obtained from the initial conditions as $(H_1, H_6):(x_{10}, x_{20}, x_{30} = \pm 4.0, 0.6, 0)$, $(H_2, H_5):(x_{10}, x_{20}, x_{30} = \pm 2.0, 0.6, 0)$, and $(H_3, H_4):(x_{10}, x_{20}, x_{30} = \pm 0.5, 0.6, 0)$. The FP is the equilibrium state of the system obtained from $(x_{10}, x_{20}, x_{30} = 0, 0, 0)$. From this Fig. 8, we conclude that our system have many interesting hidden attractors. The Lyapunov exponents were calculated for the periodic attractor where $\lambda_{1-3} = (0.00053, -0.0432, -0.00004)$ for very near to the equilibrium and $\lambda_{1-3} = (0.00052, -0.0232, -0.00022)$ far away from the equilibrium. The phase space volume was calculated as $\nabla . F = -a$ which shows that our system is purely dissipative.

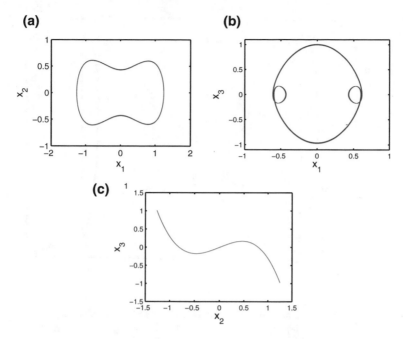

Fig. 6 Different projection of the typical phase portraits in the **a** $x_1 - x_2$, **b** $x_1 - x_3$, and **c** $x_2 - x_3$ planes of the same parameter used Fig. 5

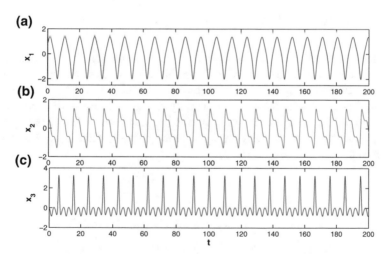

Fig. 7 Time series of **a** $(x_1(t))$, **b** $(x_2(t))$, **c** $(x_3(t))$ variables corresponding phase plot of Fig. 5

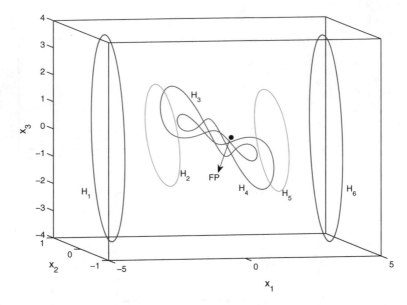

Fig. 8 Three dimensional phase plot for different coexisting hidden oscillations for different initial conditions as (H_1, H_6):$(x_{10}, x_{20}, x_{30} = \pm 4.0, 0.6, 0)$, (H_2, H_5):$(x_{10}, x_{20}, x_{30} = \pm 2.0, 0.6, 0)$, and (H_3, H_4):$(x_{10}, x_{20}, x_{30} = \pm 0.5, 0.6, 0)$

5 Controlling Hidden Attractor

In the above section the existence of the hidden oscillation is presented. In this section, the control of hidden attractor (the stabilization of fixed points) is examined. For that we take linear augmentation technique (Resmi et al. 2010). The scheme is generalized and can be applied to any other system as well. Here we coupled memristor based duffing oscillator Eq. (2) with the linear system to control the dynamics of hidden attractor in it. Initially we applied the augmentation in x_1. The coupled equation is given by:

$$\dot{x_1} = x_2 + \varepsilon u$$
$$\dot{x_2} = x_3$$
$$\dot{x_3} = -ax_3 - bx_2 - cx_1^2 x_2$$
$$\dot{u} = -ku - \varepsilon(x_1 - b) \tag{5}$$

Here ε is the coupling strength between the oscillatory and the linear systems, k is the decay parameter of the linear system u, and b is a control parameter of the augmented system. Here, we have taken only one linear system, in some cases we can take two linear systems for stabilizing both co-ordinates. We are not able to control the dynamics by augmenting the x_1 variable. The phase space plot of x_1 versus x_2 for different ε for this case is shown in Fig. 9a, b. Next, we augmented the x_2 variable

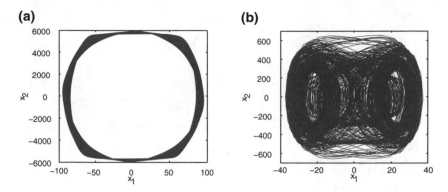

Fig. 9 Phase space trajectories of coupled memristor based duffing oscillator x_1 versus x_2 for $\varepsilon = 2.0$ **a** $k = 0.01$ and **b** $k = 1.0$

and dynamics is controlled by it. The equation is given by:

$$
\begin{aligned}
\dot{x}_1 &= x_2 \\
\dot{x}_2 &= x_3 + \varepsilon u \\
\dot{x}_3 &= -ax_3 - bx_2 - cx_1^2 x_2 \\
\dot{u} &= -ku - \varepsilon(x_2 - b).
\end{aligned}
\tag{6}
$$

Now shown in Fig. 10 is the phase space diagram in parameter space $(k - \varepsilon)$ in which A is the uncontrolled region and B is the region in which dynamics is

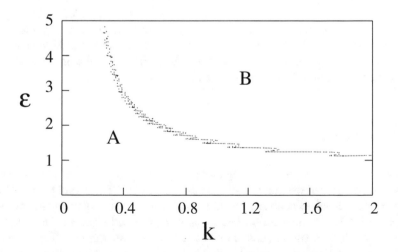

Fig. 10 Phase space diagram in the parameter space $(k - \varepsilon)$, where A and B are the uncontrolled and controlled region respectively

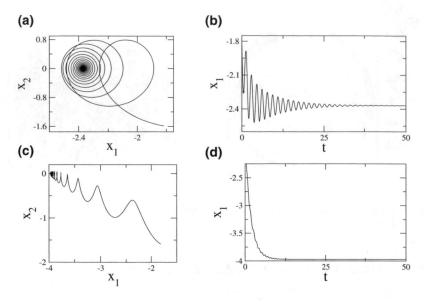

Fig. 11 Phase space trajectories of coupled memristor based duffing oscillator x_1 versus x_2 for **a** $\varepsilon = 2.0$ and **c** $\varepsilon = 5.0$. Time series for x_1 **b** $\varepsilon = 2.0$ and **d** $\varepsilon = 5.0$

controlled. The dashed line in it is demarcated when largest Lyapunov exponent changes its sign. The important thing to note here is that we are not getting the fixed point of the system, but the system is going to new fixed points created due to coupling. Now shown in Fig. 11 is the phase space (a) and (b) and time series (c) and (d) showing how the system is approaching the fixed point. Left panel is for $\varepsilon = 2$ and right panel is for $\varepsilon = 5$. It is clear from the figure that in both cases system is going to fixed point, but transient trajectory is different in both cases. So, in this section by augmentation technique we are able to control the dynamics of hidden attractor in memristor based Duffing oscillator and stabilize it to a fixed point which is not a equilibrium point of the uncoupled system.

6 Conclusion

We have observed a very interesting and emerging phenomenon of hidden oscillations in memristor based autonomous Duffing oscillator. The stability studied with the help of eigenvalues analysis. The hidden oscillations are obtained and presented in phase plot and its time series. We found our proposed system have wide range of coexisting hidden oscillations in its basin. We observed periodic form of hidden oscillation which is confirmed by Lyapunov exponents. In many applications, knowing the property of periodic oscillatory solutions is very interesting and valuable as many biological and cognitive activities require repetition. But in case of periodic hidden oscillations which leads to the new area of research by its repetition. Controlling of hidden attractors using linear augmentation technique is performed. The linear augmentation validate for all variables of our proposed system. We found that the linear augmentation will stabilize the system for only one variable mode (x_2 of Eq. (2)). The memristor emulator does not allows to stabilize the charge component (x_1 of Eq. (2)) of the system. In future, the detailed study of the hidden oscillations will be made by our proposed system with different physical situation which will be reported elsewhere (Vaibhav et al. 2017).

Acknowledgements Vaibhav Varshney, Awadhesh Prasad and acknowledge DST, Government of India for the financial support. S. Sabarathinam acknowledges the DST-SERB for the financial assistance through National Post Doctoral Fellow (NPDF) scheme. K. Thamilmaran acknowledge DST PURSE scheme for financial assisitance and Awadhesh Prasad and M. D. Shrimali also acknowledge joint project DST-RFBR for funding.

References

Ahamed AI, Lakshmanan M (2013) Nonsmooth bifurcations, transient hyperchaos and hyperchaotic beats in a memristive muralilakshmananchua circuit. Int J Bifurc Chaos 23:1350098

Bao BC, Xu JP, Liu Z (2010) Initial state dependent dynamical behaviors in a memris-tor based chaotic circuit. Chin Phys Lett 27:070504

Bar-Eli K (1985) On the stability of coupled chemical oscillators. Physica D: Nonlinear Phenom 14:242–252

Brezetskyi S, Dudkowski D, Kapitaniak T (2015) Rare and hidden attractors in Van der Pol-Duffing oscillators. Eur Phys J Spec Top 224:1459–1467

Chaudhuri U, Prasad A (2014) Complicated basins and the phenomenon of amplitude death in coupled hidden attractors. Phys Lett A 378:713–718

Cheng BB, Ping XJ, Hua ZG, Hua MZ, Ling Z (2011) Chaotic memristive cir-cuit: equivalent circuit realization and dynamical analysis. Chin Phys B 20:120502

Chen M, Li M, Yu Q, Bao B, Xu Q, Wang J (2015) Dynamics of self-excited attractors and hidden attractors in generalized memristor-based Chuas circuit. Nonlinear Dyn 81:215–226

Chen M, Yu J, Bao B-C (2015) Finding hidden attractors in improved memristor-based Chua's circuit. Electron Lett 51:462–464

Chen M, Yu J, Yu Q, Li C, Bao B (2014) A memristive diode bridge-based canonical chuas circuit. Entropy 16:6464–6476

Chua LO (1971) Memristor-the missing circuit element. IEEE Circuit Theory 18:507–519

Dudkowski D, Jafari S, Kapitaniak T, Kuznetsov NV, Leonov GA, Prasad A (2016) Hidden attractors in dynamical systems. Phys Rep 637:1–50

Dudkowski D, Prasad A, Kapitaniak T (2015) Perpetual points and hidden attractors in dynamical systems. Phys Lett A 379:2591–2596

Fujisaka H, Yamada T (1983) Stability theory of synchronized motion in coupled-oscillator systems. Prog Theor Phys 69:32–47

Ira B, Schwartz et al (1997) Tracking controlled chaos: theoretical foundations and applications. Chaos: An Interdiscip J Nonlinear Sci 7:664–679

Itoh M, Chua LO (2008) Memristor oscillators. Int J Bifurc Chaos 18:3183–3206

Jafari S, Sprott JC, Nazarimehr F (2015) Recent new examples of hidden attractors. Eur Phys J Spec Top 224:1469–1476

Javaloyes J, Perrin M, Politi A (2008) Collective atomic recoil laser as a synchronization transition. Phys Rev E 78:011108

Kapitaniak T, Leonov GA (2015) Multistability: uncovering hidden attractors. Eur Phys J Spec Top 224:1405–1408

Kim KH, Gaba S, Wheeler D, Cruz-Albrecht JM, Hussain T, Srinivasa N, Lu W (2011) A functional hybrid memristor crossbar-array/cmos system for data storage and neuromorphic applications. Nano Lett 12:389–395

Kim MY, Roy R, Aron JL, Carr TW, Schwartz IB (2005) Scaling behavior of laser population dynamics with time-delayed coupling: theory and experiment. Phys Rev Lett 94:088101

Kiseleva MA, Kuznetsov NV, Leonov GA (2016) Hidden attractors in electromechanical systems with and without equilibria. IFAC-PapersOnLine 49:51–55

Kruse D, von Cube C, Zimmermann C, Courteille PW (2003) Observation of lasing mediated by collective atomic recoil. Phys Rev Lett 91:183601

Kumar P, Prasad A, Ghosh R (2008) Stable phase-locking of an external-cavity diode laser subjected to external optical injection. J Phys B: At Mol Opt Phys 41:135402

Kumar P, Prasad A, Ghosh R (2009) Strange bifurcation and phase-locked dynamics in mutually coupled diode laser systems. J Phys B: At Mol Opt Phys 42:145401

Kuznetsov AP, Kuznetsov SP, Mosekilde E, Stankevich NV (2015) Co-existing hidden attractors in a radio-physical oscillator system. J Phys A: Math Theor 48:125101

Kuznetsov NV, Leonov GA (2014) Hidden attractors in dynamical systems: systems with no equilibria, multistability and coexisting attractors. IFAC Proceedings Volumes 47:5445–5454

Lakshmanan M, Rajaseekar S (2012) Nonlinear dynamics: integrability, chaos and patterns. Springer Science & Business Media

Leonov GA, Kuznetsov NV, Kuznetsova OA, Seledzhi SM, Vagaitsev VI (2011) Hidden oscillations in dynamical systems. Trans Syst Contr 6:54–67

Leonov GA, Kuznetsov NV, Mokaev TN (2015) Homoclinic orbits, and self-excited and hidden attractors in a Lorenz-like system describing convective fluid motion. Eur Phys J Spec Top 224:1421–1458

Leonov GA, Kuznetsov NV (2013) Hidden attractors in dynamical systems. From hidden oscillations in HilbertKolmogorov, Aizerman, and Kalman problems to hidden chaotic attractor in Chua circuits. Int J Bifurc Chaos 23:1330002

Leonov GA, Kuznetsov NV, Vagaitsev VI (2011) Localization of hidden Chuas attractors. Phys Lett A 375:2230–2233

Li C, Sprott JC (2014) Coexisting hidden attractors in a 4-D simplified Lorenz system. Int J Bifurc Chaos 24:1450034

Li Q, Zeng H, Yang XS (2014) On hidden twin attractors and bifurcation in the Chuas circuit. Nonlinear Dyn 77:255–266

Mohanty SP (2013) Memristor: from basics to deployment. IEEE Potentials 32:34–39

Mouttet B (2008) Proposal for memristors in signal processing. In: International Conference on Nano-Networks. Springer, Berlin, Heidelberg, pp 11–13

Muthuswamy B, Kokate PP (2009) Memristor-based chaotic circuits. IETE Tech Rev 26:417–429

Ott E, Grebogi C, Yorke JA (1990) Controlling chaos. Phys Rev Lett 64:1196

Pershin YV, Fontaine SL, Ventra MD (2009) Memristive model of amoeba learning. Phys Rev E 80:021926

Pershin YV, Ventra MD (2008) Spin memristive systems: spin memory effects in semiconductor spintronics. Phys Rev B 78:113309

Pershin YV, Ventra MD (2009) Frequency doubling and memory effects in the spin Hall effect. Phys Rev B 79:153307

Pham VT, Volos CK, Vaidyanathan S, Le TP, Vu VY (2015) A Memristor-based hyperchaotic system with hidden attractors: dynamics, synchronization and circuital emulating. J Eng Sci Technol Rev 8:2

Pham VT, Volos C, Jafari S, Wang X, Vaidyanathan S (2014) Hidden hyperchaotic attractor in a novel simple memristive neural network. Optoelectron Adv Mater Rapid Commun 8:1157–1163

Pikovsky AS, Rosenblum MG, Kurths J (2001) Synchronization: a universal concept in nonlinear sciences. Cambridge University Press, Cambridge

Prasad A et al (2003) Complicated basins in external-cavity semiconductor lasers. Phys Lett A 314:44–50

Prodromakis T, Peh BP, Papavassiliou C, Toumazou C (2011) A versatile memristor model with nonlinear dopant kinetics. IEEE Trans Electron Devices 58:3099–3105

Radwan AG, Zidan MA, Salama KN (2010) Hp memristor mathematical model for periodic signals and dc. In: Circuits and Systems (MWSCAS), 2010, 53rd IEEE International Midwest Symposium, pp 861–864

Rajendran J, Manem H, Karri R, Rose GS (2010) Memristor based programmable threshold logic array. In: Proceedings of the 2010 IEEE/ACM International Symposium on Nanoscale Architectures. IEEE Press, pp 5–10

Resmi V, Ambika G, Amritkar RE (2010) Synchronized states in chaotic systems coupled indirectly through a dynamic environment. Phys Rev E 81:046216

Resmi V, Ambika G, Amritkar RE, Rangarajan G (2012) Amplitude death in complex networks induced by environment. Phys Rev E 85:046211

Rosa ER, Ott E, Hess MH (1998) Transition to phase synchronization of chaos. Phys Rev Lett 80:1642

Sabarathinam S, Thamilmaran K (2017) Effect of variable memristor emulator in a Duffing nonlinear oscillator. AIP Conference Proceedings-AIP Publishing, 1832

Sabarathinam S, Volos CK, Thamilmaran K (2017) Implementation and study of the nonlinear dynamics of a memristor-based Duffing oscillator. Nonlinear Dyn 87:37–49

Saha P, Saha DC, Ray A, Chowdhury AR (2015) Memristive non-linear system and hidden attractor. Eur Phys J Spec Top 224:1563–1574

Sharma PR, Sharma A, Shrimali MD, Prasad A (2011) Targeting fixed-point solutions in nonlinear oscillators through linear augmentation. Phys Rev E 83:067201

Sharma PR, Shrimali MD, Prasad A, Feudel U (2013) Controlling bistability by linear augmentation. Phys Lett A 377:2329–2332

Sharma PR, Shrimali MD, Prasad A, Kuznetsov NV, Leonov GA (2015a) Control of multistability in hidden attractors. Eur Phys J Spec Top 224:1485–1491

Sharma PR, Shrimali M, Prasad A, Kuznetsov NV, Lenov GA (2015b) Controlling dynamics of hidden attractors. Int J Bifurc Chaos 25:1550061

Sharma PR, Singh A, Prasad A, Shrimali MD (2014) Controlling dynamical behavior of drive-response system through linear augmentation. Eur Phys J Spec Top 223:1531–1539

Sharma A, Shrimali MD, Dana SK (2012) Phase-flip transition in nonlinear oscillators coupled by dynamic environment. Chaos: An Interdiscip J Nonlinear Sci 22:023147

Sinha S et al (1990) Adaptive control in nonlinear dynamics. Physica D: Nonlinear Phenom 43:118–128

Strukov DB, Snider GS, Stewart DR, Williams RS (2008) The missing memristor found. Nature 453:80–83

Talukdar AH (2011) Nonlinear dynamics of memristor based 2nd and 3rd order oscillators. PhD thesis

Thomas A (2013) Memristor-based neural networks. J Phys D: Appl Phys 46:093001

Toth R, Taylor AF, Tinsley MR (2006) Collective behavior of a population of chemically coupled oscillators. J Phys Chem B 110:10170–10176

Tour JM, He T (2008) Electronics: the fourth element. Nature 453:42–43

Triandaf I, Schwartz IB (2000) Tracking sustained chaos: a segmentation method. Phys Rev E 62:3529

Varshney V, Sabarathinam S, Thamilmaran K, Prasad A (Submitted-2017) Hidden oscillations in a memristive autonomous Duffing oscillator-A case study. Int J Bifurc Chaos

Wang D, Hu Z, Yu X, Yu J (2009) A pwl model of memristor and its application example. In: International Conference on Communications, Circuits and Systems, (2009) ICCCAS 2009, pp 932–934

Wei Z, Zhang W (2014) Hidden hyperchaotic attractors in a modified LorenzStenflo system with only one stable equilibrium. Int J Bifurc Chaos 24:1450127

Xu C, Dong X, Jouppi NP, Xie Y (2011) Design implications of memristor-based RRAM cross-point structures. In: Design, Automation & Test in Europe Conference & Exhibition (DATE), 2011. IEEE (2011)

Zhang G, Hu J, Shen Y (2015) New results on synchronization control of delayed memristive neural networks. Nonlinear Dyn 81:1167–1178

Zhang W, Zou X (2012) Synchronization ability of coupled cell-cycle oscillators in changing environments. BMC Syst Biol 6:S13

Zhao H, Lin Y, Dai Y (2014) Hidden attractors and dynamics of a general autonomous van der PolDuffing oscillator. Int J Bifurc Chaos 24:1450080

A Novel 4-D Hyperchaotic Rikitake Dynamo System with Hidden Attractor, its Properties, Synchronization and Circuit Design

Sundarapandian Vaidyanathan, Viet-Thanh Pham, Christos Volos and Aceng Sambas

Abstract Hyperchaos has important applications in physics, chemistry, biology, ecology, secure communications, cryptosystems and many scientific branches. In this work, we propose a novel 4-D hyperchaotic Rikitake dynamo system without any equilibrium point by adding a state feedback control to the famous 3-D Rikitake two-disk dynamo system (1958). Thus, the proposed novel hyperchaotic Rikitake dynamo system exhibits hidden attractors. We describe qualitative properties of the hyperchaotic Rikitake dynamo system such as symmetry, Lyapunov exponents, Kaplan-Yorke dimension, etc. Furthermore, an adaptive integral sliding mode control scheme is proposed for the global hyperchaos synchronization of identical hyperchaotic Rikitake dynamo systems. The adaptive control mechanism helps the control design by estimating the unknown parameters. Numerical simulations using MATLAB are shown to illustrate all the main results derived in this work. Finally, the circuit experimental results of the hyperchaotic Rikitake dynamo system show agreement with the numerical simulations.

S. Vaidyanathan (✉)
Research and Development Centre, Vel Tech University,
Avadi, Chennai 600062, Tamil Nadu, India
e-mail: sundarvtu@gmail.com

V.-T. Pham
School of Electronics and Telecommunications, Hanoi University of Science
and Technology, Hanoi, Vietnam
e-mail: pvt3010@gmail.com

C. Volos
Physics Department, Aristotle University of Thessaloniki, Thessaloniki, Greece
e-mail: volos@physics.auth.gr

A. Sambas
Department of Mechanical Engineering, Universitas Muhammadiyah Tasikmalaya,
Tasikmalaya, Indonesia
e-mail: acengs@umtas.ac.id

A. Sambas
Faculty of Informatics and Computing Universiti Sultan Zainal Abidin,
Kuala Terengganu, Malaysia

© Springer International Publishing AG 2018
V.-T. Pham et al. (eds.), *Nonlinear Dynamical Systems with Self-Excited and Hidden Attractors*, Studies in Systems, Decision and Control 133,
https://doi.org/10.1007/978-3-319-71243-7_15

Keywords Chaos · Chaotic systems · Hyperchaos · Hyperchaotic systems
Dynamo system · Adaptive control · Sliding mode control · Circuit simulation

1 Introduction

A hyperchaotic system is defined as a chaotic system with at least two
positive Lyapunov exponents (Azar and Vaidyanathan 2015, 2016; Vaidyanathan
and Volos 2016a). Combined with one null exponent along the flow and one negative exponent to ensure the boundedness of the solution, the minimal dimension
for a continuous-time hyperchaotic system is four. Hyperchaotic systems have many
applications in science and engineering (Vaidyanathan and Volos 2016b, 2017; Azar
and Vaidyanathan 2017).

Some classical examples of hyperchaotic systems are hyperchaotic
Rössler system (Rössler 1979), hyperchaotic Lorenz system (Jia 2007), hyperchaotic
Chen system (Gao et al. 2006), hyperchaotic Lü system (Chen et al. 2006), etc.
Some recent examples of hyperchaotic systems are hyperchaotic
Dadras system (Dadras et al. 2012), hyperchaotic Vaidyanathan systems
(Vaidyanathan 2013, 2014b; Vaidyanathan et al. 2014, 2015a, b, d; Vaidyanathan
and Azar 2015; Vaidyanathan et al. 2015c; Vaidyanathan 2016g, b; Vaidyanathan and
Azar 2016a; Vaidyanathan 2016k, l, h; Vaidyanathan and Azar 2016b; Vaidyanathan
et al. 2016; Vaidyanathan 2016a, d, f, c, e; Vaidyanathan and Boulkroune 2016),
hyperchaotic Sampath system (Sampath et al. 2016), hyperchaotic Pham system
(Pham et al. 2016d), etc.

Recently there has been significant interest in finding and studying of infinite
number of equilibria such as equilibria located on the circle (Gotthans and Petrzela
2015), square (Gotthans et al. 2016), ellipse (Pham et al. 2016c), rounded square
Pham et al. (2016a), rounded rectangle Pham et al. (2016c), line (Jafari and Sprott
2013; Li and Sprott 2014a), two parallel lines (Li et al. 2015), two perpendicular lines
(Li et al. 2015), heart shape (Pham et al. 2017) and piecewise linear curve (Pham et al.
2016b). In addition, the chaotic system with no equilibria was also reported (Li and
Sprott 2014b, 2016; Li et al. 2016; Leonov et al. 2012, 2015).

Motivated by the above researches, a novel 4-D hyperchaotic Rikitake dynamo
system without equilibrium point is proposed in this work, which is derived by
adding a state feedback control to the famous 3-D Rikitake two-disk dynamo system
(Rikitake 1958). Thus, the proposed novel hyperchaotic Rikitake dynamo system
exhibits hidden attractors. We describe qualitative properties of the hyperchaotic
Rikitake dynamo system such as symmetry, Lyapunov exponents, Kaplan-Yorke
dimension, etc.

In this chapter, we also use adaptive integral sliding mode control for the global
hyperchaos synchronization of the identical hyperchaotic Rikitake dynamo systems.

The adaptive control mechanism helps the control design by estimating the
unknown parameters (Azar and Vaidyanathan 2015, 2016; Vaidyanathan and Volos
2016a). The sliding mode control approach is recognized as an efficient tool for

designing robust controllers for linear or nonlinear control systems operating under uncertainty conditions (Utkin 1977, 1993).

A major advantage of sliding mode control is low sensitivity to parameter variations in the plant and disturbances affecting the plant, which eliminates the necessity of exact modeling of the plant (Vaidyanathan and Sampath 2011; Sundarapandian and Sivaperumal 2011; Vaidyanathan 2011, 2012a). Sliding mode control is a popular method for the control and synchronization of chaotic systems (Vaidyanathan 2012b, 2014a; Lakhekar et al. 2016; Moussaoui et al. 2016; Vaidyanathan 2016i, j).

Next, an adaptive integral sliding mode control scheme is proposed to globally stabilize all the trajectories of the hyperchaotic two-disk dynamo system. Furthermore, an adaptive integral sliding mode control scheme is proposed for the global hyperchaos synchronization of identical hyperchaotic two-disk dynamo systems.

This work is organized as follows. Section 2 gives a review of the 3-D Rikitake dynamo chaotic system and its qualitative properties. Section 3 describes the novel 4-D hyperchaotic Rikitake dynamo system and describes its phase portraits. Section 4 details the qualitative properties of the novel 4-D hyperchaotic Rikitake dynamo system. Section 5 contains new results on the adaptive integral sliding mode controller design for the global synchronization of the novel identical 4-D hyperchaotic Rikitake dynamo systems. Section 6 contains a circuit simulation of the novel 4-D hyperchaotic Rikitake dynamo system. Section 7 contains the conclusions of this work.

2 Rikitake Two-Disk Dynamo System

In this section, we describe the two-disk dynamo chaotic system obtained by Rikitake (1958). The frequent and irregular reversals of the earth's magnetic field had inspired many early studies involving electrical currents within the earth's molten core. One of the first such dynamical models to report earth's magnetic reversals was the Rikitake two-disk dynamo model (Rikitake 1958).

The Rikitake two-disk dynamo system is modelled by the 3-D dynamics

$$
\begin{aligned}
\dot{x}_1 &= -ax_1 + x_2 x_3 \\
\dot{x}_2 &= -ax_2 + x_1(x_3 - b) \\
\dot{x}_3 &= 1 - x_1 x_2
\end{aligned}
\tag{1}
$$

where x_1, x_2, x_3 are the three states. In (1), a and b are constant, positive, parameters. It is noted that Rikitake dynamo system (1) has the same number of terms as the Lorenz chaotic system (Lorenz 1963), but with an additional quadratic nonlinearity.

In the Rikitake two-disk dynamo system (1), the parameter a stands for resistive dissipation and the parameter b stands for the difference in the angular velocities of the two disks.

In Rikitake (1958), it was established that the two-disk dynamo system depicts a *chaotic* attractor when we take the parameter values as

$$a = 1, \quad b = 1 \tag{2}$$

For MATLAB simulations, we take the initial state of the Rikitake two-disk dynamo system (1) as

$$x_1(0) = 0.4, \quad x_2(0) = 0.4, \quad x_3(0) = 0.4 \tag{3}$$

Using the parameter values (2) and the initial state (3), the Lyapunov exponents of the Rikitake two-disk dynamo system (1) are calculated using Wolf's algorithm (Wolf et al. 1985) as

$$L_1 = 0.1269, \quad L_2 = 0, \quad L_3 = -2.1269 \tag{4}$$

We note also that

$$L_1 + L_2 + L_3 = -2 < 0 \tag{5}$$

Thus, we infer that the Rikitake two-disk dynamo system (1) is chaotic and dissipative.

Also, the Kaplan-Yorke dimension of the Rikitake two-disk dynamo system (1) has been calculated as

$$D_{KY} = 2 + \frac{L_1 + L_2}{|L_3|} = 2.0597, \tag{6}$$

which is fractional.

It is easy to see that the Rikitake two-disk dynamo system (1) is invariant under the coordinates transformation

$$(x_1, x_2, x_3) \mapsto (-x_1, -x_2, x_3) \tag{7}$$

This establishes that the Rikitake two-disk dynamo system (1) has rotation symmetry about the x_3-axis. Hence, any non-trivial trajectory of the Rikitake two-disk dynamo system (1) must have a twin-trajectory.

For the parameter values $(a, b) = (1, 1)$, the Rikitake two-disk dynamo system (1) has two equilibrium points given by

$$E_1 = \begin{bmatrix} 1.2720 \\ 0.7862 \\ 1.6180 \end{bmatrix}, \quad E_2 = \begin{bmatrix} -1.2720 \\ -0.7862 \\ 1.6180 \end{bmatrix} \tag{8}$$

It is easy to verify that both equilibrium points E_1 and E_2 are marginally stable.

Figure 1 shows the phase portraits of the Rikitake two-disk dynamo chaotic system (1). It is clear that the Rikitake two-disk dynamo system exhibits a *two-scroll* chaotic attractor.

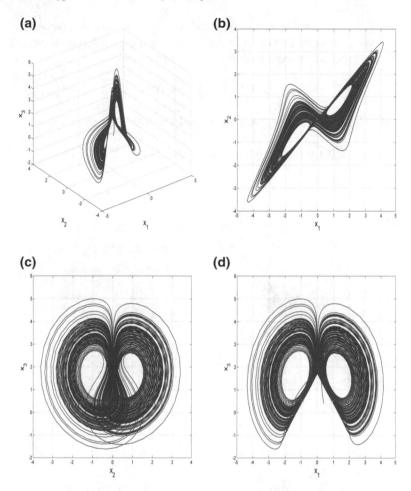

Fig. 1 Numerical simulation results of the Rikitake two-disk dynamo system (1) with two-scroll attractor for $a = 1, b = 1$, in **a** $\mathbf{R}^3$, **b** (x_1, x_2) plane, **c** (x_2, x_3) plane, **d** (x_1, x_3) plane

3 Hyperchaotic Rikitake Two-Disk Dynamo System

In this section, we derive a new 4-D hyperchaotic Rikitake two-disk dynamo system by adding a feedback control to the two-disk dynamo chaotic system (1).

Thus, our 4-D novel Rikitake two-disk dynamo system is given by the dynamics

$$
\begin{aligned}
\dot{x}_1 &= -ax_1 + x_2x_3 - x_4 \\
\dot{x}_2 &= -ax_2 + x_1(x_3 - b) - x_4 \\
\dot{x}_3 &= 1 - x_1x_2 \\
\dot{x}_4 &= cx_2
\end{aligned}
\tag{9}
$$

where x_1, x_2, x_3, x_4 are the state variables and a, b, c are positive parameters.

In this work, we show that the 4-D novel Rikitake two-disk dynamo system (9) is *hyperchaotic* when the system parameters take the values

$$a = 1, \quad b = 1, \quad c = 0.7 \tag{10}$$

For MATLAB simulations, we take the initial state of the system (9) as

$$x_1(0) = 0.4, \quad x_2(0) = 0.4, \quad x_3(0) = 0.4, \quad x_4(0) = 0.4 \tag{11}$$

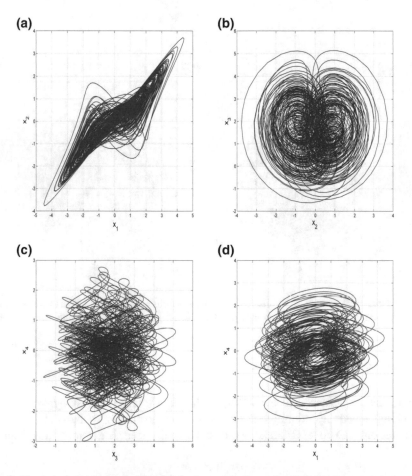

Fig. 2 Numerical simulation results of 2-D plots of the hyperchaotic Rikitake two-disk dynamo system (9) with two-scroll attractor for $a = 1, b = 1$, in **a** (x_1, x_2) plane, **b** (x_2, x_3) plane, **c** (x_3, x_4) plane and **d** (x_1, x_4) plane

For the parameter values (10) and the initial values (11), the Lyapunov exponents of the two-disk dynamo system (9) are calculated by Wolf's algorithm (Wolf et al. 1985) as

$$L_1 = 0.08175, \quad L_2 = 0.02350, \quad L_3 = 0, \quad L_4 = -2.10525 \tag{12}$$

Since there are two positive Lyapunov exponents in the LE spectrum (12), it is immediate that the 4-D novel two-disk dynamo system (9) is hyperchaotic.

Also, the Kaplan-Yorke dimension of the new 4-D hyperchaotic Rikitake two-disk dynamo system (9) is obtained as

$$D_{KY} = 3 + \frac{L_1 + L_2 + L_3}{|L_4|} = 3.05, \tag{13}$$

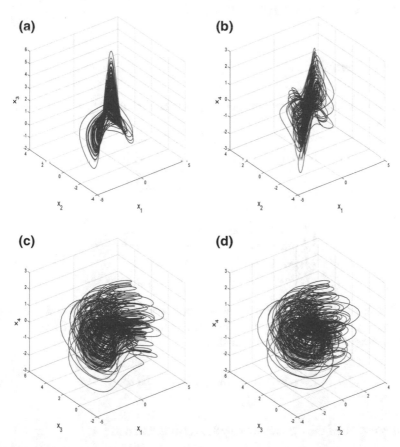

Fig. 3 Numerical simulation results of 3-D plots of the hyperchaotic Rikitake two-disk dynamo system (9) with two-scroll attractor for $a = 1, b = 1$, in **a** (x_1, x_2, x_3) space, **b** (x_1, x_2, x_4) space, **c** (x_1, x_3, x_4) space and **d** (x_2, x_3, x_4) space

which is fractional.

Figure 2 shows the 2-D phase portraits of the new hyperchaotic Rikitake two-disk dynamo system (9). Figure 3 shows the 3-D phase portraits of the new hyperchaotic Rikitake two-disk dynamo system (9). It is clear from Figs. 2 and 3 that the new hyperchaotic two-disk dynamo system (9) describes a *two-scroll* hyperchaotic attractor.

4   Analysis of the New 4-D Hyperchaotic Two-Disk Dynamo System

In this section, we give a dynamic analysis of the new 4-D hyperchaotic Rikitake two-disk dynamo system (9). We take the parameter values as in the hyperchaotic case (10), *i.e.* $a = 1$, $b = 1$ and $c = 0.7$.

4.1   Dissipativity

In vector notation, we express the new hyperchaotic Rikitake two-disk dynamo system (9) as

$$\dot{\mathbf{x}} = f(\mathbf{x}) = \begin{bmatrix} f_1(x_1, x_2, x_3, x_4) \\ f_2(x_1, x_2, x_3, x_4) \\ f_3(x_1, x_2, x_3, x_4) \\ f_4(x_1, x_2, x_3, x_4) \end{bmatrix}, \tag{14}$$

where

$$\begin{cases} f_1(x_1, x_2, x_3, x_4) = -ax_1 + x_2 x_3 - x_4 \\ f_2(x_1, x_2, x_3, x_4) = -ax_2 + x_1(x_3 - b) - x_4 \\ f_3(x_1, x_2, x_3, x_4) = 1 - x_1 x_2 \\ f_4(x_1, x_2, x_3, x_4) = cx_2 \end{cases} \tag{15}$$

Let Ω be any region in $\mathbf{R}^4$ with a smooth boundary and also, $\Omega(t) = \Phi_t(\Omega)$, where Φ_t is the flow of f. Furthermore, let $V(t)$ denote the hypervolume of $\Omega(t)$.

By Liouville's theorem, we know that

$$\dot{V}(t) = \int_{\Omega(t)} (\nabla \cdot f) \, dx_1 \, dx_2 \, dx_3 \, dx_4 \tag{16}$$

The divergence of the hyperchaotic system (14) is found as:

$$\nabla \cdot f = \frac{\partial f_1}{\partial x_1} + \frac{\partial f_2}{\partial x_2} + \frac{\partial f_3}{\partial x_3} + \frac{\partial f_4}{\partial x_4} = -(a + a) = -2 < 0 \tag{17}$$

Inserting the value of $\nabla \cdot f$ from (17) into (16), we get

$$\dot{V}(t) = \int_{\Omega(t)} (-2a)\, dx_1\, dx_2\, dx_3\, dx_4 = -2V(t) \tag{18}$$

Integrating the first order linear differential equation (18), we get

$$V(t) = \exp(-2t)V(0) \tag{19}$$

Thus, it is clear that Eq. (19) that $V(t) \to 0$ exponentially as $t \to \infty$. This shows that the new hyperchaotic Rikitake two-disk dynamo system (9) is dissipative. Hence, the system limit sets are ultimately confined into a specific limit set of zero hypervolume, and the asymptotic motion of the new hyperchaotic Rikitake two-disk dynamo system (9) settles onto a strange attractor of the system.

4.2 Equilibrium Points

We take the parameter values as in the hyperchaotic case (10).

The equilibrium points of the new hyperchaotic two-disk dynamo system (9) are obtained by solving the following system of equations.

$$-ax_1 + x_2 x_3 \, x_4 = 0 \tag{20a}$$
$$-ax_2 + (x_3 - b)x_1 - x_4 = 0 \tag{20b}$$
$$1 - x_1 x_2 = 0 \tag{20c}$$
$$cx_2 = 0 \tag{20d}$$

From (20d), $x_2 = 0$. Substituting $x_2 = 0$ in (20c), we get a contradiction.

This shows that the system of equations (20) does not admit any solution.

In other words, the new hyperchaotic two-disk dynamo system (9) has no equilibrium point. Hence, we deduce that the new hyperchaotic Rikitake two-disk dynamo system (9) exhibits hidden attractors (Li and Sprott 2014b, 2016; Li et al. 2016; Leonov et al. 2012, 2015).

4.3 Rotation Symmetry About the x_3-axis

It is easy to see that the new 4-D hyperchaotic two-disk dynamo system (9) is invariant under the change of coordinates

$$(x_1, x_2, x_3, x_4) \mapsto (-x_1, -x_2, x_3, -x_4) \tag{21}$$

Since the transformation (21) persists for all values of the system parameters, it follows that the new 4-D hyperchaotic two-disk dynamo system (9) has rotation symmetry about the x_3-axis and that any non-trivial trajectory must have a twin trajectory.

4.4 Invariance

It is easy to see that the x_3-axis is invariant under the flow of the 4-D novel hyperchaotic system (9).

The invariant motion along the x_3-axis is characterized by the scalar dynamics

$$\dot{x}_3 = 1, \tag{22}$$

which is unstable.

4.5 Lyapunov Exponents and Kaplan-Yorke Dimension

We take the parameter values of the new hyperchaotic Rikitake two-disk dynamo system (9) as in the hyperchaotic case (10), *i.e.* $a = 1, b = 1$ and $c = 0.7$.

We take the initial state of the new hyperchaotic Rikitake two-disk dynamo system (9) as (11), *i.e.* $x_i(0) = 0.4$ for $i = 1, 2, 3, 4$.

Then the Lyapunov exponents of the Rikitake two-disk dynamo system (9) are numerically obtained using MATLAB as

$$L_1 = 0.08175, \quad L_2 = 0.02350, \quad L_3 = 0, \quad L_4 = -2.10525 \tag{23}$$

Since there are two positive Lyapunov exponents in (23), the new 4-D two-disk dynamo system (9) exhibits *hyperchaotic* behavior.

The maximal Lyapunov exponent (MLE) of the new hyperchaotic two-disk dynamo system (9) is obtained as $L_1 = 0.08175$.

Since $L_1 + L_2 + L_3 + L_4 = -2 < 0$, it follows that the new hyperchaotic two-disk dynamo system (9) is dissipative.

Also, the Kaplan-Yorke dimension of the new hyperchaotic two-disk dynamo system (9) is calculated as

$$D_{KY} = 3 + \frac{L_1 + L_2 + L_3}{|L_4|} = 3.05, \tag{24}$$

which is fractional.

Figure 4 shows the Lyapunov exponents of the new hyperchaotic two-disk dynamo system (9).

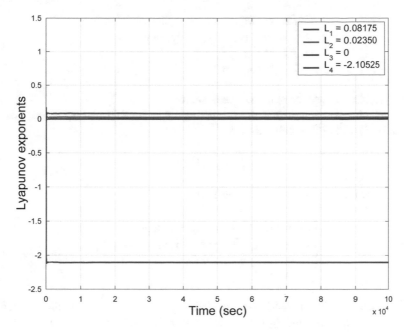

Fig. 4 Lyapunov exponents of the new hyperchaotic Rikitake two-disk dynamo system (9) for the parameter values $a = 1, b = 1$ and $c = 0.7$

5 Global Hyperchaos Synchronization of the New Hyperchaotic Rikitake Two-Wing Dynamo Systems

In this section, we use adaptive integral sliding mode control for the global hyperchaos synchronization of new hyperchaotic Rikitake two-wing dynamo systems with unknown system parameters. The adaptive control mechanism helps the control design by estimating the unknown parameters (Azar and Vaidyanathan 2015, 2016; Vaidyanathan and Volos 2016a).

As the master system, we consider the new hyperchaotic Rikitake two-wing dynamo system given by

$$\begin{cases} \dot{x}_1 = -ax_1 + x_2x_3 - x_4 \\ \dot{x}_2 = -ax_2 + (x_3 - b)x_1 - x_4 \\ \dot{x}_3 = 1 - x_1x_2 \\ \dot{x}_4 = cx_2 \end{cases} \tag{25}$$

where x_1, x_2, x_3, x_4 are the states of the system and a, b, c are unknown system parameters.

As the slave system, we consider the controlled new hyperchaotic Rikitake two-wing dynamo system given by

$$\begin{cases} \dot{y}_1 = -ay_1 + y_2y_3 - y_4 + u_1 \\ \dot{y}_2 = -ay_2 + (y_3 - b)y_1 - y_4 + u_2 \\ \dot{y}_3 = 1 - y_1y_2 + u_3 \\ \dot{y}_4 = cy_2 + u_4 \end{cases} \tag{26}$$

where y_1, y_2, y_3, y_4 are the states of the system.

The synchronization error between the hyperchaotic Rikitake two-wing dynamo systems (25) and (26) is defined as

$$e_i = y_i - x_i, \quad (i = 1, 2, 3, 4) \tag{27}$$

Then we get the error dynamics as follows:

$$\begin{cases} \dot{e}_1 = -ae_1 - e_4 + y_2y_3 - x_2x_3 + u_1 \\ \dot{e}_2 = -ae_2 - be_1 - e_4 - y_1y_3 + x_1x_3 + u_2 \\ \dot{e}_3 = -y_1y_2 + x_1x_2 + u_3 \\ \dot{e}_4 = ce_2 + u_4 \end{cases} \tag{28}$$

Based on the sliding mode control theory (Utkin 1977, 1993; Slotine and Li 1991), the integral sliding surface of each e_i $(i = 1, 2, 3, 4)$ is defined as follows:

$$s_i = \left(\frac{d}{dt} + \lambda_i \right) \left(\int_0^t e_i(\tau)d\tau \right) = e_i + \lambda_i \int_0^t e_i(\tau)d\tau, \quad i = 1, 2, 3, 4 \tag{29}$$

From Eq. (29), it follows that

$$\begin{cases} \dot{s}_1 = \dot{e}_1 + \lambda_1 e_1 \\ \dot{s}_2 = \dot{e}_2 + \lambda_2 e_2 \\ \dot{s}_3 = \dot{e}_3 + \lambda_3 e_3 \\ \dot{s}_4 = \dot{e}_4 + \lambda_4 e_4 \end{cases} \tag{30}$$

The Hurwitz condition is realized if $\lambda_i > 0$ for $i = 1, 2, 3, 4$.

We consider the adaptive feedback control given by

$$\begin{cases} u_1 = \hat{a}(t)e_1 + e_4 - y_2y_3 + x_2x_3 - \lambda_1 e_1 - \eta_1 \, \text{sgn}(s_1) - k_1 s_1 \\ u_2 = \hat{a}(t)e_2 + \hat{b}(t)e_1 + e_4 + y_1y_3 - x_1x_3 - \lambda_2 e_2 - \eta_2 \, \text{sgn}(s_2) - k_2 s_2 \\ u_3 = y_1y_2 - x_1x_2 - \lambda_3 e_3 - \eta_3 \, \text{sgn}(s_3) - k_3 s_3 \\ u_4 = -\hat{c}(t)e_2 - \lambda_4 e_4 - \eta_4 \, \text{sgn}(s_4) - k_4 s_4 \end{cases} \tag{31}$$

where $\eta_i > 0$ and $k_i > 0$ for $i = 1, 2, 3, 4$.

Substituting (31) into (28), we obtain the closed-loop error dynamics as

$$
\begin{cases}
\dot{e}_1 = -[a - \hat{a}(t)]e_1 - \lambda_1 e_1 - \eta_1 \, \mathrm{sgn}(s_1) - k_1 s_1 \\
\dot{e}_2 = -[a - \hat{a}(t)]e_2 - [b - \hat{b}(t)]e_1 - \lambda_2 e_2 - \eta_2 \, \mathrm{sgn}(s_2) - k_2 s_2 \\
\dot{e}_3 = -\lambda_3 e_3 - \eta_3 \, \mathrm{sgn}(s_3) - k_3 s_3 \\
\dot{e}_4 = [c - \hat{c}(t)]e_2 - \lambda_4 e_4 - \eta_4 \, \mathrm{sgn}(s_4) - k_4 s_4
\end{cases}
\tag{32}
$$

We define the parameter estimation errors as

$$
\begin{cases}
e_a(t) = a - \hat{a}(t) \\
e_b(t) = b - \hat{b}(t) \\
e_c(t) = c - \hat{c}(t)
\end{cases}
\tag{33}
$$

Using (33), we can simplify the closed-loop system (32) as

$$
\begin{cases}
\dot{e}_1 = -e_a e_1 - \lambda_1 e_1 - \eta_1 \, \mathrm{sgn}(s_1) - k_1 s_1 \\
\dot{e}_2 = -e_a e_2 - e_b e_1 - \lambda_2 e_2 - \eta_2 \, \mathrm{sgn}(s_2) - k_2 s_2 \\
\dot{e}_3 = -\lambda_3 e_3 - \eta_3 \, \mathrm{sgn}(s_3) - k_3 s_3 \\
\dot{e}_4 = e_c e_2 - \lambda_4 e_4 - \eta_4 \, \mathrm{sgn}(s_4) - k_4 s_4
\end{cases}
\tag{34}
$$

Differentiating (33) with respect to t, we get

$$
\begin{cases}
\dot{e}_a = -\dot{\hat{a}} \\
\dot{e}_b = -\dot{\hat{b}} \\
\dot{e}_c = -\dot{\hat{c}}
\end{cases}
\tag{35}
$$

Next, we state and prove the main result of this section.

Theorem 1 *The new hyperchaotic two-wing dynamo systems (25) and (26) are globally and asymptotically synchronized for all initial conditions $\mathbf{x}(0), \mathbf{y}(0) \in \mathbf{R}^4$ by the adaptive integral sliding mode control law (31) and the parameter update law*

$$
\begin{cases}
\dot{\hat{a}} = -s_1 e_1 - s_2 e_2 \\
\dot{\hat{b}} = -s_2 e_1 \\
\dot{\hat{c}} = s_4 e_2
\end{cases}
\tag{36}
$$

where λ_i, η_i, k_i are positive constants for $i = 1, 2, 3, 4$.

Proof We consider the quadratic Lyapunov function defined by

$$
V(s_1, s_2, s_3, s_4, e_a, e_b, e_c) = \frac{1}{2}\left(s_1^2 + s_2^2 + s_3^2 + s_4^2\right) + \frac{1}{2}\left(e_a^2 + e_b^2 + e_c^2\right)
\tag{37}
$$

Clearly, V is positive definite on $\mathbf{R}^7$.

Using (30), (34) and (35), the time-derivative of V is obtained as

$$\dot{V} = -\eta_1|s_1| - k_1s_1^2 - \eta_2|s_2| - k_2s_2^2 - \eta_3|s_3| - k_3s_3^2 - \eta_4|s_4| - k_4s_4^2$$
$$+ e_a\left(-s_1e_1 - s_2e_2 - \dot{\hat{a}}\right) + e_b\left(-s_2e_1 - \dot{\hat{b}}\right) + e_c\left(s_4e_2 - \dot{\hat{c}}\right) \tag{38}$$

Using the parameter update law (36), we obtain

$$\dot{V} = -\eta_1|s_1| - k_1s_1^2 - \eta_2|s_2| - k_2s_2^2 - \eta_3|s_3| - k_3s_3^2 - \eta_4|s_4| - k_4s_4^2 \tag{39}$$

which shows that $\dot{V}$ is negative semi-definite on $\mathbf{R}^7$.

Hence, by Barbalat's lemma (Khalil 2002), it is immediate that $\mathbf{e}(t)$ is globally asymptotically stable for all values of $\mathbf{e}(0) \in \mathbf{R}^4$.

Hence, it follows that the new hyperchaotic two-wing dynamo systems (25) and (26) are globally and asymptotically synchronized for all initial conditions $\mathbf{x}(0), \mathbf{y}(0) \in \mathbf{R}^4$.

This completes the proof. ∎

For numerical simulations, we take the parameter values of the new hyperchaotic Rikitake two-disk dynamo systems (25) and (26) as in the hyperchaotic case (10), i.e. $a = 1$, $b = 1$ and $c = 0.7$.

We take the values of the control parameters as

$$k_i = 10, \quad \eta_i = 0.1, \quad \lambda_i = 12, \quad \text{where } i = 1, 2, 3, 4 \tag{40}$$

We take the estimates of the system parameters as

$$\hat{a}(0) = 5.4, \quad \hat{b}(0) = 3.7, \quad \hat{c}(0) = 8.9 \tag{41}$$

We take the initial state of the master system (25) as

$$x_1(0) = 5.2, \quad x_2(0) = 23.7, \quad x_3(0) = 16.3, \quad x_4(0) = 7.6 \tag{42}$$

We take the initial state of the slave system (26) as

$$y_1(0) = 14.9, \quad y_2(0) = 10.4, \quad y_3(0) = 9.1, \quad y_4(0) = 12.8 \tag{43}$$

Figure 5 shows the complete synchronization between the states of the master system (25) and the slave system (26). Figure 6 shows the time-history of the synchronization errors e_1, e_2, e_3, e_4.

6 Circuit Simulation of the Hyperchaotic Rikitake Two-Disk Dynamo System

A simple electronic circuit is designed that can be used to study chaotic phenomena. In Fig. 7, the voltages of V_{C1}, V_{C2}, V_{C3} and V_{C4} are used as x_1, x_2, x_3 and x_4,

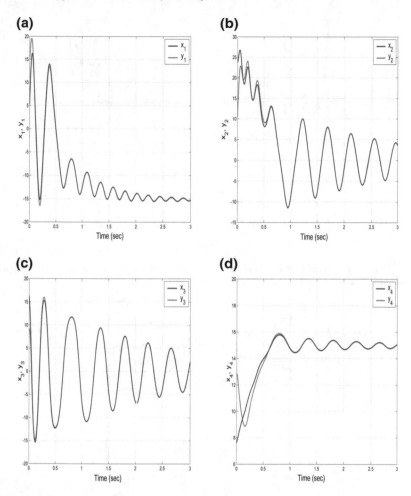

Fig. 5 Numerical simulation results of the complete synchronization of the hyperchaotic Rikitake two-disk dynamo systems (25) and (26)

respectively. By applying Kirchhoff's laws to the electronic circuit in Fig. 7, its nonlinear equations are given as follows:

$$
\begin{cases}
\dfrac{dV_{C1}}{dt} = -\dfrac{1}{C_1 R_1} V_{C1} + \dfrac{1}{10 C_1 R_2} V_{C2} V_{C3} - \dfrac{1}{C_1 R_3} V_{C4} \\[2mm]
\dfrac{dV_{C2}}{dt} = -\dfrac{1}{C_2 R_4} V_{C2} + \dfrac{1}{10 C_2 R_5} V_{C1} V_{C3} - \dfrac{1}{C_2 R_6} V_{C1} - \dfrac{1}{C_2 R_7} V_{C4} \\[2mm]
\dfrac{dV_{C3}}{dt} = \dfrac{1}{C_3 R_9} V_1 - \dfrac{1}{10 C_3 R_8} V_{C1} V_{C2} \\[2mm]
\dfrac{dV_{C4}}{dt} = \dfrac{1}{C_4 R_{10}} V_{C2}
\end{cases}
\tag{44}
$$

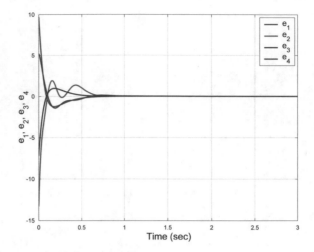

Fig. 6 Numerical simulation results of the time-history of the synchronization error between the hyperchaotic Rikitake two-disk dynamo systems (25) and (26)

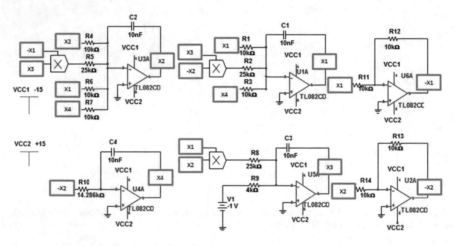

Fig. 7 Schematic of the hyperchaotic Rikitake two-disk dynamo system by using MultiSIM 10.0

We choose $R_2 = R_5 = R_8 = 25$ KΩ, $R_9 = 4$ KΩ, $R_{10} = 14.286$ KΩ, $R_1 = R_3 = R_4 = R_6 = R_7 = R_{11} = R_{12} = R_{13} = R_{14} = 10$ KΩ, $V_1 = -1V_{DC}$ and $C_1 = C_2 = C_3 = C_4 = 10$ nF. The power supplies of all active devices are $\pm$ 15 V.

The MultiSIM projections of chaotic behaviour with hidden attractor are presented in Fig. 8. The MultiSIM results also indicate that the circuit can emulate the theoretical model (9). As compared with Fig. 2, a good qualitative agreement between the numerical simulations and the MultiSIM 10.0 results of the a new 4-D hyperchaotic Rikitake two-disk dynamo system is confirmed.

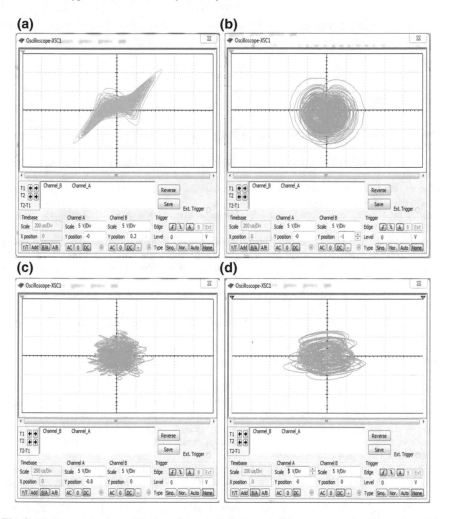

Fig. 8 Various projections of the hyperchaotic Rikitake two-disk dynamo system using MultiSIM for $a = 1, b = 1, c = 0.7$, in **a** (x_1, x_2) plane, **b** (x_2, x_3) plane, **c** (x_3, x_4) plane and **d** (x_1, x_4) plane

7   Conclusions

In this work, a hyperchaotic Rikitake two-disk dynamo system with hidden attractor was presented. The fundamental properties of the system such as dissipativity, symmetry, Lyapunov exponents and Kaplan-Yorke dimension as well as its phase portraits were described in detail. Also, an adaptive integral sliding mode control for the global hyperchaos synchronization of new hyperchaotic Rikitake two-disk dynamo systems with unknown system parameters was designed. Finally, the

MultiSIM implementation of the Hyperchaotic Rikitake two-disk dynamo system was presented for confirming the feasibility of the theoretical chaotic system.

References

Azar AT, Vaidyanathan S (2015) Chaos modeling and control systems design. Springer, Berlin, Germany

Azar AT, Vaidyanathan S (2016) Advances in chaos theory and intelligent control. Springer, Berlin, Germany

Azar AT, Vaidyanathan S (2017) Fractional order control and synchronization of chaotic systems. Springer, Berlin, Germany

Chen A, Lu J, Lü J, Yu S (2006) Generating hyperchaotic Lü attractor via state feedback control. Physica A 364:103–110

Dadras S, Momeni HR, Qi G, lin Wang Z (2012) Four-wing hyperchaotic attractor generated from a new 4D system with one equilibrium and its fractional-order form. Nonlinear Dyn 67:1161–1173

Gao T, Chen Z, Yuan Z, Chen G (2006) A hyperchaos generated from Chen's system. Int J Modern Phys C 17(4):471–478

Gotthans T, Petrzela J (2015) New class of chaotic systems with circular equilibrium. Nonlinear Dyn 81:1143–1149

Gotthans T, Sprott J, Petrzela J (2016) Simple chaotic flow with circle and square equilibrium. Int J Bifurc Chaos 26, article ID 1650137

Jafari S, Sprott J (2013) Simple chaotic flows with a line equilibrium. Chaos Solitons Fract 57:79–84

Jia Q (2007) Hyperchaos generated from the Lorenz chaotic system and its control. Phys Lett A 366(3):217–222

Khalil HK (2002) Nonlinear systems. Prentice Hall, New York, USA

Lakhekar GV, Waghmare LM, Vaidyanathan S (2016) Diving autopilot design for underwater vehicles using an adaptive neuro-fuzzy sliding mode controller. In: Vaidyanathan S, Volos C (eds) Advances and applications in nonlinear control systems. Springer, Berlin, Germany, pp 477–503

Leonov G, Kuznetsov N, Vagaitsev V (2012) Hidden attractor in smooth Chua systems. Physica D Nonlinear Phenom 241:1482–1486

Leonov G, Kuznetsov N, Mokaev T (2015) Hidden attractor and homoclinic orbit in Lorenz-like system describing convective fluid motion in rotating cavity. Commun Nonlinear Sci Numer Simul 28:166–174

Li C, Sprott J (2014a) Chaotic flows with a single nonquadratic form. Phys Lett A 378:178–183

Li C, Sprott J (2014b) Coexisting hidden attractors in a 4-D simplified Lorenz system. Int J Bifurc Chaos 24, article ID 1450034

Li C, Sprott J (2016) Variable-boostable chaotic flows. Optik 127:10,389–10,398

Li C, Sprott J, Yuan Z, Li H (2015) Constructing chaotic systems with total amplitude control. Int J Bifurc Chaos 25, article ID 1530025

Li P, Zheng T, Li C, Wang X, Hu W (2016) A unique jerk system with hidden chaotic oscillation. Nonlinear Dyn 86:197–203

Lorenz EN (1963) Deterministic nonperiodic flow. J Atmos Sci 20:130–141

Moussaoui S, Boulkroune A, Vaidyanathan S (2016) Fuzzy adaptive sliding-mode control scheme for uncertain underactuated systems. In: Vaidyanathan S, Volos C (eds) Advances and applications in nonlinear control systems. Springer, Berlin, Germany, pp 351–367

Pham V, Jafari S, Volos C, Giakoumis A, Vaidyanathan S, Kapitaniak T (2016a) A chaotic system with equilibria located on the rounded square loop and its circuit implementation. IEEE Trans Circuits Syst II Express Briefs 63:878–882

Pham V, Jafari S, Volos C, Vaidyanathan S, Kapitaniak T (2016b) A chaotic system with infinite equilibria located on a piecewise linear curve. Optik 127:9111–9117

Pham V, Jafari S, Wang X, Ma J (2016c) A chaotic system with different shapes of equilibria. Int J Bifurc Chaos 26, article ID 1650069

Pham V, Jafari S, Volos C (2017) A novel chaotic system with heart-shaped equilibrium and its circuital implementation. Optik 131:343–349

Pham VT, Vaidyanathan S, Volos C, Jafari S, Kingni ST (2016d) A no-equilibrium hyperchaotic system with a cubic nonlinear term. Optik 127(6):3259–3265

Rikitake T (1958) Oscillations of a system of disk dynamos. Math Proc Camb Philos Soc 54:89–105

Rössler O (1979) An equation for hyperchaos. Phys Lett A 71(2–3):155–157

Sampath S, Vaidyanathan S, Pham VT (2016) A novel 4-D hyperchaotic system with three quadratic nonlinearities, its adaptive control and circuit simulation. Int J Control Theory Appl 9(1):339–356

Slotine J, Li W (1991) Applied nonlinear control. Prentice-Hall, Englewood Cliffs, NJ, USA

Sundarapandian V, Sivaperumal S (2011) Sliding controller design of hybrid synchronization of four-wing Chaotic systems. Int J Soft Comput 6(5):224–231

Utkin VI (1977) Variable structure systems with sliding modes. IEEE Trans Autom Control 22(2):212–222

Utkin VI (1993) Sliding mode control design principles and applications to electric drives. IEEE Trans Ind Electr 40(1):23–36

Vaidyanathan S (2011) Analysis and synchronization of the hyperchaotic Yujun systems via sliding mode control. Adv Intel Syst Comput 176:329–337

Vaidyanathan S (2012a) Global chaos control of hyperchaotic Liu system via sliding control method. Int J Control Theory Appl 5(2):117–123

Vaidyanathan S (2012b) Sliding mode control based global chaos control of Liu-Liu-Liu-Su chaotic system. Int J Control Theory Appl 5(1):15–20

Vaidyanathan S (2013) A ten-term novel 4-D hyperchaotic system with three quadratic nonlinearities and its control. Int J Control Theory Appl 6(2):97–109

Vaidyanathan S (2014a) Global chaos synchronisation of identical Li-Wu chaotic systems via sliding mode control. Int J Model Identif Control 22(2):170–177

Vaidyanathan S (2014b) Qualitative analysis and control of an eleven-term novel 4-D hyperchaotic system with two quadratic nonlinearities. Int J Control Theory Appl 7(1):35–47

Vaidyanathan S (2016a) A non-equilibrium novel 4-D highly hyperchaotic system with four quadratic nonlinearities and its adaptive control. In: Vaidyanathan S, Volos C (eds) Adv Appl Nonlinear Control Syst. Springer, Berlin, Germany, pp 235–258

Vaidyanathan S (2016b) A novel 4-D hyperchaotic thermal convection system and its adaptive control. In: Azar AT, Vaidyanathan S (eds) Adv Chaos Theory Intel Control. Springer, Berlin, Germany, pp 75–100

Vaidyanathan S (2016c) A novel 5-D hyperchaotic system with a line of equilibrium points and its adaptive control. In: Vaidyanathan S, Volos C (eds) Adv Appl Chaotic Syst. Springer, Berlin, Germany, pp 471–494

Vaidyanathan S (2016d) A novel highly hyperchaotic system and its adaptive control. In: Vaidyanathan S, Volos C (eds) Adv Appl Chaotic Syst. Springer, Berlin, Germany, pp 513–535

Vaidyanathan S (2016e) A novel hyperchaotic hyperjerk system with two nonlinearities, its analysis, adaptive control and synchronization via backstepping control method. Int J Control Theory Appl 9(1):257–278

Vaidyanathan S (2016f) An eleven-term novel 4-D hyperchaotic system with three quadratic nonlinearities, analysis, control and synchronization via adaptive control method. Int J Control Theory Appl 9(1):21–43

Vaidyanathan S (2016g) Analysis, adaptive control and synchronization of a novel 4-D hyperchaotic hyperjerk system via backstepping control method. Arch Control Sci 26(3):311–338

Vaidyanathan S (2016h) Analysis, control and synchronization of a novel 4-D highly hyperchaotic system with hidden attractors. In: Azar AT, Vaidyanathan S (eds) Adv Chaos Theory Intel Control. Springer, Berlin, Germany, pp 529–552

Vaidyanathan S (2016i) Anti-synchronization of 3-cells cellular neural network attractors via integral sliding mode control. Int J PharmTech Res 9(1):193–205

Vaidyanathan S (2016j) Global chaos regulation of a symmetric nonlinear gyro system via integral sliding mode control. Int J ChemTech Res 9(5):462–469

Vaidyanathan S (2016k) Hyperchaos, adaptive control and synchronization of a novel 4-D hyperchaotic system with two quadratic nonlinearities. Arch Control Sci 26(4):471–495

Vaidyanathan S (2016l) Qualitative analysis and properties of a novel 4-D hyperchaotic system with two quadratic nonlinearities and its adaptive control. In: Azar AT, Vaidyanathan S (eds) Advances in chaos theory and intelligent control. Springer, Berlin, Germany, pp 455–480

Vaidyanathan S, Azar AT (2015) Analysis and control of a 4-D novel hyperchaotic system. In: Azar AT, Vaidyanathan S (eds) Chaos modeling and control systems design, studies in computational intelligence, vol 581. Springer, Germany, pp 3–17

Vaidyanathan S, Azar AT (2016a) A novel 4-D four-wing chaotic system with four quadratic nonlinearities and its synchronization via adaptive control method. In: Azar AT, Vaidyanathan S (eds) Advances in chaos theory and intelligent control. Springer, Berlin, Germany, pp 203–224

Vaidyanathan S, Azar AT (2016b) Qualitative study and adaptive control of a novel 4-D hyperchaotic system with three quadratic nonlinearities. In: Azar AT, Vaidyanathan S (eds) Advances in chaos theory and intelligent control. Springer, Berlin, Germany, pp 179–202

Vaidyanathan S, Boulkroune A (2016) A novel 4-D hyperchaotic chemical reactor system and its adaptive control. In: Vaidyanathan S, Volos C (eds) Advances and applications in chaotic systems. Springer, Berlin, Germany, pp 447–469

Vaidyanathan S, Sampath S (2011) Global chaos synchronization of hyperchaotic Lorenz systems by sliding mode control. Communications in computer and information science 205:156–164

Vaidyanathan S, Volos C (2016a) Advances and applications in chaotic systems. Springer, Berlin, Germany

Vaidyanathan S, Volos C (2016b) Advances and applications in nonlinear control systems. Springer, Berlin, Germany

Vaidyanathan S, Volos C (2017) Advances in memristors. Memristive devices and systems. Springer, Berlin, Germany

Vaidyanathan S, Volos C, Pham VT (2014) Hyperchaos, adpative control and synchronization of a novel 5-D hyperchaotic system with three positive Lyapunov exponents and its SPICE implementation. Arch Control Sci 24(4):409–446

Vaidyanathan S, Azar AT, Rajagopal K, Alexander P (2015a) Design and SPICE implementation of a 12-term novel hyperchaotic system and its synchronisation via active control. Int J Model Identif Control 23(3):267–277

Vaidyanathan S, Pham VT, Volos CK (2015b) A 5-D hyperchaotic Rikitake dynamo system with hidden attractors. Eur Phys J Special Topics 224(8):1575–1592

Vaidyanathan S, Volos C, Pham VT, Madhavan K (2015c) Analysis, adaptive control and synchronization of a novel 4-D hyperchaotic hyperjerk system and its SPICE implementation. Arch Control Sci 25(1):135–158

Vaidyanathan S, Volos CK, Pham VT (2015d) Analysis, control, synchronization and SPICE implementation of a novel 4-D hyperchaotic Rikitake dynamo system without equilibrium. J Eng Sci Technol Rev 8(2):232–244

Vaidyanathan S, Volos CK, Pham VT (2016) Hyperchaos, control, synchronization and circuit simulation of a novel 4-D hyperchaotic system with three quadratic nonlinearities. In: Azar AT, Vaidyanathan S (eds) Advances in chaos theory and intelligent control. Springer, Berlin, Germany, pp 297–325

Wolf A, Swift JB, Swinney HL, Vastano JA (1985) Determining Lyapunov exponents from a time series. Physica D 16:285–317

A Six-Term Novel Chaotic System with Hidden Attractor and Its Circuit Design

Aceng Sambas, Sundarapandian Vaidyanathan, Mustafa Mamat
and W. S. Mada Sanjaya

Abstract In this work, we propose a six-term novel 3D chaotic system with hidden attractor. The novel 3D chaotic system consists of six terms and two quadratic nonlinearities. We show that the novel chaotic system has no equilibrium point and hence it exhibits hidden attractor. A detailed qualitative analysis of the 3D chaotic system is presented such as phase portrait analysis, Lyapunov exponents, bifurcation diagram and Poincaré map. The mathematical model of the novel chaotic system is accompanied by an electrical circuit implementation, demonstrating chaotic behavior of the strange attractor. Finally, the circuit experimental results of the chaotic attractors show agreement with numerical simulations.

1 Introduction

In 1963, Lorenz constructed a 3-D model for weather prediction (Lorenz 1963). In 1976, Rössler proposed a low dimensional dissipative dynamical systems (Rössler 1976). In 1994, Sprott suggested 19 cases of simple chaotic flows (Sprott 1994). In 2000, Malasoma presented the simplest dissipative jerk equation that is parity invariant (Malasoma 2000). Some classical 3-D autonomous chaotic systems in the litera-

A. Sambas (✉)
Department of Mechanical Engineering, Universitas Muhammadiyah Tasikmalaya,
Tasikmalaya, Indonesia
e-mail: acengs@umtas.ac.id

A. Sambas · M. Mamat
Faculty of Informatics and Computing, Universiti Sultan Zainal Abidin, Kuala Terengganu,
Malaysia
e-mail: must@unisza.edu.my

S. Vaidyanathan
Research and Development Centre, Vel Tech University, Avadi, Chennai, Tamil Nadu, India
e-mail: sundarvtu@gmail.com

W. S. Mada Sanjaya
Department of Physics, Universitas Islam Negeri Sunan Gunung Djati, Bandung, Indonesia
e-mail: madasws@gmail.com

© Springer International Publishing AG 2018
V.-T. Pham et al. (eds.), *Nonlinear Dynamical Systems with Self-Excited
and Hidden Attractors*, Studies in Systems, Decision and Control 133,
https://doi.org/10.1007/978-3-319-71243-7_16

ture are Chen system (Chen and Ueta 1999), Lü system (Lü and Chen 2002), etc. In the last few decades, many new chaotic systems have been found in various applications in science and engineering (Azar and Vaidyanathan 2015, 2016; Vaidyanathan and Volos 2016a, b; Azar and Vaidyanathan 2017; Vaidyanathan and Volos 2017).

Chaos has been widely applied to many scientific disciplines such as ecology (Mada Sanjaya et al. 2012), biology (Mada Sanjaya et al. 2011), economics (Bouali et al. 2012; Tacha et al. 2016), lasers (Li et al. 2014), chemical reaction (Nakajima and Sawada 1979), robotics (Sambas et al. 2016b; Islam and Murase 2005), image encryption (Andreator and Leros 2013), voice encryption (Abdulkareem and Abduljaleel 2013), secure communication systems (Sambas et al. 2016a, 2013a, 2012, 2015b, 2013b, 2015a), etc.

Recently there has been significant interest in finding and studying of infinite number of equilibria such as equilibria located on the circle (Gotthans and Petrzela 2015), square (Gotthans et al. 2016), ellipse (Pham et al. 2016c), rounded square (Pham et al. 2016a), rounded rectangle (Pham et al. 2016c), line (Jafari and Sprott 2013; Li and Sprott 2014a), two parallel lines (Li et al. 2015), two perpendicular lines (Li et al. 2015), heart shape (Pham et al. 2017) and piecewise linear curve (Pham et al. 2016b). In addition, the chaotic system with no equilibria was also reported (Li and Sprott 2014b, 2016; Li et al. 2016; Leonov et al. 2012, 2015).

Motivated by the above researches, a novel 3-D chaotic system without equilibrium and with only two quadratic nonlinearities is proposed in this work. In Sect. 2, we present novel 3D chaotic system without equilibrium with only two quadratic nonlinearities, numerical results in evolving phase portraits, Lyapunov exponents analysis, bifurcation diagram analysis, and Poincaré map analysis. In Sect. 3, we present an electronic circuit that implements the nonlinear system. Finally, Sect. 4 contains the conclusion of this work.

2 A Six-Term Novel Chaotic System

A classical example of a chaotic system with hidden attractor is the conservative chaotic system discovered by Sprott (1994), and known as the Sprott-A hidden attractor:

$$\begin{cases} \dot{x} = y \\ \dot{y} = -x + yz \\ \dot{z} = 1 - y^2 \end{cases} \tag{1}$$

where x, y, z are the state variables. It is well-known that Sprott-A hidden attractor (1) is a special case of the Nose-Hoover oscillator (Nosé 1991) and describes many natural phenomena (Posch et al. 1986). The Sprott-A hidden attractor (1) is a conservative system and it does not have attractors. A recent study by Jafari, Sprott and Nazarimehr describes many rare flows without any equilibria (Jafari et al. 1986).

Based on system (1), we design a novel chaotic system as follows:

$$\begin{cases} \dot{x} = ay \\ \dot{y} = -x - yz \\ \dot{z} = by^2 - cx - d \end{cases} \tag{2}$$

In system (2), a, b, c and d are constant parameters. It is clear that when $d \neq 0$, the system (2) does not have any equilibrium point. The system (2) is *chaotic* when the parameter values are taken as

$$a = 0.8, \quad b = 0.5, \quad c = 0.1, \quad d = 1 \tag{3}$$

Thus, for the chaotic case (3), the system (2) does not have any equilibrium point and hence it exhibits hidden attractor.

For numerical simulation, we take the initial conditions of the system (2) as (0.1, 0.1, 0.1). The system's dynamic behavior is investigated numerically by employing a fourth order Runge-Kutta algorithm. Figure 1 shows the phase portraits of the six-term novel chaotic system (2) with hidden attractor.

The dynamics behavior of the six-term novel 3D chaotic system with hidden attractor can be characterized with its Lyapunov exponents which are computed numerically by Wolf algorithm proposed in Ref (wolf et al. 1985). The Lyapunov exponents of the six-term novel 3D chaotic system (2) are found to be $LE_1 = 0.0344, LE_2 = 0$ and $LE_3 = -0.0400$ (Fig. 2a). The system has a relatively high Kaplan-Yorke dimension of $D_{KY} = 2.86$. For $b \leq 0.5$ a strange attractor is displayed as the system has one positive Lyapunov exponent and $b \geq 0.5$ is a transition to periodic behavior (Fig. 2b). In order to get detailed view of the six-term novel 3D chaotic system with hidden attractor (2), its dynamics behavior with respect to the bifurcation parameter b is investigated. A MATLAB program was written to obtain the bifurcation diagrams for novel 3D chaotic system (2) of Fig. 2c. For the chosen value of $b \leq 0.5$ the system displays the expected chaotic behavior and $b \geq 0.5$ a periodic behavior is presented. In addition, the Poincaré map of the system (2) in Fig. 2d also reflects properties of chaos.

3 Circuit Realization of the Six-Term Novel Chaotic System

In this section, we design an electronic circuit modeling of the six-term novel 3D chaotic system with hidden attractor. The electronic circuit in Fig. 3 has been designed following an approach based on operational amplifiers (Vaidyanathan and Volos 2016a, b, 2017) where the state variables x, y, z of the system (2) are associated with the voltages across the capacitors C_1, C_2 and C_3, respectively. By applying Kirchhoffs circuit laws in to the circuit in Fig. 3, we get its circuital equations as follows:

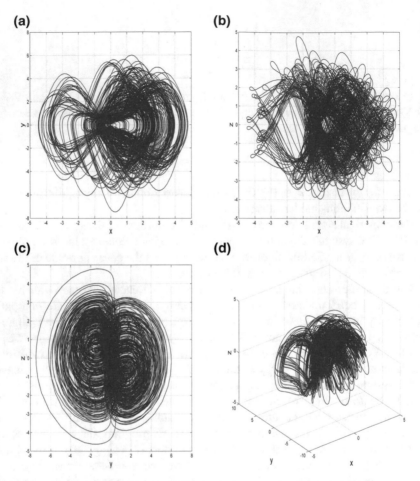

Fig. 1 Numerical simulation results of the novel chaotic system (2) with hidden attractor for $a = 0.8, b = 0.5, c = 0.1, d = 1$, in **a** $x - y$ plane, **b** $x - z$ plane, **c** $y - z$ plane, **d** $\mathbf{R}^3$

$$\begin{aligned}
\frac{dV_{C1}}{dt} &= \frac{1}{C_1 R_1} V_{C2} \\
\frac{dV_{C2}}{dt} &= -\frac{1}{C_2 R_2} V_{C1} - \frac{1}{10 C_2 R_3} V_{C2} V_{C3} \\
\frac{dV_{C3}}{dt} &= \frac{1}{10 C_3 R_4} V_{C2}^2 - \frac{1}{C_3 R_5} V_{C1} - \frac{1}{C_3 R_6} V_1
\end{aligned} \tag{4}$$

Based on known parameters of system (4), the values of the electronic components in Fig. 3 are selected as follows: $R_2 = R_7 = R_8 = 10$ KΩ, $R_1 = 12.5$ KΩ, $R_3 = 40$ KΩ, $R_4 = 80$ KΩ, $R_5 = 100$ KΩ, $R_6 = 2.5$ KΩ, $C_1 = C_2 = C_3 = C_4 = 10$ nF and $V_1 = 1 V_{DC}$.

The circuit has three integrators by using Op-amp TL082CD in a feedback loop and two multipliers IC AD633. We use the electronic simulation package MultiSIM to implement the proposed circuit. The obtained phase portraits are shown in Fig. 4.

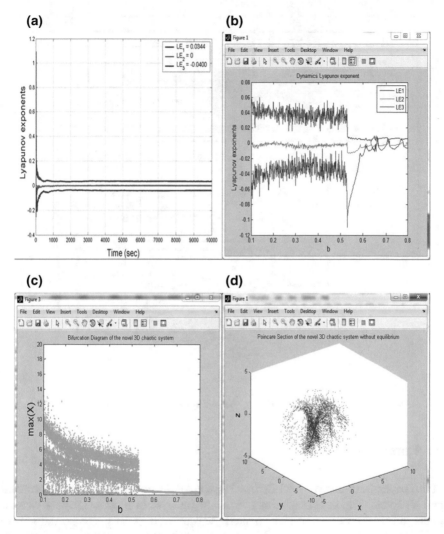

Fig. 2 Analysis of the nonlinear dynamics (2) using MATLAB 2010, for $a = 0.8, c = 0.1, d = 1$, in **a** Lyapunov exponents of the novel chaotic system for $b = 0.5$, **b** Lyapunov exponents versus the parameter control $b \in [0.1, 0.8]$ **c** Bifurcation diagram of x versus the control parameter $b \in [0.1, 0.8]$ **d** Poincaré map in the x-y-z space

A good agreement has been obtained between these circuital results and numerical simulation using MATLAB.

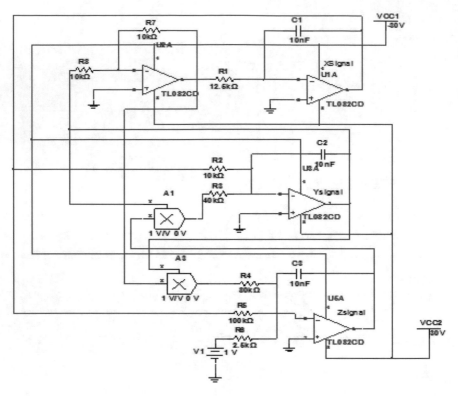

Fig. 3 Schematic of the proposed novel 3-D chaotic system by using MultiSIM 10

4 Conclusion

In this paper, a six-term novel 3-D chaotic system with hidden attractor is constructed and analyzed. The fundamental properties of the system such as Lyapunov exponents, bifurcation diagram and Poincaré map as well as its phase portraits were described in detail. By varying the value of the parameter b, the proposed system exhibits periodic and chaotic behaviors. An electronic circuit has been designed to realize the differential equations of the chaotic system proposed. Comparison of the numerical simulation MATLAB and designed circuit with MultiSIM, showed good qualitative agreement. Potential technological applications include robotics, encryption and random bit generator.

(a)

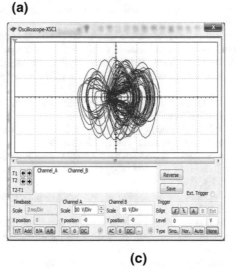

(b) **(c)**

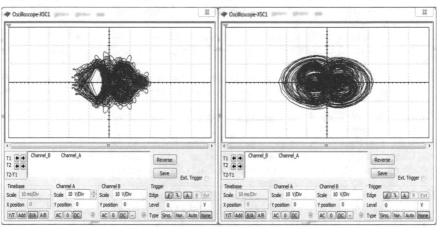

Fig. 4 Various projections of the six-term novel 3-D chaotic system with hidden attractor using MultiSIM 10.0, in **a** $x - y$ plane, **b** $x - z$ plane, and **c** $y - z$ plane

References

Abdulkareem M, Abduljaleel I (2013) Speech encryption using chaotic map and blowsh algorithms. J Basrah Res 39:68–76

Andreator A, Leros A (2013) Secure image encryption based on a Chua chaotic noise generator. J Eng Sci Technol Rev 6:90–103

Azar AT, Vaidyanathan S (2015) Chaos modeling and control systems design. Springer, Berlin, Germany

Azar AT, Vaidyanathan S (2016) Advances in chaos theory and intelligent control. Springer, Berlin, Germany

Azar AT, Vaidyanathan S (2017) Fractional order control and synchronization of chaotic systems. Springer, Berlin, Germany

Bouali S, Buscarino A, Fortuna L, Frasca M, Gambuzza L (2012) Emulating complex business cycles by using an electronic analogue. Nonlinear Anal Real World Appl 13:2459–2465

Chen G, Ueta T (1999) Yet another chaotic oscillator. Int J Bifurc Chaos 9:1465–1466

Gotthans T, Petrzela J (2015) New class of chaotic systems with circular equilibrium. Nonlinear Dyn 81:1143–1149

Gotthans T, Sprott J, Petrzela J (2016) Simple chaotic flow with circle and square equilibrium. Int J Bifurc Chaos 26, article ID 1650137

Islam M, Murase K (2005) Chaotic dynamics of a behavior-based miniature mobile robot: effects of environment and control structure. Neural Netw 18:123–144

Jafari S, Sprott J (2013) Simple chaotic flows with a line equilibrium. Chaos Solitons Fract 57:79–84

Jafari S, Sprott J, Nazarimehr F (1986) Canonical dynamics of the Nosé oscillator: stability, order, and chaos. Phys Rev A 33:4253–4265

Leonov G, Kuznetsov N, Vagaitsev V (2012) Hidden attractor in smooth Chua systems. Physica D Nonlinear Phenom 241:1482–1486

Leonov G, Kuznetsov N, Mokaev T (2015) Hidden attractor and homoclinic orbit in Lorenz-like system describing convective fluid motion in rotating cavity. Commun Nonlinear Sci Numer Simul 28:166–174

Li C, Sprott J (2014a) Chaotic flows with a single nonquadratic form. Phys Lett A 378:178–183

Li C, Sprott J (2014b) Coexisting hidden attractors in a 4-D simplified Lorenz system. Int J Bifurc Chaos 24, article ID 1450034

Li C, Sprott J (2016) Variable-boostable chaotic flows. Optik 127:10,389–10,398

Li C, Sprott J, Yuan Z, Li H (2015) Constructing chaotic systems with total amplitude control. Int J Bifurc Chaos 25, article ID 1530025

Li N, Pan W, Yan L, Luo B, Zou X (2014) Enchanced chaos synchronization and communication in cascade-coupled semiconductor ring lasers. Commun Nonlinear Sci Numer Simul 19:1874–1883

Li P, Zheng T, Li C, Wang X, Hu W (2016) A unique jerk system with hidden chaotic oscillation. Nonlinear Dynam 86:197–203

Lorenz E (1963) Deterministic nonperiodic flow. J Atmos Sci 20:130–141

LüJ., Chen G (2002) A new chaotic attractor coined. Int J Bifurc Chaos 12:659–661

Mada Sanjaya W, Mamat M, Salleh Z, Mohd I (2011) Bidirectional chaotic synchronization of Hindmarsh-Rose neuron model. Appl Math Sci 5:2685–2695

Mada Sanjaya W, Mohd IB, Mamat M, Salleh Z (2012) Mathematical model of three species food chain interaction with mixed functional response. Int J Modern Phys Conf Series 9:334–340

Malasoma J (2000) A new class of minimal chaotic flows. Phys Lett A 264:383–389

Nakajima K, Sawada Y (1979) Experimental studies on the weak coupling of oscillatory chemical reaction systems. J Chem Phys 72:2231–2234

Nosé S (1991) Constant temperature molecular dynamics methods. Progress Theor Phys Suppl 103:1–46

Pham V, Jafari S, Volos C, Giakoumis A, Vaidyanathan S, Kapitaniak T (2016a) A chaotic system with equilibria located on the rounded square loop and its circuit implementation. IEEE Trans Circuits Syst II Express Briefs 63:878–882

Pham V, Jafari S, Volos C, Vaidyanathan S, Kapitaniak T (2016b) A chaotic system with infinite equilibria located on a piecewise linear curve. Optik 127:9111–9117

Pham V, Jafari S, Wang X, Ma J (2016c) A chaotic system with different shapes of equilibria. Int J Bifurc Chaos 26, article ID 1650069

Pham V, Jafari S, Volos C (2017) A novel chaotic system with heart-shaped equilibrium and its circuital implementation. Optik 131:343–349

Posch HA, Hoover WG, Vesely FJ (1986) Canonical dynamics of the Nosé oscillator: stability, order, and chaos. Phys Rev A 33:4253–4265

Rössler O (1976) An equation for continuous chaos. Phys Lett A 57:397–398

Sambas A, Sanjaya W, Halimtaussadiyah E (2012) Unidirectional chaotic synchronization of Rossler circuit and its application for secure communication. WSEAS Trans Syst 9:506–515

Sambas A, Sanjaya W, Halimatussadiyah E (2013a) Design and analysis of bidirectional chaotic synchronization of Rossler circuit and its application for secure communication. Appl Math Sci 7:11–21

Sambas A, Sanjaya W, Mamat M (2013b) Design and numerical simulation of unidirectional chaotic synchronization and its application in secure communication system. J Eng Sci Technol Rev 6:66–73

Sambas A, Sanjaya W, Mamat M (2015a) Bidirectional coupling scheme of chaotic systems and its application in secure communication system. J Eng Sci Technol Rev 8:89–95

Sambas A, Sanjaya W, Mamat M, Salleh Z, Mohamad F (2015b) Secure communications based on the synchronization of the new Lorenz-like attractor circuit. Adv Stud Theor Phys 9:379–394

Sambas A, Sanjaya W, Mamat M, Prastio R (2016a) Mathematical modelling of chaotic jerk circuit and its application in secure communication system. In: Azar AT, Vaidyanathan S (eds) Advances in chaos theory and intelligent control, studies in fuzziness and soft computing, vol 337. Springer, Germany, pp 135–153

Sambas A, Vaidyanathan S, Mamat M, Sanjaya W, Rahayu D (2016b) A 3-D novel jerk chaotic system and its application in secure communication system and mobile robot navigation. In: Vaidyanathan S, Volos C (eds) Advances and applications in chaotic systems, studies in computational intelligence, vol 636. Springer, Germany, pp 283–310

Sprott J (1994) Some simple chaotic flows. Phys Lett E 50:647–650

Tacha O, Volos C, Kyprianidis I, Stouboulos I, Vaidyanathan S, Pham V (2016) Analysis, adaptive control and circuit simulation of a novel nonlinear finance system. Appl Math Comput 276:200–217

Vaidyanathan S, Volos C (2016a) Advances and applications in chaotic systems. Springer, Berlin, Germany

Vaidyanathan S, Volos C (2016b) Advances and applications in nonlinear control systems. Springer, Berlin, Germany

Vaidyanathan S, Volos C (2017) Advances in memristors. Memristive devices and systems. Springer, Berlin, Germany

Wolf A, Swift JB, Swinney HL, Vastano JA (1985) Determining Lyapunov exponents from a time series. Physica D: Nonlinear Phenomena, 16(3):285–317

Synchronization Phenomena in Coupled Dynamical Systems with Hidden Attractors

C. K. Volos, Viet-Thanh Pham, Ahmad Taher Azar, I. N. Stouboulos and I. M. Kyprianidis

Abstract Recently, Leonov and Kuznetsov introduced a new class of nonlinear dynamical systems, which is called systems with hidden attractors, in contrary to the well-known class of systems with self-excited attractors. In this class, dynamical systems with infinite number of equilibrium points, with stable equilibria, or without equilibrium are classified. Since then, the study of chaotic systems with hidden attractors has become an attractive research topic because this new class of dynamical systems could play an important role not only in theoretical problems but also in engineering applications. In this direction, the proposed chapter presents the bidirectional and unidirectional coupling schemes between two identical dynamical chaotic systems with no-equilibrium points. As it is observed, when the value of the coupling coefficient is increased in both coupling schemes, the coupled systems undergo a transition from desynchronization mode to complete synchronization. Also, the simulation results reveal the richness of the coupled system's dynamical behavior, especially in the bidirectional case, showing interesting nonlinear dynamics, with a transition between periodic, quasiperiodic and chaotic behavior as the coupling coefficient increases, as well as synchronization phenomena, such as complete and anti-phase synchronization. Various tools of nonlinear theory for the

C. K. Volos (✉) · I. N. Stouboulos · I. M. Kyprianidis
Physics Department, Aristotle University of Thessaloniki, 54124 Thessaloniki, Greece
e-mail: volos@physics.auth.gr

I. N. Stouboulos
e-mail: stouboulos@physics.auth.gr

I. M. Kyprianidis
e-mail: imkypr@auth.gr

V.-T. Pham
School of Electronics and Telecommunications, Hanoi University of Science and Technology, 01 Dai Co Viet, Hanoi, Vietnam
e-mail: pvt3010@gmail.com

A. T. Azar
Faculty of Computers and Information, Benha University, Benha, Egypt
e-mail: ahmad_t_azar@ieee.org

© Springer International Publishing AG 2018
V.-T. Pham et al. (eds.), *Nonlinear Dynamical Systems with Self-Excited and Hidden Attractors*, Studies in Systems, Decision and Control 133, https://doi.org/10.1007/978-3-319-71243-7_17

study of the proposed coupling method, such as bifurcation diagrams, phase portraits and Lyapunov exponents have been used.

Keywords Complete synchronization · Anti-phase synchronization
Chaos · Hidden attractors · Bifurcation diagram · Lyapunov exponent

1 Introduction

In the past three decades, the phenomenon of synchronization between coupled nonlinear systems and especially of systems with chaotic behavior has attracted the interest of the research community because it is an interesting phenomenon with a broad range of applications, such as in various complex physical, chemical and biological systems (Holstein-Rathlou et al. 2001; Mosekilde et al. 2002; Pikovsky et al. 2003; Szatmári and Chua 2008; Tognoli and Kelso 2009; Wang et al. 2009; Liu and Chen 2010), in secure and broadband communication system (Kocarev et al. 1992; Cuomo et al. 1993; Wu and Chua 1993; Feki et al. 2003; Sheng-Hai and Ke 2004; Dimitriev et al. 2006; Jafari et al. 2010) and in cryptography (Annovazzi-Lodi et al. 1997; Baptista 1998; Grassi and Mascolo 1999; Dachselt and Schwarz 2001; Klein et al. 2005; Alvarez and Li 2006; Volos et al. 2006; Banerjee 2010).

The concept of synchronization of two or more systems with chaotic behavior is the phenomenon in which the coupled systems can adjust a given of their motion property to a common behavior (equal trajectories or phase locking), due to forcing or coupling (Luo 2013). However, having two chaotic systems being synchronized, it is a major surprise, due to the exponential divergence of the nearby trajectories of the systems. Nevertheless, nowadays the phenomenon of synchronization of coupled chaotic oscillators is well-studied theoretically and proven experimentally (Ouannas et al. 2017a, b; Azar and Vaidyanathan 2015a, b, c, 2016; Vaidyanathan et al. 2015a, b, c, 2017a, b, c; Boulkroune et al. 2016a, b; Vaidyanathan and Azar 2015a, b, c, d, 2016a, b, c, d, e, f; Ouannas et al. 2016a, b).

Synchronization theory has begun studying in the 1980s and early 1990s by Fujisaka and Yamada (1983), Pikovsky (1984), Pecora and Carroll (1990). Onwards, a great number of research works based on synchronization of nonlinear systems has risen and many synchronization schemes depending on the nature of the coupling schemes and of the interacting systems have been presented. Complete or full chaotic synchronization (Maritan and Banavar 1994; Kyprianidis and Stouboulos 2003a, b; Woafo and Enjieu Kadji 2004; Kyprianidis et al. 2006a, 2008), phase synchronization (Dykman et al. 1991; Parlitz et al. 1996), lag synchronization (Rosenblum et al. 1997; Taherion and Lai 1999), generalized synchronization (Rulkov et al. 1995), antisynchronization (Kim et al. 2003; Liu et al. 2006), anti-phase synchronization (Cao and Lai 1998; Astakhov et al. 2000; Zhong et al. 2001; Blazejczuk-Okolewska et al. 2001; Kyprianidis et al. 2006b; Tsuji et al. 2007), projective synchronization (Mainieri and Rehacek 1999; Ouannas et al. 2017c), anticipating (Voss 2000), inverse lag synchronization (Li 2009) and fractional order synchronization (Tolba et al. 2017; Azar et al. 2017a, b; Pham et al. 2017c, d; Ouannas et al. 2017d, e, f, g, h, i, j, k) are the most interesting types of synchronization, which have been investigated numerically and experimentally by many research groups.

However, the most interesting and the most studied case of synchronization is the *Complete or Full synchronization*. In this case the interaction between two coupled identical nonlinear circuits leads to a perfect coincidence of their chaotic trajectories, i.e.

$$x_1(t) = x_2(t), \text{ as } t \to \infty. \tag{1}$$

Also, in 1998, another interesting type of synchronization between mutually coupled identical autonomous nonlinear systems was observed. In this new type of synchronization, which is called *Anti-phase synchronization*, each one of the uncoupled systems produces chaotic attractors (Wang et al. 2017). This synchronization phenomenon is observed when the coupled system is in a phase locked (periodic) state, depending on the coupling factor and it can be characterized by a π-phase delay. So, the periodic signals (x_1 and x_2) of each coupled circuits have a time lag τ, which is equal to $T/2$, where T is the period of the signals x_1 and x_2.

$$x_1(t) = x_2(t + \tau), \text{ where } \tau = T/2. \tag{2}$$

The anti-phase synchronization was also observed by Volos et al. (2013) in the case of two mutually coupled identical non-autonomous Duffing-type systems, which as it is known, have symmetry, because the transformation:

$$S: (x, y, t) \to (-x, -y, t + T/2) \tag{3}$$

leaves Duffing's system equations invariant.

It is well-known that chaotic dynamical systems exhibit high sensitivity on initial conditions or system's parameters and if they are identical and start from almost the same initial conditions, they follow trajectories which rapidly become uncorrelated. That is why many techniques exist to obtain chaotic synchronization. So, many of these techniques for coupling two or more nonlinear chaotic systems can be mainly divided into two classes: *unidirectional coupling* and *bidirectional* or *mutual coupling* (Gonzalez-Miranda 2004). In the first case, only the first system, the master system, drives the second one, the slave system, while in the second case, each system's dynamic behavior influences the dynamics of the other.

Recently, a great interest for dynamical systems with hidden attractors has been raised. The term *hidden attractor* is referred to the fact that in this class of systems the attractor is not associated with an unstable equilibrium and thus often remains undiscovered because it may occur in a small region of parameter space and with a small basin of attraction in the space of initial conditions (Kuznetsov et al. 2010; Leonov et al. 2011a, b, 2012; Pham et al. 2014a, b). In 2010, for the first time, a chaotic hidden attractor was discovered in the most well-known nonlinear circuit, in Chua's circuit, which is described by a three-dimensional dynamical system (Kuznetsov et al. 2010).

The problem of analyzing hidden oscillations arose for the first time in the second part of Hilbert's 16th problem (1900) for two-dimensional polynomial systems. The first nontrivial results were obtained in Bautin's works (Bautin 1939, 1952), which were devoted to constructing nested limit cycles in quadratic systems and showed the necessity of studying hidden oscillations for solving this problem. Later, in the middle of the 20th century, Kapranov studied (Kapranov 1956) the qualitative behavior of Phase-Locked Loop (PLL) systems, which are used in telecommunications and computer architectures, and estimated stability domains. In that work, Kapranov assumed that in PLL systems there were self-excited oscillations only. However, in 1961, (Gubar 1961) revealed a gap in Kapranov's work and showed analytically the possibility of the existence of hidden oscillations in two-dimensional system of PLL, thus, from a computational point of view, the system considered was globally stable, but, in fact, there was only a bounded domain of attraction.

Also, in the same period, the investigations of the widely known Markus-Yamabe (1960) and Kalman (1957) conjectures on absolute stability have led to the finding of hidden oscillations in automatic control systems with a unique stable stationary point and with a nonlinearity, which belongs to the sector of linear stability (Bernat and Llibre 1996; Fitts 1966; Leonov and Kuznetsov 2013).

Furthermore, systems with hidden attractors have received attention due to their practical and theoretical importance in other scientific branches, such as in mechanics (unexpected responses to perturbations in a structure like a bridge or in an airplane wing) (Lauvdal et al. 1997). So, the study of these systems is an interesting topic of a significant importance.

So, from the introduction of dynamical systems with hidden attractors a great number of systems belonging in this category has been reported. All these systems can be classified in three families of systems depending on the kind of systems' equilibria (Pham et al. 2017a). The first family is the systems without equilibrium points. The works of Nosé (1984) and Hoover (1985) in 1984–1985 have led the study of the aforementioned family of dynamical systems. Since then, many 3D or 4D dynamical systems of this family have been studied (Jafari et al. 2013; Wei 2011; Wang et al. 2012a; Wang and Chen 2013; Wei et al. 2014; Maaita et al. 2015; Tahir et al. 2015; Pham et al. 2016a, b; Wang et al. 2016; Zuo and Li 2016). The second family is the systems with stable equilibria (Wang and Chen 2012b; Molaie et al. 2013; Wei and Wang, 2013; Kingni et al. 2014; Lao et al. 2014; Pham et al. 2017b), while the third is the systems with an infinite number of equilibria (Jafari and Sprott 2013; Li and Sprott 2014; Gotthans and Petržela 2015; Gotthans et al. 2016; Pham et al. 2016c, d, e).

In the present chapter, the study of various synchronization phenomena between bidirectionally or unidirectionally coupled dynamical systems with hidden attractors is presented. For this reason, a no-equilibrium chaotic system, introduced by Pham et al. has been used (Pham et al. 2014c). Especially, in the case of the mutually coupled systems, except of the complete chaotic synchronization, the existence of anti-phase synchronization is also confirmed from the simulation results.

The rest of the chapter is organized as follows. Section 2 provides the mathematical model as well as the dynamics and properties of the proposed system with hidden attractors. Section 3 describes the coupling schemes of two identical no-equilibrium chaotic systems, while the simulation results of the coupled systems are thoroughly presented in Sect. 4. Finally, conclusions are drawn in Sect. 5

2 Description and Dynamics of the System Without Equilibrium

In 2013, Jafari and Sprott have introduced nine simple chaotic flows with a line equilibrium by using an exhaustive computer search (Jafari and Sprott 2013). These systems belong to the family of systems with hidden attractors because it is impossible to verify the chaotic attractor by choosing an arbitrary initial condition in the vicinity of the unstable equilibria.

As an example, the first of these systems, which is described by the following system

$$
\begin{cases}
\dot{x} = -y \\
\dot{y} = -x + yz \\
\dot{z} = -x - axy - bxz
\end{cases}
\tag{4}
$$

where a, b are real positive parameters, has a line of equilibria $E(0, 0, z)$.

In the third equation of system (5) (Pham et al. 2014c) added a real parameter c in order to obtain the following new system

$$
\begin{cases}
\dot{x} = -y \\
\dot{y} = -x + yz \\
\dot{z} = -x - axy - bxz + c
\end{cases}
\tag{5}
$$

which possesses no equilibrium points. So, it belongs to the family of dynamical systems without equilibrium.

Next, in order to discover system's (5) dynamics well-known tools of nonlinear theory, such as phase portrait, bifurcation diagram and Lyapunov spectrum, are used. For this reason the proposed system is integrated numerically using the classical fourth-order Runge-Kutta integration algorithm. For each set of parameters used in this work, the time step is always $\Delta t = 0.002$ and the calculations are

Fig. 1 **a** Bifurcation diagram
of system (5) for decreasing
values of a and **b** the graph of
the maximal Lyapunov
exponent plotted in the range
of $14 \leq a \leq 23$, with $b = 1$,
$c = 0.001$ and initial
conditions
$(x_0, y_0, z_0) = (0, 0.5, 0.5)$

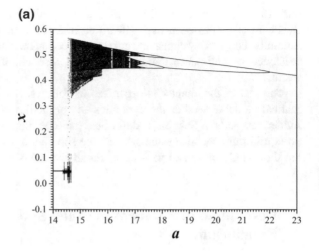

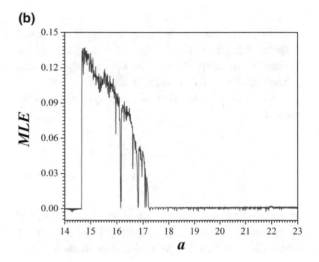

performed using variables and parameters in extended precision mode. For each
parameter settings, the system is integrated for a sufficiently long time and the
transient is discarded.

To study the type of scenario giving rise to chaos by considering the parameter
a in system (5), as the main control parameter, the bifurcation diagram in Fig. 1a is
obtained, while the other parameters remain fixed as $b = 1$ and $c = 0.001$ and the
initial conditions are chosen as $(x_0, y_0, z_0) = (0, 0.5, 0.5)$. The bifurcation diagram
is obtained by plotting the variable x when the trajectory cuts the plane $y = 0$ with
$dy/dt < 0$, as the control parameter a is decreased in tiny steps in the range of
$14 \leq a \leq 23$. From the bifurcation diagram of Fig. 1a it is possible to verify that
the system (5) is driven to chaos through a period-doubling route as the control

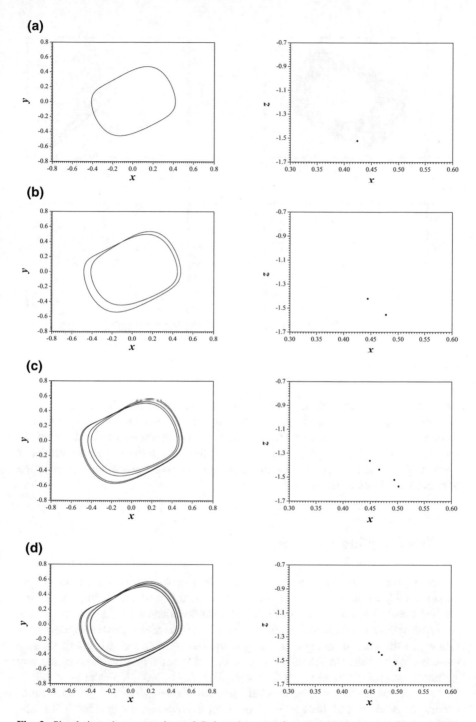

Fig. 2 Simulation phase portraits and Poincaré maps of system (5) for **a** $a = 23$ (period-1), **b** $a = 19$ (period-2), **c** $a = 17.5$ (period-4), **d** $a = 17.3$ (period-8), **e** $a = 15$ (chaos), **f** $a = 14.5$ (period-1), with $b = 1$, $c = 0.001$ and initial conditions $(x_0, y_0, z_0) = (0, 0.5, 0.5)$

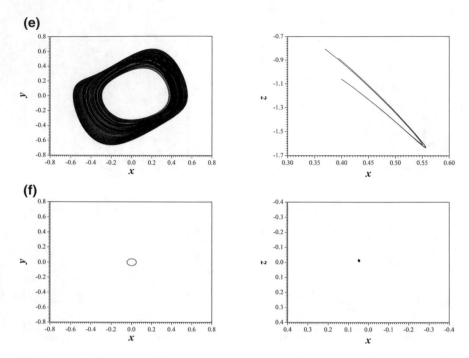

Fig. 2 (continued)

parameter is decreased and through a crisis is resulted to a period-1 steady state. Furthermore, the corresponding spectrum of the three Lyapunov exponents is shown in Fig. 1b. It can be seen that the bifurcation diagram well coincides with the spectrum of the Lyapunov exponents. Figure 2 depicts a series of phase portraits of y versus x and the respective Poincaré maps of z versus x, for various values of the parameter a, showing the route to chaos.

3   The Coupling Schemes

Generally, there are various methods of coupling between coupled nonlinear systems available in the literature. However, two are the most interesting. In the first method due to Pecora and Carroll (1990), a stable subsystem of a chaotic system could be synchronized with a separate chaotic system under certain suitable conditions. In the second method, chaos synchronization between two nonlinear systems is achieved due to the effect of coupling without requiring to construct any stable subsystem (Chua et al. 1992; Kyprianidis et al. 2005; Volos et al. 2006).

This second method can be divided into two classes: *drive-response* or *unidirectional coupling* and *bidirectional* or *mutual coupling*. In the first case, one system drives another one called the response or slave system. The system of two unidirectional coupled identical systems is described by the following set of differential equations:

$$\begin{cases} \dot{x}_1 = F(x_1) \\ \dot{x}_2 = F(x_2) + C(x_1 - x_2) \end{cases} \tag{6}$$

where $F(x)$ is a vector field in a phase space of dimension n and C a matrix of constants, which describes the nature and strength of the coupling between the oscillators. It is obvious from (6) that only the first system influences the dynamic behavior of the other.

In the second case, both the coupled systems are connected and each one influences the dynamics of the other. This is the reason for which this method is called mutual (or bidirectional). The coupled system of two mutually coupled chaotic oscillators is described by the following set of differential equations:

$$\begin{cases} \dot{x}_1 = F(x_1) + C(x_2 - x_1) \\ \dot{x}_2 = F(x_2) + C(x_1 - x_2) \end{cases} \tag{7}$$

In the last twenty years, many research groups approached the coupling methods between coupled chaotic systems, with the intention to study not only the cases of synchronization but also the various desynchronization phenomena. In this direction, the desynchronization in connection with a parameter mismatch between two coupled electronic oscillators has been studied (Astakhov et al. 1998). Furthermore, in (Yanchuk et al. 2001), the bifurcation sequence associated with desynchronization of a pair of coupled identical Rössler systems as the coupling parameter being reduced, has been followed. Starting with the transverse destabilization of a periodic orbit embedded in the fully synchronized chaotic state, this sequence proceeds via a torus bifurcation and regimes of anti-phase periodic and chaotic dynamics to asynchronous chaos.

4 Simulation Results

In this chapter, the study of the dynamic behavior of the bidirectionally and unidirectionally coupled systems with hidden attractors has been investigated numerically by employing the fourth order Runge-Kutta algorithm. Due to the fact that each one of the three system's variables and especially the variables y and z holds different order of nonlinearity the synchronization phenomena as well as the threshold for complete synchronization can be dependent on the selection of coupling variable. For this reason, in this work, the variable y has been preferred as the coupling variable because a great variety of phenomena can be observed.

So, the system of differential equations that describes the bidirectionally coupled systems' dynamics is:

$$\begin{cases} \dot{x}_1 = -y_1 \\ \dot{y}_1 = -x_1 + y_1 z_1 + \xi(y_2 - y_1) \\ \dot{z}_1 = -x_1 - ax_1 y_1 - bx_1 z_1 + c \\ \dot{x}_2 = -y_2 \\ \dot{y}_2 = -x_2 + y_2 z_2 + \xi(y_1 - y_2) \\ \dot{z}_2 = -x_2 - ax_2 y_2 - bx_2 z_2 + c \end{cases} \tag{8}$$

The first three equations of system (8) describe the first of the two coupled identical systems with hidden attractors, while the other three describe the second one. Also, the parameter ξ is the coupling coefficient and it is present in the equations of both systems, since the coupling between them is mutual.

In the case of unidirectionally coupled systems (5) the following system of differential equations is produced.

$$\begin{cases} \dot{x}_1 = -y_1 \\ \dot{y}_1 = -x_1 + y_1 z_1 \\ \dot{z}_1 = -x_1 - ax_1 y_1 - bx_1 z_1 + c \\ \dot{x}_2 = -y_2 \\ \dot{y}_2 = -x_2 + y_2 z_2 + \xi(y_1 - y_2) \\ \dot{z}_2 = -x_2 - ax_2 y_2 - bx_2 z_2 + c \end{cases} \tag{9}$$

The coupling coefficient ξ is present only in the second coupled system, since only the first system affects the dynamics of the second.

The parameters of the system are chosen as: $a = 15$, $b = 1$, $c = 0.001$. With these values each one of the coupled systems with hidden attractors are in chaotic mode.

So, by solving the coupled systems' Eqs. (8) and (9) the bifurcation diagrams of the signal's difference $(x_2 - x_1)$ versus the coupling factor ξ are produced. In details, these diagrams are produced by increasing the coupling factor ξ, from $\xi = 0$ (uncoupled systems) with step $\Delta\xi = 0.0002$, in two different ways. In the first, the initial conditions in each iteration have the same values $(x_{10}, y_{10}, z_{10}, x_{20}, y_{20}, z_{20}) = (0, 0.5, 0.5, 0.1, 0.6, 0.6)$, while in the second case the initial conditions in each iteration have different values. This occurs because the last values of the state variables in the previous iteration become the initial values for the next iteration. The second type of bifurcation diagram is more close to the experimental observation of coupled systems' dynamic behavior in many scientific fields, such as electronics, economy, biology etc.

4.1   Same Initial Conditions in Each Iteration

In the first case, as it is mentioned, the initial conditions have the same values in each iteration and the bifurcations diagrams in the cases of bidirectional and uni-directional coupling schemes have been produced (Figs. 3a and 10).

The bifurcation diagram of the bidirectionally coupling system (8) shows that the coupled system undergoes from full desynchronization, for $\xi < 0.048$, where each system is in a chaotic state and lays on its own manifold, to complete chaotic synchronization, for $\xi \geq 0.39$, where their manifolds coincide, through an inter-mediate region where the system shows a more complex dynamic behavior. This is a typical transition from full desynchronization to complete synchronization. Simulation phase portraits of x_2 versus x_1 of the bidirectionally coupled systems (8) are depicted in Fig. 4, for various values of the coupling coefficient.

Fig. 3 **a** Bifurcation diagram of $(x_2 - x_1)$ versus ξ and **b** the spectrum of Lyapunov exponents of the bidirectionally coupling system (8), with the same initial conditions in each iteration. The parameters are $a = 15$, $b = 1$, $c = 0.001$ and initial conditions $(x_{10}, y_{10}, z_{10}, x_{20}, y_{20}, z_{20}) = (0, 0.5, 0.5, 0.1, 0.6, 0.6)$

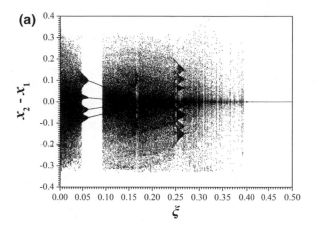

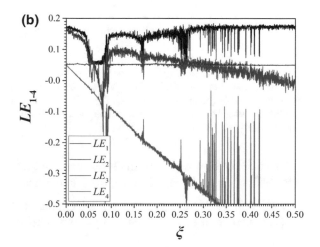

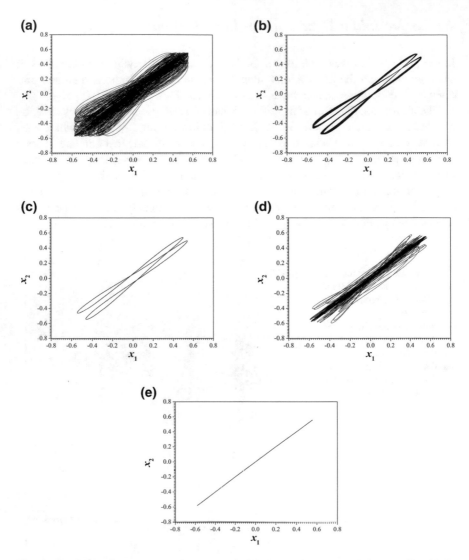

Fig. 4 Simulation phase portraits of x_2 versus x_1 of the bidirectionally coupled system (8) with the same initial conditions in each iteration, for **a** $\xi = 0.01$ (chaotic state), **b** $\xi = 0.055$ (quasiperiodic state), **c** $\xi = 0.075$ (period-4 steady state), **d** $\xi = 0.38$ (chaotic state), **e** $\xi = 0.5$ (complete chaotic synchronization). The parameters are $a = 15$, $b = 1$, $c = 0.001$ and initial conditions $(x_{10}, y_{10}, z_{10}, x_{20}, y_{20}, z_{20}) = (0, 0.5, 0.5, 0.1, 0.6, 0.6)$

The intermediate region of the bifurcation diagram of Fig. 3a is more complicated and it can be divided in three discrete regions:

- Region I: $0.048 < \xi \leq 0.063$ (Quasiperiodic state). This type of behavior is confirmed from the spectrum of Lyapunov exponents (Fig. 3b), which i.e. for

Fig. 5 Simulation phase portrait of $y_{1,2}$ versus $x_{1,2}$, for $\xi = 0.075$ (anti-phase synchronization)

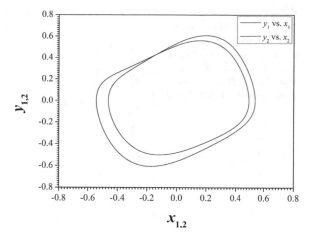

Fig. 6 Time-series of x_1, x_2, for $\xi = 0.075$

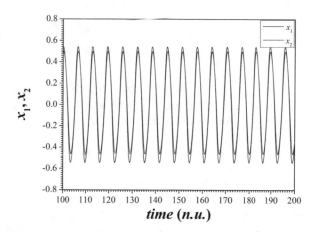

$\xi = 0.055$ are $LE_1 = 0.0094$, $LE_2 = 0.0063$, $LE_3 = -0.2398$, $LE_4 = -0.7223$, $LE_5 = -0.7590$, $LE_6 = -0.7894$.

- Region II: $0.063 < \xi \leq 0.093$ (Period-4 steady state). In this region the coupled system shows the phenomenon of anti-phase synchronization. This occurs because each one of the coupled circuits remains in the same periodic state. Figures 5 show the simulation phase portraits of $y_{1,2}$ versus $x_{1,2}$, for $\xi = 0.075$, respectively. In this figure the coincidence of circuits' attractors in the phase plain is presented. Furthermore, in Fig. 6, the time-series of the state variables x_1 and x_2 of the coupled circuits are shown. It is obvious that the two signals x_1 and x_2 are identical with a time lag.

To quantify this time lag we have used the well-known *Similarity Function S* (Rosenblum et al. 1997).

Fig. 7 The similarity
function (S) versus time (t),
for $\xi = 0.075$. $S_{min} = 0$
means lag with time shift of
$\tau_{min} = 6.39 = T/2$. So, the
phenomenon of anti-phase
synchronization is confirmed

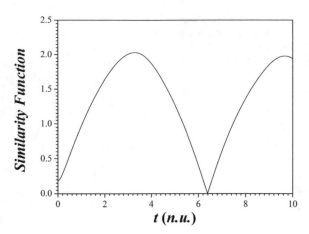

$$S(t) = \sqrt{\frac{\langle[x_2(t+\tau)-x_1(t)]^2\rangle}{\left[\langle(x_1(t))^2\rangle \cdot \langle(x_2(t))^2\rangle\right]^{1/2}}} \tag{11}$$

Let S_{min} be the minimum value of the Similarity function $S(\tau)$ and let τ_{min} be the amount of time lag, when S_{min} is achieved. The time lag τ_{min} between the variables x_1 and x_2 is found, when the conditions $S_{min} = 0$ and $\tau_{min} \neq 0$ are fulfilled. The calculation of the similarity function for $\xi = 0.075$ (Fig. 7) shows that the expected time lag $\tau_{min} = 6.39$ n.u., is equal to $T/2$, where T is the period of x_1 and x_2.

Furthermore, the same time lag is found for every value of coupling coefficient (ξ) in the Region II. So, the value of time lag remains always the same in this region and equals to the half of the period of the external voltage source. Moreover the fact that the difference of $[x_1(t) - x_2(t + T/2)]$ is equal to zero (Fig. 8), confirms that the coupled system demonstrates π phase delay, which is defined as anti-phase

Fig. 8 Time-series of
$x_1(t) - x_2(t + T/2)$, for
$\xi = 0.075$

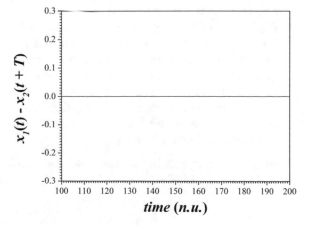

Fig. 9 Time-series of $(x_1 - x_2)$, for $\xi = 0.38$

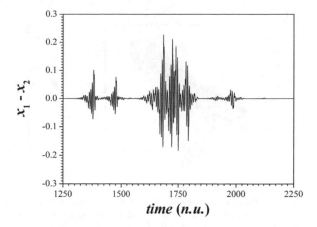

Fig. 10 a Bifurcation diagram of $(x_2 - x_1)$ versus ξ of the unidirectionally coupled system (9), with the same initial conditions in each iteration. The parameters are $a = 15$, $b = 1$, $c = 0.001$ and initial conditions $(x_{10}, y_{10}, z_{10}, x_{20}, y_{20}, z_{20}) = (0, 0.5, 0.5, 0.1, 0.6, 0.6)$

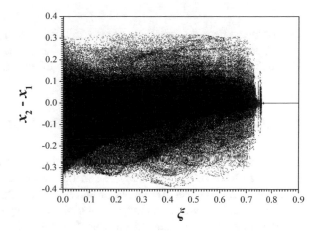

synchronization or π-lag synchronization. Finally, Fig. 9 shows the time-series of $(x_1 - x_2)$ in the case of system's intermittent behavior for $\xi = 0.38$.

- Region III: $0.093 < \xi \leq 0.39$ (Hyperchaotic state). This type of behavior is confirmed from the two positive Lyapunov exponents in Fig. 3b. For example the Lyapunov exponents for this type of behavior, for a value of the coupling coefficient $\xi = 0.2$, are $LE_1 = 0.113$, $LE_2 = 0.0521$, $LE_3 = 0$, $LE_4 = -0.2788$, $LE_5 = -0.7416$, $LE_6 = -0.8867$. Especially, in the region $0.31 < \xi \leq 0.39$ the system has an intermittent behavior as it is observed from the time-series of $x_1 - x_2$ for $\xi = 0.38$.

The bifurcation diagram of Fig. 10, in the case of unidirectionally coupling system (9), shows that the coupled system undergoes from full desynchronization, for $\xi < 0.76$ (Fig. 11a) directly to complete chaotic synchronization (Fig. 11b).

(a)
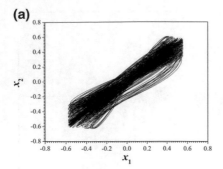

(b)
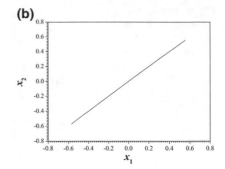

Fig. 11 Simulation phase portraits of x_2 versus x_1 of the unidirectionally coupled system (9) with the same initial conditions in each iteration, for **a** $\xi = 0.4$ (chaotic state) and **b** $\xi = 0.8$ (complete chaotic synchronization). The parameters are $a = 15$, $b = 1$, $c = 0.001$ and initial conditions $(x_{10}, y_{10}, z_{10}, x_{20}, y_{20}, z_{20}) = (0, 0.5, 0.5, 0.1, 0.6, 0.6)$

This occurred because only the first system affects the dynamics of the second. So, there is no any complex behavior and the value of the synchronization threshold ($\xi = 0.76$) is significant higher than in the case of bidirectional coupling ($\xi = 0.39$).

4.2 Different Initial Conditions in Each Iteration

In the second case of study, the initial conditions have different values in each iteration and the bifurcation diagrams in the cases of bidirectional and unidirectional coupling schemes have been produced (Figs. 12a and 14).

The bifurcation diagram in the case of bidirectionally coupling scheme (8) shows that the coupled system undergoes from full desynchronization, for $\xi < 0.048$, to complete chaotic synchronization, for $\xi \geq 0.277$, through an intermediate region where the system shows a more complex dynamic behavior than in the respective case of bidirectional coupling of the previous case. Simulation phase portraits of x_2 versus x_1 of the bidirectionally coupled systems (8) are depicted in Fig. 13, for various values of the coupling coefficient.

In the intermediate region of the bifurcation diagram of Fig. 12a, the coupled system can be characterized by three different dynamical behavior:

- *Quasiperiodic state*. This type of behavior is observed in three different distinct regions ($\xi \in (0.054, 0.061]$, $\xi \in (0.1670, 0.1684]$ and $\xi \in (0.2620, 0.2664]$) and is confirmed by the spectrum of Lyapunov exponents of Fig. 12b.
- *Periodic state*. In the following five regions of the bifurcation diagram of Fig. 13 the system is in a periodic state. In more details:

  1. For $\xi \in (0.0491, 0.0518]$ the system is in a period-12 steady state.
  2. For $\xi \in (0.061, 0.093]$ the system is in a period-4 steady state.

Fig. 12 **a** Bifurcation diagram of $(x_2 - x_1)$ versus ξ and **b** the spectrum of Lyapunov exponents of the bidirectionally coupling system (9), with different initial conditions in each iteration. The parameters are $a = 15$, $b = 1$, $c = 0.001$

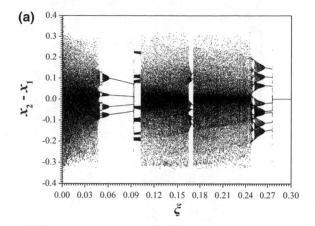

(a)

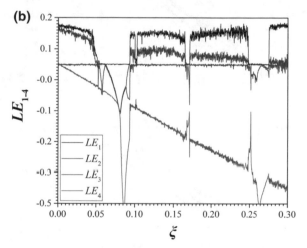

(b)

3. For $\xi \in (0.1684, 0.1710]$ the system is in a period-8 steady state.
4. For $\xi \in (0.250, 0.252]$ the system is in a period-22 steady state.
5. For $\xi \in (0.2664, 0.2760]$ the system is in a period-8 steady state.

In all these windows of periodic behavior the coupled system shows the phenomenon of anti-phase synchronization. By calculating the Similarity function $S(\tau)$ in each case we find that the expected time lag τ_{min} is equal to $T/2$, where T is the period of x_1 and x_2.

- *Hyperchaotic state.* In the rest of this intermediate region the system displays an hyperchaotic behavior, as it is observed from the respective phase portraits of Fig. 13a, f, and i, as well as from the spectrum of the Lyapunov exponents of Fig. 12b.

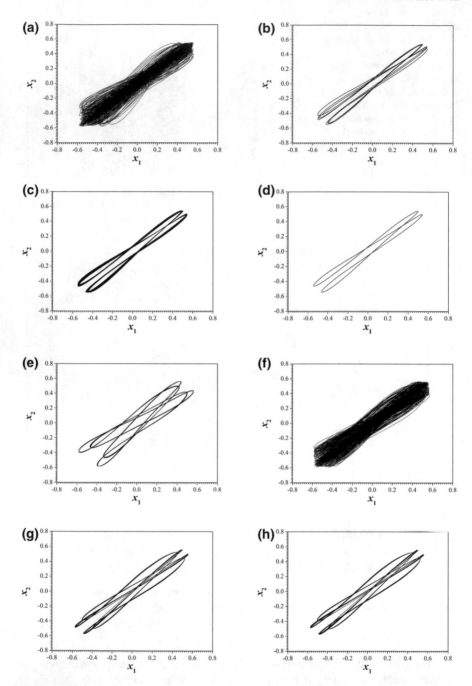

Fig. 13 Simulation phase portraits of x_2 versus x_1 of the bidirectionally coupled system (9) with different initial conditions in each iteration, for **a** $\xi = 0.03$ (hypechaotic state), **b** $\xi = 0.05$ (periodic state), **c** $\xi = 0.055$ (quasiperiodic state), **d** $\xi = 0.075$ (periodic state), **e** $\xi = 0.10$ (hyperchaotic state), **f** $\xi = 0.14$ (hyperchaotic state), **g** $\xi = 0.1678$ (quasiperiodic state), **h** $\xi = 0.17$ (periodic state), **i** $\xi = 0.2$ (chaotic state), **j** $\xi = 0.251$ (periodic state) **k** $\xi = 0.263$ (quasiperiodic state), **l** $\xi = 0.27$ (periodic state), **m** $\xi = 0.28$ (complete chaotic synchronization). The parameters are $a = 15$, $b = 1$, $c = 0.001$

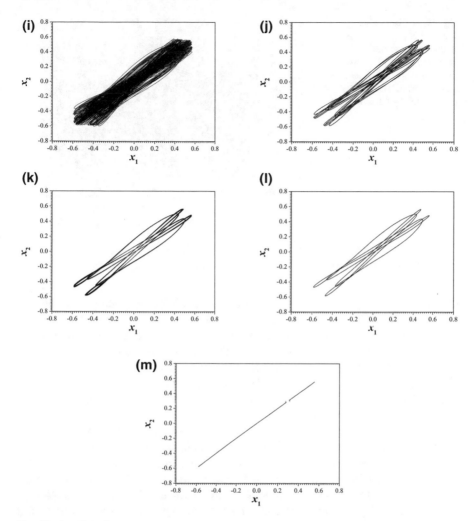

Fig. 13 (continued)

Finally, from the bifurcation diagram (Fig. 14) in the case of unidirectionally coupling system (9) we can conclude that the coupled system undergoes from full desynchronization, for $\xi < 0.55$ directly to complete chaotic synchronization, without appearing any complex dynamical behavior, while the value of the synchronization threshold ($\xi = 0.55$) is significant higher than in the case of bidirectional coupling ($\xi = 0.\ 0.277$).

Fig. 14 a Bifurcation
diagram of $(x_2 - x_1)$ versus ξ
of the unidirectionally
coupled system (10), with
different initial conditions in
each iteration. The parameters
are $a = 15$, $b = 1$, $c = 0.001$

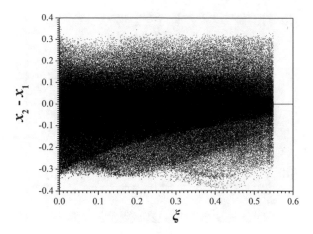

5 Conclusion

In the present chapter, a gallery of various synchronization phenomena between
resistively coupled identical nonlinear systems with hidden attractors was presented. For this reason, two coupling schemes were adopted. The first one was the
well-known bidirectional coupling while the second one was the unidirectional
coupling. In each coupling scheme two different study cases related with systems'
initial conditions were also adopted. The initial conditions in each iteration had the
same values in the first case, while in the second one the initial conditions in each
iteration had different values.

In more details, in the bidirectional coupling scheme, with the same initial
conditions in each iteration, the coupled systems undergone from full desynchronization, where each system was in a chaotic state to complete chaotic synchronization, through an intermediate region where the coupled systems were in a
periodic states showing the phenomenon of anti-phase synchronization. In the case
of unidirectionally coupled systems, the coupled system undergone from full
desynchronization directly to complete chaotic synchronization, without showing
any other complex dynamics.

Similarly, the coupling schemes (bidirectional and unidirectional), with different
initial conditions in each iteration, appeared the same route from desynchronization
to complete chaotic synchronization. However, in the bidirectional coupling
scheme, a more complex dynamics was arisen as the system had more periodic
windows where the phenomenon of anti-phase synchronization was presented.

As a future work, a more exhaustive study of coupling schemes between
identical dynamical systems with other types of hidden attractors will be done.

References

Alvarez G, Li S (2006) Some basic cryptographic requirements for chaos-based cryptosystems. Int J Bifurcat Chaos 16(08):2129–2151

Annovazzi-Lodi V, Donati S, Sciré A (1997) Synchronization of chaotic lasers by optical feedback for cryptographic applications. IEEE J Quant Electron 33(9):1449–1454

Astakhov V, Hasler M, Kapitaniak T, Shabunin A, Anishchenko V (1998) Effect of parameter mismatch on the mechanisms of chaos synchronization loss in coupled systems. Phys Rev E 58:5620–5628

Astakhov V, Shabunin A, Anishchenko V (2000) Antiphase synchronization in symmetrically coupled self-oscillators. Int J Bifurcat Chaos 10:849–857

Azar AT, Vaidyanathan S (2015a) Chaos modeling and control systems design, studies in computational intelligence, vol 581. Springer, Germany

Azar AT, Vaidyanathan S (2015b) Handbook of research on advanced intelligent control engineering and automation. Advances in computational intelligence and robotics (ACIR), Book Series. *IGI Global*, USA

Azar AT, Vaidyanathan S (2015c) Computational intelligence applications in modeling and control. In: Studies in computational intelligence, vol 575. Springer, Germany

Azar AT, Vaidyanathan S (2016) Advances in chaos theory and intelligent control. In: Studies in fuzziness and soft computing, vol 337. Springer, Germany

Azar AT, Vaidyanathan S, Ouannas A (2017a) Fractional order control and synchronization of chaotic systems. In: Studies in computational intelligence, vol 688. Springer, Germany

Azar AT, Volos C, Gerodimos NA, Tombras GS, Pham VT, Radwan AG, Vaidyanathan S, Ouannas A, Munoz-Pacheco JM (2017b) A novel chaotic system without equilibrium: dynamics, synchronization and circuit realization. Complexity, 2017: Article ID 7871467, 11 pages. https://doi.org/10.1155/2017/7871467

Banerjee S (2010) Chaos synchronization and cryptography for secure communications: applications for encryption: applications for encryption. IGI Global, USA

Baptista MS (1998) Cryptography with chaos. Phys Lett A 240(1–2):50–54

Bautin NN (1939) On the number of limit cycles generated on varying the coefficients from a focus or centre type equilibrium state. Dokl Akad Nauk SSSR 24:668–671

Bautin NN (1952) On the number of limit cycles appearing on varying the coefficients from a focus or centre type of equilibrium state. Mat Sb (N.S.) 30:181–196

Bernat J, Llibre J (1996) Counter example to Kalman and Markus-Yamabe conjectures in dimension larger than 3. Dyn Contin Discret Impul Syst 2:337–379

Blazejczuk-Okolewska B, Brindley J, Czolczynski K, Kapitaniak T (2001) Antiphase synchronization of chaos by noncontinuous coupling: two impacting oscillators. Chaos Solit Fract 2:1823–1826

Boulkroune A, Bouzeriba A, Bouden T, Azar AT (2016a). Fuzzy adaptive synchronization of uncertain fractional-order chaotic systems. In: Studies in fuzziness and soft computing, vol 337, pp 681–697. Springer, Germany

Boulkroune A, Hamel S, Azar AT (2016b). Fuzzy control-based function synchronization of unknown chaotic systems with dead-zone input. In: Studies in fuzziness and soft computing, vol 337, pp 699–718. Springer, Germany

Cao LY, Lai YC (1998) Antiphase synchronism in chaotic system. Phys Rev 58:382–386

Chua LO, Kocarev L, Eckert K, Itoh M (1992) Experimental chaos synchronization in Chua's crcuit. Int J Bifurcat Chaos 2:705–708

Cuomo KM, Oppenheim AV, Strogatz SH (1993) Synchronization of Lorenz-based chaotic circuits with applications to communications. IEEE Trans Circuits Syst II Analog Digit Signal 40(10):626–633

Dachselt F, Schwarz W (2001) Chaos and cryptography. IEEE Trans Circuits Syst I Fundam Theory, 48(12):1498–1509

Dimitriev AS, Kletsovi AV, Laktushkin AM, Panas AI, Starkov SO (2006) Ultrawideband wireless communications based on dynamic chaos. J Commun Technol Electron 51:1126–1140

Dykman GI, Landa PS, Neymark YI (1991) Synchronizing the chaotic oscillations by external force. Chaos Solit Fract 1:339–353

Feki M, Robert B, Gelle G, Colas M (2003) Secure digital communication using discrete-time chaos synchronization. Chaos Solit Fract 18(4):881–890

Fitts RE (1966) Two counter examples to Aizerman's conjecture. Trans IEEE, AC-11:553–556

Fujisaka H, Yamada T (1983) Stability theory of synchronized motion in coupled-oscillator systems. Prog Theor Phys 69:32–47

Gonzalez-Miranda JM (2004) Synchronization and control of chaos. Imperial College Press, London

Gotthans T, Petržela J (2015) New class of chaotic systems with circular equilibrium. Nonlinear Dyn 73:429–436

Gotthans T, Sportt JC, Petržela J (2016) Simple chaotic flow with circle and square equilibrium. Int J Bifurcat Chaos 26(1650):137–138

Grassi G, Mascolo S (1999) Synchronization of high-order oscillators by observer design with application to hyperchaos-based cryptography. Int J Circuit Theor Appl 27:543–553

Gubar' NA (1961) Investigation of a piecewise linear dynamical system with three parameters. J Appl Math Mech, 25:1011–1023

Holstein-Rathlou NH, Yip KP, Sosnovtseva OV, Mosekilde E (2001) Synchronization phenomena in nephron-nephron interaction. Chaos 11:417–426

Hoover W (1985) Canonical dynamics: equilibrium phase-space distributions. Phys Rev A 31:1695

Jafari S, Haeri M, Tavazoei MS (2010) Experimental study of a chaos-based communication system in the presence of unknown transmission delay. Int J Circuit Theor Appl 38:1013–1025

Jafari S, Sprott J (2013) Simple chaotic flows with a line equilibrium. Chaos Solit Fract 57:79–84

Jafari S, Sprott J, Golpayegani SMRH (2013) Elementary quadratic chaotic flows with no equilibria. Phys Lett A 377:699–702

Kalman RE (1957) Physical and mathematical mechanisms of instability in nonlinear automatic control systems. Trans ASME 79:553–566

Kapranov M (1956) Locking band for phase-locked loop. Radiofizika 2:37–52

Kim CM, Rim S, Kye WH, Rye JW, Park YJ (2003) Anti-synchronization of chaotic oscillators. Phys Lett A 320:39–46

Kingni ST, Jafari S, Simo H, Woafo P (2014) Three-dimensional chaotic autonomous system with only one stable equilibrium: Analysis, circuit design, parameter estimation, control, synchronization and its fractional-order form. Eur Phys J Plus 129:76

Klein E, Mislovaty R, Kanter I, Kinzel W (2005) Public-channel cryptography using chaos synchronization. Phys Rev E 72(1):016214

Kocarev L, Halle KS, Eckert K, Chua LO, Parlitz U (1992) Experimental demonstration of secure communications via chaotic synchronization. Int J Bifurcat Chaos 2(03):709–713

Kuznetsov NV, Leonov GA, Vagaitsev VI (2010) Analytical-numerical method for attractor localization of generalized Chua's system. IFAC Proc 4(1):29–33

Kyprianidis IM, Stouboulos IN (2003a) Synchronization of two resistively coupled nonautonomous and hyperchaotic oscillators. Chaos Solit Fract 17:314–325

Kyprianidis IM, Stouboulos IN (2003b) Synchronization of three coupled oscillators with ring connection. Chaos Solit Fract 17:327–336

Kyprianidis IM, Volos ChK, Stouboulos IN (2005) Suppression of chaos by linear resistive coupling. WSEAS Trans Circuits Syst 4:527–534

Kyprianidis IM, Volos ChK, Stouboulos IN, Hadjidemetriou J (2006a) Dynamics of two resistively coupled Duffing-type electrical oscillators. Int J Bifurcat Chaos 16:1765–1775

Kyprianidis IM, Bogiatzi AN, Papadopoulou M, Stouboulos IN, Bogiatzis GN, Bountis T (2006b) Synchronizing chaotic attractors of Chua's canonical circuit. The case of uncertainty in chaos synchronization. Int J Bifurcat Chaos 16:1961–1976

Kyprianidis IM, Volos CK, Stouboulos IN (2008) Experimental synchronization of two resistively coupled Duffing-type circuits. Nonlin Phenom Complex Syst 11:187–192

Lauvdal T, Murray R, Fossen T (1997) Stabilization of integrator chains in the presence of magnitude and rate saturations: a gain scheduling approach. IEEE Control Decis Conf, 4004–4005

Lao SK, Shekofteh Y, Jafari S, Sprott JC (2014) Cost function based on Gaussian mixture model for parameter estimation of a chaotic circuit with a hidden attractor. Int J Bifurcat Chaos 24(1450):010

Leonov G, Kuznetsov N, Vagaitsev V (2011a) Localization of hidden Chua's attractors. Phys Lett A 375:2230–2233

Leonov G, Kuznetsov N, Kuznetsova O, Seldedzhi S, Vagaitsev V (2011b) Hidden oscillations in dynamical systems. Trans Syst Control 6:54–67

Leonov G, Kuznetsov N, Vagaitsev V (2012) Hidden attractor in smooth Chua system. Physica D 241:1482–1486

Leonov G, Kuznetsov NV (2013) Analytical-numerical methods for hidden attractors' localization: the 16th Hilbert problem, Aizerman and Kalman conjectures, and Chua circuits. In: Numerical methods for differential equations, optimization, and technological problems, computational methods in applied sciences, vol 27, pp 41–64. Springer

Li GH (2009) Inverse lag synchronization in chaotic systems. Chaos Solit Fract 40:1076–1080

Li C, Sprott JC (2014) Chaotic flows with a single non quadratic term. Phys Lett A 378:178–183

Liu W, Qian X, Yang J, Xiao J (2006) Antisynchronization in coupled chaotic oscillators. Phys Lett A 354:119–125

Liu X, Chen T (2010) Synchronization of identical neural networks and other systems with an adaptive coupling strength. Int J Circ Theor Appl 38:631–648

Luo CJ (2013) Dynamical system synchronization. Springer, New York

Maaita JO, Volos CK, Stouboulos IN, Kyprianidis IM (2015) The dynamics of a cubic nonlinear system with no equilibrium point. J Nonlinear Dyn 2015:257923

Mainieri R, Rehacek J (1999) Projective synchronization in three-dimensional chaotic system. Phys Rev Lett 82:3042–3045

Maritan A, Banavar J (1994) Chaos noise and synchronization. Phys Rev Lett 72:1451–1454

Markus L, Yamabe H (1960) Global stability criteria for differential systems. Osaka Math J 12:305–317

Molaie M, Jafari S, Sprott JC, Golpayegani SMRH (2013) Simple chaotic flows with one stable equilibrium. Int J Bifurcat Chaos 23:1350

Mosekilde E, Maistrenko Y, Postnov D (2002) Chaotic synchronization: applications to living systems. World Scientific, Singapore

Nosé S (1984) A molecular dynamics method for simulations in the canonical ensemble. Mol Phys 52:255–268

Ouannas A, Azar AT, Abu-Saris R (2016a) A new type of hybrid synchronization between arbitrary hyperchaotic maps. Int J Mach Learn Cybernet. https://doi.org/10.1007/s13042-016-0566-3

Ouannas A, Azar AT, Radwan AG (2016b) On inverse problem of generalized synchronization between different dimensional integer-order and fractional-order chaotic systems. In: The 28th International Conference on Microelectronics, December 17–20, 2016. Cairo, Egypt

Ouannas A, Azar AT, Vaidyanathan S (2017a) On a simple approach for Q-S synchronization of chaotic dynamical systems in continuous-time. Int J Comput Sci Math 8(1):20–27

Ouannas A, Azar AT, Vaidyanathan S (2017b) New hybrid synchronization schemes based on coexistence of various types of synchronization between master-slave hyperchaotic systems. Int J Comput Appl Technol 55(2):112–120

Ouannas A, Azar AT, Ziar T (2017c) On inverse full state hybrid function projective synchronization for continuous-time chaotic dynamical systems with arbitrary dimensions. Diff Eq Dyn Syst. https://doi.org/10.1007/s12591-017-0362-x

Ouannas A, Azar AT, Ziar T, Vaidyanathan S (2017d) On new fractional inverse matrix projective synchronization schemes. In: Studies in computational intelligence, vol 688, pp 497–524. Springer, Germany

Ouannas A, Azar AT, Ziar T, Vaidyanathan S (2017e) Fractional inverse generalized chaos synchronization between different dimensional systems. In: Studies in computational intelligence, vol 688, pp 525–551. Springer, Germany

Ouannas A, Azar AT, Ziar T, Vaidyanathan S (2017f) A new method to synchronize fractional chaotic systems with different dimensions. In: Studies in computational intelligence, vol 688, pp 581–611. Springer, Germany

Ouannas A, Azar AT, Ziar T, Radwan AG (2017g) Study on coexistence of different types of synchronization between different dimensional fractional chaotic systems. In: Studies in computational intelligence, vol 688, pp 637–669. Springer, Germany

Ouannas A, Azar AT, Ziar T, Radwan AG (2017h) Generalized synchronization of different dimensional integer-order and fractional order chaotic systems. In: Studies in computational intelligence, vol 688, pp 671–697. Springer, Germany

Ouannas A, Azar AT, Vaidyanathan S (2017i) A robust method for new fractional hybrid chaos synchronization. Math Methods Appl Sci 40(5):1804–1812

Ouannas A, Azar AT, Vaidyanathan S (2017k) A new fractional hybrid chaos synchronization. Int J Model Identif Control, 27(4):314–322

Parlitz U, Junge L, Lauterborn W, Kocarev L (1996) Experimental observation of phase synchronization. Phys Rev E 54:2115–2217

Pecora LM, Carroll TL (1990) Synchronization in chaotic systems. Phys Rev Lett 64:521–524

Pham V-T, Volos CK, Jafari S, Wang X, Vaidyanathan S (2014a) Hidden hyperchaotic attractor in a novel simple memristive neural network. J Optoelectron Adv Mater Rapid Commun 8(11–12):1157–1163

Pham V-T, Jafari S, Volos CK, Wang X, Syed Golpayegani MRH (2014b) Is that really hidden? The presence of complex fixed-points in chaotic flows with no equilibria. Int J Bifurcat Chaos, 24(11):1450146

Pham VT, Volos C, Jafari S, Wei Z, Wang X (2014c) Constructing a novel no-equilibrium chaotic system. Int J Bifurcat Chaos 24(05):1450073

Pham V-T, Vaidyanathan S, Volos CK, Jafari S (2016a) A no-equilibrium hyperchaotic system with a cubic nonlinear term. Optik 127:3259–3265

Pham V-T, Vaidyanathan S, Volos CK, Jafari S, Kuznetsov NV, Hoang TM (2016b) A novel memristive time-delay chaotic system without equilibrium points. Eur Phys J Special Topics 225:127–136

Pham VT, Jafari S, Wang X, Ma J (2016c) A chaotic system with different shapes of equilibria. Int J Bifurcat Chaos 26:1650069

Pham V-T, Jafari S, Volos C, Giakoumis A, Vaidyanathan S, Kapitaniak T (2016d) A chaotic system with equilibria located on the rounded square loop and its circuit implementation. IEEE Trans Circuits Syst II Express Briefs 63(9):878–882

Pham V-T, Jafari S, Volos C, Vaidyanathan S, Kapitaniak T (2016e) A chaotic system with infinite equilibria located on a piecewise linear curve. Optik 127:9111–9117

Pham V-T, Volos C, Kapitaniak T (2017a) Systems with hidden attractors: from theory to realization in circuits. Springer, Switzerland

Pham V-T, Jafari S, Kapitaniak T, Volos C, Kingni ST (2017b) Generating a chaotic system with one stable equilibrium. Int J Bifurcat Chaos 27(4):1750053

Pham VT, Vaidyanathan S, Volos CK, Azar AT, Hoang TM, Yem VV (2017c) A three-dimensional no-equilibrium chaotic system: analysis, synchronization and its fractional order form. In: Studies in computational intelligence, vol 688, pp 449–470. Springer, Germany

Pham VT, Volos CK, Vaidyanathan S, Azar AT (2017d) Dynamics, synchronization and fractional order form of a chaotic system without equilibrium. In: Volos CK (ed) Nonlinear systems: design, applications and analysis

Pikovsky AS (1984) On the interaction of strange attractors. Z Phys B Condensed Matter 55: 149–154

Pikovsky AS, Rosenblum M, Kurths J (2003) Synchronization: a universal concept in nonlinear sciences. Cambridge University Press, Cambridge

Rosenblum MG, Pikovsky AS, Kurths J (1997) From phase to lag synchronization in coupled chaotic oscillators. Phys Rev Lett 78:4193–4196

Rulkov NF, Sushchik MM, Tsimring LS, Abarbanel HDI (1995) Generalized synchronization of chaos in directionally coupled chaotic systems. Phys Rev E 51:980–994

Sheng-Hai Z, Ke S (2004) Synchronization of chaotic erbium-doped fibre lasers and its application in secure communication. Chin Phys 13(8):1215

Szatmári I, Chua LO (2008) Awakening dynamics via passive coupling and synchronization mechanism in oscillatory cellular neural/nonlinear networks. Int J Circuit Theor Appl 36: 525–553

Taherion S, Lai YC (1999) Observability of lag synchronization of coupled chaotic oscillators. Phys Rev E 59:R6247–R6250

Tahir FR, Jafari S, Pham V-T, Volos CK, Wang X (2015) A novel no-equilibrium chaotic system with multiwing butterfly attractors. Int J Bifurcat Chaos 25(4):1550056

Tognoli E, Kelso JAS (2009) Brain coordination dynamics: True and false faces of phase synchrony and metastability. Prog Neurobiol 87:31–40

Tolba MF, Abdelaty AM, Soliman NS, Said LA, Madian AH, Azar AT, Radwan AG (2017) FPGA implementation of two fractional order chaotic systems. Int J Electron Commun 28 (2017):162–172

Tsuji S, Ueta T, Kawakami H (2007) Bifurcation analysis of current coupled BVP oscillators. Int J Bifurcat Chaos 17:837–850

Vaidyanathan S, Sampath S, Azar AT (2015a) Global chaos synchronisation of identical chaotic systems via novel sliding mode control method and its application to Zhu system. Int J Model Ident Control 23(1):92–100

Vaidyanathan S, Azar AT, Rajagopal K, Alexander P (2015b) Design and SPICE implementation of a 12-term novel hyperchaotic system and its synchronization via active control (2015). Int J Model Ident Control 23(3):267–277

Vaidyanathan S, Idowu BA, Azar AT (2015c) Backstepping controller design for the global chaos synchronization of Sprott's jerk systems. In: Azar AT, Vaidyanathan S (eds), Chaos modeling and control systems design, studies in computational intelligence, vol 581, pp 39–58. Springer, GmbH Berlin, Heidelberg

Vaidyanathan S, Azar AT (2015a) Anti-synchronization of identical chaotic systems using sliding mode control and an application to Vaidyanathan-Madhavan chaotic systems. In: Azar AT, Zhu Q (eds), Advances and applications in sliding mode control systems, studies in computational intelligence book Series, vol 576, pp 527–547. Springer, GmbH Berlin, Heidelberg

Vaidyanathan S, Azar AT (2015b) Hybrid synchronization of identical chaotic systems using sliding mode control and an application to Vaidyanathan chaotic systems. In: Azar AT, Zhu Q (eds), Advances and applications in sliding mode control systems, studies in computational intelligence book series, vol 576, pp 549–569. Springer, GmbH Berlin, Heidelberg

Vaidyanathan S, Azar AT (2015c) Analysis, control and synchronization of a nine-term 3-D novel chaotic system. In: Vaidyanathan S, Azar AT (eds), Chaos modeling and control systems design, studies in computational intelligence, vol 581, pp 3–17. Springer, GmbH Berlin, Heidelberg

Vaidyanathan S, Azar AT (2015d) Analysis and control of a 4-D novel hyperchaotic system. In: Vaidyanathan S, Azar AT (eds), Chaos modeling and control systems design, studies in computational intelligence, vol 581, pp 19–38. Springer, GmbH Berlin, Heidelberg

Vaidyanathan S, Azar AT (2016a) Dynamic analysis, adaptive feedback control and synchronization of an eight-term 3-D novel chaotic system with three quadratic nonlinearities. In: Studies in fuzziness and soft computing, vol 337, pp 155–178. Springer, Germany

Vaidyanathan S, Azar AT (2016b) Qualitative study and adaptive control of a novel 4-D hyperchaotic system with three quadratic nonlinearities. In: Studies in fuzziness and soft computing, vol 337, pp 179–202. Springer, Germany

Vaidyanathan S, Azar AT (2016c) A novel 4-D four-wing chaotic system with four quadratic nonlinearities and its synchronization via adaptive control method. In: Advances in chaos theory and intelligent control. Studies in fuzziness and soft computing, vol 337, pp 203–224. Springer, Germany

Vaidyanathan S, Azar AT (2016d) Adaptive control and synchronization of halvorsen circulant chaotic systems. In: Advances in chaos theory and intelligent control. Studies in fuzziness and soft computing, vol 337, pp 225–247. Springer, Germany

Vaidyanathan S, Azar AT (2016e) Adaptive backstepping control and synchronization of a novel 3-D jerk system with an exponential nonlinearity. In: Advances in chaos theory and intelligent control. Studies in fuzziness and soft computing, vol 337, pp 249–274. Springer, Germany

Vaidyanathan S, Azar AT (2016f) Generalized projective synchronization of a novel hyperchaotic four-wing system via adaptive control method. In: Advances in chaos theory and intelligent control. Studies in fuzziness and soft computing, vol 337, pp 275–296. Springer, Germany

Vaidyanathan S, Azar AT, Ouannas A (2017a) An eight-term 3-D novel chaotic system with three quadratic nonlinearities, its adaptive feedback control and synchronization. In: Studies in computational intelligence, vol 688, pp 719–746. Springer, Germany

Vaidyanathan S, Zhu Q, Azar AT (2017b) Adaptive control of a novel nonlinear double convection chaotic system. In: Studies in computational intelligence, vol 688, pp 357–385. Springer, Germany

Vaidyanathan S, Azar AT, Ouannas A (2017c) Hyperchaos and adaptive control of a novel hyperchaotic system with two quadratic nonlinearities. In: Studies in computational intelligence, vol 688, pp 773–803. Springer, Germany

Volos ChK, Kyprianidis IM, Stouboulos IN (2006a) Experimental demonstration of a chaotic cryptographic scheme. WSEAS Trans Circuit Syst 5:1654–1661

Volos ChK, Kyprianidis IM, Stouboulos IN (2006b) Designing of coupling scheme between two chaotic duffing—type electrical oscillators. WSEAS Trans Circuit Syst 5:985–992

Volos ChK, Kyprianidis IM, Stouboulos IN (2013) A gallery of synchronization phenomena in resistively coupled non-autonomous chaotic circuits. J Eng Sci Technol Rev 6(4):15–23

Voss HU (2000) Anticipating chaotic synchronization. Phys Rev E 61:5115–5119

Wang J, Che YQ, Zhou SS, Deng B (2009) Unidirectional synchronization of Hodgkin-Huxley neurons exposed to ELF electric field. Chaos Solit Fract 39:1335–1345

Wang Z, Cang S, Ochola E, Sun Y (2012) A hyperchaotic system without equilibrium. Nonlinear Dyn 69:531–537

Wang X, Chen G (2012) A chaotic system with only one stable equilibrium. Commun Nonlinear Sci Numer Simul 17:1264–1272

Wang X, Chen G (2013) Constructing a chaotic system with any number of equilibria. Nonlinear Dyn 71:429–436

Wang Z, Ma J, Cang S, Wang Z, Chen Z (2016) Simplified hyper-chaotic systems generating multi-wing non-equilibrium attractors. Optik 127:2424–2431

Wang Z, Volos C, Kingni ST, Azar AT, Pham VT (2017) Four-wing attractors in a novel chaotic system with hyperbolic sine nonlinearity. Optik Int J Light Electron Opt 131(2017):1071–1078

Wei Z (2011) Dynamical behaviors of a chaotic system with no equilibria. Phys Lett A 376:102–108

Wei Z, Wang Z (2013) Chaotic behavior and modified function projective synchronization of a simple system with one stable equilibrium. Kybernetika 49:359–374

Wei Z, Wang R, Liu A (2014) A new finding of the existence of hidden hyperchaotic attractors with no equilibria. Math Comput Simul 100:13–23

Woafo P, Enjieu Kadji HG (2004) Synchronized states in a ring of mutually coupled self sustained electrical oscillators. Phys Rev E 69:046206

Wu CW, Chua LO (1993) A simple way to synchronize chaotic systems with applications to secure communication systems. Int J Bifurcat Chaos 3(06):1619–1627

Yanchuk S, Maistrenko Yu, Mosekilde E (2001) Loss of synchronization in coupled Rössler systems. Physica D 154:26–42

Zhong GQ, Man KF, Ko KT (2001) Uncertainty in chaos synchronization. Int J Bifurcat Chaos 11:1723–1735

Zuo J, Li C (2016) Multiple attractors and dynamic analysis of a no-equilibrium chaotic system. Optik 127:7952–7959

4-D Memristive Chaotic System
with Different Families of Hidden Attractors

**Dimitrios A. Prousalis, Christos K. Volos, Viet-Thanh Pham,
Ioannis N. Stouboulos and Ioannis M. Kyprianidis**

Abstract The design of systems without equilibrium or with line of equilibrium points is a subject which has started to attract the interest of the research community the last decade. In this direction, various chaotic systems with hidden attractors, which are based on memristors or memristive systems, have been proposed. In this chapter a new 4-D memristive system is presented. The peculiarity of the model is that it displays a line of equilibrium points for a range of the parameters as well as no-equilibrium for another range of the parameters. System in both occasions presents a chaotic behavior with hidden attractors. The behavior of the proposed system is investigated through numerical simulations, by using phase portraits, Lyapunov exponents and bifurcation diagrams. The adaptive control scheme of the system is presented in order to prove that the memristive system's dynamical behavior can be controlled. Also, we have designed an electronic circuit to confirm the feasibility of the system in both cases.

Keywords Memristive system · Hidden attractor · Chaos control

D. A. Prousalis (✉) · C. K. Volos · I. N. Stouboulos · I. M. Kyprianidis
Department of Physics, Aristotle University of Thessaloniki, Thessaloniki, Greece
e-mail: dprou@physics.auth.gr

C. K. Volos
e-mail: volos@physics.auth.gr

I. N. Stouboulos
e-mail: stouboulos@physics.auth.gr

I. M. Kyprianidis
e-mail: imkypr@auth.gr

V.-T. Pham
School of Electronics and Telecommunications,
Hanoi University of Science and Technology, Hanoi, Vietnam
e-mail: thanh.phamviet@hust.edu.vn; pvt3010@gmail.com

© Springer International Publishing AG 2018　　　　　　　　　　　　　　403
V.-T. Pham et al. (eds.), *Nonlinear Dynamical Systems with Self-Excited
and Hidden Attractors*, Studies in Systems, Decision and Control 133,
https://doi.org/10.1007/978-3-319-71243-7_18

1   Introduction

The forth missing curcuit element, the memristor, was introduced for the first time in 1971 (Chua 1971). A general concept of memristive systems expanded in 1976, (Chua and Kang 1976). In 2008 the realization of a two terminal memristor was announced (Strukov et al. 2008). This announcement influenced many researchers and paved the way for various scientific fields. n 2009, other elements with memory from the nano-world, memcapacitor and meminductor was introduced (Ventra et al. 2009).

There are systems, such as thermistors, with phenomena in which internal state depends on the temperature (Sapoff and Oppenheim 1963), spintronic devices in which resistance varies according to their spin polarization (Pershin and Di Ventra 2008) and molecules in which resistance changes according to their atomic configuration (Chen et al. 2003), could be explained now with the use of the memristor. Also, electronic circuits with memory could simulate processes typical of biological systems, such as the adaptive behavior of unicellular organisms (Pershin et al. 2009) and the learning and associative memory (Pershin and Di Ventra 2010). Mem-elemets also are used in order to replace nonlinear parts of the electrical circuits.

At present, many applications of memristors based on their properties, such as memristor-based neural networks, memristor-based chaotic oscillators, memristor-based charge pump locked loops etc. have been introduced (Itoh and Chua 2008; Zhao et al. 2013; Wu et al. 2011). Research on memristor-based chaotic systems becomes a focal research topic in both the technological and the application domain (Volos et al. 2011; Yang et al. 2013; Driscoll et al. 2010; Wang et al. 2012; Shang et al. 2012; Shin et al. 2011; Cepisca et al. 2008; Cepisca and Bardis 2011; Bogdan et al. 2011; Corinto and Ascoli 2012a, b). Also, the design of memristor- based chaotic oscillators, by replacing the nonlinear part of chaotic dynamical systems with memristors has been introduced (Sabarathinam et al. 2016; Chen et al. 2015; Bao et al. 2016; Wu et al. 2016).

The last decades researchers introduced some memristor-based hyperchaotic systems, motivated by the complex dynamical behaviors of hyperchaotic systems and the special features of memristor in order to investigate whether there exists a memristor-based system that is hyperchaotic. Hyperchaos was generated by combining a memristor with its non-linear characteristics and a chaotic oscillator (Biswas et al. 2016; Ponomarenko et al. 2013; Özkaynak and Yavuz 2013; Ye and Wong 2013; Banerjee et al. 2012a, b; Banerjee and Biswas 2013).

Leonov and Kuznetsov (Kuznetsov et al. 2010; Leonov et al. 2011) in their research categorized periodic and chaotic attractors as either self-excited or hidden. A self-excited attractor has a basin of attraction that is associated with an unstable equilibrium, whereas a hidden attractor (HA) has a basin of attraction that does not intersect with small neighborhoods of any equilibrium points. The classical attractors of Lorenz, Rössler, Chen, Sprott (cases B to S), and other widely-known attractors are those excited from unstable equilibria. From a computational point of view this allows one to use a numerical method in which a trajectory started from a point on the unstable manifold in the neighborhood of an unstable equilibrium, reaches an attractor and identifies it. Hidden attractors cannot be found by this method and are important in engineering applications because they allow unexpected and potentially disastrous responses to perturbations in a structure like a bridge or an airplane wing.

Furthermore, the last two decades the subject of chaos control has attracted the interest of the research community. The control of a chaotic system aims to stabilize or regulate the system with the help of feedback control. There are many methods available for controlling a chaotic system such as active control (Sundarapandian 2010; Vaidyanathan 2011, 2016), adaptive control (Sundarapandian 2013; Vaidyanathan 2012, 2013, 2014; Azar and Vaidyanathan 2015), sliding mode control (Vaidyanathan 2012) and backstepping control (Njah and Sunday 2012; Vincent et al. 2007). Adaptive control is an active field in the design of control systems, especially of systems with hidden attractors (Vaidyanathan and Volos 2012; Wei et al. 2014; Pham et al. 2016), and deal with uncertainties. The key difference between adaptive controllers and linear controllers is the adaptive controller's ability to adjust itself in order to handle unknown model's uncertainties. Recently, much effort has been placed in adaptive control in both theory and applications. New controller design techniques are introduced to handle nonlinear and time-varying uncertainties. Broader systems with larger nonlinear uncertainties can be covered by these developments. As a result, adaptive control is used in various real world applications (Cao et al. 2012; Vaidyanathan 2015).

This research work is organized as follows. In Sect. 2 the model of the memristive system, as well as the new system are presented. In Sect. 3 the simulation results of the memristive system are also presented. The adaptive control scheme of the system is studied in Sect. 4. In Sect. 5 the circuit realization of the system is described in detail, while Sect. 6 concludes this work with a summary of the main results.

2 The Memristive System with Hidden Attractors

In this section a new memristive system with different families of hidden attractors is presented. First of all, the model of the memristive device will be analyzed, while next the mathematical description of the 4-D system will be introduced.

2.1 Model of the Memristive Device

As it is mentioned, Chua and Kang introduced the memristive device by generalizing the original definition of a memristor (Chua and Kang 1976). A memristive system can be described by:

$$
\begin{aligned}
\dot{w}_m &= F(w_m, u_m, t), \\
f_m &= G(w_m, u_m, t)u_m
\end{aligned}
\tag{1}
$$

where w_m, f_m and u_m denote the state of memristive system, output and input, respectively. The function G is a continuous and n-dimensional scalar function and F is a vector function. Based on the definition of memristive system (1), a memristive device is introduced in this section and used in our whole paper. This memristive device is described by the following equations:

$$\dot{w}_m = u_m, \tag{2a}$$

$$f_m = (1 + 0.25w_m^2 - 0.002w_m^4)u_m. \tag{2b}$$

In order to investigate the behavior of the memristive system an external sinusoidal signal u_m is applied. The form of u_m is:

$$u_m = A sin(2\pi\nu t) \tag{3}$$

where A is the amplitude and ν is the frequency. From the first equation of the system (3) we can find w_m:

$$w_m = w_m(0) + \frac{A}{2\pi t}(1 - cos(2\pi\nu t)) \tag{4}$$

where $w_m(0) = \int_{-\infty}^{0} u_m(\tau)d\tau$ is the initial condition of the internal state w_m.

Substituting Eqs. (3) and (4) into Eq. (2b) it is easy to derive the output of the memristive device. Therefore, the output f_m depends on frequency and amplitude of the applied input stimulus.

The figures below show the hysteresis loops of the proposed memristive system driven by a sinusoidal stimulus, when it is driven by a periodic signal (4).

- Figure 1 with $A = 1$, $w_0 = 0$ while $\nu = 0.1$ (green line), $\nu = 0.2$ (blue line) and $\nu = 0.5$ (red line).
- Figure 2 for $\nu = 0.1$, $w_0 = 0$ while $A = 0.5$ (green line), $A = 1$ (blue line) and $A = 1.5$ (red line).
- Figure 3 for $\nu = 0.1$, $A = 1$ while $w_0 = -1$ (green line), $w_0 = 0$ (blue line) and $w_0 = 1$ (red line).

Obviously, the proposed memristive device exhibits a pinched hysteresis loop in the input-output plane.

2.2 The New Memristive System

Finally, based on the aforementioned memristive device, the following new dynamical system can be obtained.

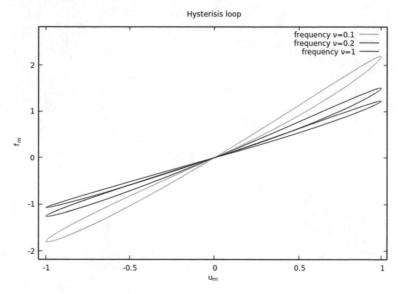

Fig. 1 Hysteresis loops of the proposed memristive device driven by a sinusoidal stimulus with $A = 1$ and $w_0 = 0$, for frequencies $\nu = 0.1$ (green line), $\nu = 0.2$ (blue line), $\nu = 0.5$ (red line)

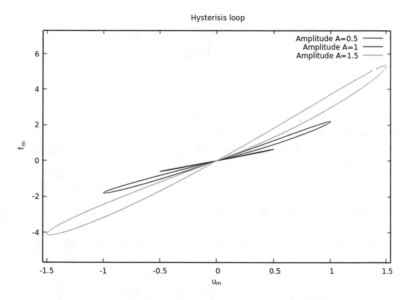

Fig. 2 Hysteresis loops of the proposed memristive device driven by a sinusoidal stimulus with $\nu = 0.1$ and $w_0 = 0$, for amplitude $A = 0.5$ (red line), $A = 1$ (blue line), $A = 1.5$ (green line)

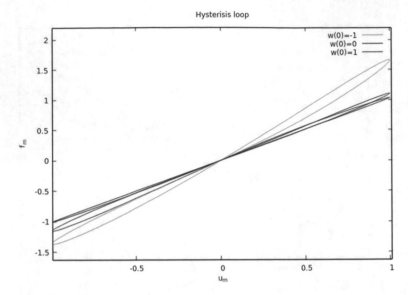

Fig. 3 Hysteresis loops of the proposed memristive device driven by a sinusoidal stimulus with $A = 1$ and $\nu = 0.1$, for $w_0 = -1$ (green line), $w_0 = 0$ (blue line) and $w_0 = 1$ (red line)

$$
\begin{aligned}
\dot{x} &= -\alpha x + \gamma f(y, w) \\
\dot{y} &= \beta x - \delta xz + \varepsilon \\
\dot{z} &= -\zeta z + xy \\
\dot{w} &= y
\end{aligned}
\tag{5}
$$

where $y = u_m$ the input, $w = w_m$ the state, $f(y, w) = f_m = (1 + 0.25w^2 - 0.002w^4)y$ the output of the memristor device and $\alpha, \beta, \gamma, \delta, \varepsilon, \zeta$ are real positive parameters. So, the fourth-order memristive system (5) is obtained and used in the following sections.

2.2.1 Analysis of the New Hyperchaotic Memristive System

The equilibria of system (5) can be derived by solving the following equations:

$$
\begin{aligned}
-\alpha x + \gamma f(y, w) &= 0 \\
\beta x - \delta xz + \varepsilon &= 0 \\
-\zeta z + xy &= 0 \\
y &= 0
\end{aligned}
\tag{6}
$$

The 4-D memristive system (5) for $\varepsilon = 0$ and for every $\alpha, \beta, \gamma, \delta, \zeta$ set of values has line of equilibrium $E(0, 0, 0, w)$. Moreover for $\varepsilon \neq 0$ and for every $\alpha, \beta, \gamma, \delta, \zeta$ has no equilibria. As a result, this memristive hyperchaotic system can be considered as a dynamical system with hidden attractors because it is impossible to verify the chaotic attractor by choosing an arbitrary initial condition in the vicinity of the unstable equilibria. This system's feature is noteworthy especially in the case of using these systems in applications, such as chaos encryption, because of its complexity.

The Jacobian of the system (5), $\mathbf{J}$ at any point is calculated as:

$$\mathbf{J} = \begin{pmatrix} -\alpha & \gamma Q & 0 & \gamma R \\ \beta - \delta z & 0 & -\delta x & 0 \\ y & x & -\zeta & 0 \\ 0 & 1 & 0 & 0 \end{pmatrix} \tag{7}$$

where,

$$Q = \frac{\partial f(y, w)}{\partial y} = 1 + 0.25w^2 - 0.002w^4$$

$$R = \frac{\partial f(y, w)}{\partial w} = 0.5wy - 0.008w^3 y$$

For the case of $\varepsilon = 0$ there are infinite equilibrium points. In this case the eigenvalues of the matrix of Eq. (7), for $\alpha = 1, \gamma = 1, \beta = 7, \delta = 1, \zeta = 1$, are:

$$\begin{aligned} \lambda_1 &= -1 \\ \lambda_2 &= 0 \\ \lambda_3 &= 0.5(-1 - (29 + 7w^2 - 0.056w^4)^{1/2}) \\ \lambda_4 &= 0.5(-1 + (29 + 7w^2 - 0.056w^4)^{1/2}) \end{aligned} \tag{8}$$

As it is clear the eigenvalue $\lambda_1 = -1$ shows that there is a stable multiplicity, $\lambda_2 = 0$ is as expected because the system has a line equilibrium and the eigenvalues λ_3 and λ_4 of the Jacobian Matrix depend on the variable w. So, it is difficult to determine the stability of the equilibrium points.

For the case of $\varepsilon \neq 0$ there are no equilibrium points. As a result there cannot be analysis of the equilibrium points.

The chaotic attractor in the (x, y, z) phase space, for $\varepsilon = 0, \alpha = 1, \gamma = 1, \beta = 8.5, \delta = 1, \zeta = 1$ is depicted in Fig. 4.

The chaotic attractor in the (x, y, z) phase space, for $\varepsilon = 0.01, \alpha = 1, \gamma = 1, \beta = 7, \delta = 2, \zeta = 1$ is depicted in Fig. 5.

According to system (5), the divergence of the system is

$$\nabla V = \frac{\partial \dot{x}}{\partial x} + \frac{\partial \dot{y}}{\partial y} + \frac{\partial \dot{z}}{\partial z} + \frac{\partial \dot{w}}{\partial w} = -\alpha - \zeta \tag{9}$$

where $\nabla V < 0$ for α and ζ positive.

The Lyapunov exponents for $\varepsilon = 0.1$ have been calculated as: $L_1 = 0.01044, L_2 = 0.05774, L_3 = 0$ and $L_4 = -2.95934$. There are two positive Lyapunov exponents, so the system is hyperchaotic. In addition the Kaplan-Yorke dimension of the system is found as:

Fig. 4 Chaotic attractor in
(x, y, z) phase space, for
$\varepsilon = 0, \alpha = 1, \gamma = 1, \beta = 8.5,$
$\delta = 1, \zeta = 1$

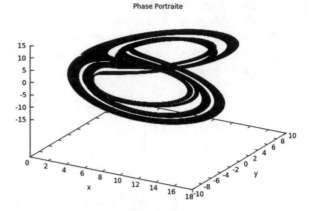

Fig. 5 Chaotic attractor in
(x, y, z) phase space, for
$\varepsilon = 0.01, \alpha = 1, \gamma = 1,$
$\beta = 7, \delta = 2, \zeta = 1$

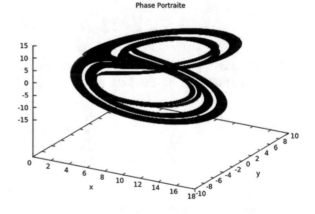

$$D_{KY} = 3 + \frac{L_1 + L_2 + L_3}{|L_4|} = 3.023038 \tag{10}$$

3   Simulation Results

In order to study the behavior of the new system, usual tools of the theory of dynamical systems such as phase portaits, bifurcation diagrams, continuation diagrams and diagram of Lyapunov exponents have been used.

Firstly, the bifurcation diagram of y versus β, for various values of the parameter ε, is obtained by plotting the variable x when the trajectory cuts the plane $w = 0$ with $dy/dt < 0$, as the control parameter β is decreased in tiny steps in the range of $7 \leq \beta \leq 10$. Also, the continuations diagrams of y versus β, in which the initial conditions in each iteration have different values, and the diagram of system's (5) Lyapunov exponents versus β are presented for different sets of values of the system's parameters. At the Lyapunov diagrams the fourth Lyapunov exponent is ignored

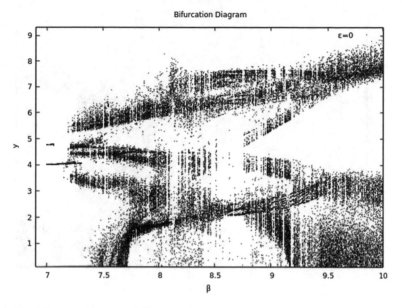

Fig. 6 The bifurcation diagram of y versus β for $\varepsilon = 0$, $\alpha = 1$, $\gamma = 1$, $\delta = 2$, $\zeta = 1$

because it takes negative values far from the zero value. Especially, the hyperchaotic behavior is shown in the Lyapunov diagrams in the region where two Lyapunov exponents become positive and one zero.

For values of the parameters $\varepsilon = 0$, $\alpha = 1$, $\gamma = 1$, $\delta = 1$, $\zeta = 1$ in Figs. 6, 7 and 8 the bifurcation diagram of y versus β, the continuation diagram of y versus β and the diagram of systems Lyapunov exponents versus β are presented.

In more details, system (5) presents the following dynamical behavior, in respect to β for $\varepsilon = 0$, $\alpha = 1$, $\gamma = 1$, $\delta = 2$, $\zeta = 1$:

- A region of periodic behavior for $\beta < 7.162$
- A region of chaotic behavior for $7.162 < \beta < 7.204$
- A region of quasi-periodic behavior for $7.204 < \beta < 7.216$
- A region of chaotic behavior for $7.216 < \beta < 7.228$
- A region of quasi-periodic behavior for $7.228 < \beta < 7.246$
- A region of chaotic behavior for $7.246 < \beta < 7.294$
- A region of quasi-periodic behavior for $7.294 < \beta < 7.306$
- A region of chaotic behavior for $7.306 < \beta < 8.134$
- A region of hyperchaotic behavior for $8.134 < \beta < 8.152$
- A region of chaotic behavior for $8.152 < \beta < 8.212$
- A region of hyperchaotic behavior for $8.212 < \beta < 8.224$
- A region of chaotic behavior for $8.224 < \beta < 8.242$
- A region of hyperchaotic behavior for $8.242 < \beta < 8.254$
- A region of chaotic behavior for $8.254 < \beta < 9.472$
- A region of hyperchaotic behavior for $9.472 < \beta < 10$.

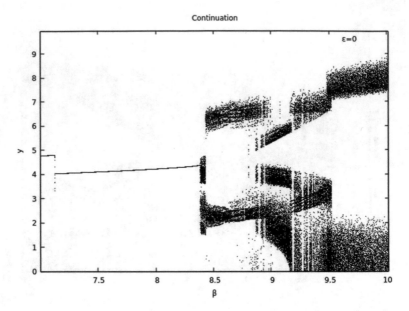

Fig. 7 The continuation diagram of y versus β for $\varepsilon = 0$, $\alpha = 1$, $\gamma = 1$, $\delta = 2$, $\zeta = 1$

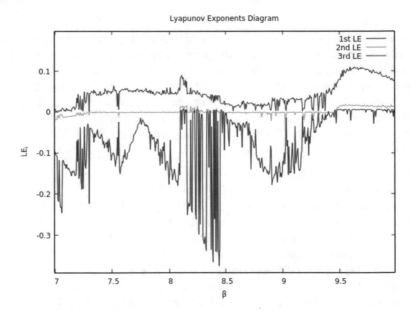

Fig. 8 The Lyapunov diagram of Lyapunov exponents versus β for $\varepsilon = 0$, $\alpha = 1$, $\gamma = 1$, $\delta = 2$, $\zeta = 1$

Bifurcation Diagram

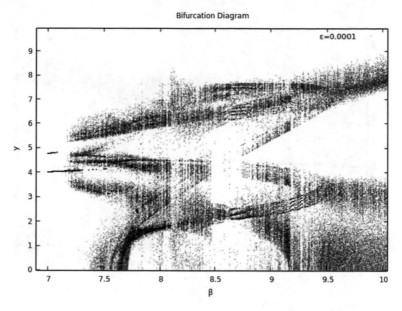

Fig. 9 The bifurcation diagram of y versus β for $\varepsilon = 0.0001$, $\alpha = 1$, $\gamma = 1$, $\delta = 1$, $\zeta = 1$

For values of the parameters $\varepsilon = 0.0001$, $\alpha = 1$, $\gamma = 1$, $\delta = 1$, $\zeta = 1$ in Figs. 9, 10 and 11 the bifurcation diagram of y versus β, the continuation diagram of y versus β and the diagram of systems Lyapunov exponents versus β are presented.

In more details, system (5) presents the following dynamical behavior, in respect to β for $\varepsilon = 0.0001$, $\alpha = 1$, $\gamma = 1$, $\delta = 1$, $\zeta = 1$:

- A region of periodic behavior for $\beta < 7.216$
- A region of chaotic for $7.216 < \beta < 8.11$
- A region of hyperchaotic behavior for $8.11 < \beta < 8.158$
- A region of chaotic behavior for $8.158 < \beta < 9.49$
- A region of hyperchaotic behavior for $9.49 < \beta < 10$.

For values of the parameters $\varepsilon = 0.001$, $\alpha = 1$, $\gamma = 1$, $\delta = 1$, $\zeta = 1$ in Figs. 12, 13 and 14 the bifurcation diagram of y versus β, the continuation diagram of y versus β and the diagram of systems Lyapunov exponents versus β are presented.

In more details, system (5) presents the following dynamical behavior, in respect to β for $\varepsilon = 0.001$, $\alpha = 1$, $\gamma = 1$, $\delta = 1$, $\zeta = 1$:

- A region of periodic behavior for $\beta < 7.138$
- A region of quasi-periodic behavior for $7.144 < \beta < 7.204$
- A region of chaotic behavior for $7.204 < \beta < 7.234$
- A region of quasi-periodic behavior for $7.234 < \beta < 7.246$
- A region of chaotic behavior for $7.246 < \beta < 8.164$
- A region of hyperchaotic behavior for $8.164 < \beta < 8.254$
- A region of chaotic behavior for $8.254 < \beta < 9.502$
- A region of hyperchaotic behavior for $9.502 < \beta < 10$.

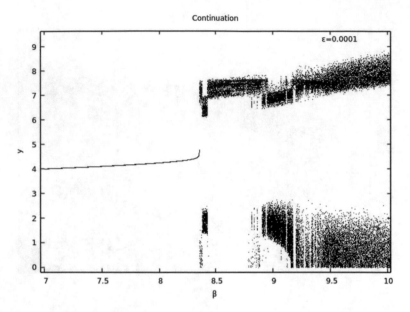

Fig. 10 The continuation diagram of y versus β for $\varepsilon = 0.0001, \alpha = 1, \gamma = 1, \delta = 1, \zeta = 1$

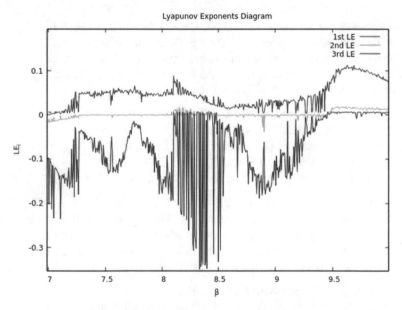

Fig. 11 The Lyapunov Diagram of Lyapunov exponents versus β for $\varepsilon = 0.0001, \alpha = 1, \gamma = 1, \delta = 1, \zeta = 1$

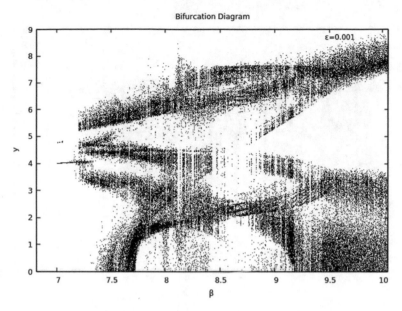

Fig. 12 The bifurcation diagram of y versus β for $\varepsilon = 0.001$, $\alpha = 1$, $\gamma = 1$, $\delta = 1$, $\zeta = 1$

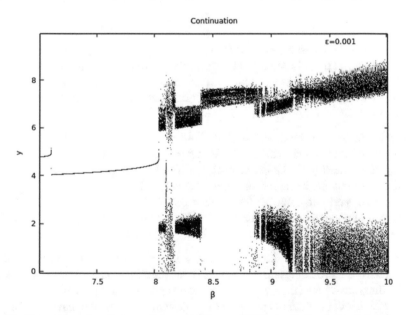

Fig. 13 The continuation diagram of y versus β for $\varepsilon = 0.001$, $\alpha = 1$, $\gamma = 1$, $\delta = 1$, $\zeta = 1$

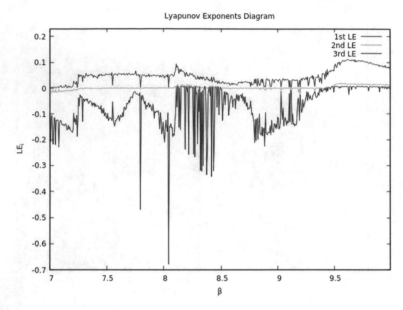

Fig. 14 The Lyapunov Diagram of Lyapunov exponents versus β for $\varepsilon = 0.001$, $\alpha = 1$, $\gamma = 1$, $\delta = 1$, $\zeta = 1$

For values of the parameters $\varepsilon = 0.01$, $\alpha = 1$, $\gamma = 1$, $\delta = 1$, $\zeta = 1$ in Figs. 15, 16 and 17 the bifurcation diagram of y versus β, the continuation diagram of y versus β and the diagram of systems Lyapunov exponents versus β are presented.

In more details, system (5) presents the following dynamical behavior, in respect to β for $\varepsilon = 0.01$, $\alpha = 1$, $\gamma = 1$, $\delta = 2$, $\zeta = 1$:

- A region of periodic behavior for β < 7.048
- A region of chaotic behavior for 7.048 < β < 7.06
- A region of quasi-periodic behavior for 7.06 < β < 7.066
- A region of chaotic behavior for 7.066 < β < 7.732
- A region of periodic behavior for 7.732 < β < 7.75
- A region of chaotic behavior for 7.732 < β < 9.508
- A region of hyperchaotic behavior for 9.508 < β < 10.

For values of the parameters $\varepsilon = 0.1$, $\alpha = 1$, $\gamma = 1$, $\delta = 1$, $\zeta = 1$ in Figs. 18, 19 and 20 the bifurcation diagram of y versus β, the continuation diagram of y versus β and the diagram of systems Lyapunov exponents versus β are presented.

In more details, system (5) presents the following dynamical behavior, in respect to β for $\varepsilon = 0.1$, $\alpha = 1$, $\gamma = 1$, $\delta = 2$, $\zeta = 1$.:

- A region of periodic behavior for 7.108 < β < 7.126
- A region of quasi-periodic behavior for 7.126 < β < 7.156
- A region of periodic behavior for 7.156 < β < 7.258
- A region of chaotic behavior for 7.258 < β < 8.83

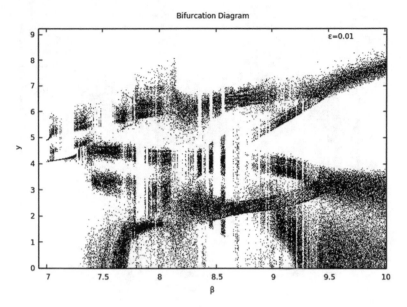

Fig. 15 The bifurcation diagram of y versus β for $\varepsilon = 0.01$, $\alpha = 1$, $\gamma = 1$, $\delta = 2$, $\zeta = 1$

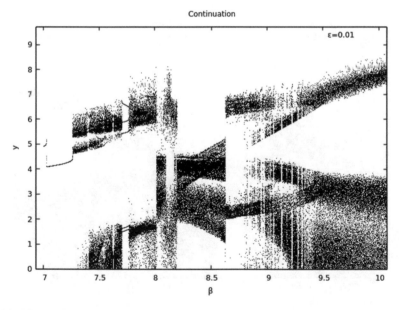

Fig. 16 The continuation diagram of y versus β for $\varepsilon = 0.01$, $\alpha = 1$, $\gamma = 1$, $\delta = 2$, $\zeta = 1$

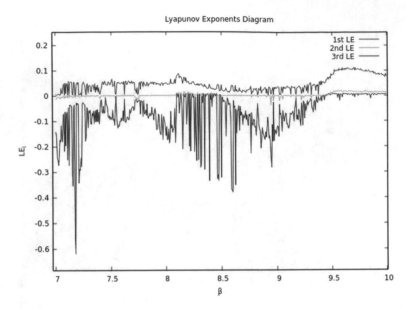

Fig. 17   The Lyapunov Diagram of Lyapunov exponents versus β for $\varepsilon = 0.01, \alpha = 1, \gamma = 1, \delta = 2,$ $\zeta = 1$

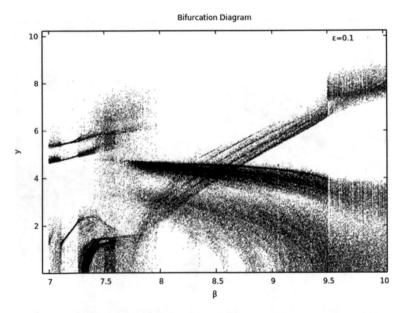

Fig. 18   The bifurcation diagram of y versus β for $\varepsilon = 0.1, \alpha = 1, \gamma = 1, \delta = 2, \zeta = 1$

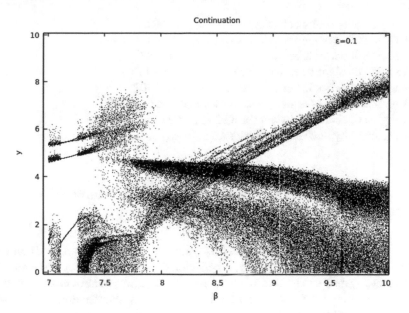

Fig. 19 The continuation diagram of y versus β for $\varepsilon = 0.1$, $\alpha = 1$, $\gamma = 1$, $\delta = 2$, $\zeta = 1$

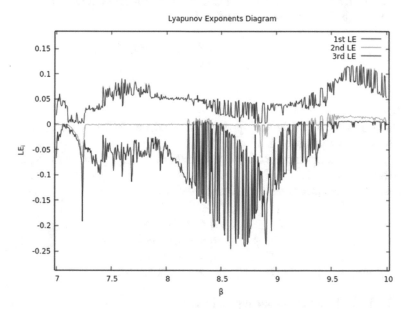

Fig. 20 The Lyapunov diagram of Lyapunov exponents versus β for $\varepsilon = 0.1$, $\alpha = 1$, $\gamma = 1$, $\delta = 2$, $\zeta = 1$

- A region of periodic behavior for $8.83 < \beta < 8.842$
- A region of quasi-periodic behavior for $8.842 < \beta < 8.854$
- A region of periodic behavior for $8.854 < \beta < 8.872$
- A region of chaotic behavior for $8.872 < \beta < 8.896$
- A region of periodic behavior for $8.896 < \beta < 8.902$
- A region of quasi-periodic behavior for $8.902 < \beta < 8.92$
- A region of chaotic behavior for $8.92 < \beta < 9.46$
- A region of hyperchaotic behavior for $9.46 < \beta < 10$.

4 Adaptive Control of the 4-D Hyperchaotic Memristive Dynamical System

From the results of the simulations it is shown that the memristor adds an extra complexity to the system's dynamical behavior. So it is useful to see if the new 4-D memristive system can be controlled by using the adaptive control method, in order to derive an adaptive feedback control law for globally stabilization of the system with unknown parameters.

The controlled 4-D hyperchaotic memristive dynamical system given by following state equilibrium for $\gamma = 1$, $\varepsilon = 0, \zeta = 1$:

$$
\begin{aligned}
\dot{x} &= -\alpha x + f(y, w) + u_1 \\
\dot{y} &= \beta x - \delta xz + u_2 \\
\dot{z} &= -z + xy + u_3 \\
\dot{w} &= y + u_4
\end{aligned}
\tag{11}
$$

where x, y, z, w are the states and u_1, u_2, u_3, u_4 are the adaptive controls and α, β and δ are the unknown parameters of the system.

The problem is finding the adaptive controls u_1, u_2, u_3, u_4 so as to regulate the variables x, y, z, w.

Consider the adaptive feedback control law:

$$
\begin{aligned}
u_1 &= \hat{\alpha}(t)x - f(y, w) - k_1 x \\
u_2 &= -\hat{\beta}(t)x + \hat{\delta}(t)xz - k_2 y \\
u_3 &= z - xy - k_3 z \\
u_4 &= -y - k_4 w
\end{aligned}
\tag{12}
$$

where k_1, k_2, k_3, k_4 are the positive gain constants.

Substituting Eq. (12) into Eq. (11), the closed-loop plant dynamics is given as:

$$\dot{x} = -(\alpha - \hat{\alpha}(t))x - k_1 x$$
$$\dot{y} = (\beta - \hat{\beta}(t))x - (\delta - \hat{\delta}(t))xz - k_2 y$$
$$\dot{z} = -k_3 z$$
$$\dot{w} = -k_4 w$$

(13)

The parameter estimation errors are defined as:

$$e_\alpha = \alpha - \hat{\alpha}(t)$$
$$e_\beta = \beta - \hat{\beta}(t)$$
$$e_\delta = \delta - \hat{\delta}(t)$$

(14)

Differentiating the Eq. (14) with respect to t

$$\dot{e}_\alpha = -\dot{\hat{\alpha}}(t)$$
$$\dot{e}_\beta = -\dot{\hat{\beta}}(t)$$
$$\dot{e}_\delta = -\dot{\hat{\delta}}(t)$$

(15)

In the view of Eq. (15) the plant dynamics can be simplified as:

$$\dot{x} = -e_\alpha x - k_1 x$$
$$\dot{y} = e_\beta x - e_\delta xz - k_2 y$$
$$\dot{z} = -k_3 z$$
$$\dot{w} = -k_4 w$$

(16)

Next the adaptive control theory is used in order to find an update law for the parameter estimates. Consider the quadratic candidate Lyapunov function defined by

$$V(x, y, z, w, e_\alpha, e_\beta, e_\delta) =$$
$$= \frac{1}{2}(x^2 + y^2 + z^2 + w^2) + \frac{1}{2}(e_\alpha^2 + e_\beta^2 + e_\delta^2)$$

(17)

Differentiating the Eq. (17) with respect to t

$$\dot{V} = x\dot{x} + y\dot{y} + z\dot{z} + w\dot{w} + e_\alpha \dot{e}_\alpha + e_\beta \dot{e}_\beta + e_\delta \dot{e}_\delta$$

(18)

Finally,

$$\dot{V} = -k_1 x^2 - k_2 y^2 - k_3 z^2 - k_4 w^2 +$$
$$+ e_\alpha(x^2 - \dot{\hat{\alpha}}) + e_\beta(xy - \dot{\hat{\beta}}) - e_\delta(zxy \dot{\hat{\delta}})$$

(19)

From Eq. (19) the parameter update law is

$$\dot{\hat{\alpha}}(t) = -x^2$$
$$\dot{\hat{\beta}}(t) = xy$$
$$\dot{\hat{\delta}}(t) = -zxy$$

(20)

Theorem 1 *The states x, y, z, w of the 4-D hyperchaotic memristive dynamical system (5) with unknown system parameters are globaly and exponentially regulated for all initial conditions to the desired constant values* α, β, δ *by the adaptive control law (11) and the parameter update law (19), where* k_1, k_2, k_3 *and* k_4 *are positive gain constants.*

Proof This result will be prooved by applying Lyapunov stability theory (Khalil 2001).

The quadratic Lyapunov function defined by Eq. (17), which is a positive definite function on $\mathfrak{R}^7$, is considered.

By substituting the Eq. (15) into Eq. (14) the time derivative of *V* is obtained as:

$$\dot{V} = -k_1 x^2 - k_2 y^2 - k_3 z^2 - k_4 w^2 \tag{21}$$

From the above equation (21) it is obvious that the derivative of *V* respect to *t*, $\frac{dV}{dt} < 0$ is a negative semi-definite function on $\mathfrak{R}^7$. So the state vector *x(t)* and the parameter estimation error can be concluded that are globally bounded, i.e.

$$[x\, y\, z\, w\, e_\alpha(t)\, e_\beta(t)\, e_\delta(t)]^T \in L_\infty$$

where the function space L_∞ consists of all functions of the form *h(t)* that satisfies $| h(\cdot, t) | < \infty$ for all *t*.

If $k = min\{k_1, k_2, k_3, k_4\}$, then it follows from the Eq. (16) that

$$\dot{V} \leq -k||\mathbf{x}(t)||^2 \tag{22}$$

Thus

$$k||\mathbf{x}(t)||^2 \leq \dot{V} \tag{23}$$

Integrating the inequality (23)

$$k \int_0^t ||\mathbf{x}(t)||^2 d\tau \leq V(0) - V(t) \tag{24}$$

From Eq. (24) it follows that $x, y, z, w \in L_2$, where the function space L_2 consists of all functions *h(t)* with properties such that the integral $\int_0^\infty \sqrt{h(t)^2}$ exists for all *t*. By using Barbalat's lemma (Khalil 2001), the $x, y, z, w \to 0$ exponentially as $t \to \infty$ for all initial conditions $x(0), y(0), z(0), w(0) \in \mathfrak{R}^4$. Then it follows that ths states x, y, z, w of the system with the unknown parameters α, β, δ are globally exponentially regulated for all the initial conditions, by the adaptive control laws (12) and the parameter update law (20).

Here the proof is completed.

For the numerical simulations the parameter values are $\alpha = 1$, $\beta = 8$, $\delta = 2$ as used before. Also the positive gain constants are chosen $k_1 = k_2 = k_3 = k_4 = 5$.

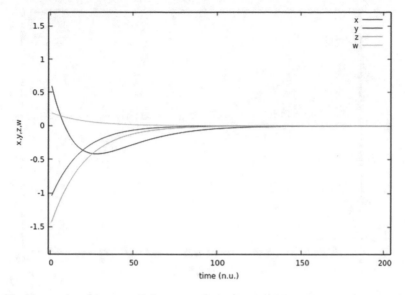

Fig. 21 Time-series of the controlled states x, y, z, w

Futhermore the initial conditions are $x(0) = -1.1$, $y(0) = 0.6$, $z(0) = -1.5$, $w(0) = 0.2$, and $\hat{\alpha}(0) = -0.5$, $\hat{\beta}(0) = -0.2$, $\hat{\delta}(0) = -0.1$. In Fig. 21 the exponential convergence of the controlled states of the system, is depicted.

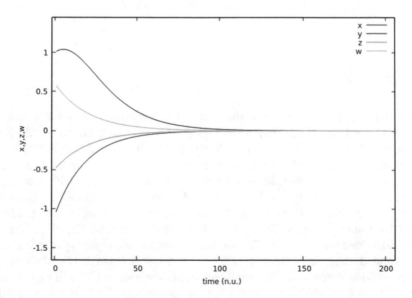

Fig. 22 Time-series of the controlled states x, y, z, w

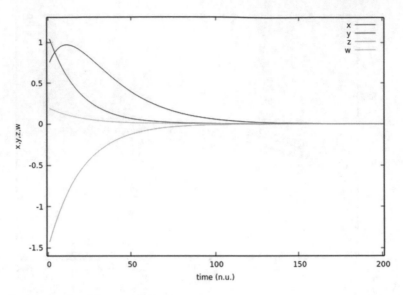

Fig. 23 Time-series of the controlled states x, y, z, w

In Fig. 22 the parameter values are $\alpha = 1, \beta = 8, \delta = 2$, while the initial conditions are $x(0) = -1.1$, $y(0) = 1$, $z(0) = -0.5$, $w(0) = 0.7$, and $\hat{\alpha}(0) = -0.5$, $\hat{\beta}(0) = -0.2$, $\hat{\delta}(0) = -0.1$.

In Fig. 23 the parameter values are $\alpha = 1, \beta = 8, \delta = 2$, while the initial conditions are $x(0) = 1.1$, $y(0) = 0.8$, $z(0) = -1.5$, $w(0) = 0.2$, and $\hat{a}(0) = -0.5$, $\hat{b}(0) = -0.2$, $\hat{\delta}(0) = -0.1$.

5 Circuit Realization

The classical approach for the verification of the feasibility of theoretical chaotic models is the physical realization through electronic circuits (Borah et al. 2016; Bouali et al. 2012; Kingni et al. 2016; Wu et al. 2015; Zhou et al. 2015). Furthermore, the circuital realization of chaotic systems has been applied in numerous engineering applications, for example in secure communications (Banerjee 2010; Cicek et al. 2010), liquid mixing (Sahin and Guzelic 2013), robotics (Volos et al. 2012), image encryption process (Volos et al. 2013), audio encryption scheme (Liu et al. 2016), target detection (Wang et al. 2015) or random signal generation (Fatemi-Behbahani et al. 2016; Yalcin et al. 2004). For this reason, analog and digital approaches have been applied to realize chaotic oscillators by using different kinds of electronic devices such as common off-the-shelf electronic components (Elwakil and Ozoguz 2003; Piper and Sprott 2010), integrated circuit technology (Trejo-Guerra et al. 2012,

2013), microcontroller (Pano-Azucena et al. 2017) or field-programmable gate array (FPGA) (Koyuncu et al. 2014; Tlelo-Cuautle et al. 2015).

Therefore, in this section, we will confirm the feasibility of the proposed memristive system by discussing its circuital realization by using the general operational amplier–based approach. The third state variable (z) of the memristive system has been rescaled as $Z = z/2$, in order to avoid the limitations problems of the components of our electronic circuit. Therefore, the memristive system is transformed into the following equivalent system:

$$
\begin{aligned}
\dot{X} &= -X + F(Y, W) \\
\dot{Y} &= \beta X - 2\delta XZ + \varepsilon \\
\dot{Z} &= -Z + \tfrac{1}{2}XY \\
\dot{W} &= Y
\end{aligned}
\tag{25}
$$

where $F(Y, W) = (1 + 0.25W^2 - 0.002W^4)Y$ the output of the memristive device. Figure 24 shows the schematic of the circuit for realizing the system (5). As shown in this figure, the circuit includes sixteen resistors, four capacitors, seven operational amplifiers (TL081) and five analog multipliers (AD633). By applying Kirchhoffs circuit laws into the designed circuit, we get the following circuital equation:

$$
\begin{aligned}
\dot{x} &= \tfrac{1}{R \cdot C}[-X + F(Y, W)]y + \tfrac{R}{10V \cdot R_1}y \cdot z] \\
\dot{y} &= \tfrac{1}{R \cdot C}[\tfrac{R}{R_\beta}X - \tfrac{R}{10V \cdot R_\delta}XZ + V + \varepsilon] \\
\dot{z} &= \tfrac{1}{R \cdot C}[-Z + \tfrac{R}{10V \cdot R_1}X \cdot Y] \\
\dot{w} &= \tfrac{1}{R \cdot C}Y
\end{aligned}
\tag{26}
$$

where

$$
F(Y, W) = [\tfrac{R}{10V \cdot R_a}V_f + \tfrac{R}{(10V)^2 \cdot R_b}W^2 - \tfrac{R}{(10V)^4 \cdot R_c}W^4]y
\tag{27}
$$

is the output of the memristive circuit in the dotted frame of the schematic in Fig. 16, which implements the opposite of the memristive function of Eq. (2).

In system (26), X, Y, Z and W correspond to the voltages on the integrators (U1–U4), respectively, while the power supply is $\pm 15V_{DC}$. System (26) is normalized by using $\tau = t/RC$. It can thus be suggested that system (26) is equivalent to system (5), with $a = \tfrac{R}{10V \cdot R_a}$, $b = \tfrac{R}{(10V)^2 \cdot R_b}$, $c = \tfrac{R}{(10V)^4 \cdot R_c}$, $d = R/R_\delta$, $2e = \tfrac{R}{10V \cdot R_e}$, $m = V_m$ and $\tfrac{R}{10V \cdot R_1} = 0.5$. So, the values of circuit components are: $R = 10\,\text{k}\Omega$, $R_a = 1\,\text{k}\Omega$, $R_b = 0.4\,\text{k}\Omega$, $R_c = 0.5\,\text{k}\Omega$, $R_\delta = 1\,\text{k}\Omega$, $Re = 0.5\,\text{k}\Omega$, $R_1 = 2\,\text{k}\Omega$, $C = 10\,\text{nF}$, $V_f = 1\,\text{V}$ and $V_\varepsilon = 0\,\text{V}$ (for the case of $\varepsilon = 0$). The designed circuit has been implemented in Multisim and PSpice results are reported in Fig. 24. It is easy to see the good agreement between the circuit's simulation results (Figs. 25, 26 and 27) and numerical results (Fig. 2).

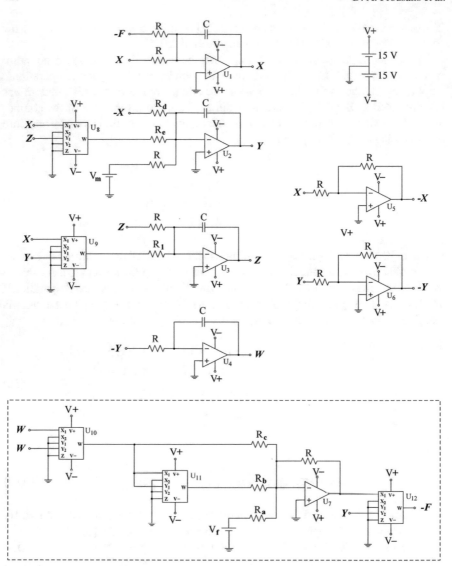

Fig. 24 Schematic of the circuit including sixteen resistors, four capacitors, seven operational amplifiers and five analog multipliers. The power supplies of all operational amplifiers and analog multipliers are $\pm 15 V_{DC}$

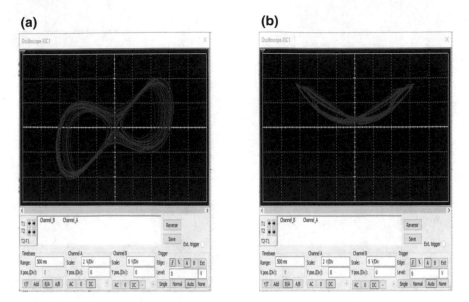

Fig. 25 **a** PSpice chaotic attractors of the designed circuit in (**a**) $X - Y$ plane, **b** $X - Z$ plane for $\varepsilon = 0$

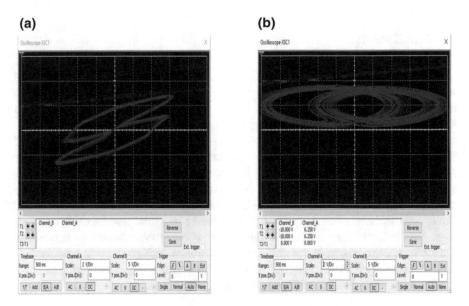

Fig. 26 **a** PSpice chaotic attractors of the designed circuit in (**a**) $X - W$ plane, **b** $Y - Z$ plane for $\varepsilon = 0$

(a)   **(b)**

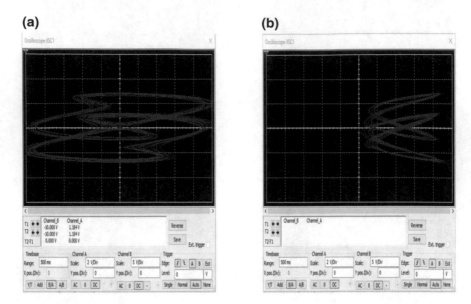

Fig. 27 **a** PSpice chaotic attractors of the designed circuit in (**a**) $Y - W$ plane, **b** $Z - W$ plane for $\varepsilon = 0$

6 Conclusion

The existence of a memristor-based hyperchaotic system with line of equibria and with no equilibria has been studied in this paper. Although 4-D memristive systems often only generate chaos, the presence of a memristive device leads the proposed system to a hyperchaotic system with hidden attractors. The system has rich dynamical behavior as confirmed by the reported example of attractor and by the presented numerical bifurcation diagrams and Lyapunov exponents. It is worth noting that the possibilities of control of such system with unknown parameters is verified by constructing an adaptive controller. Also, the designed circuit emulates very well the proposed hyperchaotic memristive system. Because there is little knowledge about the special features of such systems, future works will continue focusing on their dynamical behaviors, as well as the possibility of synchronization of such systems. Furthermore, the robustness of the control technique with respect to noise is very crucial especially in practical applications. For this reason, the investigation of noise effect on the control scheme will be taken as a future work.

References

Azar AT, Vaidyanathan S (2015) Analysis and control of a 4-D novel hyperchaotic system. Stud Comput Intell 581:3–17

Bogdan M, Buga M, Medianu R, Cepisca C, Bardis N (2011) Obtaining a model of photovoltaic cell with optimized quantum efficiency. Sci Bull Electr Eng Fac 2(16):1843–6188

Banerjee T, Biswas D, Sarkar BC (2012a) Design and analysis of a first order time-delayed chaotic system. Nonlinear Dyn 70(1):721–734

Banerjee T, Biswas D, Sarkar BC (2012b) Design of chaotic and hyperchaotic time-delayed electronic circuit. Bonfring Int J Power Syst Integr Circuits 2(4):13

Banerjee T, Biswas D (2013) Theory and experiment of a first-order chaotic delay dynamical system. Int J Bifurc Chaos 23(06):1330020

Banerjee S (2010) Chaos synchronization and cryptography for secure communications. Applications for encryption: applications for encryption IGI Global2010

Bao B, Jiang T, Xu Q, Chen M, Wu H, Hu Y (2016) Coexisting infinitely many attractors in active band-pass filter-based memristive circuit. Nonlinear Dyn 1–13

Biswas D, Banerjee T (2016) A simple chaotic and hyperchaotic time-delay system: design and electronic circuit implementation. Nonlinear Dyn 83(4):2331–2347

Borah M, Singh PP, Roy BK (2016) Improved chaotic dynamics of a fractional-order system, its chaos-suppressed synchronisation and circuit implementation. Circuits Syst Signal Process 35:1871–1907

Bouali S, Buscarino A, Fortuna L, Frasca M, Gambuzza LV (2012) Emulating complex business cycles by using an electronic analogue. Nonlinear Anal Real World Appl 13:2459–2465

Cao C, Ma L, Xu Y (2012) Adaptive control theory and applications. J Control Sci Eng 827353:2

Cepisca C, Grigorescu SD, Ganatsios S, Bardis NG (2008) Passive and active compensations for current transformers. Metrologie 5–10

Cepisca C, Bardis NG (2011) Measurement and control in street lighting. Electra publication

Chen Y, Jung GY, Ohlberg DA, Li XM, Stewart DR, Jeppesen JO et al (2003) Nanoscale molecular-switch crossbar circuits. Nanotechology 14;462–468

Chen M, Li M, Yu Q, Bao B, Xu Q, Wang J (2015) Dynamics of self-excited attractors and hidden attractors in generalized memristor-based Chuas circuit. Nonlinear Dyn 81(1–2):215–226

Chua LO (1971) Memristors-the missing circuit element. In: IEEE transactions on circuit theory, vol CT-18, pp 507–519

Chua LO, Kang SM (1976) Memristive devices and systems. In: Proceeding of the IEEE no 23, pp 209–223. SIAM

Cicek S, Uyaroglu Y, Pehlivan I (2010) Simulation and circuit implementation of sprott case H chaotic system and its synchronization application for secure communication systems. J Circuits Syst Comput 22:1350022

Corinto F, Ascoli A (2012a) Memristor based elements for chaotic circuits. PIEICE Nonlinear Theory Appl 3(3):336–356

Corinto F, Ascoli A (2012b) Analysis of current-voltage characteristics for memristive elements in pattern recognition systems. Int J Circuit Theory Appl 40(12):1277–1320

Driscoll T, Quinn J, Klein S, Kim HT, Kim BJ, Pershin YV, Di Ventra M, Basov DN (2010) Memristive adaptive filters. Appl Phys Lett 97:093502-1–3

Elwakil A, Ozoguz S (2003) Chaos in pulse-excited resonator with self feedback. Electron Lett 39:831–3

Fatemi-Behbahani E, Ansari-Asl K, Farshidi E (2016) A new approach to analysis and design of chaos-based random number generators using algorithmic converter. Circuits Syst Signal Process 35:3830–3846

Itoh M, Chua LO (2008) Memristor oscillators. Int J Bifurc Chaos 18:3183–3206

Khalil HK (2001) Nonlinear systems, 3rd edn. Prentice Hall, New Jersey

Kingni ST, Pham VT, Jafari S, Kol GR, Woafo P (2016) Three-dimensional chaotic autonomous system with a circular equilibrium: analysis, circuit implementation and its fractional-order form. Circuits, Syst Signal Process 35:1933–1948

Koyuncu I, Ozcerit AT, Pehlivan I (2014) Implementation of FPGA-based real time novel chaotic oscillator. Nonlinear Dyn 77:49–59

Kuznetsov NV, Leonov GA, Vagaitsev VI (2010) Analytical-numerical method for attractor localization of generalized Chuas system. in: IFAC proceedings volumes (IFAC-Papers Online), p 4

Leonov GA, Kuznetsov NV, Vagaytsev VI (2011) Localization of hidden Chuasattractors. Phys Lett A 375(23):2230–2233

Liu H, Kadir A, Li Y (2016) Audio encryption scheme by confusion and diffusion based on multi-scroll chaotic system and one-time keys. Opt-Int J Light Electron Opt 127:7431–8

Njah AN, Sunday OD (2012) Generalization on the chaos control of 4-D chaotic systems using recursive backstepping nonlinear controller. Chaos Solitons Fractals 41(5):2371–2376

Özkaynak F, Yavuz S (2013) Designing chaotic S-boxes based on time-delay chaotic system. Nonlinear Dyn 74(3):551–557

Pano-Azucena AD, de Jesus R-MJ, Tlelo-Cuautle E, de Jesus Q-VA (2017) CArduino-based chaotic secure communication system using multi-directional multi-scroll chaotic oscillators. Nonlinear Dyn 87:2203–2217

Pershin YV, Di Ventra M (2008) Spin memristive systems: spin memory effects in semiconductor spintronics. Phys Rev B 78:113309/1–113309/4

Pershin YV, La Fontaine S, Di Ventra M (2009) Memristive model of amoeba learning. Phys Rev E 80:021926/1–021926/6

Pershin YV, Di Ventra M (2010) Experimental demonstration of associative memory with memristive neural networks. Neural Netw 23:881

Ponomarenko VI, Prokhorov MD, Karavaev AS, Kulminskiy DD (2013) An experimental digital communication scheme based on chaotic time-delay system. Nonlinear Dyn 74(4):1013–1020

Piper JR, Sprott JC (2010) Simple autonomous chaotic circuits. IEEE Trans Circuits Syst II: Express Briefs 57:730–734

Pham VT, Vaidyanathan S, Volos CK, Hoang TM, Van Yem V (2016) Dynamics, synchronization and SPICE implementation of a memristive system with hidden hyperchaotic attractor. In: Advances in chaos theory and intelligent control. Springer International Publishing, pp 35–52

Sabarathinam S, Volos CK, Thamilmaran K (2016) Implementation and study of the nonlinear dynamics of a memristor-based Duffing oscillator. Nonlinear Dyn 1–13

Sahin S, Guzelic C (2013) A dynamical state feedback chaotification method with application on liquid mixing. J Circuits Syst Comput 22:1350059

Sapoff M, Oppenheim RM (1963) Theory and application of self-heated thermistors. In: Proceedings of the IEEE, vol 51, pp 1292–1305

Shang Y, Fei W, Yu H (2012) Analysis and modeling of internal state variables for dynamic effects of nonvolatile memory devices. IEEE Trans Circuits Syst I: Regul Pap 59:1906–1918

Shin S, Kim K, Kang SM (2011) Memristor applications for programmable analog ICs. IEEE Trans Nanotechnol 10:266–274

Sundarapandian V (2010) Output regulation of the Lorenz attractor. Far East J Math Sci 42(2):289–299

Sundarapandian V (2013) Adaptive control and synchronization design for the Lu-Xiao chaotic system. Lect Notes Electr Eng 131:319–327

Strukov DB, Snider GS, Stewart GR, Williams RS (2008) The missing memristor found. Nature 453:80–83

Tlelo-Cuautle E, Rangel-Magdaleno J, Pano-Azucena A, Obeso-Rodelo P, Nuez-Perez JC (2015) CFPGA realization of multi-scroll chaotic oscillators. Commun Nonlinear Sci Numer Simul 27:66–80

Tlelo-Cuautle E, Pano-Azucena A, Rangel-Magdaleno J, Carbajal-Gomez V, Rodriguez-Gomez G (2016) Generating a 50-scroll chaotic attractor at 66 MHz by using FPGAs. Nonlinear Dyn 85:2143–57

Trejo-Guerra R, Tlelo-Cuautle E, Jimenez-Fuentes J, Snchez-Lpez C, Muoz-Pacheco J, Espinosa-Flores-Verdad G et al (2012) Integrated circuit generating 3-and 5-scroll attractors. Commun Nonlinear Sci Numer Simul 17:4328–4335

Trejo-Guerra R, Tlelo-Cuautle E, Jimnez-Fuentes M, Muoz-Pacheco J, Snchez-Lpez C (2013) Multiscroll floating gatebased integrated chaotic oscillator. Int J Circuit Theory Appl 41:831–843

Vaidyanathan S (2011) Output regulation of Arneodo-Coullet chaotic system. Commun Comput Inf Sci 133:98–107

Vaidyanathan S (2016) Analysis, control and synchronization of a novel 4-D highly hyperchaotic system with hidden attractors. In: Advances in chaos theory and intelligent control. Springer International Publishing, pp 529–552

Vaidyanathan S (2012) Adaptive controller and syncrhonizer design for the Qi-Chen chaotic system. Advances in computer science and information technology. Comput Sci Eng 84:73–82

Vaidyanathan S (2013) A ten-term novel 4-D hyperchaotic system with three quadratic nonlinearities and its control. Int J Control Theory Appl 6(2):97–109

Vaidyanathan S (2014) Qualitative analysis and control of an eleven-term novel 4-D hyperchaotic system with two quadratic nonlinearities. Int J Control Theory Appl 7:35–47

Vaidyanathan S, Volos C (2012) Analysis and adaptive control of a novel 3-D conservative no-equilibrium chaotic system. Arch Control Sci 25(3):333–353

Vaidyanathan S (2012) Global chaos control of hyperchaotic Liu system via sliding control method. Int J Control Theory Appl 5(2):117–123

Vaidyanathan S (2015) Adaptive control of Rikitake two-disk dynamo system. Int J ChemTech Res 8(8):121–133

Ventra MD, Pershin YV, Chua LO (2009) Circuit elements with memory: memristors, memcapacitors, and meminductors. Proc IEEE 97(10):1717–1724

Vincent UE, Njah AN, Laoye JA (2007) Controlling chaos and deterministic directed transport in inertia ratchets using backstepping control. Phys D 231(2):130–136

Volos CK, Kyprianidis IM, Stouboulos IN (2011) The memristor as an electric synapse synchronization phenomena In: Proceedings of the International Conference on DSP2011 (Confu, Greece), pp 1–6

Volos CK, Kyprianidis IM, Stouboulos IN (2012) A chaotic path planning generator for autonomous mobile robots. Robot Auton Syst 60:651–656

Volos CK, Kyprianidis IM, Stouboulos IN (2013) Image encryption process based on chaotic synchronization phenomena. Signal Process 93:1328–40

Wang L, Zhang C, Chen L, Lai J, Tong J (2012) A novel memristor-based rSRAM structure for multiple-bit upsets immunity. IEICE Electron Express 9:861–867

Wang B, Xu H, Yang P, Liu L, Li J (2015) Target detection and ranging through lossy media using chaotic radar. Entropy 17:2082–93

Wei Z, Moroz I, Liu A (2014) Degenerate Hopf bifurcations, hidden attractors, and control in the extended Sprott E system with only one stable equilibrium. Turkish J Math 38(4):672–687

Wu AL, Zhang JN, Zeng ZG (2011) Dynamical behaviors of a class of memristor-based Hopfield networks. TPhys Lett A 375:1661–1665

Wu H, Bao B, Liu Z, Xu Q, Jiang P (2016) Chaotic and periodic bursting phenomena in a memristive Wien-bridge oscillator. Nonlinear Dyn 83(1–2):893–903

Wu X, He Y, Yu W, Yin B (2015) A new chaotic attractor and its synchronization implementation. Circuits Syst Signal Process 34:1747–1768

Yang JJ, Strukov DB, Stewart DR (2013) Memristive devices for computing. Nat Nanotechnol 8:13–24

Yalcin ME, Suykens JA, Vandewalle J (2004) True random bit generation from a double-scroll attractor. EEE Trans Circuits Syst I: Regul Pap 51:1395–1404

Ye G, Wong KW (2013) An image encryption scheme based on time-delay and hyperchaotic system. Nonlinear Dyn 71(1–2):259–267

Zhao YB, Tse CK, Feng JC, Guo YC (2013) Application of memristor-based controller for loop filter design in charge-pump phase-locked loop. Circuits Syst Signal Process 32:1013–1023

Zhou W-j, Wang Z-p, Wu M-w, Zheng W-h, Weng J-f (2015) Dynamics analysis and circuit implementation of a new three-dimensional chaotic system. Opt-Int J Light Electron Opt 126:765–768

Hidden Chaotic Path Planning and Control of a Two-Link Flexible Robot Manipulator

Kshetrimayum Lochan, Jay Prakash Singh, Binoy Krishna Roy
and Bidyadhar Subudhi

Abstract Robotics is an emerging and interesting area in many fields of technical science. In general, a robot manipulator (rigid/flexible) is a more focused research direction in comparison with other areas of robotics. Specifically, flexible manipulators are more applicable in many fields when compared with its rigid counterparts because of many advantages like lightweight, more workspace, lower energy consumption, smaller in size, mobility, etc. These advantages give rise to many control challenges like underactuation, nonminimum phase, noncollocation, control spillover, uncertainties, nonlinearities, complex dynamical behaviours, etc. Path planning or trajectory tracking problem is considered as an interesting and challenging control problem for a flexible manipulator in comparison with the regulation problem. In recent decades, the theory of chaos is used in various technical fields. Aperiodic long time, highly sensitive to initial conditions, unpredictable behaviours, etc. are the fundamental properties of a chaotic signal arising out of a deterministic nonlinear system. Many continuous/discrete/fractional order autonomous and non-autonomous chaotic dynamical systems are available in the literature. In the recent past, more attention has been given to the design and applications of hidden chaotic dynamical systems. The path planning problem of a flexible manipulator requires a reference signal. Various reference signals are used in the literature. Recently, a chaotic signal is used as a reference signal for path planning. However, we have not found any paper wherein a reference signal using a hidden chaotic system is used

K. Lochan (✉) · J. P. Singh · B. K. Roy
Department of Electrical Engineering, National Institute of Technology Silchar,
Silchar, India
e-mail: lochan.nits@gmail.com

J. P. Singh
e-mail: jayprakash1261@gmail.com

B. K. Roy
e-mail: bkrnits@gmail.com

B. Subudhi
Department of Electrical Engineering, National Institute of Technology Rourkela,
Rourkela, India
e-mail: bidyadhar@nitrkl.ac.in

© Springer International Publishing AG 2018   433
V.-T. Pham et al. (eds.), *Nonlinear Dynamical Systems with Self-Excited
and Hidden Attractors*, Studies in Systems, Decision and Control 133,
https://doi.org/10.1007/978-3-319-71243-7_19

for path planning. The use of a signal from a hidden chaotic attractor for path planning of a flexible manipulator can provide a new domain of research. Hidden chaotic path planning/trajectory tracking of a two-link flexible manipulator is the aim of this chapter. Use of hidden chaotic attractors as a path/trajectory reference creates extra challenges and complexity in controlling the flexible manipulator. Thus, controlling a flexible manipulator in such a scenario is a challenging task. The dynamics of a two-link flexible manipulator is first modelled using assumed modes method and divided into two parts using two-time scale separation principle (singular perturbation). One subsystem is called as the slow subsystem involving with the rigid parts and another subsystem is called as the fast subsystem which incorporates the flexible dynamics. Separate control techniques are applied to each subsystem. An adaptive sliding mode control technique is designed for the slow subsystem which tackles the uncertainties and helps in fast tracking of the desired hidden chaotic trajectory. A backstepping controller is designed for the fast subsystem system for quick suppression of tip deflections and vibration suppressions. The proposed control techniques are validated using a reference chaotic signal generated from a 3-D hidden attractors chaotic system in MATLAB simulation environment and results are demonstrated. The results reveal that the objective of the chapter is achieved successfully by the proposed control techniques.

Keywords Hidden chaotic attractors · Chaotic path planning · Singular perturbation · Adaptive SMC · Backstepping control · Two-link flexible manipulator

1 Introduction

Aperiodic long time behaviour in a nonlinear deterministic dynamical system is considered as the chaotic phenomenon (Pham et al. 2016a). In the last two decades, many chaotic systems have been reported in the literature based on their various behaviours and characteristics (Pham et al. 2014a, b, 2016a, c, e, g, 2017a, b; Vaidyanathan et al. 2015; Wang et al. 2017). Recently, dynamical chaotic systems are used for various applications like secure communication (Tlelo-Cuautle et al. 2015), information theory (Esteban et al. 2016), image processing (Tlelo-Cuautle et al. 2015), structural engineering (Nichols et al. 2003), security (Pham et al. 2014b), economics (Andrievskii and Fradkov 2004), biomedical (Andrievskii and Fradkov 2004), robotics (Tlelo-Cuautle et al. 2014; Lochan et al. 2016c), etc. But, the use of these chaotic systems in many areas of technical sciences are still less explored and require consideration. The use of a chaotic signal as the desired trajectory for path planning/trajectory tracking of a flexible manipulator is the motivation of this chapter.

The nature of an equilibrium point of a chaotic system plays an important role in its classification. Various chaotic systems are reported based on the different nature of its equilibrium points (Pham et al. 2014a, b, 2016a, e, g, 2017a; Vaidyanathan et al. 2015; Wang et al. 2017). In recent years, chaotic systems are mainly classified into two groups: (i) chaotic systems (Vaidyanathan and Volos 2016; Hoppensteadt 2000; Azar et al. 2017) with self-excited attractors and (ii) chaotic systems with hidden attractors (Leonov et al. 2011a, b, 2012, 2014, 2015; Leonov and Kuznetsov 2013). Chaotic system with (a) stable equilibrium points (Pham et al. 2016d; Chen et al. 2017; Molaie et al. 2013), (b) no equilibrium point (Kiseleva et al. 2016; Pham et al. 2016f; Leonov et al. 2011a; Kingni et al. 2016; Singh and Roy 2017a) and (c) infinitely many equilibrium points (Pham et al. 2016b; Jafari and Sprott 2013; Singh and Roy 2017b) satisfy the requirement of the hidden attractors chaotic systems and hence, belong to the category of hidden attractors. The conventional chaotic systems like Lorenz (1963), Chen (1999), Lü (2002), Sprott (1994), systems in Singh and Roy (2015a, b, 2016a, b, c), Singh et al. (2014) are classified under the category of self-excited chaotic system. Hidden attractors are also seen in many electromechanical systems like in induction motor (Leonov et al. 2011a), drilling system (Leonov et al. 2014), and many others.

Flexible manipulators are used in many applications like aerospace (Sabatini et al. 2012), industry (Lochan et al. 2016a), medical science (Bruno et al. 2014; Arora et al. 2014), home (Nakamura et al. 2003), education (Lochan et al. 2014, 2016a; Suklabaidya et al. 2015), etc., (Lochan et al. 2016a; Kiang et al. 2014). The above-stated applications of flexible manipulators (Suklabaidya et al. 2014a, b) are gradually increasing as compared with their rigid counterpart (Lochan and Roy 2015a) because of their inherent advantages (Lochan et al. 2016a). Many control problems are considered in the literature for a flexible manipulator. The most commonly used control problems are path planning/trajectory tracking (Lochan and Roy 2016) for the hub angle and path planning/trajectory tracking for the tip position (Lochan et al. 2016a). Various desired paths/trajectories are considered in the literature. The available desired signals used in the literature are listed in Table 1. It is seen from Table 1 that the use of a signal generated from a hidden attractors chaotic system as the desired trajectory for a flexible manipulator is not found in the literature. Hence, this work uses a hidden attractors chaotic signal as the desired trajectory for the path planning/trajectory tracking control of a two-link flexible manipulator.

Table 1 Type of desired trajectories used for the trajectory tracking control of a flexible manipulator

Sl. no.	Desired trajectories	References of papers
1.	Bang-bang	Aoustin and Formal'sky (1999)
		Pradhan and Subudhi (2014)
2.	Circular	Masoud et al. (2010)
		Zhang and Liu (2012)
3.	Exponentially varying	Lee and Lee (2002)
		Pradhan and Subudhi (2012)
4.	Straight Link	Li et al. (2005)
5.	Chaotic signal	Lochan et al. (2016b, c)
6.	Hidden attractors chaotic signal	This work

The design of a controller largely depends on the modelling method used to model a system. Many modelling methods are available in the literature for modelling of a flexible manipulator. The commonly used modelling methods are assumed modes method (Subudhi and Morris 2002), lumped parameter method (Lochan et al. 2016c; Lochan and Roy 2015b) and finite element method (Lochan et al. 2016a; Korayem and Haghpanahi 2009). Among these three modelling methods, assumed modes method is widely used (Lochan et al. 2016a). It gives the desirable response of the dynamics by selecting a suitable choice of the number of modes. Another interesting modelling method is the singular perturbation (SP) technique (Lochan et al. 2016a). In singular perturbation, a two-time scale separation principle is used. Using this, dynamics is divided into two parts: slow and fast subsystem dynamics (Subudhi and Morris 2002; Siciliano and Book 1986). In the case of flexible manipulators, the singular perturbation is used to divide the system dynamics into a slow subsystem consisting of the rigid dynamics and a fast subsystem consisting of the flexible dynamics (Siciliano and Book 1986). Thus, it is easy to design the control inputs separately for each subsystem and the desired performances can be achieved.

Link deflection is another important control problem considered in the field of flexible manipulators. Various control techniques are reported in the literature for the quick suppression of link deflection (Özer and Semercigil 2010; Chu and Cui 2015; Karagulle et al. 2015). The use of a singular perturbation (SP) modelling method for the design of a control technique for quick suppression of link deflection is more worthy as compared with other methods. Because in SP, the dynamics of the fast subsystem representing the flexible dynamics of the FM can be used to design a separate control technique for the quick suppression of the links deflection. Different types of flexible manipulators are available in the literature like single link FM, two-link FM and flexible joint FM, multi-link FM (Lochan et al. 2016a). But, two-link flexible manipulators are mainly used and considered in this paper for the design of a controller (Lochan et al. 2016a).

Many control techniques including classical and robust are reported in the literature for controlling two-link flexible manipulators like backstepping control, state feedback control (Lochan et al. 2016a), observer based control (Zhang and Liu 2012), extended state observer (Yu et al. 2015), adaptive control (Pradhan and Subudhi 2014), sliding mode control (SMC) (Lochan and Roy 2015b; Lochan et al. 2015a, b, 2016c), adaptive SMC (Lochan et al. 2015b, 2016b), hybrid control technique (Pradhan and Subudhi 2012), intelligent control techniques like fuzzy logic control (Subudhi and Morris 2009), artificial neural network (Subudhi and Morris 2009), genetic algorithm (Subudhi and Morris 2009), etc. Most of the reported control techniques on TLFMs use assumed modes modelling method for designing their controllers. But, designing of a controller for a two-link flexible manipulator with singular perturbation modelling method is found to be predominately less. The reported papers using SP on the dynamics of a TLFM are classified in Table 2. It is seen from Table 2, that the use of singular perturbation for designing a controller of a TLFM is still less explored. Motivated with the above discussion, this chapter attempts to broaden the literature by the use of singular perturbation for designing a controller for a TLFM.

Table 2 Categorisation of controllers applied on a singular perturbation model of two-link flexible manipulators

References of paper	Modelling method	Types of control (Slow subsystem)	Types of control (Fast subsystem)
Khorrami et al. (1994)	AMM	PID feedback control	PID
Khorrami and Jain (1993)	AMM	Feedback linearisation	Linear quadratic regulator (LQR)
Subudhi and Morris (2002)	FEM	Computed torque control	LQR based state feedback
Bo and Bakakawa (2004)	AMM	PD	Feedback control
Lee and Lee (2002)	AMM	VSC	Virtual force control
Li et al. (2005)	AMM without gravitational force	PID+ANN	H_∞ control
Zhang et al. (2005)	AMM	Adaptive Normal SMC with H_∞	LQR
Matsuno and Yamamoto (1994)	AMM	PID Feedback control	PID
Wang et al. (2008)	PDE model	Fuzzy non-singular TSMC	Reduced order-observer based LQR
Ashayeri and Farid (2008)	AMM without Disturbance matrix	PD type inverse dynamic based control	Lyapunov based control
Li et al. (2010)	AMM	Normal SMC	H_∞
Mirzaee et al. (2010)	AMM	VSC	Lyapunov based controller
Yue-jiao et al. (2010)	AMM without gravitational force	NN	LQR
Wang et al. (2014)	AMM	Continuous non-singular TMC	Reduced order-observer based LQR
This work	AMM	Adaptive SMC	Backstepping

In this chapter, an adaptive SMC is designed for the path planning/tracking control of the desired signal generated from a hidden attractors chaotic system and a backstepping control for tip deflection suppression of a two-link flexible manipulator. An adaptive SMC and backstepping control techniques are designed for the slow subsystem and fast subsystem, respectively.

The contributions of the chapter are given below:

1. A signal of a hidden chaotic attractors system is considered as the desired signal for path planning/ trajectory tracking of a two-link flexible manipulator.
2. The designed controllers (adaptive-SMC and backstepping) offer faster path planning/trajectory tracking and quicker suppression of link deflection.

3. The robustness of the proposed controller is evaluated in the presence of parameter uncertainties and variation of payloads.
4. The performance of the designed controller is compared with the controller reported in Mirzaee et al. (2010). The proposed controller is found to be better than the controller in Mirzaee et al. (2010).

The organisation of the chapter is as follows. The concept of the dynamic modelling is briefly discussed in Sect. 2. Section 3 introduces the design of composite control. In Sect. 4, a hidden attractor chaotic system is used to generate the desired signal for tracking. The results and discussion are presented in Sect. 5 followed by conclusions in Sect. 6.

2 Modelling of a TLFM

The schematic representation of a planar two-link flexible manipulator (TLFM) is shown in Fig. 1. In Fig. 1, (X_0, Y_0) represents the generalised coordinated frame. $(\hat{X}_i, \hat{Y}_i)$ is the inertial frame and (X_i, Y_i) gives the rigid body moving frame associated with the ith link. W_{hi} and J_{hi} are the mass and inertia, respectively of the ith hub whereas τ_i represents the actuated torque. $u_i(x_i, t)$ represents the elastic deflection of the ith link. M_p and J_p are the mass and inertia, respectively of the payload attached at the end of the final link. θ_i is the ith joint angle of the ith link.

The rigid body motion is described with the help of joint angle θ_i and flexible motion is described with $u_i(l_i, t)$. The dynamical model of a planar TLFM is derived by using Euler-Bernoulli beam theory in the form of a partial differential equation (PDE) along with the boundary conditions representing the motion of the

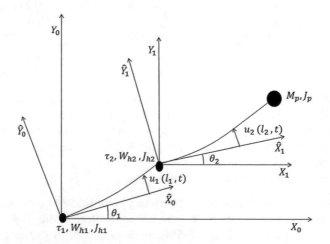

Fig. 1 Physical representation of a two-link flexible manipulator

links. Equation of system energies in the dynamics of a TLFM is obtained by using the Lagrangian formation approach along with assumed modes modelling (AMM) method (Pradhan and Subudhi 2014; Subudhi and Morris 2002). The Lagrangian dynamics of the flexible motion is described as

$$\frac{d}{dt}\frac{\delta((E_{KE})_i - (E_{PE})_i)}{\delta q_i} - \frac{\delta((E_{KE})_i - (E_{PE})_i)}{\delta q_i} = \tau_i \tag{1}$$

where $(E_{KE})_i$ and $(E_{PE})_i$ are the total kinetic and potential energy, respectively, of the ith link and q_i is the generalised coordinate consists of joint angles, joint velocities and modal coordinates (Subudhi and Morris 2002). The total kinetic energy $(E_{KE})_i$ can be obtained as $(E_{KE})_i =$ (total KE due to i^{th} joint) + (total KE due to ith link) + (total KE due M_p) in the absence of gravity. The partial differential equation of the link deflection can be modelled as

$$(EI)_i\frac{\delta^4 u_i(l_i, t)}{\delta l_i^4} + \rho_i\frac{\delta^2 u_i(l_i, t)}{\delta l_i^2} = 0 \tag{2}$$

where

i	ith *link*
$(EI)_i$	*Flexural rigidity*
l_i	*Length*
ρ_i	*Density*
t	*Time*
$u_i(l_i, t)$	*Deflection*

A solution of (2) can be obtained by applying proper boundary conditions. The commonly used different boundary conditions are shown in Fig. 2. The third boundary condition in Fig. 2, also called as pseudo pinned, is locked in the vertical direction but is free to move in the angular direction (DeLuca and Siciliano 1991).

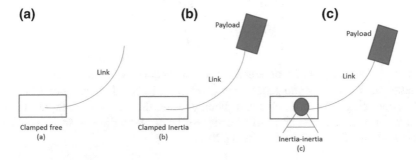

Fig. 2 Different boundary conditions

Considering that the mass of the links is negligible compared with the mass of the payload, we can write as in (DeLuca and Siciliano (1991), Subudhi et al. (2011), Subudhi and Pradhan (2016)):

$$
\begin{cases}
(EI)_i \frac{\delta^4 u_i(l_i,t)}{\delta l_i^4} = -J_{eq_i} \frac{d^2}{dt^2}\left(\frac{\delta u_i(l_i,t)}{\delta l_i}\right) \\
(EI)_i \frac{\delta^3 u_i(l_i,t)}{\delta l_i^3} = -M_{eq_i} \frac{d^2}{dt^2}(u_i(l_i,t))
\end{cases}
\tag{3}
$$

where J_{eq_i} and M_{eq_i} are the moment of inertia and mass, respectively, at the end of the ith link. Finite dimensional expression of the links flexibility $u_i(l_i,t)$ can be written using AMM (DeLuca and Siciliano 1991) as

$$
u_i(l_i,t) = \sum_{j=1}^{n} \phi_{ij}(l_i)\delta_{ij}(t)
\tag{4}
$$

where

ϕ_{ij} jth mode shapes (spatial coordinates)
δ_{ij} jth modal coordinates (time coordinates)
n Number of assumed modes.

The general solution of (1) can be obtained using (4) in the form of time harmonic function and space eigen function as in first equation of (5) and second equation of (5), respectively (DeLuca and Siciliano 1991).

$$
\begin{cases}
\delta_{ij}(t) = e^{j\phi_{ij}t} \\
\phi_{ij} = C_{1,i}\sin(\alpha_i,l_i) + C_{2,i}\cos(\alpha_i,l_i) + C_{3,i}\sinh(\alpha_i,l_i) + C_{4,i}\cosh(\alpha_i,l_i)
\end{cases}
\tag{5}
$$

where ϕ_i is the natural frequency and $(\alpha)_i^4 = \frac{(\phi)_i^4 \rho_i}{(EI)_i}$. Using the first and the second boundary conditions shown in Fig. 2a, b, the constants in (5) can be obtained as

$$
\begin{cases}
C_{3,i} = -C_{1,i}, C_{4,i} = -C_{2,i} \\
[f(\alpha_i,l_i)]\begin{pmatrix} C_{1,i} \\ C_{2,i} \end{pmatrix} = 0
\end{cases}
\tag{6}
$$

The values of α_i can be obtained by solving second equation of (6), using first equation of (6), and second equation of (5). Similarly, the finite solution of the link deflection using (2) can be obtained. Finally, using the Lagrangian expression (1), dynamic model equation of a TLFM using AMM can be written as in (Subudhi and Morris (2002), DeLuca and Siciliano (1991))

$$
B(\theta_i,\delta_i)\begin{pmatrix} \ddot{\theta}_i \\ \ddot{\delta}_i \end{pmatrix} + \begin{pmatrix} H_1(\theta_i,\delta_i,\dot{\theta}_i,\dot{\delta}_i) \\ H_2(\theta_i,\delta_i,\dot{\theta}_i,\dot{\delta}_i) \end{pmatrix} + K\begin{pmatrix} 0 \\ \delta_i \end{pmatrix} + D\begin{pmatrix} \dot{\theta}_i \\ \dot{\delta}_i \end{pmatrix} = \begin{pmatrix} \tau_i \\ 0 \end{pmatrix}
\tag{7}
$$

where

τ_i	Actuated torques
$\delta_i, \dot{\delta}_i$	Modal displacements and velocities
$\theta_i, \dot{\theta}_i$	Joint angle and velocity
B	Positive definite mass inertia matrix
H_1, H_2	Vectors of coriolis and centrifugal forces
K	Positive definite stiffness matrix
D	Positive definite damping matrix

2.1 Singular Perturbation Modelling of a TLFM

This subsection describes the dynamic decomposition of a TLFM dynamics into a slow and a fast subsystem. This is achieved by using the singular perturbation technique using two-time scale property. The slow subsystem consists of the rigid body dynamics of the manipulator and the fast subsystem consists of the flexible mode dynamics of the manipulator. Therefore, the dynamics of a two-link flexible manipulator (TLFM) given in (8) is separated into a slow subsystem (18) which corresponds to rigid dynamics of the manipulator and a fast subsystem (21) which represents the flexible dynamics of the manipulator. It is seen that two separate controllers can be designed for these two subsystems to obtain the desired performances.

The dynamics model (7) of a TLFM can be rewritten as

$$B \begin{pmatrix} \ddot{\theta} \\ \ddot{\delta} \end{pmatrix} + \begin{pmatrix} H_r + D_r \dot{\theta} \\ H_f + D_f \dot{\delta} + K\delta \end{pmatrix} = \begin{pmatrix} \tau_i \\ 0 \end{pmatrix} \tag{8}$$

or in simplified form as

$$\ddot{\theta} = -M_{11}(H_r + D_r\dot{\theta}) - M_{12}(H_f + D_f\dot{\delta} + K\delta) + M_{11}\tau_i \tag{9}$$

$$\ddot{\delta} = -M_{21}(H_r + D_r\dot{\theta}) - M_{22}(H_f + D_f\dot{\delta} + K\delta) + M_{21}\tau_i \tag{10}$$

where

$\theta = [\theta_1, \theta_2]^T] \epsilon R^2$	Vector of joint angle
$\delta = [\delta_{11}, \delta_{12}, \delta_{21}, \delta_{22}]^T \epsilon R^4$	Vector of flexible modes
$D_r \epsilon R^{2X2}, D_f \epsilon R^{4X4}$	Damping matrices
$K \epsilon R^{4X4}$	Stiffness matrix
$H_r \epsilon R^2, H_f \epsilon R^4$	Matrices of gravitational, coriolis and centripetal forces
$B \epsilon R^{6X6}$	Inertia matrix

The inertial matrix B can be represented as

$$M = \begin{pmatrix} M_{11} & M_{12} \\ M_{21} & M_{22} \end{pmatrix} = \begin{pmatrix} B_r & B_{rf} \\ (B_{rf})^T & B_f \end{pmatrix}^{-1} \tag{11}$$

where $M_{11} \epsilon R^{2X2}, M_{12} \epsilon R^{2X4}, M_{21} \epsilon R^{4X2}, M_{22} \epsilon R^{4X4}$ and

$$B_r = [M_{11} - M_{12}(M_{22})^{-1}M_{21}]^{-1} \tag{12}$$

Considering new state variables $\delta = \epsilon q$ and $K_s = \epsilon K$, where ϵ is the singular perturbation parameter which is defined as $\epsilon = 1/K_m$, K_m is the value of smallest stiffness. Using new state variables, the singularly perturbed model of the flexible manipulator dynamics (8) can be written as

$$\ddot{\theta} = -M_{11}(H_r + D_r\dot{\theta}) - M_{12}(H_f + D_f\epsilon\dot{q} + K_s q) + M_{11}\tau_i \tag{13}$$

$$\epsilon\ddot{q} = -M_{21}(H_r + D_r\dot{\theta}) - M_{22}(H_f + D_f\epsilon\dot{q} + K\delta) + M_{21}\tau_i \tag{14}$$

A composite control τ_i is described as

$$\tau_i = \tau_s + \tau_f \tag{15}$$

where τ_s and τ_f are the control inputs for the slow and fast subsystems, respectively.

2.2 Dynamics of a Slow Subsystem

The slow subsystem dynamics of the manipulator can be obtained by considering $\epsilon = 0$ in (14), and solving for q as

$$\bar{q} = K_s^{-1}(\bar{M}_{22})^{-1}(\bar{M}_{21}\bar{D}_r\dot{\bar{\theta}} + \bar{M}_{21}\bar{H}_r + \bar{M}_{22}\bar{H}_f - \bar{M}_{21}\tau_s) \tag{16}$$

where over bar represents the quantity with $\epsilon = 0$.

Substituting (16) in (13), we can write as

$$\ddot{\bar{\theta}} = (\bar{M}_{11} - \bar{M}_{12}(\bar{M}_{22})^{-1}\bar{M}_{21})(-\bar{H}_r - \bar{D}_r\dot{\bar{\theta}} + \tau_s) \tag{17}$$

which represents the rigid body dynamics of the manipulator. Using (12), the slow subsystem dynamics can be written as

$$\ddot{\bar{\theta}} = (\bar{B}_r)^{-1}(-\bar{H}_r - \bar{D}_r\dot{\bar{\theta}} + \tau_s) \tag{18}$$

The state space dynamics of the slow subsystem (18) represents the linear system with $\bar{\theta}$ as the parameter. In order to obtain the dynamics of the fast subsystem, two-time scale method is used. Consider a fast time scale $t = \tau\sqrt{\varepsilon}$ and boundary correction terms $y_1 = q - \bar{q}$ and $y_2 = \sqrt{\varepsilon}\dot{q}$. Thus, using (14), the boundary layer system can be written as

$$\begin{cases} \frac{dy_1}{d\tau} = y_2 \\ \frac{dy_2}{d\tau} = -M_{21}(H_r + D_r\dot{\theta}) - M_{22}(H_f + D_f\varepsilon\dot{q} + K\delta) + M_{21}\tau_i \end{cases} \tag{19}$$

2.3 Dynamics of a Fast Subsystem

Using the property of two-time scale separation, the slow dynamics variables can be treated as frozen parameters (Siciliano and Book 1986), thus $\frac{d\bar{q}}{d\tau} = \sqrt{\varepsilon}\dot{q} = 0$. Using (16) into (19) with $\varepsilon = 0$, we can write as

$$\frac{dy_2}{d\tau} = -\bar{M}_{22}K_s y_1 + \bar{M}_{21}\tau_f \tag{20}$$

Now, the dynamics of the fast subsystem in state space form can be written as

$$\dot{y} = A_f y + B_f \tau_f \tag{21}$$

where $y = [y_1, y_2]^T \varepsilon \mathcal{R}^0$

$$A_f = \begin{pmatrix} 0 & 1 \\ -\bar{M}_{22}K_s & 0 \end{pmatrix}, B_f = \begin{pmatrix} 0 \\ \bar{M}_{21} \end{pmatrix} \tag{22}$$

3 Design of a Composite Control

This section describes the design of a composite control input $\tau_i = \tau_s + \tau_f$ for the TLFM dynamics (11). This is achieved by designing separate controllers τ_s and τ_f for the fast and the slow subsystems, respectively.

3.1 Adaptive Sliding Mode Control for the Slow Subsystem

In order to track the desired trajectory by the TLFM dynamics (11) in the presence of bounded matched disturbances, an adaptive sliding mode control (A-SMC) is designed. The controller is designed with the help of slow subsystem dynamics of the TLFM (18).

The designing of A-SMC is achieved in two steps. The first step is the design of a suitable sliding surface and the second step is the determination of sliding mode control law.

Suppose, the slow subsystem dynamics of the TLFM is affected by the bounded and matched disturbances $\Delta f(\theta)$.

$$\ddot{\bar{\theta}} = -(\bar{B})^{-1}(-\bar{H}_r - \bar{D}_r\dot{\bar{\theta}} + \tau_s) + \Delta f(\theta) \tag{23}$$

It is assumed that the system uncertainties $\Delta f(\theta)$ is bounded as $|\Delta f(\theta)| \leq L$. Suppose, θ_d be a twice differentiable desired trajectory. Here, the desired signal θ_d is considered as a signal generated from a hidden attractors chaotic system. The tracking error for the slow subsystem is defined as

$$e_s = -\theta_d + \bar{\theta} \tag{24}$$

where $\bar{\theta}$ is the states of the slow subsystem. The sliding surface is defined as

$$s = \dot{e}_s + c_s e_s \tag{25}$$

where c_s is a positive definite constant gain matrix. When the system operates on the sliding mode, it satisfies the conditions $s = \dot{s} = 0$, i.e.

$$\begin{cases} s(t) = \dot{e}_s + c_s e_s = 0 \\ \dot{s}(t) = \ddot{e}_s + c_s \dot{e}_s = 0 \end{cases} \tag{26}$$

Therefore, the equivalent sliding mode dynamics can be written as

$$\ddot{e}_s = -c_s\dot{e}_s, \quad e_{sy} = \dot{e}_s, \quad \dot{e}_{sy} = -c_s e_{sy} \tag{27}$$

Now, we show the stability analysis of the equivalent sliding mode dynamics (26) using Lyapunov stability theory. A Lyapunov function candidate is selected as $V_{1s}(e_{sy}) = \frac{1}{2}(e_{sy}(e_{sy})^T)$. The time derivative of the Lyapunov function candidate using (26) can be written as $\dot{V}_{1s}(t) = -c_s(e_{sy}^2)$. Thus, according to Lyapunov stability theory, we can say that sliding motion on the sliding surface is stable and ensures the asymptotical convergence of error dynamics to zero.

After stabilising the sliding surface designed in (25), next step is to design sliding mode control to drive the system trajectory onto the sliding mode $s = 0$.

When the system is in the sliding mode, it satisfies $\dot{s} = 0$ and during the reaching phase, it satisfies

$$\dot{s} < -\rho \tanh(s) \tag{28}$$

Gain ρ is selected such that the reaching condition is satisfied and the sliding mode motion occurs. Now the control torque τ_s is designed to guarantee that the system trajectory hits $s = 0$. Using (18) and (27), the control torque τ_s can be obtained as

$$\tau_s = (\bar{B}_r)[-c_s \dot{e}_s + \ddot{\theta}_d + (\bar{B}_r)^{-1}(\bar{H}_r + \bar{D}_r \dot{\theta}) - \rho \tanh(s)] \tag{29}$$

The ability of the control torque τ_s defined in (29) to drive the system (19) to the sliding mode $s = 0$, can be expressed using Theorem 1.

Theorem 1 *Consider the uncertain slow system dynamics of the TLFM given in (18) and is controlled by τ_s in (29). Then, the trajectory of the slow subsystem (18) converges to the sliding surface, $s = 0$.*

Proof Consider another Lyapunov function candidate as $V(s) = \frac{1}{2} ss^T$, then its time derivative can be written as

$$\dot{V}(s) = s^T \dot{s} = s^T \{(\bar{B}_r)^{-1}(-\bar{H}_r - \bar{D}_r \dot{\theta} + \tau_s) + \Delta f(\theta) - \ddot{\theta}_d + c_s \dot{e}_s\} \tag{30}$$

Using control input (29), (30) can be written as

$$\dot{V}(s) = s^T \dot{s} = s^T \{-\rho \tanh(s) + \Delta f(\theta)\} \tag{31}$$

Now, using the boundness of the uncertainties $|\Delta f(\theta)| \leq L$, (31) can be written as,

$$\dot{V}(s) \leq -\rho |s| + L|s| \leq -|s|(\rho - L) \tag{32}$$

With a suitable choice of $\rho > L$, we can ensure the negative definiteness of the Lyapunov function $V(s)$. Therefore, the closed loop system (slow subsystem of the TLFM (18)) is asymptotically stable.

Now, it is assumed that the system uncertainties $\Delta f(\theta)$ are unknown and satisfy the condition $|\Delta f(\theta)| \leq L < \rho$. In real-life, the upper bound of the uncertainties is unknown and challenging to determine. Then, the control law (29) is modified as:

$$\tau_s = (\bar{B}_r)[-c_s \dot{e}_s + \ddot{\theta}_d + (\bar{B}_r)^{-1}(\bar{H}_r + \bar{D}_r \dot{\theta}) - \hat{\rho} \tanh(s)] \tag{33}$$

where $\hat{\rho}$ is the estimate of ρ. To calculate the parameter $\hat{\rho}$, following adaptation law is defined

$$\dot{\hat{\rho}} = k^{-1}|s| \tag{34}$$

where $k > 0$ is the adaptation gain matrix. Considering $\tilde{\rho} = \hat{\rho} - \rho$, above adaptation law can be written as

$$\dot{\tilde{\rho}} = \dot{\hat{\rho}} = k^{-1}|s| \tag{35}$$

The stability of the adaptation law of the parameter $\hat{\rho}$ can be proved by considering the following Lyapunov function candidate as

$$V_{2s}(s, \rho) = \frac{1}{2}(ss^T + k\tilde{\rho}\tilde{\rho}^T) \tag{36}$$

Taking time derivative of (36) and using (33) and (35), we can write as

$$\dot{V}_{2s}(s,\rho) = s^T\dot{s} + k\tilde{\rho}^T\dot{\tilde{\rho}} = s^T\{-\hat{\rho}|s| + \Delta f(\theta)\} + \tilde{\rho}|s| \tag{37}$$

Using the boundness condition $|\Delta f(\theta)| \le L < \rho$, we can write as

$$\dot{V}(s) = -\rho|s| + s^T\{\Delta f(\theta)\} \le L|s| - \rho|s| \le -(\rho - L)|s| \tag{38}$$

Since $L > 0$, we can say that (38) is a negative definite function. Therefore, we can say that the trajectory of the slow subsystem dynamics converge towards the sliding surface and remains on it.

3.2 Design of Backstepping Control for the Fast Subsystem

In this subsection, a backstepping control technique is designed for controlling the fast subsystem. Dynamics of the fast subsystem of the flexible manipulator can be written as:

$$\begin{cases} \dot{y}_1 = y_2 \\ \dot{y}_2 = -A_{f3}y_1 + B_{f2}\tau_f \end{cases} \tag{39}$$

where $A_{f3} = \bar{M}_{22}K_s$ and $B_{f2} = \bar{M}_{21}$. Suppose, y_d is a twice differentiable desired link deflection, and v is a virtual control variable. The link deflection error is defined as

$$e_{1f} = y_d - y_1 \tag{40}$$

$$e_{2f} = v - y_2 \tag{41}$$

The error dynamics can be obtained as

$$\dot{e}_{1f} = \dot{y}_d - y_2 \tag{42}$$

$$\dot{e}_{2f} = \dot{v} + A_{f3}y_1 - B_{f2}\tau_f \tag{43}$$

The control law designed for controlling the fast subsystem is obtained by using Theorem 2.

Theorem 2 *Suppose the backstepping control law is defined in (44) using the error variable (42) and (43), then the fast subsystem of the manipulator dynamics (39) follows the desired trajectory y_d, i.e. the links deflection of the manipulator are suppressed to zero properly.*

$$\tau_f = (B_{f2})^{-1}(\dot{v} + A_{f3}y_1 + k_2e_{2f}) \tag{44}$$

Proof Designing of a backstepping controller for the fast subsystem of the TLFM (8) is achieved using the following steps:

Step 1: Consider a Lyapunov function candidate as

$$V_{1f} = \frac{1}{2}e_{1f}^2 \tag{45}$$

Time derivative of (45) using (42) and (43), we get

$$\dot{V}_{1f} = e_{1f}(\dot{y}_d + e_{2f} - v) = e_{1f}\dot{y}_d - e_{1f}v + e_{1f}e_{2f} \tag{46}$$

Now, consider the virtual control variable v as

$$v = \dot{y}_d + k_1 e_{1f} + e_{2f} \tag{47}$$

where $k_1 > 0$ is a positive definite matrix. Using (47), the time derivative of Lyapunov function candidate (46) can be written as

$$\dot{V}_{1f} = -k_1 e_{1f}^2 \tag{48}$$

It is seen from (48) that the time derivative of Lyapunov function $\dot{V}_{1f}$ is a negative definite. Thus, the first state variable of the fast subsystem (39) is stabilised. The next step is to show the stability of the second state variable and to obtain the control input τ_f for the fast subsystem.

Step 2: Consider another Lyapunov function candidate as

$$V_{2f} = V_{1f} + \frac{1}{2}e_{2f}^2 \tag{49}$$

Using (43) and (48) in time derivative of (49), we can write as

$$\dot{V}_{2f} = -k_1 e_{1f}^2 + e_{2f}(\dot{v} + A_{f3}y_1 - B_{f2}\tau_f) \tag{50}$$

Now, we can obtain the actual torque input as

$$\tau_f = (B_{f2})^{-1}(\dot{v} + A_{f3}y_1 + k_2 e_{2f}) \tag{51}$$

Using (51), $\dot{V}_{2f}$ in (50) is written as

$$\dot{V}_{2f} = -(k_1 e_{1f}^2 + k_2 e_{2f}^2) \tag{52}$$

Since k_1, k_2 are the positive definite constant matrices, then using Lyapunov stability theory, we can say that (52) is a negative definite function. Thus, the error variables e_{1f} and e_{2f} asymptotically converge to the origin with a suitable choice of constant matrices k_1, k_2. Therefore, the link deflections of the fast subsystem of the TLFM are suppressed to their desired values, i.e. at origin.

The structure of the composite control technique designed for the hidden chaotic path planning is shown in Fig. 3.

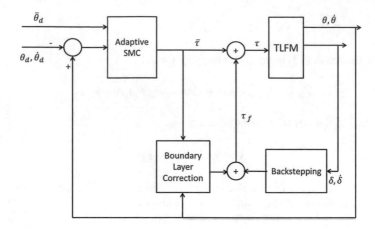

Fig. 3 Structure of the composite control technique

4 Hidden Chaotic Attractor as a Desired Signal

The dynamics of a hidden attractors chaotic system whose signal is used as the desired trajectory for the TLFM is described as Molaie et al. (2013)

$$\begin{cases} \dot{x}_1 = x_2 \\ \dot{x}_2 = x_3 \\ \dot{x}_3 = -x_1 - 2.9x_3^2 + x_1x_2 + 1.1x_1x_3 - 1 \end{cases} \tag{53}$$

System (53) is chaotic with initial conditions $x(0) = (-2.2, 0.6, 0)^T$ where Lyapunov exponents of the system are $L_i = (0.0638, 0, -1.0638)$. The system has the equilibrium points at $E = (-1, 0, 0)$ and stable eigenvalues corresponding to this equilibrium points. Thus, the system given in (53) qualifies to be in the category of hidden

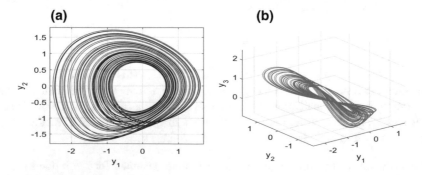

Fig. 4 Hidden chaotic attractors of system (53)

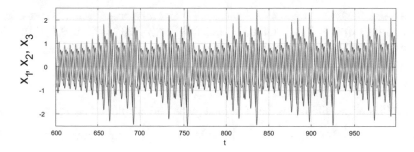

Fig. 5 Hidden chaotic signals of system (53)

chaotic attractors (Molaie et al. 2013). The chaotic attractors and chaotic signals of system (53) with initial conditions $x(0) = (-2.2, 0.6, 0)^T$ are shown in Fig. 4 and Fig. 5, respectively.

In this chapter, the signals x_1, x_2, x_3 are used as the desired trajectories for the TLFM as $\theta_{d1} = \theta_{d2} = x_1$, $\dot{\theta}_{d1} = \dot{\theta}_{d2} = x_2$ and $\ddot{\theta}_{d1} = \ddot{\theta}_{d2} = x_3$.

5 Results and Discussion

This section discusses the results and discussion for the tracking control of the hidden chaotic desired signal. The parameters of the physical TLFM used for simulating the flexible manipulator dynamics (7) are given in Table 3.

All the simulations in this chapter are carried out using ode-45 solver in MATLAB 14-a simulation environment. The initial conditions used for the simulation

Table 3 Symbols and its descriptions

Mass of link-1, $m_1 = 0.15268$ kg	Coefficients of viscous damping, $B_{eq1} = 4$ Nms/rad, $B_{eq2} = 1.5$ Nms/rad
Mass of link-2, $m_2 = 0.0535$ kg	Efficiency of gear boxes, $\eta_{g1} = 0.85$, $\eta_{g2} = 0.9$
Length of link-1, $L_1 = 0.202$ m	Efficiency of motors, $\eta_{m1} = 0.85$, $\eta_{m2} = 0.85$
Length of link-2, $L_2 = 0.2018$	Constants of back e.m.f, $K_{m1} = 0.119$ v/rad, $K_{m2} = 0.0234$ v/rad
Resistance of armatures, $R_{m1} = 11.5\Omega$, $R_{m2} = 2.32\Omega$	Gear ratio, $K_{g1} = 100$, $K_{g2} = 50$
Equivalent MI at load, $J_{eq1} = 0.17043$ kgm^3	Motor torque constants $K_{t1} = 0.119$ Nm/A, $K_{t2} = 0.0234$ Nm/A
Equivalent MI at load, $J_{eq2} = 0.0064387$ kgm^3	Stiffness of the links, $K_{s1} = 22$ Nm/rad, $K_{s2} = 2.5$ Nm/rad
Link-1 MI, $J_{arm1} = 0.002035$ kgm^2	Link-2 MI, $J_{arm2} = 0.0007204$ kgm^2

of two-link flexible manipulator dynamics (7) and adaptation laws (34) are considered as $\theta(0), \delta(0) = (0.1, 0.1, 0.0, 0.0, 0.0, 0.0)^T$, $\dot{\theta}(0), \dot{\delta}(0) = (0, 0, 0, 0, 0, 0)^T$. The system uncertainties added in the manipulator dynamics are given as

$$\Delta f(\theta) = [0.1 \sin(\theta) \sin(\dot{\theta}), 0.1 \sin(\theta) \sin(\dot{\theta})]^T \tag{54}$$

The values of the sliding surface constant gains and adaptation constant gains used for adaptive SMC are $c_s = \begin{pmatrix} 20 & 0 \\ 0 & 20 \end{pmatrix}$ and $k = \begin{pmatrix} 14 & 0 \\ 0 & 14 \end{pmatrix}$, respectively, and the gains of the backstepping controller are considered as $k_1 = k_2 = \begin{pmatrix} 10 & 0 & 0 & 0 \\ 0 & 10 & 0 & 0 \\ 0 & 0 & 10 & 0 \\ 0 & 0 & 0 & 10 \end{pmatrix}$. These gains are considered in a manner to achieve better tracking performances and require less control efforts.

5.1   Simulation Results with a Nominal Payload (0.145 kg)

Hidden chaotic path planning for the two-link flexible manipulator (7) is first discussed here with a nominal payload (0.145 kg) and in the presence of system uncertainties (54). The trajectory tracking errors for both the links are shown in Fig. 6. It is observed from Fig. 6 that the chaotic trajectory tracking for both the links are achieved within 6 s. The modes of the first and second flexible link with 0.145 kg payload are shown in Fig. 7 and Fig. 8, respectively. It is apparent from Figs. 7 and 8 that the modes of the links are suppressed within values $[10^{-3}, 10^{-5}]$ and $[10^{-4}, 10^{-6}]mm$, respectively. The behaviour of tip deflection for both the links are shown in Fig. 9. It is noted from Fig. 9 that the tip deflections of the first and second link are suppressed within the values 10^{-2} mm and 10^{-3} mm, respectively. Responses

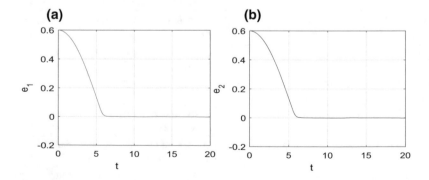

Fig. 6 Chaotic path following errors of the joint angles with nominal payload: **a** link-1 and **b** link-2

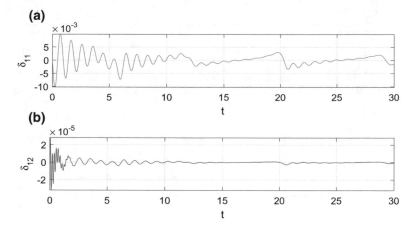

Fig. 7 Modes with nominal payload condition: **a** mode-1 for link-1 and **b** mode-2 for link-1

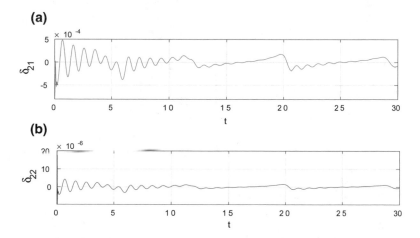

Fig. 8 Modes with nominal payload condition: **a** mode-1 for link-2 and **b** mode-2 for link-2

of the sliding surfaces designed for the slow subsystem of both the links are shown in Fig. 10. The required control torque inputs in the slow subsystem are shown in Fig. 11. It is seen from Figs. 10 and 11 that the chattering is not present in the sliding surfaces and control inputs. Responses of the control input required in the fast subsystem are shown in Fig. 12. Responses of the composite control input designed for the two-link flexible manipulator are shown in Fig. 13. It is noted from Fig. 13 that the required control torques using the composite control for both the links are initially high but after sometime, these torques decay within small values. It is also seen from Fig. 13 that the variation in control inputs is low.

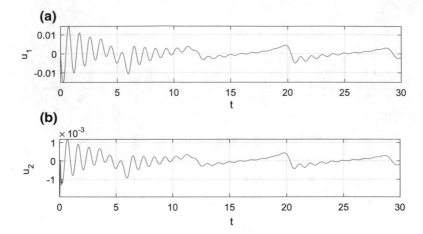

Fig. 9 Response of tip deflections of both the links for chaotic path planning with nominal payload

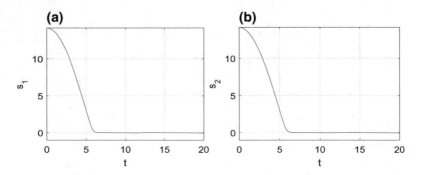

Fig. 10 Sliding surfaces with nominal payload

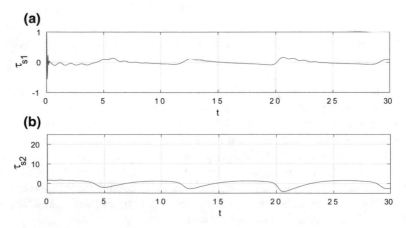

Fig. 11 Required control torque during chaotic trajectory tracking for the slow subsystem with the nominal payload

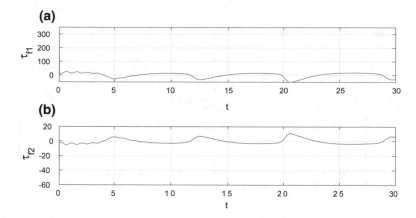

Fig. 12 Required control torque for the fast subsystem with the nominal payload

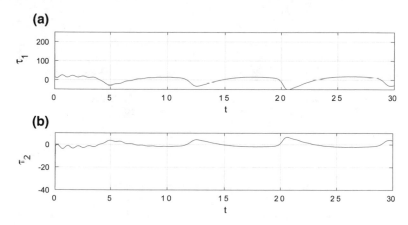

Fig. 13 Composite control torque for the two-link flexible manipulator during chaotic trajectory tracking with the nominal payload

5.2 Simulation Results with 0.3 kg Payload

This subsection describes the robustness of the controllers for chaotic tracking results of two-link flexible manipulator (7) with a payload of 0.3 kg. The chaotic path planning errors for both the links with 0.3 kg payload are shown in Fig. 14. It is apparent from Fig. 14 that with the addition of payload, the settling time (8 s) of the tracking error increases as compared with the nominal payload (6 s). Behaviours of the modes of both the links are shown in Figs. 15 and 16, respectively. The responses of the tip deflection of both the links are shown in Fig. 17. It is noted from Fig. 17 that with an increase of payload there is a small increase in the magnitude of deflection

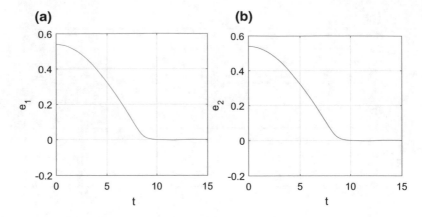

Fig. 14 Chaotic path tracking errors of the joint angles with 0.3 kg payload: **a** link-1 and **b** link-2

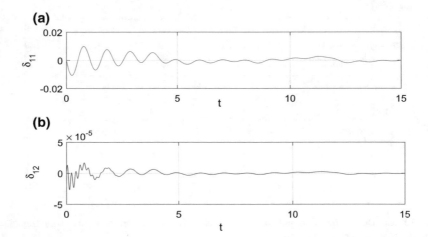

Fig. 15 Modes with 0.3 kg payload condition: **a** mode-1 for link-1 and **b** mode-2 for link-2

for both the links. The behaviours of the sliding surfaces designed for the slow subsystem with 0.3 *kg* payload are shown in Fig. 18. It is seen from Fig. 18 that there is an increase of reaching time (3 *s*) for the sliding surfaces with 0.3 *kg* payload as compared with the nominal payload. Responses of the composite control inputs generated for the two-link flexible manipulator with 0.3 *kg* payload are shown in Fig. 19. It is visible from Fig. 19 that due to increase in the payload on the tip of the second link, it requires more control efforts as compared with the nominal payload.

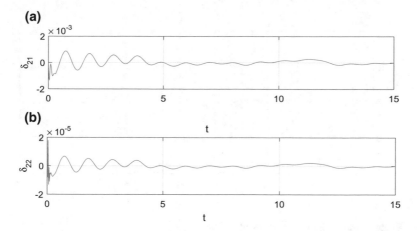

Fig. 16 Modes with 0.3 kg payload condition: **a** mode-1 for link-2 and **b** mode-2 for link-2

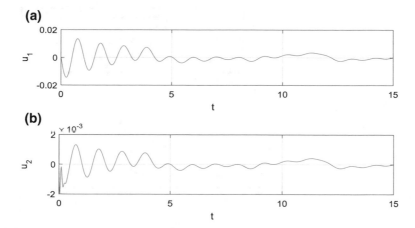

Fig. 17 Responses of the tip deflections of link-1 and link-2 for chaotic path planning with 0.3 kg payload

5.3 Comparison of the Proposed Composite Controller with the Controller of (Mirzaee et al. 2010)

Chaotic trajectory tracking performances of the proposed composite controller with a nominal payload for the TLFM given in (7) are compared with the composite controller available (Mirzaee et al. 2010). Tracking errors using the proposed composite controller and the controller of (Mirzaee et al. 2010) for both the links are shown in Fig. 20. It is apparent from Fig. 20 that the reaching time for the proposed controller is greater than (Mirzaee et al. 2010). But, the steady state error using the

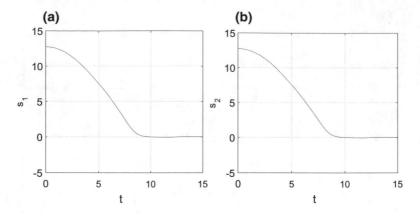

Fig. 18 Response of sliding surfaces with 0.3 kg payload

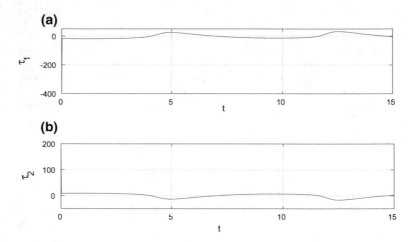

Fig. 19 Composite control torque for the two-link flexible manipulator during chaotic trajectory tracking with 0.3 kg payload

proposed composite controller is lower as compared with the controller of (Mirzaee et al. 2010). The behaviour of the tip deflections using both the composite controllers are shown in Fig. 21. It is observed from Fig. 21 that tip deflection using the proposed controller is lower as compared with controller of (Mirzaee et al. 2010). The natures of the proposed composite control technique and composite controller of (Mirzaee et al. 2010) are shown in Fig. 22. It is noted from Fig. 22 that the required control efforts using the proposed composite controller is comparatively less than that of the controller of (Mirzaee et al. 2010) in link-2 and is comparatively same in link-1. The comparison on the performances of the controller are given in Table 4.

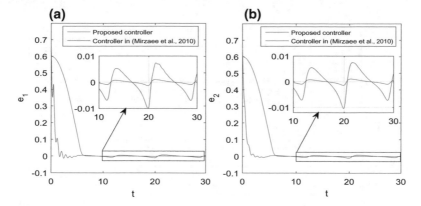

Fig. 20 Chaotic path tracking errors of the joint angles with nominal payload: **a** link-1 and **b** link-2 using the proposed composite controller and controller of (Mirzaee et al. 2010)

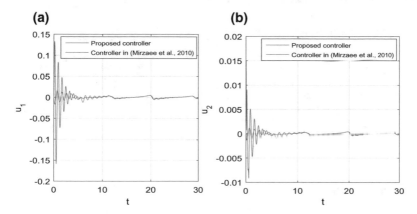

Fig. 21 Responses of the tip deflections of both the links for chaotic path planning with nominal payload using the proposed composite controller and controller of (Mirzaee et al. 2010)

The 2-norm and integral square error (ISE) are the two performance indices considered for comparison. It is seen from the Table 4 that the proposed control technique required less control efforts and has less ISE in the case of links deflection. Moreover, from the results of Figs. 20, 21 and 22, we can say that the proposed composite controller performs better than the controller of (Mirzaee et al. 2010) in terms of steady state tracking error, tip deflection and control efforts.

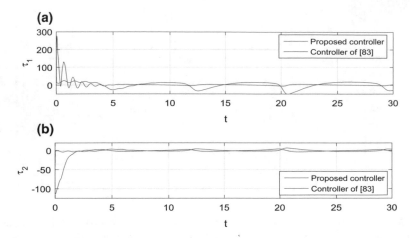

Fig. 22 Comparison of nature of composite control torques for the two-link flexible manipulator during chaotic trajectory tracking with nominal payload

Table 4 Comparison on performances of controllers

	Performances of controllers			
Control techniques	Proposed controller		Controller in (Mirzaee et al. 2010)	
Performance Indices	Link-1	Link-2	Link-1	Link-2
Control energy ($\|u\|_2$)	98.0413	45.7030	176.6112	555.5637
ISE of tip deflections	0.058	0.0047	0.164	0.011

6 Conclusions

In this chapter, hidden chaotic trajectory tracking of a two-link flexible manipulator is proposed. A chaotic system with hidden chaotic attractor is used for the generation of the desired signals. Since, the dynamics of a TLFM is very nonlinear and complex, trajectory tracking of such a system is a challenging task. To meet the challenge, the dynamics of the TLFM is decomposed into two subsystems consisting of the slow and the fast subsystems. The slow subsystem consists of the rigid dynamics of the manipulator and the fast subsystem consists of the flexible mode dynamics of the manipulator. This is achieved by the singular perturbation method. Then, a composite controller consisting of controller of both subsystems is designed. The composite control technique consists of adaptive sliding mode control (A-SMC) for slow subsystem and a backstepping controller for the fast subsystem. It is shown using MATLAB simulation that the proposed composite controller works effectively. The performances of the proposed composite controller are compared with the controller available in (Mirzaee et al. 2010). The simulation results confirm that the proposed composite controller has better performances than the controller given in

(Mirzaee et al. 2010). The proposed controller has smaller steady state errors, quick and smaller tip deflection and required lesser control efforts when compared with the controller in (Mirzaee et al. 2010).

References

Andrievskii BR, Fradkov AL (2004) Control of chaos : methods and applications. II. applications. Autom Remote Control 65(4):505–533

Aoustin Y, Formal'sky A (1999) On the feedforward torques and reference trajectory for flexible two-link arm. Multibody Syst Dyn 3:241–265

Arora A, Ambe Y, Kim TH, Ariizumi R, Matsuno F (2014) Development of a maneuverable flexible manipulator for minimally invasive surgery with varied stiffness. Artif Life Robot 19(4):340–346

Ashayeri A, Farid M (2008) Trajectory tracking for two-link flexible arm via two-time scale and boundary control methods. In: Proceedings of IMECE2008 2008 ASME international mechanical engineering congress and exposition, pp 1–9

Azar AT, Vaidyanathan S, Ouannas A (2017) In: Fractional order control and synchronization of chaotic systems, 688

Bo XU, Bakakawa Y (2004) Control two-link flexible manipulators using controlled Lagrangian method. In: SICE annual conference in Sapporo, pp 289–294

Bruno D, Calinon S, Caldwell DG (2014) Null space redundancy learning for a flexible surgical robot. In: IEEE international conference on robotics and automation, Hong Kong Convention and Exhibition Center, pp 2443–2448. https://doi.org/10.1109/ICRA.2014.6907199

Chen G, Ueta T (1999) Yet another chaotic attractor. Int J Bifurcat Chaos 9:1465–1999

Chen M, Xu Q, Lin Y, Bao B (2017) Multistability induced by two symmetric stable node-foci in modified canonical Chua's circuit. Nonlinear Dyn 87(2):789–802

Chu Z, Cui J (2015) Experiment on vibration control of a two-link flexible manipulator using an input shaper and adaptive positive position feedback. Adv Mech Eng 7(10):1–13

DeLuca A, Siciliano B (1991) Closed-form dynamic model of planar multilink lightweight robots. IEEE Trans Syst Man Cybern 21(4):826–839

Esteban T, Rangel M, de la Fraga L (2016) Engineering applications of FPGAs : chaotic systems, artificial neural networks, random number generators, and secure communication systems. Springer, Switzerland

Hoppensteadt FC (2000) In: analysis and simulation of chaotic systems, 94

Jafari S, Sprott JC (2013) Simple chaotic flows with a line equilibrium. Chaos, Solitons Fractals 57:79–84

Karagulle H, Malgaca L, Dirilmis M, Akdag M, Yavuz S (August 2015) Vibration control of a two-link flexible manipulator. J Vib Control

Khorrami F, Jain S (1993) Non-linear control with end-point acceleration feedback for a two-link flexible manipulator: experimental results. J Robot Syst 10(4):505–530

Khorrami F, Jain S, Tzes A (1994) Experiments on rigid body-based controllers with input preshaping for a two-link flexible manipulator. IEEE Trans Robot Autom 10(11):55–65

Kiang CT, Spowage A, Yoong CK (2014) Review of control and sensor system of flexible manipulator. J Intell Robot Syst Theory Appl 77(1):187–213

Kingni ST, Pham VT, Jafari S, Kol GR, Woafo P (2016) Three-dimensional chaotic autonomous system with a circular equilibrium: analysis, circuit implementation and its fractional-order form. Circ Syst Sig Process 35(6):1807–1813

Kiseleva MA, Kuznetsov NV, Leonov GA (2016) Hidden and self-excited attractors in electromechanical systems with and without equilibria. pp. 1–10. arXiv:1601.069091601.06909

Korayem MH, Haghpanahi M (2009) Finite element method and optimal control theory for path planning of elastic manipulators. New Adv Intell Decis Technol 199(2009):117–126

Lee SH, Lee CW (2002) Hybrid control scheme for robust tracking of two-link flexible manipulator. J Intell Robot Syst 34(4):431–452

Leonov GA, Kuznetsov NV (2013) Hidden attractors in dynamical systems: from hidden oscillations in Hilbert-Kolmogorov, Aizerman, and Kalman problems to hidden chaotic attractor in Chua circuits. Int J Bifurcat Chaos 23(01):1330,002–130,071

Leonov GA, Kuznetsov NV, Kuznestova OA, Seledzhi SM, Vagaitsev VI (2011a) Hidden oscillations in dynamical systems system. Trans Syst Control 6(2):1–14

Leonov GA, Kuznetsov NV, Vagaitsev VI (2011b) Localization of hidden Chuas attractors. Phy Lett A 375(23):2230–2233

Leonov GA, Kuznetsov NV, Vagaitsev VI (2012) Hidden attractor in smooth Chua systems. Phy D 241(18):1482–1486

Leonov GA, Kuznetsov NV, Kiseleva MA, Solovyeva EP, Zaretskiy AM (2014) Hidden oscillations in mathematical model of drilling system actuated by induction motor with a wound rotor. Nonlinear Dyn 77(1–2):277–288

Leonov GA, Kuznetsov NV, Mokaev TN (2015) Hidden attractor and homoclinic orbit in Lorenz-like system describing convective fluid motion in rotating cavity. Commun Nonlinear Sci Numer Simul 28(1–3):166–174. https://doi.org/10.1016/j.cnsns.2015.04.0071412.7667

Li Y, Liu G, Hong T, Liu K (2005) Robust control of a two-link flexible manipulator with quasi-static deflection compensation using neural networks. J Intell Robot Syst 44(3):263–276

Li YC, Tang BJ, Shi ZX, Lu YF (2010) Experimental study for trajectory tracking of a two-link flexible manipulator. Int J Syst Sci 31(1):3–9

Lochan K, Roy BK (2015a) Control of two link 2-DOF robot manipulator using fuzzy logic techniques: a review. In: Proceedings of fourth international conference on soft computing for problem solving, pp 499–511

Lochan K, Roy BK (2015b) Position control of two-link flexible manipulator using low chattering SMC techniques. Int J Control Theory Appl 8(3):1137–1146

Lochan K, Roy BK (2016) Trajectory tracking control of an AMM modelled TLFM using backstepping method. Int J Control Theory Appl 9(39):239–246

Lochan K, Suklabaidya S, Roy BK (2014) Dynamic modelling, simulation of single-link flexible robot manipulator and its control. J Control Instrum 5(3):15–21

Lochan K, Suklabaidya S, Roy BK (June 2015a) Comparison of chattering in single link flexible manipulator with sliding mode controllers. In: 2015 international conference on energy, power and environment: towards sustainable growth (ICEPE)

Lochan K, Suklabaidya S, Roy BK (2015b) Sliding mode and adaptive sliding mode control approaches of two link flexible manipulator. In: Proceedings of 2nd advances in robotics, Goa, India, 1, pp 2–7

Lochan K, Roy BK, Subudhi B (2016a) A review on two-link flexible manipulators. Annu Rev Control 42:346–367

Lochan K, Roy BK, Subudhi B (2016b) Generalized projective synchronization between controlled master and multiple slave TLFMs with modified adaptive SMC. Trans Inst Measur Control. https://doi.org/10.1177/0142331216674067

Lochan K, Roy BK, Subudhi B (2016c) SMC controlled chaotic trajectory tracking of two-link flexible manipulator with PID sliding surface. IFAC-PapersOnLine 49(1):219–224

Lorenz EN (1963) Deterministic nonperiodic flow. J Atmos Sci 20(2):130–141

Lü J, Chen G, Cheng D, Celikovsky S (2002) Bridge the gap between the Lorenz system and the Chen system. Int J Bifurcat Chaos 12(12):2917–2926

Masoud M, Mostafa G, Mostafa SN (2010) Observer based tip tracking control of two-link flexible manipulator. Int Conf Control Autom Syst 2010:9–13

Matsuno F, Yamamoto K (1994) Dynamic hybrid force/position control of two degrre of freedom flexible manipulator. J Robot Syst 11(5):355–366

Mirzaee E, Eghtesad M, Fazelzadeh SA (2010) Maneuver control and active vibration suppression of a two-link flexible arm using a hybrid variable structure/Lyapunov control design. Acta Astronaut 67(9–10):1218–1232

Molaie M, Jafari S, Sprott JC, Golpayegani S, Hashemi MR (2013) Simple chaotic flows with one stable equilibrium. Int J Bifurcat Chaos 23(11):1350,188

Nakamura T, Saga N, Nakazawa M, Kawamura T (2003) Development of a soft manipulator using a smart flexible joint for safe contact with humans. In: IEEE/ASME international conference on advanced intelligent mechatronics, AIM, pp 441–446

Nichols JM, Todd MD, Wait JR (2003) Using state space predictive modeling with chaotic interrogation in detecting joint preload loss in a frame structure. Smart Mater Struct 12:580–601

Özer A, Semercigil SE (2010) Effective vibration supression of maneuvering two-link flexible arm with an event-based stiff. Proceedings of the IMAC-XXVIII 3:323–330

Pham V, Jafari S, Volos C, Giakoumis A, Vaidyanathan S, Kapitaniak T (2016a) A chaotic system with equilibria located on the rounded square loop and its circuit implementation. IEEE Trans Circ Syst II Express Briefs 63(9):878–882

Pham VT, Volos C, Jafari S, Wang X, Vaidyanathan S (2014a) Hidden hyperchaotic attractor in a novel simple memristive neural network. Optoelectron Adv Mater Rapid Commun 8(11–12):1157–1163

Pham VT, Volos C, Jafari S, Wei Z, Wang X (2014b) Constructing a novel no equilibrium chaotic system. Int J Bifurcat Chaos 24(05):1450,073–1450,087

Pham VT, Jafari S, Volos C, Giakoumis A, Vaidyanathan S, Kapitaniak T (2016b) A chaotic system with equilibria located on the rounded square loop and its circuit implementation. IEEE Trans Circ Syst II Express Briefs 63(9):878–882

Pham VT, Jafari S, Volos C, Vaidyanathand S, Kapitaniake T (2016c) A chaotic system with infinite equilibria located on a piecewise linear curve 127:9111–9117

Pham VT, Jafari S, Wang X (2016d) A chaotic system with different shapes of equilibria. Int J Bifurcat Chaos 26(04):1650,069–1650,074

Pham VT, Sundarapandian V, Volos CK, Jafari S, Kuznetsov NV, Hoang TM (2016e) A novel memristive time-delay chaotic system without equilibrium points. Eur Phy J Spec Top 225(1):127–136

Pham VT, Volos C, Jafari S, Kapitaniak T (2016f) Coexistence of hidden chaotic attractors in a novel no-equilibrium system. Nonlinear Dyn pp 1–10

Pham VT, Volos C, Vaidyanathan S, Wang X (2016g) A chaotic system with an infinite number of equilibrium points dynamics, horseshoe, and synchronization. Adv Math Phy 2016:1–9

Pham VT, Akgul A, Volos C, Jafari S, Kapitaniak T (2017a) Dynamics and circuit realization of a no-equilibrium chaotic system with a boostable variable. AEU Int J Electron Commun 78:134–140

Pham VT, Takougang S, Volos C, Jafari S, Kapitaniak T (2017b) A simple three-dimensional fractional-order chaotic system without equilibrium: Dynamics, circuitry implementation, chaos control and synchronization. AEÜ Int J Electron Commun (In Press)

Pradhan SK, Subudhi B (2012) Real-time adaptive control of a flexible manipulator using reinforcement learning. IEEE Trans Autom Sci Eng 9(2):237–249

Pradhan SK, Subudhi B (2014) Nonlinear adaptive model predictive controller for a flexible manipulator: an experimental study. IEEE Trans Control Syst Technol 22(5):1754–1768

Sabatini M, Gasbarri P, Monti R, Palmerini GB (2012) Vibration control of a flexible space manipulator during on orbit operations. Acta Astronaut 73(1):109–121

Siciliano B, Book WJ (1986) A singular perturbation approach to control of lightweight flexible manipulators. PhD thesis

Singh JP, Roy BK (2015a) A novel asymmetric hyperchaotic system and its circuit validation. Int J Control Theory Appl 8(3):1005–1013

Singh JP, Roy BK (2015b) Analysis of an one equilibrium novel hyperchaotic system and its circuit validation. Int J Control Theory Appl 8(3):1015–1023

Singh JP, Roy BK (2014) (2016a) Comment on "Theoretical analysis and circuit verification for fractional-order chaotic behavior in a new hyperchaotic system". Math Probl Eng 1:1–4

Singh JP, Roy BK (2016b) Crisis and inverse crisis route to chaos in a new 3-D chaotic system with saddle , saddle foci and stable node foci nature of equilibria. Optik 127(24):11,982–12,002

Singh JP, Roy BK (2016c) The nature of Lyapunov exponents is (+, +, -, -). Is it a hyperchaotic system?. Chaos Solitons Fractals 92:73–85

Singh JP, Roy BK (2017a) Multistability and hidden chaotic attractors in a new simple 4-D chaotic system with chaotic 2-torus behaviour. Int J Dyn Control (In Press). https://doi.org/10.1007/s40435-017-0332-8

Singh JP, Roy BK (2017b) The simplest 4-D chaotic system with line of equilibria, chaotic 2-torus and 3-torus behaviour. Nonlinear Dyn. https://doi.org/10.1007/s11071-017-3556-4

Singh PP, Singh JP, Roy BK (2014) Synchronization and anti-synchronization of Lu and Bhalekar-Gejji chaotic systems using nonlinear active control. Chaos Solitons Fractals 69:31–39

Sprott JC (1994) Some simple chaotic flow. Phy Rev E 50(2):647–650

Subudhi B, Morris AS (2002) Dynamic modelling, simulation and control of a manipulator with flexible links and joints. Robot Auton Syst 41(4):257–270

Subudhi B, Morris AS (2009) Soft computing methods applied to the control of a flexible robot manipulator. Appl Soft Comput 9(1):149–158

Subudhi B, Pradhan SK (2016) A flexible robotic control experiment for teaching nonlinear adaptive control. Int J Electr Eng Educ. http://ije.sagepub.com/lookup/doi/10.1177/0020720916631159

Subudhi B, Ranasingh S, Swain AK (2011) Evolutionary computation approaches to tip position controller design for a two-link flexible manipulator. Arch Control Sci 21(3):269–285

Suklabaidya S, Lochan K, Roy BK (2014a) Dynamic modelling and state-feedback control of a rotational base flexible link manipulator. J Electron Des Technol 5(3):15–21

Suklabaidya S, Lochan K, Roy BK (2014b) Modeling and sliding mode control of flexible link flexible joint robot manipulator. In: Proceedings of the 2015 conference on advances in robotics, 59. https://doi.org/10.1145/2783449.2783509

Suklabaidya S, Lochan K, Roy BK (2015) Control of rotational base single link flexible manipulator using different SMC techniques for variable payloads. In: 2015 international conference on energy, power and environment, towards sustainable growth (ICEPE), pp 1–6

Tlelo-Cuautle E, Ramos-López HC, Sánchez-Sánchez M, Pano-Azucena AD, Sánchez-Gaspariano LA, Núñez-Pérez JC, Camas-Anzueto JL (2014) Application of a chaotic oscillator in an autonomous mobile robot. J Electr Eng 65(3):157–162

Tlelo-Cuautle E, Carbajal-Gomez VH, Obeso-Rodelo PJ, Rangel-Magdaleno JJ, Núñez-Pérez JC (2015) FPGA realization of a chaotic communication system applied to image processing. Nonlinear Dyn 82(4):1879–1892

Vaidyanathan S, Volos C (2016) In: advances and applications in chaotic systems, 636

Vaidyanathan S, Pham VT, Volos CK (2015) A 5-D hyperchaotic Rikitake dynamo system with hidden attractors. Eur Phy J Spec Top 224(8):1575–1592

Wang X, Vaidyanathan S, Volos C, Pham VT, Kapitaniak T (2017) Dynamics, circuit realization, control and synchronization of a hyperchaotic hyperjerk system with coexisting attractors. Nonlinear Dyn (In Press)

Wang Y, Feng Y, Yu X (2008) Fuzzy terminal sliding mode control of two-link flexible manipulators. vol 2, pp 1620–1625. http://ieeexplore.ieee.org/lpdocs/epic03/wrapper.htm?arnumber=4758196

Wang Y, Han F, Feng Y, Hongwei X (2014) Hybrid continuous nonsingular terminal sliding mode control of uncertain flexible manipulators. In: 40th IEEE annual conference of the industrial electronics society (IECON), 51307035, pp 190–196. https://doi.org/10.1109/IECON.2009.5415316

Yu Y, Yuan Y, Fan X, Yang H (2015) Back-stepping control of two-link flexible manipulator based on extended state observer. Adv Space Res 56:2312–2322

Yue-jiao D, Xi C, Ming Z, Jun R (2010) Anti-windup for two-link flexible arms with actuator saturation using neural network. In: 2010 international conference on e-product e-service and e-entertainment (ICEEE), 06, pp 6–9

Zhang L, Liu J (2012) Observer-based partial differential equation boundary control for a flexible two-link manipulator in task space. IET Control Theory Appl 6(13):2120–2133

Zhang Y, Mi Y, Zhu M, Lu F (Aug 2005) Adaptive sliding mode control for two-link flexible manipulator with H infinity tracking, pp 702–707

5-D Hyperchaotic and Chaotic Systems with Non-hyperbolic Equilibria and Many Equilibria

Jay Prakash Singh and Binoy Krishna Roy

Abstract In the present decade, chaotic systems are used and appeared in many fields like in information security, communication systems, economics, bioengineering, mathematics, etc. Thus, developing of chaotic dynamical systems is most interesting and desirable in comparison with dynamical systems with regular behaviour. The chaotic systems are categorised into two groups. These are (i) system with self-excited attractors and (ii) systems with hidden attractors. A self-excited attractor is generated depending on the location of its unstable equilibrium point and in such case, the basin of attraction touches the equilibria. But, in the case of hidden attractors, the basin of attraction does not touch the equilibria and also finding of such attractors is a difficult task. The systems with (i) no equilibrium point and (ii) stable equilibrium points belong to the category of hidden attractors. Recently chaotic systems with infinitely many equilibria/a line of equilibria are also considered under the cattegory of hidden attractors. Higher dimensional chaotic systems have more complexity and disorders compared with lower dimensional chaotic systems. Recently, more attention is given to the development of higher dimensional chaotic systems with hidden attractors. But, the development of higher dimensional chaotic systems having both hidden attractors and self-excited attractors is more demanding. This chapter reports three hyperchaotic and two chaotic, 5-D new systems having the nature of both the self-excited and hidden attractors. The systems have non-hyperbolic equilibria, hence, belong to the category of self-excited attractors. Also, the systems have many equilibria, and hence, may be considered under the category of a chaotic system with hidden attractors. A systematic procedure is used to develop the new systems from the well-known 3-D Lorenz chaotic system. All the five systems exhibit multistability with the change of initial conditions. Various theoretical and numerical tools like phase portrait, Lyapunov spectrum, bifurcation diagram, Poincaré map, and frequency spectrum are used to confirm the chaotic nature of the new systems.

J. P. Singh (✉) · B. K. Roy
National Institute of Technology Silchar, Silchar 788010, Assam, India
e-mail: jayprakash1261@gmail.com

B. K. Roy
e-mail: bkr_nits@yahoo.co.in

© Springer International Publishing AG 2018
V.-T. Pham et al. (eds.), *Nonlinear Dynamical Systems with Self-Excited and Hidden Attractors*, Studies in Systems, Decision and Control 133,
https://doi.org/10.1007/978-3-319-71243-7_20

465

The MATLAB simulation results of the new systems are validated by designing their circuits and realising the same.

Keywords Non-hyperbolic equilibria · Many equilibria · A line of equilibria Hyperchaotic system · Chaotic system · 5-D chaotic system 5-D hyperchaotic system · Hidden attractors

1 Introduction

Recently, the development and applications of chaotic systems are seen in many fields like in communication theory (Xiong et al. 2016), image processing (Tlelo-Cuautle et al. 2015a, b), information theory (Esteban et al. 2016; Valtierra et al. 2016), robotics (Lochan and Roy 2015, 2016; Lochan et al. 2016a, b, c; Tlelo-Cuautle et al. 2014; Singh et al. 2017a, b; Andrievskii and Fradkov 2004), etc. Based on the desired behaviours and responses, many hyperchaotic/chaotic systems are reported in the last decade (Pham et al. 2014, 2016g; Vaidyanathan et al. 2015). An equilibrium point plays an important role in the generation of the desired behaviour and responses. Recently, many hyperchaotic/chaotic systems are reported based on different nature of equilibrium points (Pham et al. 2016e, f, h, 2017a, b; Wang et al. 2017; Sharma et al. (2015)). Higher dimensional (4-D/5-D) hyperchaotic/chaotic systems are more important from the application point of view as compared with the lower dimensional systems (Pham et al. 2016b; Shen et al. 2014a, b). This is because of their more complex and disorder behaviour as compared with the lower dimensional systems (Shen et al. 2014a, b). Thus, development of higher dimensional (5-D) hyperchaotic/chaotic systems with unique and interesting nature of equilibrium points is the motivational background of this work.

Many control techniques are proposed in the literature and used in the last decade for the applications of hyperchaotic/chaotic systems. Some of these are sliding mode control (SMC) (Singh and Roy 2015a), backstepping control (Yu et al. 2012), feedback control (Pang and Liu 2011), nonlinear active control (Singh et al. 2014a, 2017a, b), adaptive control (Effati et al. 2014), H_∞ (Wang et al. 2013), sampled data control (Lam and Li 2014), etc.

The reported hyperchaotic/chaotic systems can be classified into two major categories. These are: (i) hyperchaotic/chaotic systems with hidden attractors and (ii) self-excited attractors hyperchaotic/chaotic systems (Leonov and Kuznetsov 2013; Leonov et al. 2011a, b, 2012, 2014; Singh and Roy 2017a; Singh and Roy (2016a); Singh et al. (2015)). Some of the conventional chaotic systems like Lorenz system (Lorenz 1963), Rössler (1976), Chen and Ueta (1999), Lü et al. (2002), Bhalekar-Gejji systems (Singh et al. 2014a) and systems in Singh and Roy (2015a, b, 2016a, b), etc., are grouped under the category of self-excited attractors. The family of the hyperchaotic/chaotic systems with hidden attractors is grouped with the systems having (i) only stable equilibrium points (Kingni et al. 2014), (ii) no equilibrium point (Lin et al. 2016; Singh and Roy 2017a) and (iii) an infinite number of equilibria (Jafari and Sprott 2013; Wang and Chen 2012). The chaotic systems with an infinite/line/many equilibria belong to the category of hidden attractors (Leonov et al. 2014, 2015; Pham et al. 2016a, b, c, d, e, f, g, h). In a hyperchaotic/chaotic system with hidden attractors, the basin of attraction does not intersect with small neighbourhoods of its equilibria (Leonov et al. 2011a, b, 2012). However, in a chaotic system with infinitely many equilibria, the basin of attraction may intersect the equilibrium surface in some sections. Since there are usually uncountable sections/points on the surface of equilibria which are outside the basin of attraction and from a computational point of view, these attractors of hyperchaotic/chaotic systems with many equilibria also belong to the family of hidden attractors (Barati et al. 2016; Pham et al. 2016a, b). Because the knowledge about the locations of equilibria in such systems does not help in the generation of attractors.

Very less attention is given to the development of 5-D hyperchaotic/chaotic systems (Kemih et al. 2013; Ojoniyi and Njah 2016; Vaidyanathan et al. 2014, 2015, 2016). Recently, many hyperchaotic/chaotic systems with an infinite number of equilibria are reported. The systems with infinitely many equilibria are the systems with a line of equilibria (Singh and Roy 2017b), plane of equilibria (Jafari et al. 2016a), surface of equilibria (Jafari et al. 2016b), sphere of equilibria (Qi and Chen 2015), square shaped equilibria (Qi and Chen 2015; Gotthans et al. 2016; Pham et al. 2016a, b, c, d, e, f, g, h), etc. The reported systems with an infinite number of equilibria are classified in Table 1.

It is seen from Table 1 that very few 5-D hyperchaotic/chaotic systems are reported with infinitely many equilibria. Motivated by this finding, an attempt is made in this chapter to construct five new 5-D hyperchaotic/chaotic systems with infinitely many equilibria. Multistability in a hyperchaotic/chaotic system is defined as the coexistence of various possible steady states/attractors of the system (Pisarchik and Feudel 2014; Sharma et al. 2015; Kiseleva et al. 2017). The occurrence of multistability is governed by the choice of initial conditions, hence, creates a complicated basin of attraction (Pisarchik and Feudel 2014; Sharma et al. 2015). Multistability is seen in several areas (Sharma et al. 2015), like in an electronic circuit, a laser system, chaotic/hyperchaotic system, etc., (Chen et al. 2017; Chudzik et al. 2011; Leonov and Kuznetsov 2013; Pisarchik and Feudel 2014; Sharma et al. 2015).

Most of the reported chaotic systems have hyperbolic nature of equilibria. Very few hyperchaotic/chaotic systems are reported with non-hyperbolic nature of equilibria (Sprott 2015; Wei et al. 2015a, b; Yang et al. 2010; Li and Xiong 2017). Higher dimensional hyperchaotic/chaotic systems with non-hyperbolic nature of equilibria are rare in the literature. Thus, developing higher dimensional hyperchaotic/chaotic systems with some fascinating attributes like non-hyperbolic equilibria and multistability is also a worthy motivation of this chapter.

In this chapter, three new 5-D hyperchaotic and two chaotic systems are reported. Out of these five systems, four of them have many equilibria and thus qualify to be chaotic systems with hidden attractors. Again, all the five systems exhibit non-hyperbolic equilibria and hence behave like a chaotic system with self-attractors. Therefore, four new chaotic systems have both the self-attractor and

Table 1 Categorisation of the reported chaotic and hyperchaotic systems with an infinite number of equilibria

Sl. no.	3-D/4-D system	Nature of systems	References of papers
1.	3-D chaotic system	Line of equilibria	Jafari and Sprott (2015, 2013), Kingni et al. (2016a, b)
		Many equilibria	Wang and Chen (2012)
		Circle of equilibria	Gotthans and Petržela (2015), Gotthans et al. (2016), Kingni et al. (2016a, b), Pham et al. (2016d, f)
		Surface of equilibria	Jafari et al. (2016b)
		Curve of equilibria	Barati et al. (2016), Pham et al. (2016c)
		Square shaped equilibria	Gotthans et al. (2016), Pham et al. (2016b, d, f)
		Ellipse shaped equilibria	Pham et al. (2016d)
		Sphere of equilibria	Qi and Chen (2015)
2.	4-D chaotic system	Plane of equilibria	Jafari et al. (2016a, b)
		Line of equilibria	Singh and Roy (2017b), Pham et al. (2016c)
3.	4-D hyperchaotic system	Line of equilibria	Li et al. (2014a, b), Zhou and Yang (2014)
		Curve of equilibria	Chen and Yang (2015)
4.	4-D memristive hyperchaotic system	Line of equilibria	Li et al. (2014a, b), Ma et al. (2015)
5.	5-D hyperchaotic/ chaotic system	Line of equilibria	Vaidyanathan (2016)
		Line of equilibria with coexistence of attractors	This work
		Hyperbolic curve of equilibria with coexistence of attractors	This work

hidden attractor. The three new systems have hyperbolic curve of equilibria and one system has a line of equilibria. All the five new systems depict multistability. The systems have various dynamical behaviours like hyperchaotic, chaotic, periodic, quasi-periodic, etc. Various numerical tools are used to find the different dynamical behaviour of the systems like phase portrait, Lyapunov spectrum, bifurcation diagram, Poincaré map, frequency spectrum. Chaotic natures of the systems are validated by circuit design and implementation. Circuit implementation results of the two systems have good agreement with the MATLAB simulation results.

The rest part of the chapter is organised as follows. Section 2 describes the development of the new systems. The findings of different dynamic behaviour of the systems are shown in Sect. 3. Circuit design and implementations of the systems are discussed in Sect. 4. Section 5 presents the conclusions of the chapter.

2 Development of the Systems with Non-hyperbolic Equilibria and a Line of Equilibria

The dynamics of the 3-D Lorenz chaotic system with linear control inputs is described as (Lorenz 1963):

$$\begin{cases} \dot{x}_1 = a(x_2 - x_1) + u_1 \\ \dot{x}_2 = rx_1 - x_2 - x_1 x_3 + u_2 \\ \dot{x}_3 = x_1 x_2 + cx_3 + u_3 \end{cases} \tag{1}$$

Table 2 Three hyperchaotic and two chaotic systems with non-hyperbolic and many equilibria

Case	System dynamics	LEs and nature	D_{KY}	Initial conditions
NHE1	$\dot{x}_1 = a(x_3 - x_1)$ $\dot{x}_2 = b - x_1 x_3 + x_4$ $\dot{x}_3 = x_1 x_2 + x_5$ $\dot{x}_4 = cx_3$ $\dot{x}_5 = -cx_2 x_3$ $a = 10, b = 45, c = 0.0183$	$LE =$ $\begin{pmatrix} 0.1632, \\ 0.0124, \\ 0, \\ -0.0069, \\ -11.6351 \end{pmatrix}$ and hyperchaotic	4.014	$x(0) =$ $\begin{pmatrix} 0.001, \\ 0.002, \\ 0.003, \\ 0.001, \\ 0.001 \end{pmatrix}^T$
NHE2	$\dot{x}_1 = a(x_3 - x_1)$ $\dot{x}_2 = b - x_1 x_3 + x_4$ $\dot{x}_3 = x_1 x_2 + x_5$ $\dot{x}_4 = cx_3$ $\dot{x}_5 = -cx_2 x_5$ $a = 10, b = 45, c = 0.0105$	$LE =$ $\begin{pmatrix} 0.9470, \\ 0.0011, \\ 0, \\ -0.0129, \\ -10.9316 \end{pmatrix}$ and hyperchaotic	4.085	$x(0) =$ $\begin{pmatrix} 0.001, \\ 0.002, \\ 0.003, \\ 0.001, \\ 0.001 \end{pmatrix}^T$
NHE3	$\dot{x}_1 = a(x_3 - x_1)$ $\dot{x}_2 = b - x_1 x_3 + x_4$ $\dot{x}_3 = x_1 x_2 + x_5$ $\dot{x}_4 = -cx_2$ $\dot{x}_5 = -cx_2 x_3$ $a = 10, b = 45, c = 0.0198$	$LE =$ $\begin{pmatrix} 2.0149, \\ 0.0120, \\ 0, \\ -0.0012, \\ -12.0229 \end{pmatrix}$ and hyperchaotic	4.168	$x(0) =$ $\begin{pmatrix} 0.001, \\ 0.002, \\ 0.003, \\ 0.001, \\ 0.001 \end{pmatrix}^T$
NHE4	$\dot{x}_1 = a(x_3 - x_1)$ $\dot{x}_2 = b - x_1 x_3 + x_4$ $\dot{x}_3 = x_1 x_2 + x_5$ $\dot{x}_4 = -cx_2$ $\dot{x}_5 = -dx_3 x_5$ $a = 10, b = 45, c = 0.01, d = 0.001$	$LE =$ $\begin{pmatrix} 1.0288, \\ 0.0, \\ -0.0004, \\ -0.0007, \\ -11.0212 \end{pmatrix}$ and chaotic	4.093	$x(0) =$ $\begin{pmatrix} 0.001, \\ 0.002, \\ 0.003, \\ 0.001, \\ 0.001 \end{pmatrix}^T$
NHE5	$\dot{x}_1 = a(x_3 - x_1)$ $\dot{x}_2 = b - x_1 x_3 + x_4$ $\dot{x}_3 = x_1 x_2 + x_5$ $\dot{x}_4 = -cx_2$ $\dot{x}_5 = -cx_1 x_2$ $a = 10, b = 45, c = 0.001$	$LE =$ $\begin{pmatrix} 1.0513, \\ 0.0, \\ 0.0, \\ 0.0, \\ -11.0461 \end{pmatrix}$ and chaotic	4.095	$x(0) =$ $\begin{pmatrix} 0.001, \\ 0.002, \\ 0.003, \\ 0.001, \\ 0.001 \end{pmatrix}^T$

Table 3 Stability analysis of equilibrium points of the systems

System	Equilibrium points	Shape of equilibria	Eigenvalues	Nature
NHE1	$E1 = (0,0,0,-45,0)$	Constant	$\lambda = (-10,0,0,0,0)$	Non-hyperbolic
	$E2 = (0,x_2,0,-45,0)$	Line of equilibria	The system has non-hyperbolic, stable focus and saddle nature of eigenvalues for different values of state variables x_2	
NHE2	$E1 = (0,0,0,-45,0)$	Constant	$\lambda = (0.2653, \pm 0.2653i, -0.2653, -10)$	Non-hyperbolic
NHE3	$E1 = (0,0,0,-45,0)$	Constant	$\lambda = (-10,0,0,\pm 0.1407i)$	Non-hyperbolic
	$E2 = (x_1,0,x_1,x_1^2-45,0)$	Hyperbolic curve of equilibria	The system has non-hyperbolic and saddle nature of eigenvalues for different values of state variables x_1	
NHE4	$E1 = (0,0,0,-45,0)$	Constant	$\lambda = (-10,0,0,\pm 0.10i)$	
	$E2 = (x_1,0,x_1,x_1^2-45,0)$	Hyperbolic curve of equilibria	The system has non-hyperbolic and saddle nature of eigenvalues for different values of state variables x_1	
NHE5	$E1 = (0,0,0,-45,0)$	Constant	$\lambda = (-10,0,0,\pm 0.0316i)$	
	$E2 = (x_1,0,x_1,x_1^2-45,0)$	Hyperbolic curve of equilibria	The system has non-hyperbolic and saddle nature of eigenvalues for different values of state variables x_1	

where a, r, c are the parameters, x_1, x_2, x_3 are the state variables and u_1, u_2, u_3 are the control inputs. Selecting control inputs u_1, u_2, u_3 as in the form of (2) (Yuhua et al. 2010), we get (3).

$$\begin{cases} u_1 = a(x_3 - x_2) \\ u_2 = b - rx_1 + x_2 \\ u_3 = -cx_3 \end{cases} \tag{2}$$

Using (2), the dynamics of the Lorenz system can be written as in (3).

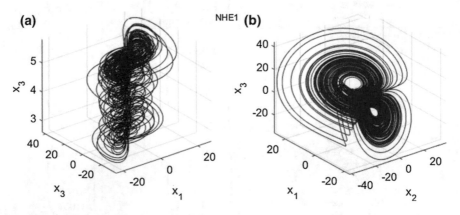

Fig. 1 Hyperchaotic attractors of the NHE1 system with $a = 10, b = 45, c = 0.0183$

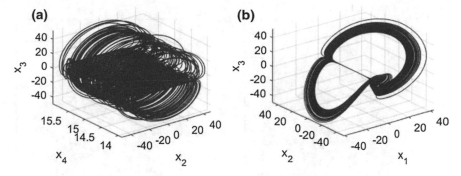

Fig. 2 Hyperchaotic attractors of the NHE2 system with $a = 10, b = 45, c = 0.0105$

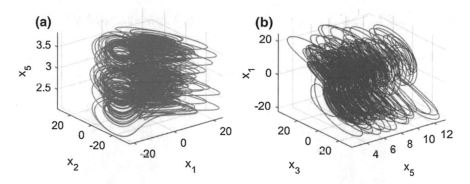

Fig. 3 Hyperchaotic attractors of the NHE3 system with $a = 10, b = 45, c = 0.0198$

$$\begin{cases} \dot{x}_1 = a(x_3 - x_1) \\ \dot{x}_2 = b - x_1 x_3 \\ \dot{x}_3 = x_1 x_2 \end{cases} \tag{3}$$

System (3) is chaotic with $a = 10, b = 45$ (Yuhua et al. 2010).

Using the above system (3), this chapter presents five new 5-D self-attractor/ hidden attractor hyperchaotic/chaotic systems with non-hyperbolic and many equilibria. A known and widely used systematic search procedure is used to develop the systems as used in the paper (Munmuangsaen et al. 2011; Pham et al. 2016f; Sprott 1993, 2000, 2010). The procedure considers various combinations of states to generate hyperchaotic/chaotic systems with largest Lyapunov exponents at least greater than 0.9. The general expression of the new 5-D hyperchaotic or chaotic systems is considered as:

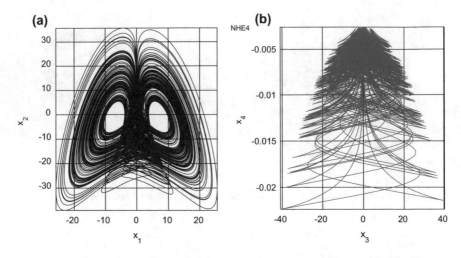

Fig. 4 Chaotic attractors of the NHE4 system with $a = 10, b = 45, c = 0.01, d = 0.001$

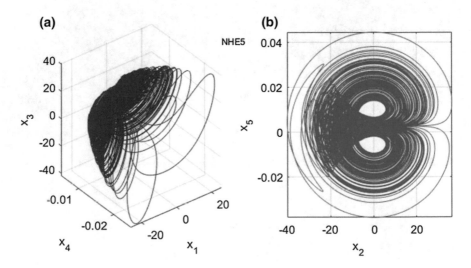

Fig. 5 Chaotic attractors of the NHE5 system with $a = 10, b = 45, c = 0.001$

$$\begin{cases} \dot{x}_1 = a(x_3 - x_1) \\ \dot{x}_2 = b - x_1 x_3 + x_4 \\ \dot{x}_3 = x_1 x_2 + x_5 \\ \dot{x}_4 = f_1(x_1, x_2, x_3) \\ \dot{x}_5 = f_2(x_1, x_2, x_3, x_4, x_5) \end{cases} \tag{4}$$

where $f_1(x_1, x_2, x_3)$ and $f_2(x_1, x_2, x_3, x_4, x_5)$ are linear and nonlinear functions, respectively. Different choices of $f_1(x_1, x_2, x_3)$ and $f_2(x_1, x_2, x_3, x_4, x_5)$ lead to systems with various type of equilibria.

With suitable choices of $f_1(x_1, x_2, x_3)$ and $f_2(x_1, x_2, x_3, x_4, x_5)$, five different types of hyperchaotic or chaotic systems are developed. There are named as NHE1 to NHE5 and the details are shown in Table 2. The first three systems (NHE1 to NHE3) have hyperchaotic behaviour and the rest two systems (NHE4 and NHE5) have chaotic behaviour. Table 2 describes the dynamics of the systems, Lyapunov exponents (LEs), nature of the systems, Lyapunov dimension/Kaplan-Yorke dimension (D_{KY}) and initial conditions used for simulation of these systems.

Stability analysis of the equilibrium points of the systems given in Table 2 is discussed in Table 3. It is seen from Table 3 that all the systems have non-hyperbolic nature of equilibria. All the systems have many equilibria except the system NHE2.

3 Numerical Findings of the Proposed Systems Given in Table 2

This section discusses various numerical tools like time series plot, phase portrait, Lyapunov spectrum, bifurcation diagram, frequency spectrum, Poincaré maps used for finding different dynamical behaviour of the new systems given in Table 2.

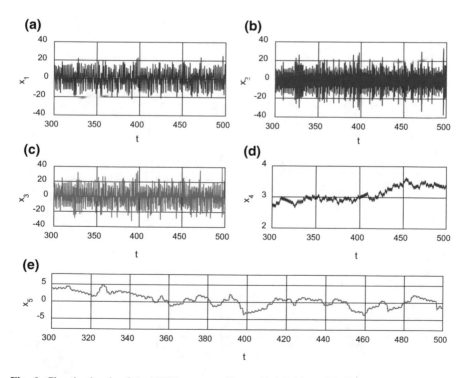

Fig. 6 Chaotic signals of the NHE1 system with $a = 10, b = 45, c = 0.0183$

3.1   Time Series and Phase Portrait

Chaotic behaviour of the systems given in Table 2 is confirmed by plotting their time responses and phase portraits. Figures 1, 2, 3, 4 and 5 show the hyperchaotic and chaotic attractors of the new systems. The irregular shape of the phase portraits of the systems in Figs. 1, 2, 3, 4 and 5 depicts their chaotic behaviours. Time responses of the systems NHE1 and NHE3 are shown in Figs. 6 and 7, respectively. Aperiodic nature of the responses confirms the chaotic behaviour of the systems (Singh and Roy 2015a, b, 2016a, b, 2017a, b, c; Singh et al. 2017a, b). All the time responses and phase portraits of the systems are generated using the fixed initial conditions and value of the parameters which are given in Table 2.

3.2   Lyapunov Spectrum and Bifurcation Diagram

Different dynamical behaviour of the systems given in Table 2 are calculated using Lyapunov spectrum and bifurcation diagram. Lyapunov spectrums of all the systems are calculated by finding Lyapunov exponents using Wolf algorithm (Wolf et al. 1985) with the observation time $T = 20000$, step size $\Delta t = 0.01$ and fixed

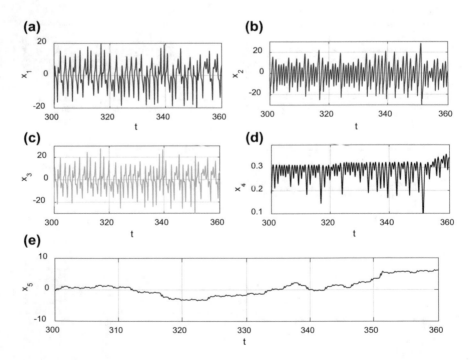

Fig. 7 Chaotic signals of the NHE3 system with $a = 10, b = 45, c = 0.0198$

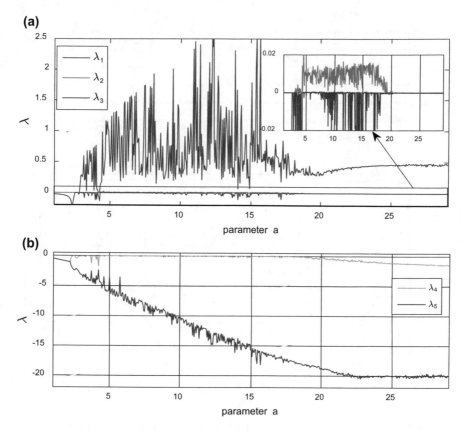

Fig. 8 Lyapunov spectrum of the NHE1 system with $b = 45, c = 0.0183$ and $a \in [1, 30]$

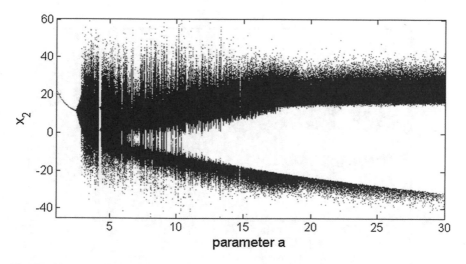

Fig. 9 Bifurcation diagram of the NHE1 system with $b = 45, c = 0.0183$ and $a \in [1, 30]$

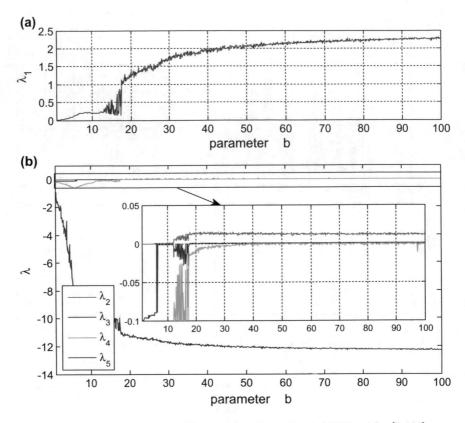

Fig. 10 Lyapunov spectrum of the NHE1 system with $a = 10, c = 0.0183$ and $b \in [5, 100]$

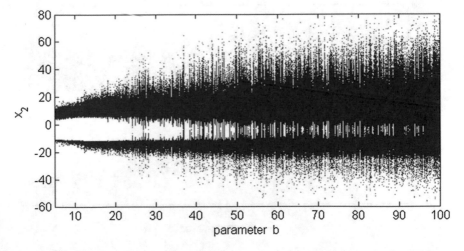

Fig. 11 Bifurcation diagram of the NHE1 system with $a = 10, c = 0.0183$ and $b \in [5, 100]$

(a)

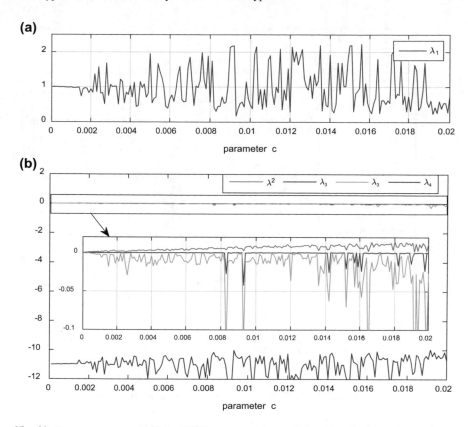

(b)

Fig. 12 Lyapunov spectrum of the NHE1 system with $a = 10, b = 45$ and $c \in [0.0001, 0.02]$

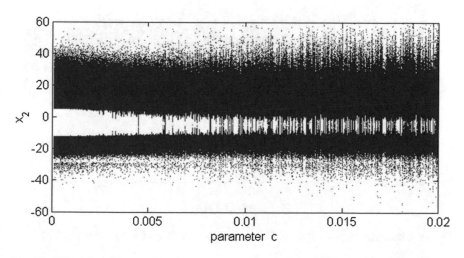

Fig. 13 Bifurcation diagram of the NHE1 system with $a = 10, b = 45$ and $c \in [0.0001, 0.02]$

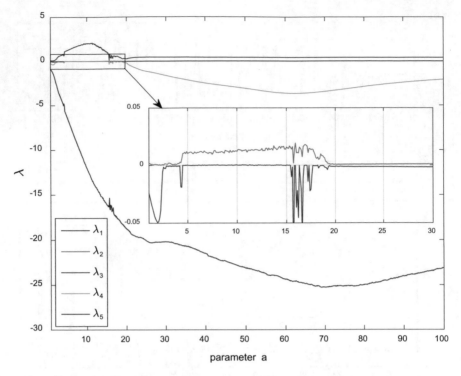

Fig. 14 Lyapunov spectrum of the NHE3 system with $b = 45, c = 0.0198$ and $a \in [1, 100]$

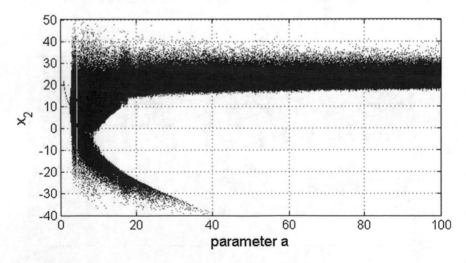

Fig. 15 Bifurcation diagram of the NHE3 system with $b = 45, c = 0.0198$ and $a \in [5, 100]$

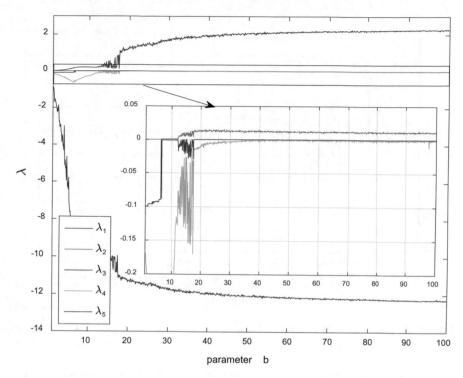

Fig. 16 Lyapunov spectrum of the NHE3 system with $a = 10, c = 0.0198$ and $b \in [5, 100]$

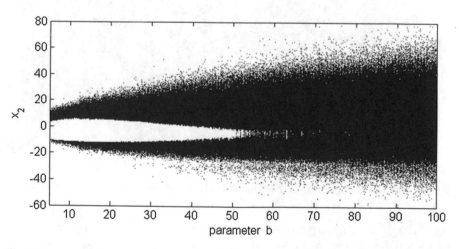

Fig. 17 Bifurcation diagram of the NHE3 system with $a = 10, c = 0.0198$ and $b \in [5, 100]$

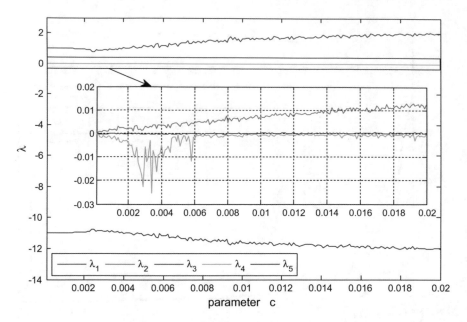

Fig. 18 Lyapunov spectrum of the NHE3 system with $a = 10, b = 45$ and $c \in [0.0001, 0.02]$

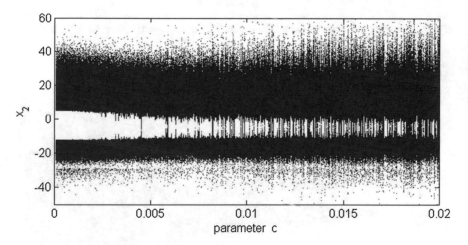

Fig. 19 Bifurcation diagram of the NHE3 system with $a = 10, b = 45$ and $c \in [0.0001, 0.02]$

initial conditions $x(0) = (0.001, 0.002, 0.003, 0.001, 0.001)^T$. In MATLAB, the time variable is selected as $T = 0 : \Delta t : 1000$, where Δt is the step size and 1000 is the total observation time. It may be noted that T does not reflect the actual time of calculation. Lyapunov spectrum and bifurcation diagram of the systems are shown with the variation of one parameter and keeping other fixed. Here, Lyapunov

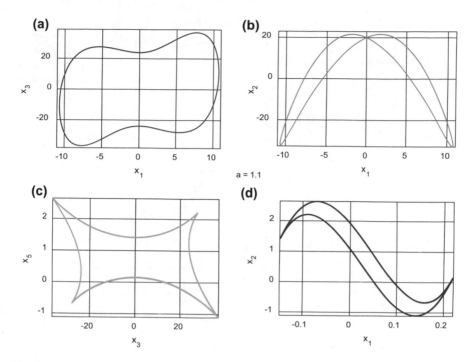

Fig. 20 Periodic attractors of the NHE1 system with $a = 1.1, b = 45, c = 0.0183$ and $x(0) = (0.001, 0.002, 0.003, 0.001, 0.001)^T$

spectrums and bifurcation diagrams of systems NHE1 and NHE3 are only shown. These figures for the other systems can also be calculated in a similar manner and are not shown here to avoid the repetition. However, Lyapunov exponents of systems NHE2, NHE4 and NHE5 are given in Table 2.

Lyapunov spectrum and bifurcation diagram of the NHE1 system with the variation of one parameter, out of a, b or c, and keeping the rest two fixed are shown in Figs. 8, 9, 10, 11, 12 and 13. Similarly, Lyapunov spectrum and bifurcation diagram of the NHE3 system are shown in Figs. 14, 15, 16, 17, 18 and 19. It is observed from Figs. 8, 10, 12, 14, 16 and 18 that NHE1 and NHE3 systems, respectively, have different dynamical behaviours like hyperchaotic, chaotic, periodic and quasi-periodic. It is also observed from Figs. 9, 11, 13, 15, 17 and 19 that NHE1 and NHE3 systems, respectively, have various dynamical behaviours like chaotic and periodic.

Periodic nature of the NHE1 system with $a = 1.1, b = 45, c = 0.0183$ and $a = 4.25, b = 45, c = 0.0183$ is shown in Figs. 20 and 21, respectively. Periodic nature of the NHE3 system with $a = 1.1, b = 45, c = 0.0198$ is shown in Fig. 22. The NHE2 system shows transient chaotic behaviour with trajectory going to infinity for smaller values of parameter b. The transient chaotic behaviour of the NHE2 system with $a = 10, b = 10, c = 0.0105$ is shown in Fig. 23. It is apparent from Fig. 23 that

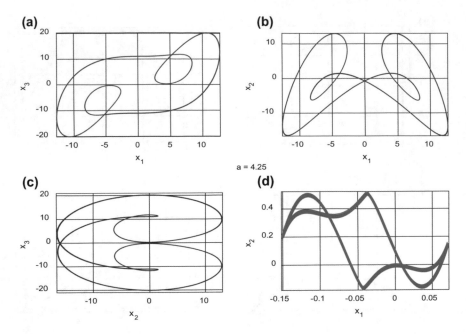

Fig. 21 Periodic attractors of the NHE1 system with $a = 4.25, b = 45, c = 0.0183$ and $x(0) = (0.001, 0.002, 0.003, 0.001, 0.001)^T$

the NHE2 system has chaotic behaviour approximately for $t < 1500$ and trajectory going to infinity at $t > 1600$ approximately.

3.3 Coexistences of Attractors

All the proposed systems show multistability (i.e. coexistences of attractors) with the change of initial conditions. Coexistence of chaotic attractors of NHE1, NHE2 and NHE3 systems are shown in Figs. 24, 26 and 27, respectively. Coexistences of the quasi-periodic behaviour of the NHE2 system is shown in Fig. 25. Other two systems, i.e. NHE4 and NHE5 also show the coexistences of attractors with the changes of initial conditions. Their results are not shown here to avoid the repetition.

3.4 Frequency Spectrum and Poincaré Maps

Frequency spectra of $x_2(t)$ and $x_3(t)$ signals of NHE1 and NHE3 systems are shown in Fig. 28 and Fig. 29, respectively. Aperiodic continuous natures of the spectra

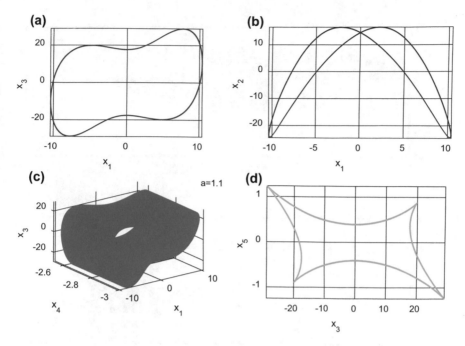

Fig. 22 Periodic attractors of the NHE3 system with $a=1.1, b=45, c=0.0198$ and $x(0) = (0.001, 0.002, 0.003, 0.001, 0.001)^T$

(Figs. 28 and 29) indicate the chaotic behaviour of the systems. Poincaré maps across different section of planes of NHE1 and NHE3 systems are shown in Fig. 30 and Fig. 31, respectively. Random locations of dots in the maps indicate the chaotic behaviour of the systems (Singh and Roy 2015a, b, 2016a, b, 2017a, b, c; Singh et al. 2017a, b). Frequency spectra and Poincaré maps of other systems can also be shown in a similar way but avoided here.

4 Circuit Implementation

This section describes the circuit design and realisation of NHE1 and NHE3 systems. Circuit realisations of other systems can also be done in a similar manner and are not shown here to avoid repetition.

Circuit realisation of a chaotic system represents its practical applicability (Trejo-Guerra et al. 2011, 2012; Nunez et al. 2015; Valtierra et al. 2015; Tlelo-Cuautle et al. 2016a). Circuit realisation of various chaotic/hyperchaotic systems are achieved by FPGA tool (Tlelo-Cuautle et al. 2015a, b, 2016b; Esteban et al. 2016), Cadence OrCAD (Trejo-Guerra et al. 2011) and NI Multisim (Ruo-Xun and Shi-ping 2010; Lao et al. 2014; Xiong et al. 2016) software. In this

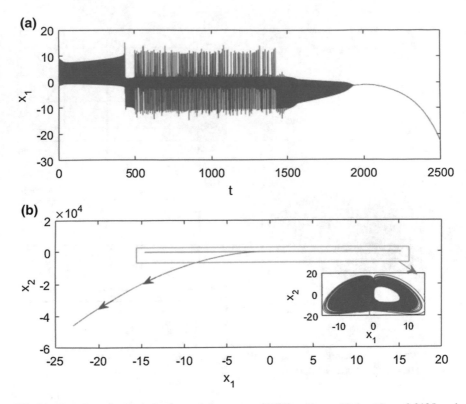

Fig. 23 Transient chaotic behaviour of the system NHE2 with $a = 10, b = 10, c = 0.0105$ and $x(0) = (0.001, 0.002, 0.003, 0.001, 0.001)^T$

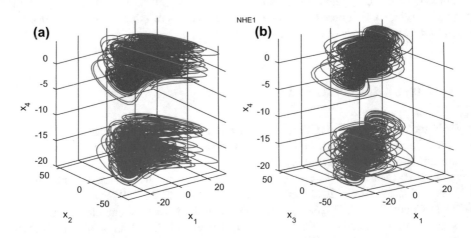

Fig. 24 Coexistences of chaotic attractors of the NHE1 system with $a = 10, b = 100, c = 0.0183$ and $x(0) = (\pm 0.001, \pm 0.002, \pm 0.003, \pm 0.001, \pm 0.001)^T$

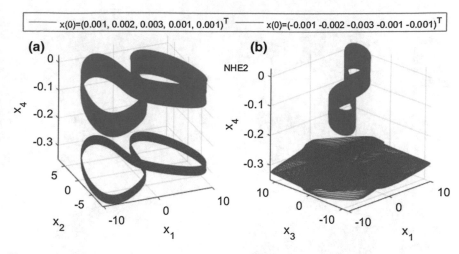

Fig. 25 Coexistences of the quasi-periodic behaviour of the NHE2 system with $a = 10, b = 25, c = 0.0105$ and $x(0) = (\pm 0.001, \pm 0.002, \pm 0.003, \pm 0.001, \pm 0.001)^T$

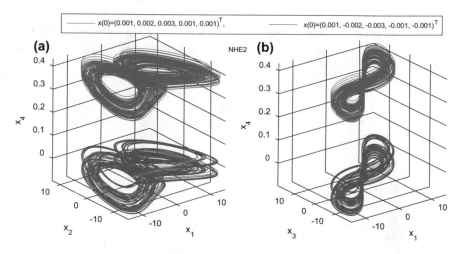

Fig. 26 Coexistences of chaotic attractors of NHE2 system with $a = 10, b = 35, c = 0.0105$ and $x(0) = (\pm 0.001, \pm 0.002, \pm 0.003, \pm 0.001, \pm 0.001)^T$

chapter, the circuit realisation of NHE1 and NHE3 systems are achieved using NI Multisim v12 software. Chaotic attractors of the NHE1 system obtained using the circuit implementation are shown in Figs. 32 and 33. The circuit designed for the implementation of the NHE1 system is shown in Fig. 34. The circuit which is shown in Fig. 34 has five integrators (U9A, U1A, U3A, U5A, and U7A) and use to realise the five states of the NHE1 system. The circuit consists of capacitors

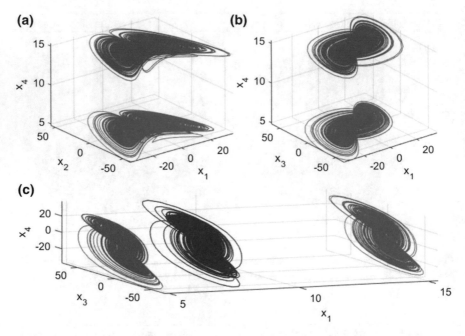

Fig. 27 Coexistences of chaotic attractors of NHE3 system with $a = 100, b = 45, c = 0.0198$, $x(0) = (\pm 0.001, \pm 0.002, \pm 0.003, \pm 0.001, \pm 0.001)^T$ (blue, brown) and $x(0) = (0.001, 0.002, 0.003, -0.001, -0.001)^T$ (red)

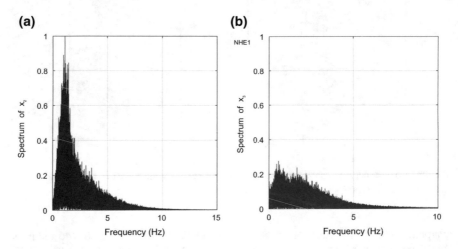

Fig. 28 The frequency spectrum of the NHE1 system with $a = 10, b = 45$ and $c = 0.0183$

(C1, C2, C3, C4, C5), resistances (R1,..., R19), Op-Amp (LF353D) and multipliers (AD633). The circuit equations of the NHE1 system can be written by using Kirchhoff's laws as:

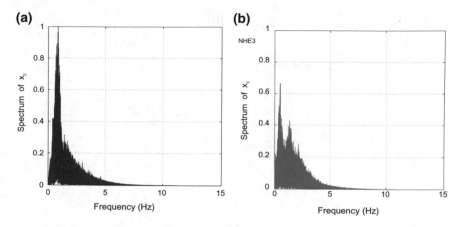

Fig. 29 The frequency spectrum of the NHE3 system with $a = 10, b = 45$ and $c = 0.0198$

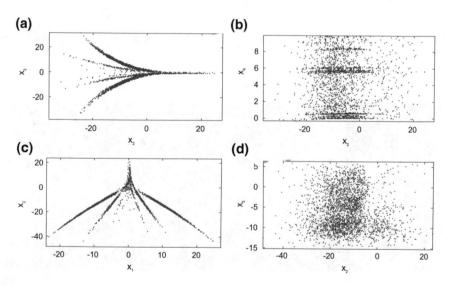

Fig. 30 Poincaré maps of the NHE1 system with $a = 10, b = 45$ and $c = 0.0183$ for: $x_1 = 0$ in (**a**), (**b**) and $x_3 = 0$ in (**c**), (**d**)

$$\begin{cases} \dot{x}_1 = \frac{1}{RC1} \left[\frac{R}{R1} x_3 - \frac{R}{R1} x_1 \right] \\ \dot{x}_2 = \frac{1}{RC2} \left[\frac{R}{R7} V1 - \frac{0.1R}{R6} x_1 x_3 + \frac{R}{R5} x_4 \right] \\ \dot{x}_3 = \frac{1}{RC3} \left[\frac{0.1R}{R11} x_1 x_2 + \frac{R}{R10} x_5 \right] \\ \dot{x}_4 = \frac{1}{RC4} \left[\frac{R}{R14} x_3 \right] \\ \dot{x}_5 = \frac{1}{RC5} \left[- \frac{0.1R}{R17} x_2 x_3 \right] \end{cases} \quad (5)$$

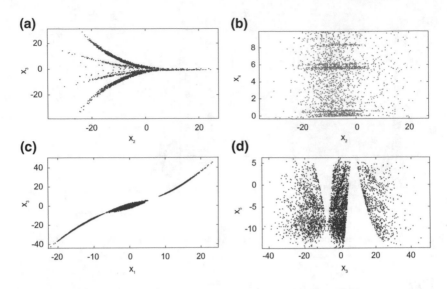

Fig. 31 Poincaré maps of the NHE3 system with $a = 10, b = 45$ and $c = 0.0198$ for: $x_1 = 0$ in (**a**), (**b**) and $x_2 = 0$ in (**c**), (**d**)

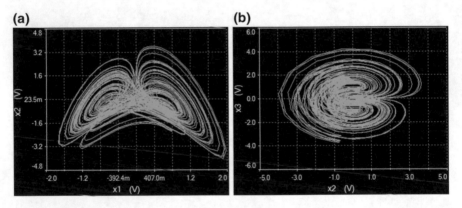

Fig. 32 Chaotic attractors of the NHE1 system obtained using circuit implementation with $a = 10, b = 45$ and $c = 0.0183$

where the variables x_1, x_2, x_3, x_4 and x_5 are the outcome of U9A, U1A, U3A, U5A and U7A, respectively. The system in (5) is equivalent to the NHE1 system with $\tau = t/RC$, $R1 = R2 = bR = 40\,k\Omega$, $R6 = R11 = 40\,k\Omega$, $R5 = R10 = 400\,k\Omega$, $R7 = bR = 8.88\,k\Omega$, $R14 = cR = 21857.92\,k\Omega$, $R17 = 0.1Rc = 2185.79\,k\Omega$, $C1 = C2 = C3 = C4 = C5 = 10\,nF$, $a = 10, b = 45, c = 0.0183$.

The circuit designed for implementation of the NHE3 system is shown in Fig. 35. The circuit in Fig. 35 consists of five integrators (U9A, U1A, U3A, U5A and U7A) which are used to realise the five states of the NHE3 system. The circuit

(a) **(b)**

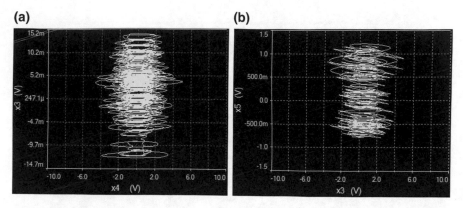

Fig. 33 Chaotic attractors of the NHE1 system obtained using circuit implementation with $a = 10, b = 45$ and $c = 0.0183$

consists of capacitors (C1, C2, C3, C4, C5), resistors (R1,…, R19), Op-Amp (LF353D) and multipliers (AD633). The circuit equations of the NHE3 system can be written as:

$$\begin{cases} \dot{x}_1 = \frac{1}{RC1} \left[\frac{R}{R1} x_3 - \frac{R}{R1} x_1 \right] \\ \dot{x}_2 = \frac{1}{RC2} \left[\frac{R}{R7} V1 - \frac{0.1R}{R6} x_1 x_3 + \frac{R}{R5} x_4 \right] \\ \dot{x}_3 = \frac{1}{RC3} \left[\frac{0.1R}{R11} x_1 x_2 + \frac{R}{R10} x_5 \right] \\ \dot{x}_4 = \frac{1}{RC4} \left[- \frac{R}{R14} x_2 \right] \\ \dot{x}_5 = \frac{1}{RC5} \left[- \frac{0.1R}{R17} x_2 x_3 \right] \end{cases} \qquad (6)$$

where the variables x_1, x_2, x_3, x_4 and x_5 are the outcome of U9A, U1A, U3A U5A and U7A, respectively. The system in (6) is equivalent to the NHE3 system with $\tau = t/RC$, $R1 = R2 = bR = 40\,\text{k}\Omega$, $R6 = R11 = 40\,\text{k}\Omega$, $R5 = R10 = 400\,\text{k}\Omega, R7 = bR = 8.88\,\text{k}\Omega$, $R14 = cR = 20202.02\,\text{k}\Omega$, $R17 = 0.1Rc = 2020.20\,\text{k}\Omega$, $C1 = C2 = C3 = C4 = C5 = 10\,\text{nF}$, $a = 10, b = 45, c = 0.0198$. The chaotic attractors of the NHE3 system are shown in Figs. 36 and 37.

It is apparent from Figs. 31, 33, 36 and 37 that the attractors of NHE1 and NHE3 systems obtained using circuit implementation match with the MATLAB simulation results. It is visible from Figs. 32 and 33 that the ranges of state variables are different from the MATLAB simulation results. This is because of difference in time constants considered. Relation between the time constant of system for MATLAB simulation and time used for circuit implementation is $\tau = \frac{t}{RC}$, where $R = 400\,\text{k}\Omega, C = 10\,\text{nF}$.

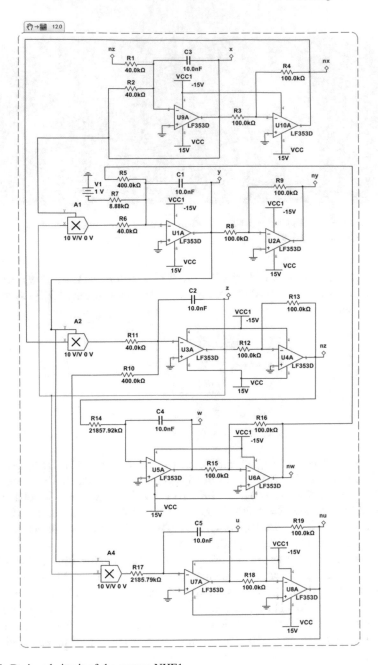

Fig. 34 Designed circuit of the system NHE1

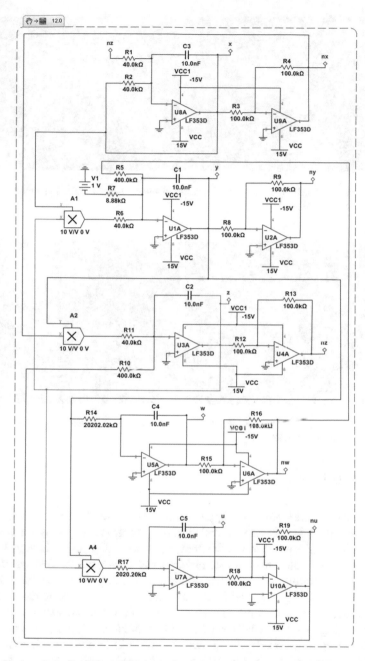

Fig. 35 Designed circuit of the NHE3 system

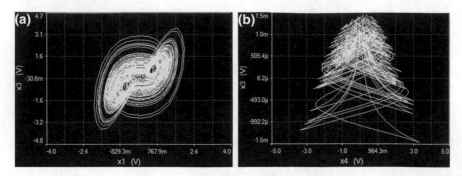

Fig. 36 Chaotic attractors of the NHE3 system obtained using circuit implementation with $a = 10, b = 45$ and $c = 0.0198$

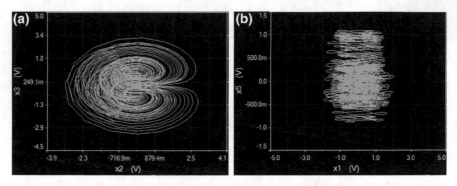

Fig. 37 Chaotic attractors of the NHE3 system obtained using circuit implementation with $a = 10, b = 45$ and $c = 0.0198$

5   Conclusions

In this chapter, three new 5-D hyperchaotic systems and two new 5-D chaotic systems with the nature of self-excited attractors are reported. Four of these systems may behave as hidden chaotic attractors. Such chaotic systems having both the self-excited and hidden attractors are rare in the literature. All the five systems have non-hyperbolic equilibria and hence belong to the category of self-excited attractors. NHE1, NHE3, NHE4 and NHE5 systems have many equilibria along with non-hyperbolic nature of equilibria. Hence, these four systems may be considered under the category of both self-excited and hidden attractors chaotic systems. The new systems are developed from the well-known 3-D Lorenz chaotic system with some transformation. All the five systems exhibit multistability. Various numerical tools like phase portrait, Lyapunov spectrum, bifurcation diagram, Poincaré map, and frequency spectrum are used to find different dynamic behaviour of the new systems. These behaviours confirm the chaotic nature of the proposed systems. The

results obtained using MATLAB simulations are validated by using circuit realisation. The proposed 5-D systems can have better application in the field of secure
communications.

References

Andrievskii BR, Fradkov AL (2004) Control of chaos : methods and applications. II. Applications.
    Autom Remote Control 65(4):505–533
Barati K, Jafari S, Sprott JC, Pham V (2016) Simple chaotic flows with a curve of equilibria. Int J
    Bifurcat Chaos 26(12):1630034–1630040
Chen G, Ueta T (1999) Yet another chaotic attractor. Int J Bifurcat Chaos 9:1465–1999
Chen Y, Yang Q (2015) A new Lorenz-type hyperchaotic system with a curve of equilibria. Math
    Comput Simul 112:40–55
Chen M, Xu Q, Lin Y, Bao B (2017) Multistability induced by two symmetric stable node-foci in
    modified canonical Chua's circuit. Nonlinear Dyn 87(2):789–802
Chudzik A, Perlikowski P, Stefański A, Kapitaniak T (2011) Multistability and rare attractors in
    van der Pol–Duffing oscillator. Int J Bifurcat Chaos 21(7):1907–1912
Effati S, Saberi-Nadjafi J, Saberi Nik H (2014) Optimal and adaptive control for a kind of 3D
    chaotic and 4D hyper-chaotic systems. Appl Math Model 38(2):759–774
Esteban T-C, de Jesus Rangel MJ, de la Fraga LG (2016) Engineering applications of FPGAs :
    chaotic systems, artificial neural networks, random number generators, and secure commu-
    nication systems. Springer, Switzerland
Gotthans T, Petržela J (2015) New class of chaotic systems with circular equilibrium. Nonlinear
    Dyn 81:1143–1149
Gotthans T, Sprott JC, Petrzela J (2016) Simple chaotic flow with circle and square equilibrium.
    Int J Bifurcat Chaos 26(8):1650137–1650145
Jafari S, Sprott JC (2013) Simple chaotic flows with a line equilibrium. Chaos Solitons Fractals
    57:79–84. https://doi.org/10.1016/j.chaos.2013.08.018
Jafari S, Sprott JC (2015) Erratum: simple chaotic flows with a line equilibrium (Chaos Solitons
    Fractals (2013) 57:79–84). Chaos Solitons Fractals 77:341–342 (2016a)
Jafari S, Sprott JC, Molaie M (2016a) A simple chaotic flow with a plane of equilibria. Int J
    Bifurcat Chaos 26(6):1650098–1650104
Jafari S, Sprott JC, Pham V, Volos C, Li C (2016b) Simple chaotic 3D flows with surfaces of
    equilibria. Nonlinear Dyn 86(2):1349–1358
Kemih K, Bouraoui H, Messadi M, Ghanes M (2013) Impulsive control and synchronization of a
    new 5D hyperchaotic system. Acta Phys Pol A 123(2):193–195
Kingni ST, Jafari S, Simo H, Woafo P (2014) Three-dimensional chaotic autonomous system with
    only one stable equilibrium: analysis, circuit design, parameter estimation, control, synchro-
    nization and its fractional-order form. Eur Phys J Plus 129(76):1–16
Kingni ST, Jafari S, Pham V-T, Woafo P (2016a) Constructing and analysing of a unique
    three-dimensional chaotic autonomous system exhibiting three families of hidden attractors.
    Math Comput Simul 132:172–182
Kingni ST, Pham V-T, Jafari S, Kol GR, Woafo P (2016b) Three-dimensional chaotic autonomous
    system with a circular equilibrium: analysis, circuit implementation and its fractional-order
    form. Circ Syst Signal Process 35(6):1807–1813

Kiseleva M, Kondratyeva N, Kuznetsov N, Leonov G (2017) Hidden oscillations in electrome-chanical systems. In: Dynamics and Control of Advanced Structures and Machines. Springer, pp 119–124

Lam HK, Li H (2014) Synchronization of chaotic systems using sampled-data polynomial controller. J Dyn Syst Meas Control 136(3):31006. https://doi.org/10.1115/1.4026304

Lao S, Tam L, Chen H, Sheu L (2014) Hybrid stability checking method for synchronization of chaotic fractional-order systems. Abstr Appl Anal 2014:1–11

Leonov GA, Kuznetsov NV (2013) Hidden attractors in dynamical systems: from hidden oscillations in Hilbert-Kolmogorov, Aizerman, and Kalman problems to hidden chaotic attractor in Chua circuits. Int J Bifurcat Chaos 23(1):130071–1330002

Leonov GA, Kuznetsov NV, Vagaitsev VI (2011a) Localization of hidden Chuas attractors. Phys Lett A 375(23):2230–2233

Leonov GA, Kuznetsov NV, Kuznestova OA, Seledzhi SM, Vagaitsev VI (2011b) Hidden oscillations in dynamical systems system. Trans Syst Control 6(2):1–14

Leonov GA, Kuznetsov NV, Vagaitsev VI (2012) Hidden attractor in smooth Chua systems. Phys D 241(18):1482–1486

Leonov GA, Kuznetsov NV, Kiseleva MA, Solovyeva EP, Zaretskiy AM (2014) Hidden oscillations in mathematical model of drilling system actuated by induction motor with a wound rotor. Nonlinear Dyn 77(1–2):277–288

Leonov GA, Kuznetsov NV, Mokaev TN (2015) Hidden attractor and homoclinic orbit in Lorenz-like system describing convective fluid motion in rotating cavity. Commun Nonlinear Sci Numer Simul 28(1–3):166–174

Li CL, Xiong J Bin (2017) A simple chaotic system with non-hyperbolic equilibria. Optik 128:42–49

Li C, Sprott JC, Thio W (2014a) Bistability in a hyperchaotic system with a line equilibrium. J Exp Theor Phys 118(3):494–500

Li Q, Hu S, Tang S, Zeng G (2014b) Hyperchaos and horseshoe in a 4D memristive system with a line of equilibria and its implementation. Int J Circuit Theory Appl 42(11):1172–1188

Lin Y, Wang C, He H, Zhou LL (2016) A novel four-wing non-equilibrium chaotic system and its circuit implementation. Pramana 86(4):801–807

Lochan K, Roy BK (2015) Control of two-link 2-DOF robot manipulator using fuzzy logic techniques: a review. Advances in Intelligent Systems and Computing Proceedings of Fourth International Conference on Soft Computing for Problem Solving, vol 336, pp 205–14

Lochan K, Roy BK (2016) Trajectory tracking control of an AMM modelled TLFM using backstepping method. Int J Control Theory Appl 9(39):241–248

Lochan K, Roy BK, Subudhi B (2016a) Generalized projective synchronization between controlled master and multiple slave TLFMs with modified adaptive SMC. Trans Inst Meas Control, 1–23. http://doi.org/10.1177/0142331216674067

Lochan K, Roy BK, Subudhi B (2016b) SMC controlled chaotic trajectory tracking of two-link flexible manipulator with PID sliding. IFAC-PapersOnLine 49(1):219–224

Lochan K, Roy BK, Subudhi B (2016c) A review on two-link flexible manipulators. Annu Rev Control 42:346–367

Lorenz EN (1963) Deterministic nonperiodic flow. J Atmos Sci 20(2):130–141

Lü J, Chen G, Cheng D, Celikovsky S (2002) Bridge the gap between the Lorenz system and the Chen system. Int J Bifurcat Chaos 12(12):2917–2926

Ma J, Chen Z, Wang Z, Zhang Q (2015) A four-wing hyper-chaotic attractor generated from a 4-D memristive system with a line equilibrium. Nonlinear Dyn 81(3):1275–1288

Munmuangsaen B, Srisuchinwong B, Sprott JC (2011) Generalization of the simplest autonomous chaotic system. Phys Lett A 375(12):1445–1450

Nunez JC, Tlelo E, Ramirez C, Jimenez JM (2015) CCII+Based on QFGMOS for implementing chua's chaotic oscillator. IEEE Latin Am Trans 13(9):2865–2870

Ojoniyi OS, Njah AN (2016) A 5D hyperchaotic Sprott B system with coexisting hidden attractors. Chaos Solitons Fractals 87

Pang S, Liu Y (2011) A new hyperchaotic system from the Lu system and its control. J Comput Appl Math 235(8):2775–2789

Pham V-T, Volos C, Jafari S, Wei Z, Wang X (2014) Constructing a novel no-equilibrium chaotic system. Int J Bifurcat Chaos 24(5):1450073

Pham V-T, Jafari S, Kapitaniak T (2016a) Constructing a chaotic system with an infinite number of equilibrium points. Int J Bifurcat Chaos 26(13):1650225–1650232

Pham V-T, Jafari S, Volos C, Giakoumis A, Vaidyanathan S, Kapitaniak T (2016b) A chaotic system with equilibria located on the rounded square loop and its circuit implementation. IEEE Trans Circuits Syst II Express Briefs 63(9):878–882

Pham V-T, Jafari S, Volos C, Vaidyanathand S, Kapitaniake T (2016c) A chaotic system with infinite equilibria located on a piecewise linear curve. Optik 127:9111–9117

Pham V-T, Jafari S, Wang X (2016d) A chaotic system with different shapes of equilibria. Int J Bifurcat Chaos 26(4):1650069–1650075

Pham V, Jafari S, Volos C, Kapitaniak T (2016e) A gallery of chaotic systems with an infinite number of equilibrium points. Chaos Solitons Fractals 93:58–63

Pham V-T, Volos C, Vaidyanathan S, Wang X (2016f) A chaotic system with an infinite number of equilibrium points dynamics, horseshoe, and synchronization. Adv Math Phys 2016

Pham VT, Sundarapandian V, Volos CK, Jafari S, Kuznetsov NV, Hoang TM (2016g) A novel memristive time-delay chaotic system without equilibrium points. Eur Phys J Spec Top 225(1): 127–136

Pham V, Volos C, Jafari S, Vaidyanathan S, Kapitaniak T, Wang X (2016h) A chaotic system with different families of hidden attractors. Int J Bifurcat Chaos 26(8):1650139–1650148

Pham V-T, Akgul A, Volos C, Jafari S, Kapitaniak T (2017a) Dynamics and circuit realization of a no-equilibrium chaotic system with a boostable variable. AEU-Int J Electron Commun 78:134–140

Pham V, Kingni ST, Volos C, Jafari S, Kapitaniak T (2017) A simple three-dimensional fractional-order chaotic system without equilibrium: dynamics, circuitry implementation, chaos control and synchronization. AEÜ-Int J Electron Commun (In Press)

Pisarchik AN, Feudel U (2014) Control of multistability. Phys Rep 540(4):167–218

Qi G, Chen G (2015) A spherical chaotic system. Nonlinear Dyn 81(3):1381–1392

Rössler OE (1976) An equation for continuous chaos. Phys Lett A 57(5):397–398

Ruo-Xun Z, Shi-ping Y (2010) Adaptive synchronisation of fractional-order chaotic systems. Chin Phys B 19(2):1–7

Sánchez Valtierra, de la Vega JL, Tlelo-Cuautle E (2015) Simulation of piecewise-linear one-dimensional chaotic maps by Verilog-A. IETE Tech Rev 32(4):304–310

Sharma PR, Shrimali MD, Prasad A, Kuznetsov NV, Leonov GA (2015) Control of multistability in hidden attractors. Eur Phys J Spec Top 224(8):1485–1491

Shen C, Yu S, Lu J, Chen G (2014a) Designing hyperchaotic systems with any desired number of positive Lyapunov exponents via a simple model. IEEE Trans Circuits Syst I Regul Pap 61 (8):2380–2389

Shen C, Yu S, Lü J, Chen G (2014b) A systematic methodology for constructing hyperchaotic systems with multiple positive Lyapunov exponents and circuit implementation. IEEE Trans Circuits Syst I Regul Pap 61(3):854–864

Singh JP, Roy BK (2015a) A novel asymmetric hyperchaotic system and its circuit validation. Int J Control Theory Appl 8(3):1005–1013

Singh JP, Roy BK (2015b) Analysis of an one equilibrium novel hyperchaotic system and its circuit validation. Int J Control Theory Appl 8(3):1015–1023

Singh JP, Roy BK (2016a) A Novel hyperchaotic system with stable and unstable line of equilibria and sigma-shaped Poincare map. IFAC-PapersOnLine 49(1):526–531

Singh JP, Roy BK (2016b) Comment on "Theoretical analysis and circuit verification for fractional-order chaotic behavior in a new hyperchaotic system". Math Prob Eng 2014(1):1–4. https://doi.org/10.1155/2014/682408

Singh JP, Roy BK (2016c) The nature of Lyapunov exponents is (+, +, −, −). Is it a hyperchaotic system? Chaos Solitons Fractals 92:73–85

Singh JP, Roy BK (2017a) Multistability and hidden chaotic attractors in a new simple 4-D chaotic system with chaotic 2-torus behaviour. Int J Dyn Control (In Press). http://doi.org/10.1007/s40435-017-0332-8

Singh JP, Roy BK (2017b) The simplest 4-D chaotic system with line of equilibria, chaotic 2-torus and 3-torus behaviour. Nonlinear Dyn. http://doi.org/10.1007/s11071-017-3556-4

Singh JP, Roy BK (2017c) Coexistence of asymmetric hidden chaotic attractors in a new simple 4-D chaotic system with curve of equilibria. Optik 2017(145):209–217

Singh PP, Singh JP, Roy BK (2014) Synchronization and anti-synchronization of Lu and Bhalekar-Gejji chaotic systems using nonlinear active control. Chaos Solitons Fractals 69:31–39

Singh JP, Singh PP, Roy BK (2015) PI based sliding mode control for hybrid synchronization of Chen and Liu-Yang chaotic systems with circuit design and simulation. In: 1st IEEE Indian control conference, Chennai, India, pp 257–262

Singh JP, Lochan K, Kuznetsov NV, Roy BK (2017a) Coexistence of single- and multi-scroll chaotic orbits in a single-link flexible joint robot manipulator with stable spiral and index-4 spiral repellor types of equilibria. Nonlinear Dyn. https://doi.org/10.1007/s11071-017-3726-4

Singh PP, Singh JP, Roy BK (2017b) NAC-based synchronisation and anti-synchronisation between hyperchaotic and chaotic systems, its analogue circuit design and application. IETE J Res 1–17. http://doi.org/10.1080/03772063.2017.1331758

Sprott JC (1993) Automatic generation of strange attractors. Comput Graph 17(3):325–332

Sprott JC (2000) Simple chaotic systems and circuits. Am J Phys 68(8):758–763

Sprott JC (2010) Elegant chaos, algebraically simple chaotic flows. World Scientific Publishing Co. Pte. Ltd

Sprott JC (2015) Review strange attractors with various equilibrium types. Eur Phys J Spec Top 224:1409–1419

Tlelo-Cuautle E, Ramos-López HC, Sánchez-Sánchez M, Pano-Azucena AD, Sánchez-Gaspariano LA, Núñez-Pérez JC, Camas-Anzueto JL (2014) Application of a chaotic oscillator in an autonomous mobile robot. J Electr Eng 65(3):157–162

Tlelo-Cuautle E, Carbajal-Gomez VH, Obeso-Rodelo PJ, Rangel-Magdaleno JJ, Núñez-Pérez JC (2015a) FPGA realization of a chaotic communication system applied to image processing. Nonlinear Dyn 82(4):1879–1892

Tlelo-Cuautle E, Rangel-Magdaleno JJ, Pano-Azucena AD, Obeso-Rodelo PJ, Nunez-Perez JC (2015b) FPGA realization of multi-scroll chaotic oscillators. Commun Nonlinear Sci Numer Simul 27(1–3):66–80

Tlelo-Cuautle E, Pano-Azucena AD, Rangel-Magdaleno JJ, Carbajal-Gomez VH, Rodriguez-Gomez G (2016a) Generating a 50-scroll chaotic attractor at 66 MHz by using FPGAs. Nonlinear Dyn 85(4):2143–2157

Tlelo-Cuautle E, Quintas-Valles ADJ, De La Fraga LG, Rangel-Magdaleno JDJ (2016b) VHDL descriptions for the FPGA implementation of PWL-function-based multi-scroll chaotic oscillators. PLoS ONE 11. http://doi.org/10.1371/journal.pone.0168300

Trejo-Guerra R, Tlelo-Cuautle E, Jiménez-Fuentes M, Muñoz-Pacheco JM, Sánchez-López C (2011) Multiscroll floating gate-based integrated chaotic oscillator. Int J Circuit Theory Appl 38(7):689–708

Trejo-Guerra R, Tlelo-Cuautle E, Jiménez-Fuentes JM, Sánchez-López C, Muñoz-Pacheco JM, Espinosa-Flores-Verdad G, Rocha-Pérez JM (2012) Integrated circuit generating 3- and 5-scroll attractors. Commun Nonlinear Sci Numer Simul 17(11):4328–4335

Vaidyanathan S (2016) A novel 5-D hyperchaotic system with a line of equilibrium points and its adaptive control. In: Advances and Applications in Chaotic Systems Studies in Computational Intelligence, vol 636

Vaidyanathan S, Volos C, Pham V-T (2014) Hyperchaos, adaptive control and synchronization of a novel 5-D hyperchaotic system with three positive Lyapunov exponents and its SPICE implementation. Arch Control Sci 24(4):409–446

Vaidyanathan S, Pham VT, Volos CK (2015) A 5-D hyperchaotic Rikitake dynamo system with hidden attractors. Eur Phys J Spec Top 224(8):1575–1592

Valtierra JL, Tlelo-Cuautle E, Rodríguez-Vázquez Á (2016) A switched-capacitor skew-tent map implementation for random number generation. Int J Circuit Theory Appl 45(2):305–315

Wang X, Chen G (2012) Constructing a chaotic system with any number of equilibria. Nonlinear Dyn 71(3):429–436

Wang B, Shi P, Karimi HR, Song Y, Wang J (2013) Robust H_∞ synchronization of a hyper-chaotic system with disturbance input. Nonlinear Anal Real World Appl 14(3): 1487–1495

Wang X, Vaidyanathan S, Volos C, Pham V-T, Kapitaniak T (2017) Dynamics, circuit realization, control and synchronization of a hyperchaotic hyperjerk system with coexisting attractors. Nonlinear Dyn (In Press). http://doi.org/10.1007/s11071-017-3542-x

Wei Z, Sprott JC, Chen H (2015a) Elementary quadratic chaotic flows with a single non-hyperbolic equilibrium. Phys Lett A 379:2184–2187

Wei Z, Zhang W, Yao M (2015b) On the periodic orbit bifurcating from one single non-hyperbolic equilibrium in a chaotic jerk system. Nonlinear Dyn 82(3):1251–1258. https://doi.org/10.1007/s11071-015-2230-y

Wolf A, Swift JB, Swinney HL, Vastano JA (1985) Determining Lyapunov exponents from a time series. Phys D 16(3):285–317

Xiong L, Lu Y-J, Zhang Y-F, Zhang X-G, Gupta P (2016) Design and hardware implementation of a new chaotic secure communication technique. PLoS ONE 11(8):1–19

Yang Q, Wei Z, Chen Gua (2010) An unusual 3D autonomous quadratic chaotic system with two stable node-foci. Int J Bifurcat Chaos 20(4):1061–1083

Yu H, Wang J, Deng B, Wei X, Che Y, Wong YK, Chan WL, Tsang KM (2012) Adaptive backstepping sliding mode control for chaos synchronization of two coupled neurons in the external electrical stimulation. Commun Nonlinear Sci Numer Simul 17(3):1344–1354

Yuhua XYX, Wuneng ZWZ, Jianan FJF (2010) On dynamics analysis of a new symmetrical five-term chaotic attractor. In: Proceedings of the 29th Chinese Control Conference, Beijing, China, pp 610–614

Zhou P, Yang F (2014) Hyperchaos, chaos, and horseshoe in a 4D nonlinear system with an infinite number of equilibrium points. Nonlinear Dyn 76(1):473–480

Printed in the United States
By Bookmasters